中华人民共和国工程建设标准目录

(2016年版)

住房和城乡建设部标准定额研究所 编

中国建筑工业出版社

图书在版编目（CIP）数据

中华人民共和国工程建设标准目录：2016年版/住房和城乡建设部标准定额研究所编. —北京：中国建筑工业出版社，2017.9
ISBN 978-7-112-20996-5

Ⅰ.①中… Ⅱ.①住… Ⅲ.①建筑工程-标准-中国-目录 Ⅳ.①TU-65

中国版本图书馆CIP数据核字（2017）第169098号

责任编辑：孙玉珍
责任校对：李欣慰　赵　力

中华人民共和国工程建设标准目录
（2016年版）
住房和城乡建设部标准定额研究所　编
*
中国建筑工业出版社出版、发行（北京海淀三里河路9号）
各地新华书店、建筑书店经销
北京科地亚盟排版公司制版
北京同文印刷有限责任公司印刷
*
开本：880×1230毫米　1/16　印张：31¾　字数：1315千字
2017年9月第一版　2017年9月第一次印刷
定价：**86.00**元
ISBN 978-7-112-20996-5
（30643）

版权所有　翻印必究
如有印装质量问题，可寄本社退换
（邮政编码 100037）

前　言

为满足广大工程建设管理人员、技术人员的需求，适时了解现行工程建设标准的情况，我们在原《中华人民共和国工程建设标准目录》（2007年版）的基础上，结合近年来批准发布的国家标准和已经备案的行业标准、地方标准的情况，组织修编了本目录。

本目录的内容包括现行工程建设国家标准、行业标准、地方标准以及中国工程建设标准化协会标准。截至2016年底，各类工程建设标准总计7059项，其中，国家标准1143项，行业标准2641项，地方标准2904项，中国工程建设标准化协会标准371项；局部修订公告72项。此外，我们将近年翻译出版的工程建设标准英文版目录也一并附上，供使用者参考。

参加本《目录》修编和审查的主要人员有：杨瑾峰、李铮、张红彦、李艳海、董博雅、林常青、倪知之、姚涛、杨申武、汤亚军、杜刘艺。

本《目录》由于修编的时间紧，错漏之处在所难免，请在使用过程中批评指正，并将意见函告我们。

联系地址：北京市三里河路九号院，邮编：100835。

住房和城乡建设部标准定额研究所

2017年6月

目　录

一、工程建设国家标准 ·· 1

二、工程建设标准局部修订公告 ·· 68

三、工程建设行业标准 ·· 70
  建筑工程 ··· 70
  城镇建设工程 ·· 93
  电力工程 ·· 107
  石油天然气工程 ·· 141
  海洋石油工程 ··· 158
  石油化工工程 ··· 166
  化工工程 ·· 186
  煤炭工业工程 ··· 204
  水利工程 ·· 205
  铁路工程 ·· 210
  公路工程 ·· 217
  水路工程 ·· 224
  航空工业工程 ··· 234
  冶金工业工程 ··· 235
  有色金属工业工程 ·· 237
  民航工程 ·· 241
  建材工业工程 ··· 244
  机械工业工程 ··· 245
  信息产业（邮政、电信、电子）工程 ··· 246
  广播电影电视工程 ·· 247
  国内贸易工程 ··· 250
  林业工程 ·· 251
  纺织业工程 ·· 252
  轻工业工程 ·· 253
  船舶工业工程 ··· 255

四、中国工程建设标准化协会标准 ·· 256

五、工程建设地方标准 ·· 282
  北京市 ··· 282
  天津市 ··· 294
  河北省 ··· 306

山西省	317
内蒙古自治区	323
辽宁省	325
吉林省	333
黑龙江省	338
上海市	343
江苏省	364
福建省	375
浙江省	388
安徽省	395
江西省	404
山东省	406
河南省	416
湖南省	421
湖北省	424
广东省	427
广西壮族自治区	433
重庆市	437
四川省	450
云南省	458
贵州省	463
陕西省	466
甘肃省	471
宁夏回族自治区	477
青海省	479
新疆维吾尔自治区	481
海南省	486
西藏自治区	489

附：标准英文版目录 ································ 490

一、工程建设国家标准

序号	标准编号	标准名称	被代替标准编号	主编单位
1	GB/T 50001-2010	房屋建筑制图统一标准 Unified standard for building drawings	GB/T 50001-2001	中国建筑标准设计研究院
2	GB/T 50002-2013	建筑模数协调标准 Standard for modular coordination of building	GBJ 2-86 GB/T 50100-2001	中国建筑标准设计研究院 中国建筑设计研究院
3	GB 50003-2011	砌体结构设计规范 Code for design of masonry structures	GB 50003-2001	中国建筑东北设计研究院有限公司
4	GB 50005-2003 （2005年版）	木结构设计规范 Code for design of timber structures	GBJ 5-88	中国建筑西南设计研究院 四川省建筑科学研究院
5	GB/T 50006-2010	厂房建筑模数协调标准 Standard for modular coordination of industrial buildings	GBJ 6-86	中国联合工程公司
6	GB 50007-2011	建筑地基基础设计规范 Code for design of building foundation	GB 50007-2002	中国建筑科学研究院
7	GB 50009-2012	建筑结构荷载规范 Load code for the design of building structures	GB 50009-2001 （2006年版）	中国建筑科学研究院
8	GB 50010-2010 （2015年版）	混凝土结构设计规范 Code for design of concrete structures	GB 50010-2001	中国建筑科学研究院
9	GB 50011-2010 （2016年版）	建筑抗震设计规范 Code for seismic design of buildings	GB 50011-2001	中国建筑科学研究院
10	GB 50012-2012	Ⅲ、Ⅳ级铁路设计规范 Code for design of class Ⅲ、Ⅳ railway	GBJ 12-87	中铁第四勘察设计院集团有限公司
11	GB 50013-2006	室外给水设计规范 Code for design of outdoor water supply engineering	GBJ 13-86 （1997年版）	上海市政工程设计研究院
12	GB 50014-2006 （2016年版）	室外排水设计规范 Code for design of outdoor wastewater engineering	GBJ 14-87 （1997年版）（2011年版）（2014年版）	上海市政工程设计研究总院（集团）有限公司
13	GB 50015-2003 （2009年版）	建筑给水排水设计规范 Code for design of building water supply and drainage		上海现代建筑设计（集团）有限公司
14	GB 50016-2014	建筑设计防火规范 Code for fire protection design of buildings	GB 50016-2006 GB 50045-95 （2005年版） GBJ 16-87	公安部天津消防研究所 公安部四川消防研究所
15	GB 50017-2003	钢结构设计规范 Code for design of steel structures	GBJ 17-88	北京钢铁设计研究总院
16	GB 50018-2002	冷弯薄壁型钢结构技术规范 Technical code of cold-formed thin-wall steel structures	GBJ 18-87	中南建筑设计院
17	GB 50019-2015	工业建筑供暖通风与空气调节设计规范 Design code for heating ventilation and air conditioning of industrial buildings	GBJ 19-87 （2000年版） GB 50019-2003	中国有色工程有限公司 中国恩菲工程技术有限公司
18	GB 50021-2001 （2009年版）	岩土工程勘察规范 Code for investigation of geotechnical engineering		建设综合勘察研究设计院

工程建设国家标准

序号	标准编号	标准名称	被代替标准编号	主编单位
19	GBJ 22-87	厂矿道路设计规范	TJ 22-77	交通部公路规划设计院
20	GB 50023-2009	建筑抗震鉴定标准 Standard for seismic appraisal of buildings	GB 50023-95	中国建筑科学研究院
21	GB 50025-2004	湿陷性黄土地区建筑规范 Code for building construction in collapsible loess regions	GBJ 25-90	陕西省建筑科学研究设计院
22	GB 50026-2007	工程测量规范 Code for engineering surveying	GB 50026-93	中国有色金属工业西安勘察设计研究院
23	GB 50027-2001	供水水文地质勘察规范 Standard for hydrogeological investigation of water-supply	GBJ 27-88	中国冶金建设集团武汉勘察研究总院
24	GB 50028-2006	城镇燃气设计规范 Code for design of city gas engineering	GB 50028-93	中国市政工程华北设计研究院
25	GB 50029-2014	压缩空气站设计规范 Code for design of compressed air station	GB 50029-2003	中机国际工程设计研究院有限责任公司（原机械工业部第八设计研究院） 红五环集团股份有限公司
26	GB 50030-2013	氧气站设计规范 Code for design of oxygen station	GB 50030-91	中国中元国际工程有限公司 中国电子工程设计院
27	GB 50031-91	乙炔站设计规范 Norm of acetylene plant design	TJ 31-78	机械电子工业部设计研究院
28	GB 50032-2003	室外给水排水和燃气热力工程抗震设计规范 Code for earthquake-resistant design of outdoor water supply, sewerage, gas and heating engineering	TJ 32-78	北京市政工程设计研究总院
29	GB 50033-2013	建筑采光设计标准 Standard for daylighting design of buildings	GB/T 50033-2001 GB 50033-91	中国建筑科学研究院
30	GB 50034-2013	建筑照明设计标准 Standard for lighting design of buildings	GB 50034-2004	中国建筑科学研究院
31	GB 50037-2013	建筑地面设计规范 Code for design of building ground	GB 50037-96 TJ 37-79	中国联合工程公司 深圳中邦（集团）建设总承包有限公司
32	GB 50038-2005	人民防空地下室设计规范 Code for design of civil air defence basement		
33	GB 50039-2010	农村防火规范 Code for fire protection and prevention of rural area	GBJ 39-90	山西省公安消防总队
34	GB 50040-96	动力机器基础设计规范 Code for design of dynamic machine foundation	GBJ 40-79	机械工业部设计研究院
35	GB 50041-2008	锅炉房设计规范 Code for design of boiler plant	GB 50041-92	中国联合工程公司
36	GB 50046-2008	工业建筑防腐蚀设计规范 Code for anticorrosion design of industrial constructions	GB 50046-95	中国寰球工程公司
37	GB 50049-2011	小型火力发电厂设计规范 Code for design of small fossil fired power plant	GB 50049-94	河南省电力勘测设计院
38	GB 50050-2007	工业循环冷却水处理设计规范 Code for design of industrial recirculating cooling water treatment	GB 50050-95	中国寰球工程公司

工程建设国家标准

序号	标准编号	标准名称	被代替标准编号	主编单位
39	GB 50051-2013	烟囱设计规范 Code for design of chimneys	GB 50051-2002	中冶东方工程技术有限公司
40	GB 50052-2009	供配电系统设计规范 Code for design electric power supply systems	GB 50052-95	中国联合工程公司
41	GB 50053-2013	20kV及以下变电所设计规范 Code for design of 20kV and below substation	GB 50053-94	中机中电设计研究院有限公司
42	GB 50054-2011	低压配电设计规范 Code for design of low voltage electrical installations	GB 50054-95	中机中电设计研究院有限公司
43	GB 50055-2011	通用用电设备配电设计规范 Code for design of electric distribution of general-purpose utilization equipment	GB 50055-93 GBJ 55-83	中国新时代国际工程公司
44	GB 50056-93	电热设备电力装置设计规范 Code for design of electrical equipment of electroheat installations	GBJ 56-83	机械工业部设计研究院
45	GB 50057-2010	建筑物防雷设计规范 Code for design protection of structures against lightning	GB 50057-94 （2000年版）	中国中元国际工程公司
46	GB 50058-2014	爆炸危险环境电力装置设计规范 Code for design of electrical installations in explosive atmospheres	GB 50058-92	中国寰球工程公司
47	GB 50059-2011	35kV～110kV变电站设计规范 Code for design of 35kV～110kV substation	GB 50059-92	华东电力设计院 上海电力设计院有限公司
48	GB 50060-2008	3～110kV高压配电装置设计规范 Code for design of 3～110kV high voltage electrical installations	GB 50060-92	中国电力工程顾问集团西北电力设计院
49	GB 50061-2010	66kV及以下架空电力线路设计规范 Code for design of 66kV or under overhead electrical power transmission line	GB 50061-97	辽宁电力勘测设计院
50	GB/T 50062-2008	电力装置的继电保护和自动装置设计规范 Code for design of relaying protection and automatic device of electric power installation	GB 50062-92	中国电力工程顾问集团东北电力设计院
51	GB/T 50063-2008	电力装置的电测量仪表装置设计规范 Code for design of electrical measuring device of power system	GBJ 63-90	中国电力工程顾问集团西南电力设计院
52	GB/T 50064-2014	交流电气装置的过电压保护和绝缘配合设计规范 Code for design of overvoltage protection and insulation coordination for AC electrical installations	GBJ 64-83	中国电力科学研究院
53	GB/T 50065-2011	交流电器装置的接地设计规范 Code for design of ac electrical installations earthing	GBJ 65-83	中国电力科学研究院
54	GB 50067-2014	汽车库、修车库、停车场设计防火规范 Code for fire protection design of garage, motor repair shop and parking area	GB 50067-97	上海市公安消防总队
55	GB 50068-2001	建筑结构可靠度设计统一标准 Unified standard for reliability design of building structures	GBT 68-84	中国建筑科学研究院

工程建设国家标准

序号	标准编号	标准名称	被代替标准编号	主编单位
56	GB 50069-2002	给水排水工程构筑物结构设计规范 Structural design code for special structures of water supply and waste water engineering	GBT 69-84	北京市政工程设计研究总院
57	GB 50070-2009	矿山电力设计规范 Code for design of electric power in mine	GB 50070-94	中煤国际工程集团北京华宇工程有限公司
58	GB 50071-2014	小型水力发电站设计规范 Design code for small hydropower station	GB 50071-2002	水利部水利水电规划设计总院 四川省水利水电勘测设计研究院
59	GB 50072-2010	冷库设计规范 Code for design of cold store	GB 50072-2001	国内贸易工程设计研究院
60	GB 50073-2013	洁净厂房设计规范 Code for design of clean room	GB 50073-2001	中国电子工程设计院
61	GB 50074-2014	石油库设计规范 Code for design of oil depot	GB 50074-2002	中国石化工程建设有限公司
62	GB/T 50076-2013	室内混响时间测量规范 Code for measurement of the reverberation time in rooms	GBJ 76-84	清华大学建筑学院
63	GB 50077-2003	钢筋混凝土筒仓设计规范 Code for design of reinforced concrete silos	GBJ 77-85	中煤国际工程设计研究总院
64	GB 50078-2008	烟囱工程施工及验收规范 Code for construction and acceptance of chimney engineering	GBJ 78-85	中冶京唐建设有限公司
65	GB/T 50080-2016	普通混凝土拌合物性能试验方法标准 Standard for test method of performance on ordinary fresh concrete mixture	GBJ 80-85 GB/T 50080-2002	中国建筑科学研究院
66	GB/T 50081-2002	普通混凝土力学性能试验方法标准 Standard for test method of mechanical properties of ordinary concrete	GBJ 81-85	中国建筑科学研究院
67	GB/T 50082-2009	普通混凝土长期性能和耐久性能试验方法标准 Standard for test methods of long-term performance and durability of ordinary concrete	GBJ 82-85	中国建筑科学研究院
68	GB/T 50083-2014	工程结构设计基本术语标准 Standard for general terms used in design of engineering structures	GB/T 50083-97	中国建筑科学研究院
69	GB 50084-2001 (2005年版)	自动喷水灭火系统设计规范 Code of design for sprinkler systems	GBJ 84-85	公安部天津消防科学研究所
70	GB/T 50085-2007	喷灌工程技术规范 Technical code for sprinkling engineering	GBJ 85-85	北京工业大学继续教育学院 中国水利水电科学研究院
71	GB 50086-2015	岩土锚杆与喷射混凝土支护工程技术规范 Technical code for engineering of ground anchoragea and shotcrete support	GBJ 86-85 GB 50086-2001	中冶建筑研究总院有限公司
72	GB/T 50087-2013	工业企业噪声控制设计规范 Code for design of noise control of industrial enterprises	GBJ 87-85	北京市劳动保护科学研究所
73	GB 50089-2007	民用爆破器材工程设计安全规范 Safety code for design of engineering of civil expolosives materials	GB 50089-98	五洲工程设计研究院
74	GB 50090-2006	铁路线路设计规范 Code for design of railway line	GB 50090-99	铁道第一勘察设计院

工程建设国家标准

序号	标准编号	标准名称	被代替标准编号	主编单位
75	GB 50091-2006	铁路车站及枢纽设计规范 Code for design of railway stations and railway terminals	GB 50091-99	铁道第一勘察设计院
76	GB 50092-96	沥青路面施工及验收规范 Code for construction and acceptance of asphalt pavement	GBJ 92-86	交通部公路科学研究所
77	GB 50093-2013	自动化仪表工程施工及质量验收规范 Code for construction and quality acceptance of automation instrumentation engineering	GB 50093-2002 GB 50131-2007	中国化学工程第十一建设有限公司 全国化工施工标准化管理中心站
78	GB 50094-2010	球形储罐施工规范 Code for construction of spherical tanks	GB 50094-98	中国石油天然气第一建设公司
79	GB/T 50095-2014	水文基本术语和符号标准 Standard for essential terminology and symbol in hydrology	GB/T 50095-98	水利部水文局
80	GB 50096-2011	住宅设计规范 Design code for residential buildings	GB 50096-1999 (2003年版)	中国建筑设计研究院
81	GBJ 97-87	水泥混凝土路面施工及验收规范 Code for construction and acceptance of cement concrete pavement		浙江省交通厅
82	GB 50098-2009	人民防空工程设计防火规范 Code for fire protection design of civil air defence works	GB 50098-98	总参工程兵第四设计研究院
83	GB 50099-2011	中小学校设计规范 Code for design of school	GBJ 99-86	北京市建筑设计研究院 天津市建筑设计院
84	GB/T 50102-2014	工业循环水冷却设计规范 Code for design of cooling for industrial recirculating water	GB/T 50102-2003	中国电力工程顾问集团东北电力设计院
85	GB/T 50103-2010	总图制图标准 Standard for general layout drawings	GB/T 50103-2001 GBJ 103-87	中国建筑标准设计研究院
86	GB/T 50104-2010	建筑制图标准 Standard for architectural drawings	GB/T 50104-2001 GBJ 104-87	中国建筑标准设计研究院
87	GB/T 50105-2010	建筑结构制图标准 Standard for structural drawings	GB/T 50105-2001 GBJ 105-87	中国建筑标准设计研究院
88	GB/T 50106-2010	建筑给水排水制图标准 Standard for building water supply and drainage drawings	GB/T 50106-2001 GBJ 106-87	中国建筑标准设计研究院
89	GB/T 50107-2010	混凝土强度检验评定标准 Standard for evaluation of concrete compressive strength	GBJ 107-87	中国建筑科学研究院
90	GB 50108-2008	地下工程防水技术规范 Technical code for waterproofing of underground works	GB 50108-2001 GBJ 108-87 GBJ 208-83	总参工程兵科研三所
91	GB/T 50109-2014	工业用水软化除盐设计规范 Design code for softening and demineralization of industrial water	GB/T 50109-2006	中国电力工程顾问集团西北电力设计院
92	GBJ 110-87	卤代烷1211灭火系统设计规范 Specifications for design of haloalkane 1211 fire-fighting systems		公安部天津消防科学研究所

工程建设国家标准

序号	标准编号	标准名称	被代替标准编号	主编单位
93	GB 50111-2006（2009年版）	铁路工程抗震设计规范 Code for seismic design of railway engineering	GBJ 111-87	中铁第一勘察设计院集团有限公司
94	GB 50112-2013	膨胀土地区建筑技术规范 Technical code for buildings in expansive soil regions	GBJ 112-87	中国建筑科学研究院
95	GB 50113-2005	滑动模板工程技术规范 Technical code slipform engineering	GBJ 113-87	中冶集团建筑研究总院
96	GB/T 50114-2010	暖通空调制图标准 Standard for heating, ventilation and air conditioning drawings	GB/T 50114-2001 GBJ 114-88	中国建筑标准设计研究院
97	GB 50115-2009	工业电视系统工程设计规范 Code for design of industrial television system	GBJ 115-87	中冶京诚工程技术有限公司
98	GB 50116-2013	火灾自动报警系统设计规范 Code for design of automatic fire alarm system	GB 50116-98 GBJ 116-88	公安部沈阳消防研究所
99	GB 50117-2014	构筑物抗震鉴定标准 Standard for seismic appraisal of special structures	GBJ 117-88	中冶建筑研究总院有限公司
100	GBJ 118-2010	民用建筑隔声设计规范 Code for design of sound insulation of civil buildings	GBJ 118-88	中国建筑科学研究院
101	GB 50119-2013	混凝土外加剂应用技术规范 Code for concrete admixture application	GB 50119-2003	中国建筑科学研究院
102	GB/T 50121-2005	建筑隔声评价标准 Standard for assessment of building sound insulation	GBJ 121-88	北京市建筑设计研究院
103	GBJ 122-88	工业企业噪声测量规范 Code for measurement of noise in industrial enterprises		北京市劳动保护科学研究所
104	GB/T 50123-1999	土工试验方法标准 Standard for soil test method	GBJ 123-88	南京水利科学研究院
105	GBJ 124-88	道路工程术语标准 Standard for terms used in road engineering		交通部公路规划设计院
106	GB/T 50125-2010	给水排水工程基本术语标准 Standard for basic terms of water and wastewater engineering	GBJ 125-89	上海市政工程设计研究总院 腾达建设集团股份有限公司
107	GB 50126-2008	工业设备及管道绝热工程施工规范 Code for construction of industrial equipment and pipeline insulation engineering	GBJ 126-89	全国化工施工标准化管理中心站
108	GB 50127-2007	架空索道工程技术规范 Technical standard for aerial ropeway engineering	GBJ 127-89	昆明有色冶金设计研究院
109	GB 50128-2014	立式圆筒形钢制焊接储罐施工规范 Code for construction of vertical cylindrical steel welded storage tanks	GB 50128-2005	中国石油天然气第一建设公司
110	GB/T 129-2011	砌体基本力学性能试验方法标准 Standard for test method of basic mechanics properties of masonry	GBJ 129-90	四川省建筑科学研究院 山西四建集团有限公司
111	GBJ 130-90	钢筋混凝土升板结构技术规范 Technical code for reinforced concrete lift-slab structure		中国建筑科学研究院

工程建设国家标准

序号	标准编号	标准名称	被代替标准编号	主编单位
112	GB/T 50132-2014	工程结构设计通用符号标准 Standard for general symbols used in design of engineering structures	GBJ 132-90	中国建筑科学研究院
113	GB 50134-2004	人民防空工程施工及验收规范 Code for construction and acceptance of civil air defence works	GBJ 134-90	辽宁省人防建筑设计研究院
114	GB 50135-2006	高耸结构设计规范 Code for design of high-rising structures	GBJ 135-90	同济大学
115	GB 50136-2011	电镀废水治理设计规范 Code for design of electroplating wastewater processing	GBJ 136-90	中国新时代国际工程公司
116	GB 50137-2011	城市用地分类与规划建设用地标准 Code for classification of urban land use and planning standards of development land	GBJ 137-90 GBJ 139-90	中国城市规划设计研究院
117	GB/T 50138-2010	水位观测标准 Standard for stage observation	GBJ 138-90	水利部长江水利委员会水文局
118	GB 50139-2014	内河通航标准 Navigation standard of inland waterway	GB 50139-2004	长江航道局
119	GB 50140-2005	建筑灭火器配置设计规范 Code for design of extinguisher distribution in buildings	GBJ 140-90	公安部上海消防研究所
120	GB 50141-2008	给水排水构筑物工程施工及验收规范 Code for construction and acceptance of water and sewerage structures	GBJ 141-90	北京市政建设集团有限责任公司
121	GBJ 142-90	中、短波广播发射台与电缆载波通信系统的防护间距标准 Standard for protective specing between medium, short wave broadcast transmitting station and electric cable carrier telecommunication systems		广播电影电视部标准化规划研究所
122	GBJ 143-90	架空电力线路、变电所对电视差转台、转播台无线电干扰防护间距标准 Standard for protective spacing of radio interference from overhead electric lines, substations to television transposer station and retransmitting station		广播电影电视部标准化规划研究所
123	GB 50144-2008	工业建筑可靠性鉴定标准 Standard for appraisal of reliability of industrial buildings and structures	GBJ 144-90	中冶建筑研究总院有限公司（原冶金工业部建筑研究总院）
124	GB/T 50145-2007	土的工程分类标准 Standard for engineering classification of soil	GBJ 145-90	南京水利科学研究院
125	GB/T 50146-2014	粉煤灰混凝土应用技术规范 Technical code for application of fly ash concrete	GBJ 146-90	中国水利水电科学研究院
126	GB 50147-2010	电气装置安装工程 高压电器施工及验收规范 Code for construction and acceptance of high-voltage electric equipment installation engineering	GBJ 147-90 GBJ 232-82（部分）	中国电力科学研究院（原国电电力建设研究所） 广东省输变电工程公司
127	GB 50148-2010	电气装置安装工程 电力变压器、油浸电抗器、互感器施工及验收规范 Code for construction and acceptance of power transformers oil reactor and mutual inductor	GBJ 148-90 GBJ 232-82（部分）	中国电力科学研究院（原国电电力建设研究所） 广东省输变电工程公司

工程建设国家标准

序号	标准编号	标准名称	被代替标准编号	主编单位
128	GB 50149-2010	电气装置安装工程 母线装置施工及验收规范 Code for construction and acceptance of busbar installation of electric equipment installation engineering	GBJ 149-90 GBJ 232-82（部分）	中国电力科学研究院 江苏电力建设第一工程公司
129	GB 50150-2016	电气装置安装工程 电气设备交接试验标准 Standard for hand-over test of electric equipment electric equipment installation engineering	GB 50150-91 GBJ 232-82（部分） GB 50150-2006	国网北京电力建设研究院
130	GB 50151-2010	泡沫灭火系统设计规范 Code for design of foam extinguishing systems	GB 50151-92 （2000年版） GB 50196-93 （2002年版）	公安部天津消防研究所
131	GB/T 50152-2012	混凝土结构试验方法标准 Standard for test method of concrete structures	GB 50152-92	中国建筑科学研究院 中建国际建设有限公司
132	GB 50153-2008	工程结构可靠性设计统一标准 Unified standard for reliability design of engineering structures	GB 50153-92	中国建筑科学研究院
133	GB 50154-2009	地下及覆土火药炸药仓库设计安全规范 Safety code for design of underground and earth covered magazine of powders and explosives	GB 50154-92	兵器工业安全技术研究所
134	GB/T 50155-2015	供暖通风与空气调节术语标准 Standard for terminology of heating, ventilation and air conditioning	GB 50155-92	亚太建设科技信息研究院有限公司 中国建筑设计院有限公司
135	GB 50156-2012 （2014年版）	汽车加油加气站设计与施工规范 Code for design and construction of filling station	GB 50156-2002 （2006年版）	中国石化工程建设有限公司
136	GB 50157-2013	地铁设计规范 Code for design of metro	GB 50157-2003	北京城建设计研究总院有限责任公司 中国地铁工程咨询有限责任公司
137	GB 50158-2010	港口工程结构可靠性设计统一标准 Unified standard for reliability design of port engineering structures	GB 50158-92	中国交通建设股份有限公司
138	GB/T 50159-2015	河流悬移质泥沙测验规范 Code for measurements of suspended load in open channels	GB 50159-92	水利部水文局（水利部水利信息中心）
139	GB 50160-2008	石油化工企业设计防火规范 Fire prevention code of petrochemical enterprise design	GB 50160-92 （1999年）	中国石化集团洛阳石油化工工程公司 中国石化工程建设公司
140	GB 50161-2009	烟花爆竹工程设计安全规范 Safety code for design of engineering of fireworks and firecracker	GB 50161-92	兵器工业安全技术研究所 国家安全生产宜春烟花爆竹检测检验中心
141	GB 50162-92	道路工程制图标准 Drawing standards of road engineering		交通部公路规划设计院
142	GB 50163-92	卤代烷1301灭火系统设计规范 Code for design of halon 1301 fire extinguishing systems		公安部天津消防科学研究所
143	GB 50164-2011	混凝土质量控制标准 Standard for quality control of concrete	GB 50164-92	中国建筑科学研究院 北京中关村开发建设有限公司
144	GB 50165-92	古建筑木结构维护与加固技术规范 Technical code for maintenance and strengthing of ancient timber buildings		四川省建筑科学研究院

工程建设国家标准

序号	标准编号	标准名称	被代替标准编号	主编单位
145	GB 50166-2007	火灾自动报警系统施工及验收规范 Code for installation and acceptance of fire alarm system	GB 50166-92	公安部沈阳消防研究所
146	GB 50167-2014	工程摄影测量规范 Code for engineering photogrammetry	GB 50167-92	中国有色金属工业西安勘察设计研究院
147	GB 50168-2006	电气装置安装工程 电缆线路施工及验收规范 Code for construction and acceptance of cable system electric equipment installation engineering	GB 50168-92	国网北京电力建设研究院
148	GB 50169-2016	电气装置安装工程 接地装置施工及验收规范 Code for construction and acceptance of grounding connection electric equipment installation engineering	GB 50169-92 GB 50169-2006	中国电力科学研究院
149	GB 50170-2006	电气装置安装工程 旋转电机施工及验收规范 Code for construction and acceptance of rotating electrical machines electric equipment installation engineering	GB 50170-92	国网北京电力建设研究院
150	GB 50171-2012	电气装置安装工程 盘、柜及二次回路接线施工及验收规范 Code for construction and acceptance of switchbord outfit complete cubicle and secondary circuit electric equipment installation engineering	GB 50171-92	广东火电工程总公司 中国电力企业联合会
151	GB 50172-2012	电气装置安装工程 蓄电池施工及验收规范 Code for construction and acceptance of battery electric equipment installation engineering	GB 50172-92	湖南省火电建设公司 中国电力企业联合会
152	GB 50173-2014	电气装置安装工程 66kV及以下架空电力线路施工及验收规范 Code for construction and acceptance of 66kV and under overhead electric power transmission line	GB 50173-92	中国电力科学研究院 葛洲坝集团电力有限责任公司
153	GB 50174-2008	电子信息系统机房设计规范 Code for design of electronic information system room	GB 50174-93	中国电子工程设计院
154	GB 50175-2014	露天煤矿工程质量验收规范 Code for acceptance of surface coal mine engineering	GB 50175-93	中煤平朔集团有限公司
155	GB 50176-2016	民用建筑热工设计规范 Thermal design code for civil building	GB 50176-93	中国建筑科学研究院
156	GB 50177-2005	氢气站设计规范 Design code for hydrogen station	GB 50177-93	中国电子工程设计院
157	GB 50178-93	建筑气候区划标准 Standard for climatic regionalization for architecture		中国建筑科学研究院
158	GB 50179-2015	河流流量测验规范 Code for liquid flow measurment in open channels	GB 50179-93	水利部水文局（水利部水利信息中心）
159	GB 50180-93 (2016年版)	城市居住区规划设计规范 Code of urban residential areas planning & design	GB 50180-93 (2002年版)	中国城市规划设计研究院

工程建设国家标准

序号	标准编号	标准名称	被代替标准编号	主编单位
160	GB 50181-93 (1998年版)	蓄滞洪区建筑工程技术规范 Technical code for building and civil engineering in retarding basin		中国建筑科学研究院
161	GB 50183-2004	石油天然气工程设计防火规范 Code for fire protection design of crude oil and natural gas works	GB 50183-93	中国石油天然气股份有限责任公司西南分公司
162	GB 50184-2011	工业金属管道工程施工质量验收规范 Code for acceptance of construction quality of industrial metallic piping	GB 50184-93	中国石油和化工勘察设计协会 中国化学工程第三建设有限公司
163	GB 50185-2010	工业设备及管道绝热工程施工质量验收规范 Code for acceptance of construction quality of industrial equipment and pipeline insulation engineering	GB 50185-93	中国石油化工勘察设计协会 全国化工施工标准化管理中心站
164	GB/T 50186-2013	港口工程基本术语标准 Standard for fundamental terms of port engineering	GB 50186-93	中交水运规划设计院有限公司
165	GB 50187-2012	工业企业总平面设计规范 Code for design of general layout of industrial enterprises	GB 50187-93	中冶南方工程技术有限公司
166	GB 50188-2007	镇规划标准 Standard for planning of town	GB 50188-93 GB 50188-2006	中国建筑设计研究院
167	GB 50189-2015	公共建筑节能设计标准 Design standard for energy efficiency of public buildings	GB 50189-2005 GB 50189-93	中国建筑科学研究院
168	GB 50190-93	多层厂房楼盖抗微振设计规范 Code for design of anti-microvibration of multistory factory floor		机械工业部设计研究院
169	GB 50191-2012	构筑物抗震设计规范 Code for seismic design of special structures	GB 50191-93	中冶建筑研究总院有限公司
170	GB 50193-93 (2010年版)	二氧化碳灭火系统设计规范 Code for design of carbon dioxide fire extinguishing systems	GB 50193-93 (1999年版)	公安部天津消防研究所
171	GB 50194-2014	建设工程施工现场供用电安全规范 Code for safety of power supply and consumption for construction site	GB 50194-93	中国电力企业联合会 河南省第二建设集团有限公司
172	GB 50195-2013	发生炉煤气站设计规范 Design code for producer gas station	GB 50195-94	中国中元国际工程公司
173	GB 50197-2015	煤炭工业露天矿设计规范 Code for design of open pit mine of coal industry	GB 50197-2005	中国煤炭建设协会勘察设计委员会 中煤科工集团沈阳设计研究院有限公司
174	GB 50198-2011	民用闭路监视电视系统工程技术规范 Technical code for project of civil closed circuit monitoring television system	GB 50198-94	武汉市广播影视局
175	GB 50199-2013	水利水电工程结构可靠性设计统一标准 Unified standard for reliability design of hydraulic engineering structures	GB 50199-94	水电水利规划设计总院
176	GB 50200-94	有线电视系统工程技术规范 Technical code for regunation of CATV system		武汉广播电视局
177	GB 50201-2014	防洪标准 Standard for flood control	GB 50201-94	水利部水利水电规划设计总院 黄河勘测规划设计有限公司

工程建设国家标准

序号	标准编号	标准名称	被代替标准编号	主编单位
178	GB 50201-2012	土方与爆破工程施工及验收规范 Code for construction and acceptance of earthwork and blasting engineering	GB 50201-83	中国华西企业股份有限公司四川省建筑机械化工程公司
179	GB 50202-2002	建筑地基基础工程施工质量验收规范 Code for acceptance of construction quality of building foundation	GBJ 201-83 GBJ 202-83 两个规范中有关土方工程	上海市基础工程公司
180	GB 50203-2011	砌体结构工程施工质量验收规范 Code for acceptance of constructional quality of masonry structures	GB 50203-2002	陕西省建筑科学研究院 陕西建工集团总公司
181	GB 50204-2015	混凝土结构工程施工质量验收规范 Code for quality acceptance of concrete structure construction	GB 50204-2002 （2010年版）	中国建筑科学研究院
182	GB 50205-2001	钢结构工程施工质量验收规范 Code for acceptance of construction quality of steel structure	GB 50205-95 GB 50221-95	冶金工业部建筑研究总院
183	GB 50206-2012	木结构工程施工质量验收规范 Code for acceptanc of construction quality of timber structures	GB 50206-2002	哈尔滨工业大学 中建新疆建工（集团）有限公司
184	GB 50207-2012	屋面工程质量验收规范 Code for acceptance of construction quality of roof	GB 50207-2002	山西建筑工程（集团）总公司 上海市第二建筑有限公司
185	GB 50208-2011	地下防水工程质量验收规范 Code for acceptance of construction quality of underground waterproof	GB 50208-2002	山西建筑工程（集团）总公司 福建省闽南建筑工程（集团）有限公司
186	GB 50209-2010	建筑地面工程施工质量验收规范 Code for acceptance of construction quality of building ground	GB 50209-2002	江苏省建筑工程集团有限公司 江苏省华建建设股份有限公司
187	GB 50210-2001	建筑装饰装修工程质量验收规范 Code for construction quality acceptance of building decoration	GBJ 210-83 JGJ 73-91 GBJ 301-88中第十、第十一章	中国建筑科学研究院
188	GB 50211-2014	工业炉砌筑工程施工与验收规范 Code for construction and acceptance of industrial furnaces building	GB 50211-2004	中冶武汉冶金建筑研究院有限公司 宏盛建业投资集团有限公司
189	GB 50212-2014	建筑防腐蚀工程施工规范 Code for construction of building anticorrosive engineering	GB 50212-2002	中国石油和化工勘察设计协会 全国化工施工标准化管理中心站
190	GB 50213-2010	煤矿井巷工程质量验收规范 Code for acceptance of shaft sinking and drifting of coal mine	GBJ 213-90	中煤第五建设有限公司
191	GB/T 50214-2013	组合钢模板技术规范 Technical code for composite steel-form	GB 50214-2001 GBJ 214-89	中冶建筑研究总院有限公司 中国模板协会
192	GB 50215-2015	煤炭工业矿井设计规范 Code for design of mine of coal industry	GB 50215-94 GB 50215-2005	中国煤炭科工集团南京设计研究院有限公司 中国煤炭建设协会勘察设计委员会
193	GB 50216-94	铁路工程结构可靠度设计统一标准 Unified design standard for reliability of railway engineering structures		铁道部科学研究院

工程建设国家标准

序号	标准编号	标准名称	被代替标准编号	主编单位
194	GB 50217-2007	电力工程电缆设计规范 Code for design of cables of electric engineering	GB 50217-94	中国电力工程顾问集团西南电力设计院
195	GB/T 50218-2014	工程岩体分级标准 Standard for engineering classification of rock mass	GB/T 50218-94	长江水利委员会长江科学院
196	GB 50219-2014	水喷雾灭火系统技术规范 Technical code for water spray fire protection systems	GB 50219-95	公安部天津消防研究所
197	GB 50220-95	城市道路交通规划设计规范 Code for planning design of transport on urban road		同济大学
198	GB 50222-95 (2001年版)	建筑内部装修设计防火规范 Code for fire prevention design of interior decoration of buildings		中国建筑科学研究院
199	GB 50223-2008	建筑工程抗震设防分类标准 Standard for classification of seismic protection of buildings constructions	GB 50223-2004	中国建筑科学研究院
200	GB 50224-2010	建筑防腐蚀工程施工质量验收规范 Code for acceptance of construction quality of anticorrosive engineering of buildings	GB 50224-95	全国化工施工标准化管理中心站
201	GB 50225-2005	人民防空工程设计规范 Code for design of civil air defence works	GB 50225-95	
202	GB 50226-2007 (2011年版)	铁路旅客车站建筑设计规范 Code for design of railway passenger station buildings	GB 50226-95	铁道第三勘察设计院集团有限公司
203	GB 50227-2008	并联电容器装置设计规范 Code for design of installation of shunt capacitors	GB 50227-95	中国电力工程顾问集团西南电力设计院 济南迪生电子电气有限公司
204	GB/T 50228-2011	工程测量基本术语标准 Standard for foundational terminology of engineering survey	GB/T 50228-96	中国有色金属工业西安勘察设计研究院
205	GB 50229-2006	火力发电厂与变电站设计防火规范 Code for design of fire protection for foossil fuel power plants and substations	GB 50229-96	中国电力工程顾问集团东北电力设计院
206	GB 50231-2009	机械设备安装工程施工及验收通用规范 General code for construction and acceptance of mechanical equipment installation engineering	GB 50231-98	中国机械工业建设总公司
207	GB 50233-2014	110kV~750kV架空输电线路施工及验收规范 Code for construction and acceptance of 110kV~750kV overhead transmission line	GB 50233-2005 GB 50389-2006	中国电力企业联合会 中国电力科学研究院
208	GB 50235-2010	工业金属管道工程施工规范 Code for construction of industrial metallic piping engineering	GB 50235-1997	中国石油和化工勘察设计协会 中国化学工程第三建设有限公司
209	GB 50236-2011	现场设备、工业管道焊接工程施工规范 Code for construction of field equipment, industrial pipe welding engineering	GB 50236-98	中国石油和化工勘察设计协会 中油吉林化建工程股份有限公司
210	GB 50242-2002	建筑给水排水及采暖工程施工质量验收规范 Code for acceptance of construction quality of water supply drainage and heating works	GBJ 242-82 和 GBJ 302-88 中"采暖卫生工程"部分	沈阳市城乡建设委员会

工程建设国家标准

序号	标准编号	标准名称	被代替标准编号	主编单位
211	GB 50243-2016	通风与空调工程施工质量验收规范 Code for acceptance for construction quality of ventilation and air conditioning works	GBJ 304-88 GB 50243-97 GB 50243-2002	上海市安装工程有限公司
212	GB 50251-2015	输气管道工程设计规范 Code for design of gas transmission pipeline engineering	GB 50251-2003	中国石油集团工程设计有限责任公司西南分公司
213	GB 50252-2010	工业安装工程施工质量验收统一标准 Unified standard for constructional quality acceptance of industrial installation engineering	GB 50252-94	全国化工施工标准化管理中心站
214	GB 50253-2014	输油管道工程设计规范 Code for design of oil transportation pipeline engineering	GB 50253-2003	中国石油天然气管道工程有限公司
215	GB 50254-2014	电气装置安装工程 低压电器施工及验收规范 Code for construction and acceptance of low-voltage apparatus electric equipment installation engineering	GB 50254-96	中国电力企业联合会 北京建工集团有限责任公司
216	GB 50255-2014	电气装置安装工程 电力变流设备施工及验收规范 Code for construction and acceptance of power conversion equipment electric equipment installation engineering	GB 50255-96	葛洲坝集团电力有限责任公司 中国电力科学研究院
217	GB 50256-2014	电气装置安装工程 起重机电气装置施工及验收规范 Code for construction and acceptance of electric device of crane electrical equipment installation engineering	GB 50256-96	中国电力企业联合会 国核工程有限公司
218	GB 50257-2014	电气装置安装工程 爆炸和火灾危险环境电气装置施工及验收规范 Code for construction and acceptance of electric equipment on fire and explosion hazard electrical equipment installation engineering	GB 50257-96	中国电力企业联合会 国核工程有限公司
219	GB 50260-2013	电力设施抗震设计规范 Code for seismic design of electrical installations	GB 50260-96	中国电力工程顾问集团西北电力设计院
220	GB 50261-2005	自动喷水灭火系统施工及验收规范 Code for installation and commissioning of sprinkler systems	GB 50261-96	公安部四川消防研究所
221	GB/T 50262-2013	铁路工程基本术语标准 Standard for basic terms of railway engineering	GB/T 50262-97	铁道第三勘察设计集团有限公司
222	GB 50263-2007	气体灭火系统施工及验收规范 Code for installation and acceptance of gas fire extinguishing systems	GB 50263-97	公安部天津消防研究所
223	GB 50264-2013	工业设备及管道绝热工程设计规范 Code for design of industrial equipment and pipeline insulation engineering	GB 50264-97	中国石油和化工勘察设计协会 中国成达工程有限公司
224	GB 50265-2010	泵站设计规范 Design code for pumping station	GB/T 50265-97	湖北省水利水电勘测设计院
225	GB/T 50266-2013	工程岩体试验方法标准 Standard for test methods of engineering rock mass	GB/T 50266-99	中国水电顾问集团成都勘测设计研究院 水利水电规划设计总院 中国电力企业联合会

工程建设国家标准

序号	标准编号	标准名称	被代替标准编号	主编单位
226	GB 50267-97	核电厂抗震设计规范 Code for seismic design of unclear power plants		国家地震局工程力学研究所
227	GB 50268-2008	给水排水管道工程施工及验收规范 Code for construction and acceptance of water and sewerage pipeline works	GB 50268-97 CJJ 3-90	北京市政建设集团有限公司
228	GB/T 50269-2015	地基动力特性测试规范 Code for measurement method of dynamic properties of subsoil	GB/T 50269-97	机械工业勘察设计研究院有限公 中国机械工业集团有限公司
229	GB 50270-2010	输送设备安装工程施工及验收规范 Code for construction and acceptance of conveyor equipment installation engineering	GB 50270-98	中国机械工业建设总公司 中国机械工业机械化施工公司 北京市工业设计研究院
230	GB 50271-2009	金属切削机床安装工程施工及验收规范 Code for construction and acceptance of metal-cutting machine installation engineering	GB 50271-98	中国机械工业建设总公司
231	GB 50272-2009	锻压设备安装工程施工及验收规范 Code for construction and acceptance of forging-press equipment installation engineering	GB 50272-98	中国机械工业建设总公司
232	GB 50273-2009	锅炉安装工程施工及验收规范 Code for construction and acceptance of boiler installation engineering	GB 50273-98	中国机械工业建设总公司
233	GB 50274-2010	制冷设备、空气分离设备安装工程施工及验收规范 Code for construction and acceptance of refrigeration plant and air seperation plant installation	GB 50274-98	中国机械工业建设总公司 中国机械工业第四建设工程公司 北京市工业设计研究院
234	GB 50275-2010	风机、压缩机、泵安装工程施工及验收规范 Code for construction and acceptance of fan, compressor and pump installation	GB 50275-98	中国机械工业建设总公司 中国机械工业第四建设工程公司 北京市工业设计研究院
235	GB 50276-2010	破碎、粉磨设备安装工程施工及验收规范 Code for construction and acceptance of crusher and grinding equipment installation engineering	GB 50276-98	中国机械工业建设总公司 中国机械工业第五建设工程公司 北京市工业设计研究院
236	GB 50277-2010	铸造设备安装工程施工及验收规范 Code for construction and acceptance of casting plant installation	GB 50277-98	中国机械工业建设总公司 中国机械工业第二建设工程公司 北京市工业设计研究院
237	GB 50278-2010	起重设备安装工程施工及验收规范 Code for construction and acceptance of crane installation engineering	GB 50278-98	中国机械工业建设总公司 中国机械工业机械化施工公司 北京市工业设计研究院
238	GB/T 50279-2014	岩土工程基本术语标准 Standard for fundamental terms of geotechnical engineering	GB/T 50279-98	水利部水利水电规划设计总院 南京水利科学研究院
239	GB/T 50280-98	城市规划基本术语标准 Standard for basic terminology of urban planning		中国城市规划设计研究院
240	GB 50281-2006	泡沫灭火系统施工及验收规范 Code for installation and acceptance of foam fire extinguishing systems	GB 50281-98	公安部天津消防研究所
241	GB 50282-2016	城市给水工程规划规范 Code for urban water supply engineering planning	GB 50282-98	浙江省城乡规划设计研究院
242	GB/T 50283-1999	公路工程结构可靠度设计统一标准 Unified standard of reliability design of highway engineering structures		交通部公路规划设计院

工程建设国家标准

序号	标准编号	标准名称	被代替标准编号	主编单位
243	GB 50284-2008	飞机库设计防火规范 Code for fire protection design of aircraft hangar	GB 50284-98	中国航空工业规划设计研究院
244	GB 50285-98	调幅收音台和调频电视转播台与公路的防护间距标准 Standad for protection distance from highway to AM, FM and TV rebroadcast stations		广电总局标准化规划研究所
245	GB 50286-2013	堤防工程设计规范 Code for design of levee project	GB 50286-98	水利部水利水电规划设计总院
246	GB 50287-2016	水力发电工程地质勘察规范 Code for water resources and hydropower engineering gelolgical investigation	GB 50287-99 GB 50287-2006	水利水电规划设计总院
247	GB 50288-99	灌溉与排水工程设计规范 Code for design of irrigation and drainage		水利部农田灌溉研究所 华北水利水电学院北京研究生部 水利部水利水电规划设计总院
248	GB 50289-2016	城市工程管线综合规划规范 Code for urban engineering pipeline comprehensive planning	GB 50289-98	沈阳市规划设计研究院
249	GB/T 50290-2014	土工合成材料应用技术规范 Technical code for application of geosynthetics	GB 50290-98	水利部水利水电规划设计总院 中国水利水电科学研究院
250	GB/T 50291-2015	房地产估价规范 Code for real estate appraisal	GB/T 50291-1999	中国房地产估价师与房地产经纪人学会
251	GB 50292-2015	民用建筑可靠性鉴定标准 Standard for appraiser of reliability of civil buildings	GB 50292-1999	四川省建筑科学研究院
252	GB/T 50293-2014	城市电力规划规范 Code for planning of urban electric power	GB 50293-99	中国城市规划设计研究院
253	GB/T 50294-2014	核电厂总平面及运输设计规范 Design code for general plan and transportation of nuclear power plants	GB/T 50294-1999	中国核电工程有限公司
254	GB 50295-2016	水泥工厂设计规范 Code for design of cement plant	GB 50295-2008	天津水泥工业设计研究院有限公司
255	GB 50296-2014	管井技术规范 Technical code for tube well	GB 50296-99	中冶集团武汉勘察研究院有限公司
256	GB/T 50297-2006	电力工程基本术语标准 Standard for basic terms of electric power engineering	GB/T 50297-1999	中国电力企业联合会
257	GB 50298-1999	风景名胜区规划规范 Code for scenic area planning		中国城市规划设计研究院
258	GB 50299-1999 (2003年版)	地下铁道工程施工及验收规范 Code for construction and acceptance of metro engineering		北京城建集团有限责任公司
259	GB 50300-2013	建筑工程施工质量验收统一标准 Unified standard for constructional quality acceptance of building engineering	GB 50300-2001	中国建筑科学研究院
260	GB 50303-2015	建筑电气工程施工质量验收规范 Code of acceptance of construction quality of building electrical engineering	GBJ 303-88 GB 50258-96 GB 50259-96 GB 50303-2002	浙江省工业设备安装集团有限公司

工程建设国家标准

序号	标准编号	标准名称	被代替标准编号	主编单位
261	GB 50307-2012	城市轨道交通岩土工程勘察规范 Code for geotechnical investigations of urban rail transit	GB 50307-1999	北京城建勘测设计研究院有限责任公司
262	GB 50308-2008	城市轨道交通工程测量规范 Code for urban rail transit engineering survey	GB 50308-1999	北京城建勘察设计研究院
263	GB 50309-2007	工业炉砌筑工程质量验收规范 Code for quality inspection and acceptance of industrial furnaces building	GB 50309-92	武汉冶金建筑研究院
264	GB 50310-2002	电梯工程施工质量验收规范 Code for acceptance of installation quality of lifts, escalators and passenger conveyors	GBJ 310-88 GB 50182-93	中国建筑科学研究院建筑机械研究分院
265	GB 50311-2016	综合布线系统工程设计规范 Code for engineering design of generic cabling system	GB/T 50311-2000 GB 50311-2007	中国移动通信集团设计院
266	GB 50312-2016	综合布线系统工程验收规范 Code for engineering acceptance of generic cabling system	GB/T 50312-2000 GB 50312-2007	中国移动通信集团设计院
267	GB 50313-2013	消防通信指挥系统设计规范 Code for design of fire communication and command system	GB 50313-2000	公安部沈阳消防研究所
268	GB 50314-2015	智能建筑设计标准 Standard for design of intelligent building	GB/T 50314-2006	上海现代建筑设计（集团）有限公司
269	GB/T 50315-2011	砌体工程现场检测技术标准 Technical standard for site testing of masonry engineering	GB/T 50315-2000	四川省建筑科学院 成都建筑工程集团总公司
270	GB 50316-2000 （2008年版）	工业金属管道设计规范 Design code for industrial metallic piping	GB 50316-2000	中国寰球化学工程公司
271	GB 50317-2009	猪屠宰与分割车间设计规范 Code for design of pig's slaughtering and cutting rooms	GB 50317-2000	国内贸易工程设计研究院
272	GB 50318-2000	城市排水工程规划规范 Code of urban wastewater engineering planning		陕西省城乡规划设计研究院
273	GB/T 50319-2013	建设工程监理规范 Code of construction project management	GB 50319-2000	中国建设监理协会
274	GB 50320-2014	粮食平房仓设计规范 Code for design of grain storehouses	GB 50320-2001	河南工业大学 河南创达建设工程管理有限公司
275	GB 50322-2011	粮食钢板筒仓设计规范 Code for design of grain steel silos	GB 50322-2001	郑州粮油食品工程建筑设计院 郑州市第一建筑工程集团有限公司
276	GB/T 50323-2001	城市建设档案著录规范 Code for urban construction archives description		建设部城建档案工作办公室
277	GB 50324-2014	冻土工程地质勘察规范 Code for engineering geological investigation of frozen ground	GB 50324-2001	内蒙古筑业工程勘察设计有限公司 中国科学院寒区旱区环境与工程研究所
278	GB 50325-2010 （2013年版）	民用建筑工程室内环境污染控制规范 Code for indoor environmental pollution control of civil building engineering	GB 50325-2001	河南省建筑科学研究院有限公司 泰宏建设发展有限公司
279	GB/T 50326-2006	建设工程项目管理规范 The code of construction project management	GB/T 50326-2001	中国建筑业协会工程项目管理专业委员会

工程建设国家标准

序号	标准编号	标准名称	被代替标准编号	主编单位
280	GB 50327-2001	住宅装饰装修工程施工规范 Code for construction of decoration of housings		中国建筑装饰协会
281	GB/T 50328-2014	建设工程文件归档规范 Code for putting construction project documents into records	GB/T 50328-2001	住建部城建档案办公室 住建部科技与产业化发展中心
282	GB/T 50329-2012	木结构试验方法标准 Standard for test methods of timber structures	GB/T 50329-2002	重庆大学 中国新兴保信建设总公司
283	GB 50330-2013	建筑边坡工程技术规范 Technical code for building slope engineering	GB 50330-2002	重庆市设计院 中国建筑技术集团有限公司
284	GB/T 50331-2002	城市居民生活用水量标准 The standard of water quantity for city's residential use		住建部城市建设局
285	GB 50332-2002	给水排水工程管道结构设计规范 Structural design code for pipelines of water supply and waste water engineering	GBJ 69-84	北京市政工程设计研究总院
286	GB 50333-2013	医院洁净手术部建筑技术规范 Architectural technical code for hospital clean operating department	GB 50333-2002	中国建筑科学研究院
287	GB 50334-2002	城市污水处理厂工程质量验收规范 Quality acceptance code for municipal sewage treatment plant engineering		中国市政工程协会 天津市政工程局
288	GB 50335-2016	城镇污水再生利用工程设计规范 Code for design of wastewater reclamation and reuse	GB 50335-2002	中国市政工程东北设计研究院
289	GB 50336-2002	建筑中水设计规范 Code for design of building reclaimed water system		中国人民解放军总后勤部建筑设计研究院
290	GB 50337-2003	城市环境卫生设施规划规范 Code for planning of urban environmental sanitation facilities		成都市规划设计研究院
291	GB 50338-2003	固定消防炮灭火系统设计规范 Code of design for fixed fire monitor extinguishing systems		公安部上海消防研究所
292	GB 50339-2013	智能建筑工程质量验收规范 Code for acceptance of quality of intelligent building systems	GB 50339-2003	同方股份有限公司
293	GB 50340-2016	老年人居住建筑设计规范 Code for design of residential building for the aged	GB/T 50340-2003	中国建筑设计研究院 民政部社会福利和社会事务司
294	GB 50341-2014	立式圆筒形钢制焊接油罐设计规范 Code for design of vertical cylindrical welded steel oil tanks	GB 50341-2003	中国石油天然气管道工程有限公司
295	GB 50342-2003	混凝土电视塔结构技术规范 Technical code for concrete structure of TV tower		国家广播电影电视总局设计院
296	GB 50343-2012	建筑物电子信息系统防雷技术规范 Technical code for protection of building electronic information system against lightning	GB 50343-2004	中国建筑标准设计研究院 四川中光防雷科技股份有限公司
297	GB/T 50344-2004	建筑结构检测技术标准 Techical standard for inspection of building structure	GB/T 50344-2004	中国建筑科学研究院

工程建设国家标准

序号	标准编号	标准名称	被代替标准编号	主编单位
298	GB 50345-2012	屋面工程技术规范 Technical code for roof engineering	GB 50345-2004	山西建筑工程（集团）总公司 浙江省长城建设集团股份有限公司
299	GB 50346-2011	生物安全实验室建筑技术规范 Architectural and technical code for biosafety laboratories	GB 50346-2004	中国建筑科学研究院 江苏双楼建设集团有限公司
300	GB 50347-2004	干粉灭火系统设计规范 Code of design for powder extinguishing systems		公安部天津消防研究所
301	GB 50348-2004	安全防范工程技术规范 Technical code for engineering of security and protection system		公安部科技局 国家安全防范报警系统标准化技术委员会
302	GB 50349-2015	气田集输设计规范 Code for design of gas gathering and transportation system in gas field		中国石油集团工程设计有限责任公司西南分公司
303	GB 50350-2015	油田油气集输设计规范 Code for design of oil-gas gathering and transportation systems of oilfield	GB 50350-2005	大庆油田工程有限公司
304	GB 50351-2014	储罐区防火堤设计规范 Code for design of fire dike in storage tank farm	GB 50351-2005	中国石油天然气管道工程有限公司
305	GB 50352-2005	民用建筑设计通则 Code for design of civil buildings	JGJ 37-87	中国建筑设计研究院 中国建筑标准设计研究院
306	GB/T 50353-2013	建筑工程建筑面积计算规范 Calculation code for construction area of building	GB/T 50353-2005	住房和城乡建设部标准定额研究所
307	GB 50354-2005	建筑内部装修防火施工及验收规范 Code for fire prevention installation and acceptance in construction of interior decoration engineering of buildings		中国建筑科学院研究院
308	GB/T 50355-2005	住宅建筑室内振动限值及其测量方法标准 Standard of limit and measurement method of vibration in the room of residential buildings		住建部标准定额研究所
309	GB/T 50356-2005	剧场、电影院和多用途厅堂建筑声学设计规范 Code for architectural acoustical design of theater, cinema and multi-use auditorium		同济大学
310	GB 50357-2005	历史文化名城保护规划规范 Code of conservation planning for historic cities		中国城市规划设计研究院
311	GB/T 50358-2005	建设项目工程总承包管理规范 Code for management of engineering contracting projects		中国勘察设计协会
312	GB 50359-2016	煤炭洗选工程设计规范 Code for design of coal cleaning engineering	GB 50359-2005	中煤国际工程集团北京华宇工程有限公司
313	GB 50360-2016	水煤浆工程设计规范 Code for design of coal water mixture engineering	GB 50360-2005	中国煤炭建设协会勘察设计委员会 中煤科工集团北京华宇工程有限公司
314	GB/T 50361-2005	木骨架组合墙体技术规范 Technical code for partitions with timber frame work		国家建筑材料工业局标准定额中心站 中国建筑西南设计研究院
315	GB/T 50362-2005	住宅性能评定技术标准 Technical standard for performance assessment of residential buildings		住建部住宅产业化促进中心 中国建筑科学研究院
316	GB/T 50363-2006	节水灌溉工程技术规范 Technical specification for water-saving irrigation engineering		水利部农村水利司 中国灌溉排水发展中心

工程建设国家标准

序号	标准编号	标准名称	被代替标准编号	主编单位
317	GB 50364-2005	民用建筑太阳能热水系统应用技术规范 Technical code for solar water heating system of civil buildings		中国建筑设计研究院
318	GB 50365-2005	空调通风系统运行管理规范 Code for operation and management of central air conditioning system		中国建筑科学研究院 中国疾病预防控制中心
319	GB 50366-2005 (2009年版)	地源热泵系统工程技术规范 Technical code for ground source heat pump system	GB 50366-2005	中国建筑科学研究院
320	GB 50367-2013	混凝土结构加固设计规范 Code for design of strengthening concrete structure	GB 50367-2006	四川省建筑科学研究院 山西八建集团有限公司
321	GB 50368-2005	住宅建筑规范 Residential building code		中国建筑科学研究院
322	GB 50369-2014	油气长输管道工程施工及验收规范 Code for construction and acceptance of oil and gas long-distance transmission pipeline engineering	GB 50369-2006	中国石油天然气管道局
323	GB 50370-2005	气体灭火系统设计规范 Code for design of gas fire extinguishing systems		公安部天津消防研究所
324	GB 50371-2006	厅堂扩声系统设计规范 Code for sound reinforcement system design of auditorium		中广电广播电影电视设计研究院
325	GB 50372-2006	炼铁机械设备工程安装验收规范 Code for installation acceptance of metallurgical machinery ironmaking equipment engineering		上海宝冶建设有限公司
326	GB 50373-2006	通信管道与通道工程设计规范 Design code for communication conduit and passage engineering		中讯邮电咨询设计院
327	GB 50374-2006	通信管道工程施工及验收规范 Code of construction and acceptance for communication conduit engineering		中讯邮电咨询设计院
328	GB/T 50375-2016	建筑工程施工质量评价标准 Evaluating standard for excellent quality of building engineering	GB/T 50375-2006	中国建筑业协会
329	GB 50376-2015	橡胶工厂节能设计规范 Design code for energy saving of rubber factory	GB 50376-2006	中国石油和化工勘察设计协会 中国化学工业桂林工程有限公司
330	GB 50377-2006	选矿机械设备工程安装验收规范 Code for acceptance of mineral processing equipment installation engineering		中国第三冶金建设公司
331	GB/T 50378-2014	绿色建筑评价标准 Assessment standard for green building	GB/T 50378-2006	中国建筑科学研究院 上海市建筑科学研究院
332	GB/T 50379-2006	工程建设勘察企业质量管理规范 Code for quality management of geotechnical investigation enterprises		北京市勘察设计研究院
333	GB/T 50380-2006	工程建设设计企业质量管理规范 Code for quality management of engineering design enterprises		中国天辰化学工程公司
334	GB 50381-2010	城市轨道交通自动售检票系统工程质量验收规范 Code for constructional quality acceptance of urban rail transit automatic fare collection system engineering	GB 50381-2006	上海地铁咨询监理科技有限公司

工程建设国家标准

序号	标准编号	标准名称	被代替标准编号	主编单位
335	GB 50382-2016	城市轨道交通通信工程质量验收规范 Code for constructional quality acceptance of urban rail transit communication engineering	GB 50382-2006	中国铁路通信信号上海工程公司
336	GB 50383-2016	煤矿井下消防、洒水设计规范 Code for design of the fire protecting, sprinkling system in underground coalmine	GB 50383-2006	中煤邯郸设计工程有限责任公司
337	GB 50384-2016	煤矿立井井筒及硐室设计规范 Code for design of coal mine shaft and chamber	GB 50384-2007	中国煤炭建设协会勘察设计委员会 中煤科工集团南京设计研究院有限公司
338	GB 50385-2006	矿山井架设计规范 Code for design of the mine headframes		煤炭工业邯郸设计研究院
339	GB 50386-2016	轧机机械设备工程安装验收规范 Code for acceptance of enginneering installation for mechanical equipment of rolling mill	GB 50386-2006	中国二十冶集团有限公司
340	GB 50387-2006	冶金机械液压、润滑和气动设备工程安装验收规范 Code for installation acceptance of metal mechanical hydromatic, lubricating and air-driven system equipment		中国第一冶金建设有限责任公司
341	GB 50388-2016	煤矿井下机车车辆运输信号设计规范 Code for design of underground locomotive vehicle transport signal of coal mine	GB 50388-2006	中煤国际工程集团沈阳设计研究院
342	GB 50390-2006	焦化机械设备工程安装验收规范 Code for installation acceptance of coking and chemical mechanical equipment engineering		中国第五冶金建设公司
343	GB 50391-2014	油田注水工程设计规范 Code for design of oilfield water injection engineering	GB 50391-2006	大庆油田工程有限公司
344	GB/T 50392-2016	机械通风冷却塔工艺设计规范 Code for design of cooling tower for mechanical ventilation	GB/T 50392-2006	全国化工给排水设计技术中心站
345	GB 50393-2008	钢质石油储罐防腐蚀工程技术规范 Technical code for anticorrosive engineering of the steel petroleum tank		中国石化集团洛阳石油化工工程有限公司
346	GB 50394-2007	入侵报警系统工程设计规范 Code for design for intrusion alarm systems engineering		全国安全防范报警系统标准化技术委员会
347	GB 50395-2007	视频安防监控系统工程设计规范 Code of design for video monitoring system		全国安全防范报警系统标准化技术委员会
348	GB 50396-2007	出入口控制系统工程设计规范 Code of design for access control systems engineering		全国安全防范报警系统标准化技术委员会
349	GB 50397-2007	冶金电气设备工程安装验收规范 Code for acceptance of electrical construction installation in metallurgy		中国第三冶金建设公司
350	GB 50398-2006	无缝钢管工艺设计规范 Code for design of rolling process on seamless steel tubes		中冶东方工程技术有限公司

工程建设国家标准

序号	标准编号	标准名称	被代替标准编号	主编单位
351	GB 50399-2006	煤炭工业小型矿井设计规范 Code for design of small capacity mine of coal industry		中国煤炭建设协会勘察设计委员会
352	GB 50400-2016	建筑与小区雨水控制及利用工程技术规范 Engineering technical code for rain utilization in building and sub-district	GB 50400-2006	中国建筑设计研究院
353	GB 50401-2007	消防通信指挥系统施工及验收规范 Code for installation and acceptance of fire communication and command system		公安部沈阳消防研究所
354	GB 50402-2007	烧结机械设备工程安装验收规范 Code for installation acceptance of sintering mechanical equipment engineering		中国第十三冶金建设公司
355	GB 50403-2007	炼钢机械设备工程安装验收规范 Code for engineering installment acceptance of steel-making mechanical equipment		中国第一冶金建设有限责任公司
356	GB 50404-2007	硬泡聚氨酯保温防水工程技术规范 Technical code for rigid polyurethane foam insulation and waterproof engineering		烟台同化防水保温工程有限公司
357	GB 50405-2007	钢铁工业资源综合利用设计规范 Code for design of comprehensive utilization of iron and steel industry resources		中冶京诚工程技术有限公司
358	GB 50406-2007	钢铁工业环境保护设计规范 Code for design of environmental protection of iron and steel industry		中冶京诚工程技术有限公司
359	GB 50408-2015	烧结厂设计规范 Code for design of sintering plant	GB 50408-2007	中冶长天国际工程有限责任公司
360	GB 50410-2014	型钢轧钢工程设计规范 Code for design of section steel hot rolling mills	GB 50410-2007	中冶南方工程技术有限公司 中冶赛迪工程技术股份有限公司
361	GB 50411-2007	建筑节能工程施工质量验收规范 Code for acceptance of energy efficient building construction		中国建筑科学研究院
362	GB/T 50412-2007	厅堂音质模型试验规范 Code for test of scale acoustic model for auditorium		清华大学
363	GB 50413-2007	城市抗震防灾规划标准 Standard for urban planning on earthquake resistance and hazardous prevention		北京工业大学抗震减灾研究所 河北省地震工程研究中心
364	GB 50414-2007	钢铁冶金企业设计防火规范 Code of design for fire protection and prevention for iron & steel metallurgy enterprises		中冶京诚工程技术有限公司 首安工业消防有限公司
365	GB 50415-2007	煤矿斜井井筒及硐室设计规范 Code for design of inclind shafts and chambers of coal mine		中煤国际工程集团武汉设计研究院
366	GB 50416-2007	煤矿井底车场硐室设计规范 Code for design of chambers around pit-bottom of coal mine		中煤国际工程集团武汉设计研究院
367	GB 50417-2007	煤矿井下供配电设计规范 Code for design of electric power supply of under the coal mine		中煤国际工程集团武汉设计研究院

工程建设国家标准

序号	标准编号	标准名称	被代替标准编号	主编单位
368	GB 50418-2007	煤矿井下热害防治设计规范 Code for design of prevention and elimination of thermal disaster in coal mines		中煤国际工程集团武汉设计研究院
369	GB 50419-2007	煤矿巷道断面和交岔点设计规范 Code for design of roadway section and junction of coal mine		中煤西安设计工程有限责任公司
370	GB 50420-2007 （2016年版）	城市绿地设计规范 Code for the design of urban green space	GB 50420-2007	上海市园林设计院有限公司
371	GB 50421-2007	有色金属矿山排土场设计规范 Code for waste dump design of nonferrous metal mines		长沙有色冶金设计研究院
372	GB 50422-2007	预应力混凝土路面工程技术规范 Technical code for engineerings of prestressed concrete pavement		东南大学
373	GB 50423-2013	油气输送管道穿越工程设计规范 Code for design of oil and gas transportation pipeline crossing engineering	GB 50423-2007	中国石油天然气管道工程有限公司
374	GB 50424-2015	油气输送管道穿越工程施工规范 Code for construction of oil and gas transmission pipeline crossing engineering	GB 50424-2007	中国石油天然气管道局
375	GB 50425-2008	纺织工业企业环境保护设计规范 Code for design of environmental protection of textile industry enterprise		上海纺织建筑设计研究院
376	GB 50426-2016	印染工厂设计规范 Code for design of dyeing and printing plant	GB 50426-2007	中国纺织工业联合会 浙江省省直建筑设计院
377	GB 50427-2015	高炉炼铁工程设计规范 Code for design of blast furnace ironmaking plant	GB 50427-2008	中冶赛迪集团有限公司
378	GB 50428-2015	油田采出水处理设计规范 Code for design of oil field produced water treatment	GB 50428-2007	大庆油田工程有限公司
379	GB 50429-2007	铝合金结构设计规范 Code for design of aluminium structures		同济大学 现代建筑设计集团上海建筑设计研究院有限公司
380	GB/T 50430-2007	工程建设施工企业质量管理规范 Code for quality management of engineering construction enterprises		中国建筑业协会
381	GB 50431-2008	带式输送机工程设计规范 Code for design of belt conveyor engineering		中煤国际工程集团沈阳设计研究院
382	GB 50432-2007	炼焦工艺设计规范 Code for design of coking technology		中冶焦耐工程技术有限公司
383	GB 50433-2008	开发建设项目水土保持技术规范 Technical code on soil and water conservation of development and construction projects		水利部水土保持监测中心
384	GB 50434-2008	开发建设项目水土流失防治标准 Control standards for soil and water loss on development and construction projects		水利部水土保持监测中心
385	GB 50435-2016	平板玻璃工厂设计规范 Code for design of flat glass plant	GB 50435-2007	蚌埠玻璃工业设计研究院

工程建设国家标准

序号	标准编号	标准名称	被代替标准编号	主编单位
386	GB 50436-2007	线材轧钢工艺设计规范 Code for design of technology of wire rod mill		中冶东方工程技术有限公司
387	GB 50437-2007	城镇老年人设施规划规范 Code for planning of city and town facilities for the aged		南京市规划设计研究院
388	GB/T 50438-2007	地铁运营安全评价标准 Standard for the operation safety assessment of existing metro		北京市地铁运营有限公司
389	GB 50439-2015	炼钢工程设计规范 Code for design of steelmaking engineering	GB 50439-2008	中冶京诚工程技术有限公司
390	GB 50440-2007	城市消防远程监控系统技术规范 Technical code for remote-monitoring system of urban fire protection		公安部沈阳消防研究所
391	GB/T 50441-2016	石油化工设计能耗计算标准 Standard for calculation of energy consumption in petrochemical engineering design	GB/T 50441-2007	中国石化集团洛阳石油化工公司
392	GB 50442-2008	城市公共设施规划规范 Code for urban public facilities planning		天津市城市规划设计研究院
393	GB 50443-2016	水泥工厂节能设计规范 Code for design of energy conservation of cement plant	GB 50443-2007	中国水泥协会 天津水泥工业设计研究院有限公司
394	GB 50444-2008	建筑灭火器配置验收及检查规范 Code for acceptance and inspection of extinguisher distribution in buildings		公安部上海消防研究所
395	GB 50445-2008	村庄整治技术规范 Technique code for village rehabilitation		中国建筑设计研究院
396	GB 50446-2008	盾构法隧道施工与验收规范 Code for construction and acceptance of shield tunnelling method		住建部科技发展促进中心
397	GB 50447-2008	实验动物设施建筑技术规范 Architectural and technical code for laboratory animal facility		中国建筑科学研究院
398	GB/T 50448-2015	水泥基灌浆材料应用技术规范 Technical code for application of cementitious grout	GB/T 50448-2008	中冶建筑研究总院有限公司 鲲鹏建设集团有限公司
399	GB 50449-2008	城市容貌标准 Standard for urban appearance	CJ/T 12-1999	上海市容环境卫生管理局
400	GB 50450-2008	煤矿主要通风机站设计规范 Code for design of main ventilating fan station of coal mine		中煤邯郸设计工程有限责任公司 （原煤炭工业邯郸设计研究院）
401	GB 50451-2008	煤矿井下排水泵站及排水管路设计规范 Code for design of pumping station and pipeline of under coal mine		中煤邯郸设计工程有限责任公司 （原煤炭工业邯郸设计研究院）
402	GB/T 50452-2008	古建筑防工业振动技术规范 Technical specifications for protection of historic buildings against man-made vibration		五洲工程设计研究院
403	GB 50453-2008	石油化工建（构）筑物抗震设防分类标准 Standard for classification of seismic protection of buildings and special structures in petrochemical engineering		中国石化工程建设公司

工程建设国家标准

序号	标准编号	标准名称	被代替标准编号	主编单位
404	GB 50454-2008	航空发动机试车台设计规范 Code for design of aero-engine test cell		中国航空工业规划设计研究院
405	GB 50455-2008	地下水封石洞油库设计规范 Code for design of underground oil storage in rock caverns		青岛英派尔化学工程有限公司
406	GB 50457-2008	医药工业洁净厂房设计规范 Code for design of pharmaceutical industry clean room		中国石化集团上海工程有限公司
407	GB 50458-2008	跨座式单轨交通设计规范 Code for design of straddle monorail transit		重庆市轨道交通总公司
408	GB 50459-2009	油气输送管道跨越工程设计规范 Code for design of oil and gas transportation pipeline aerial crossing engineering		中国石油集团工程设计有限公司西南分公司
409	GB 50460-2015	油气输送管道跨越工程施工规范 Code for construction of oil and gas transmission pipeline aerial crossing engineering	GB 50460-2008	四川石油天然气建设工程有限责任公司
410	GB 50461-2008	石油化工静设备安装工程施工质量验收规范 Code for quality acceptance of static equipment installation in petrochemical engineering		中国石化集团第四建设公司
411	GB 50462-2015	数据中心基础设施工及验收规范 Code for construction and acceptance of data center infrastructure	GB 50462-2008	工业和信息化部电子工业标准化研究院电子工程标准定额站 中国机房设施工程有限公司
412	GB 50463-2008	隔振设计规范 Code for design of vibration isolation		中国中元国际工程公司
413	GB 50464-2008	视频显示系统工程技术规范 Code for technical of video display system engineering		中国电子科技集团公司第三研究所 北京奥特维科技开发总公司
414	GB 50465-2008	煤炭工业矿区总体规划规范 Code for general planning of mining area of coal industry		中国煤炭建设协会勘察设计委员会
415	GB/T 50466-2008	煤炭工业供热通风与空气调节设计规范 Code for design of heating ventilation and air conditioning of coal industry		中煤国际工程集团北京华宇工程公司
416	GB 50467-2008	微电子生产设备安装工程施工及验收规范 Code for construction and acceptance of micro-electronics manufacturing equipment installation engineering		中国电力系统工程第二建设公司
417	GB 50468-2008	焊管工艺设计规范 Code for design of welded-pipe process		中冶赛迪工程技术股份有限公司
418	GB 50469-2016	橡胶工厂环境保护设计规范 Code for design of environmental protection of rubber factory	GB 50469-2008	中国石油和化工勘察设计协会 全国橡胶塑料设计技术中心
419	GB 50470-2008	油气输送管道线路工程抗震技术规范 Seismic technical code for oil and gas transmission pipeline engineering		中国石油天然气管道局
420	GB 50471-2008	煤矿瓦斯抽采工程设计规范 Code for design of the gas drainage engineering of coal mine		中煤国际工程集团重庆设计研究院

工程建设国家标准

序号	标准编号	标准名称	被代替标准编号	主编单位
421	GB 50472-2008	电子工业洁净厂房设计规范 Code for design of electronic industry clean room		中国电子工程设计院
422	GB 50473-2008	钢制储罐地基基础设计规范 Code for design of steel tank foundation		中国石化工程建设公司
423	GB 50474-2008	隔热耐磨衬里技术规范 Technical code for heat-insulation and wear-resistant linings		天津金耐达筑炉衬里有限公司 中国石化集团洛阳石化工程公司
424	GB 50475-2008	石油化工全厂性仓库及堆场设计规范 Code for design of general warehouse and lay down area of petrochemical industry		镇海石化工程有限责任公司
425	GB/T 50476-2008	混凝土结构耐久性设计规范 Code for durability design of concrete structures		清华大学
426	GB 50477-2009	纺织工业企业职业安全卫生设计规范 Code of design of occupational safety and health for textile industry enterprises		北京维拓时代建筑设计有限公司
427	GB 50478-2008	地热电站岩土工程勘察规范 Code for investigation of geotechnical engineering of geothermal power plant		中国电力工程顾问集团西南电力设计院
428	GB/T 50479-2011	电力系统继电保护及自动化设备柜（屏）工程技术规范 Technical code of cabinet (panel) for protection and automation equipments of electric power system		国电南京自动化股份有限公司
429	GB/T 50480-2008	冶金工业岩土勘察原位测试规范 Code for insitu tests of geotechnical engineering investigation of metallurgical industry		中冶沈勘工程技术有限公司
430	GB 50481-2009	棉纺织工厂设计规范 Code for design of cotton spinning and weaving factory		河南省纺织建筑设计院有限公司
431	GB 50482-2009	铝加工厂工艺设计规范 Code for design of aluminium processing plant		中色科技股份有限公司（中国铝业公司洛阳有色金属加工设计研究院）
432	GB 50483-2009	化工建设项目环境保护设计规范 Code for design of environmental protectin of chemical industry projects		中国石油和化工勘察设计协会
433	GB 50484-2008	石油化工建设工程施工安全技术规范 Code for technical of construction safety in petrochemical engineering		中国石化集团第五建设公司
434	GB/T 50485-2009	微灌工程技术规范 Technical code for microirrigation engineering		中国灌溉排水发展中心 中国农业大学
435	GB 50486-2009	钢铁厂工业炉设计规范 Code for design of industrial furnaces in iron & steel works		中冶华天工程技术有限公司
436	GB 50487-2008	水利水电工程地质勘察规范 Code for engineering geological investigation of water resources and hydropower		水利部水利水电规划设计总院 长江水利委员会长江勘测规划设计研究院
437	GB 50488-2009	腈纶工厂设计规范 Code for design of acrylic fibres plant		上海纺织建筑设计研究院
438	GB 50489-2009	化工企业总图运输设计规范 Code for design of general plot plan and transportation of chemical industrial enterprises		中国石油和化工勘察设计协会 全国化工总图运输设计技术中心站

工程建设国家标准

序号	标准编号	标准名称	被代替标准编号	主编单位
439	GB 50490-2009	城市轨道交通技术规范 Technical code of urban rail transit		中国城市规划设计研究院
440	GB 50491-2009	铁矿球团工程设计规范 Code for design of iron pellet engineering		中冶长天国际工程有限责任公司 中冶北方工程技术有限公司
441	GB 50492-2009	聚酯工厂设计规范 Code for design of PET plant		中国纺织工业设计院
442	GB 50493-2009	石油化工可燃气体和有毒气体检测报警设计规范 Code for design of combustible gas and toxic gas detection and alarm for petrochemical industry		中国石化集团洛阳石油化工工程公司
443	GB 50494-2009	城镇燃气技术规范 Technical code for city gas		住建部标准定额研究所 中国市政工程华北设计研究院
444	GB 50495-2009	太阳能供热采暖工程技术规范 Technical code for solar heating system		中国建筑科学研究院
445	GB 50496-2009	大体积混凝土施工规范 Code for construction of mass concrete		中冶建筑研究总院有限公司
446	GB 50497-2009	建筑基坑工程监测技术规范 Technical code for monitoring of building excavation engineering		济南大学 莱西市建筑总公司 山东省工程建设标准造价协会
447	GB 50498-2009	固定消防炮灭火系统施工与验收规范 Code for installation and acceptance of fixed fire monitor extinguishing systems		公安部上海消防研究所
448	GB 50499-2009	麻纺织工厂设计规范 Code for design of bast textile mill		黑龙江省纺织工业设计院
449	GB 50500-2013	建筑工程工程量清单计价规范 Code of bills of quantities and valuation for construction works	GB 50500-2008	住建部标准定额研究所 四川省建设工程造价管理总站
450	GB 50501-2007	水利工程工程量清单计价规范 Code of valuation with bill quantity of water conservancy construction works		北京峡光经济技术咨询有限责任公司 长江流域水利建设工程造价（定额）管理站
451	GB/T 50502-2009	建筑施工组织设计规范 Code for construction organization plan of building engineering		中国建筑技术集团有限公司
452	GB/T 50503-2014	兵器工业工厂废水监控规范 Monitoring and control code for waste water of ordnance industry factory		北京北方节能环保有限公司
453	GB/T 50504-2009	民用建筑设计术语标准 Standard for terminology of civil architectural design		同济大学建筑设计研究院 中国建筑标准设计研究院
454	GB 50505-2009	高炉煤气干法袋式除尘设计规范 Code for design of BFG dry bag filter		北京首钢国际工程技术有限公司（原北京首钢设计院）
455	GB 50506-2009	钢铁企业节水设计规范 Code for design of water saving for iron and steel enterprises		中冶京诚工程技术有限公司
456	GB 50507-2010	铁路罐车清洗设施设计规范 Code of design of railway tank wagon cleaning facilities		中国石化集团洛阳石油化工工程公司
457	GB 50508-2010	涤纶工厂设计规范 Code for design of polyester fiber plant		中国纺织工业设计院

工程建设国家标准

序号	标准编号	标准名称	被代替标准编号	主编单位
458	GB/T 50509-2009	灌区规划规范 Code for irrigation areas programming		山东省水利勘测设计院
459	GB/T 50510-2009	泵站更新改造技术规范 Technical code for renewal and renovation of pumping station		中国灌溉排水发展中心 武汉大学
460	GB 50511-2010	煤矿井巷工程施工规范 Code for construction of shaft and roadway of coal mine		中煤第一建设公司
461	GB 50512-2009	冶金露天矿准轨铁路设计规范 Code for design standard-gauge railway of metallurgical open-pit mine		中冶北方工程技术有限公司
462	GB 50513-2009 （2016年版）	城市水系规划规范 Code for plan of urban water system		武汉城市规划设计研究院
463	GB 50514-2009	非织造布工厂设计规范 Code for design of nonwovens factory		辽宁天维纺织研究建筑设计有限公司（原辽宁省建筑纺织设计院）
464	GB 50515-2010	导（防）静电地面设计规范 Code for design of conductive or anti-static ground surface and floor		北方设计研究院
465	GB 50516-2010	加氢站技术规范 Technical code for hydrogen fuelling station		中国电子工程设计院
466	GB 50517-2010	石油化工金属管道工程施工质量验收规范 Code for construction quality acceptance of metallic piping in petrochemical engineering		中国石化集团第十建设公司
467	GB/T 50518-2010	矿井通风安全装备标准 Standard for the equipment of ventilative safety of coal colliery		中煤国际工程集团重庆设计研究院
468	GB 50520-2009	核工业铀水冶厂尾矿库、尾渣库安全设计规范 Code for safety design of uranium mills tailings pond in nuclear industry		核工业第四研究设计院
469	GB 50521-2009	核工业铀矿冶工程设计规范 Code for design of uranium mining and metallurgy engineering in nuclear industry		核工业第四研究设计院
470	GB/T 50522-2009	核电厂建设工程监理规范 Code of construction project management for unclear power plants		核工业四达建设监理公司
471	GB 50523-2010	电子工业职业安全卫生设计规范 Code for design of occupational safety and hygiene in electronics industry		中国电子工程设计院
472	GB 50524-2010	红外线同声传译系统工程技术规范 Technical code for infrared simultaneous interpretation system		中国电子科技集团公司第三研究所 深圳市台电实业有限公司
473	GB/T 50525-2010	视频显示系统工程测量规范 Code of measurement for video display system		中国电子科技集团公司第三研究所
474	GB 50526-2010	公共广播系统工程技术规范 Technical code for public address system engineering		中国电子学会声频工程分会 广州市迪士普音响科技有限公司
475	GB 50527-2009	平板玻璃工厂节能设计规范 Code for design of energy conservation of flat glass plant		中国新型建筑材料工业杭州设计研究院 国家建筑材料工业标准定额总站

工程建设国家标准

序号	标准编号	标准名称	被代替标准编号	主编单位
476	GB 50528-2009	烧结砖瓦工厂节能设计规范 Code for design of energy conservation of fired brick and tile plant		中国建材西安墙体材料研究设计院
477	GB 50529-2009	维纶工厂设计规范 Code for design of vinylon fiber plant		福建省建筑轻纺设计院
478	GB 50530-2010	氧化铝厂工艺设计规范 Code for design of alumina refinery process		中铝国际工程有限公司沈阳分公司（沈阳铝镁设计研究院）
479	GB/T 50531-2009	建设工程计价设备材料划分标准 Standard of valuated building services and components' classification		国家建筑材料工业标准定额总站
480	GB 50532-2009	煤炭工业矿区机电设备修理设施设计规范 Coal industry design code of repairing facilities for electro-mechanical equipment of mining district		中煤国际工程集团南京设计研究院 中国煤炭建设协会勘察设计委员会
481	GB 50533-2009	煤矿井下辅助运输设计规范 Code for design of underground auxiliary haulage system of coal mine		煤炭工业济南设计研究院 中国煤炭建设协会勘察设计委员会
482	GB 50534-2009	煤矿采区车场和硐室设计规范 Code for design of district stations and chambers of coal mine		中煤国际工程集团沈阳设计研究院 中国煤炭建设协会勘察设计委员会
483	GB 50535-2009	煤矿井底车场设计规范 Code for design of pit bottom of coal mine		中煤国际工程集团沈阳设计研究院 中国煤炭建设协会勘察设计委员会
484	GB 50536-2009	煤矿综采采区设计规范 Code for design of coal mine full-mechanized mining district		煤炭工业济南设计研究院 中国煤炭建设协会勘察设计委员会
485	GB/T 50537-2009	油气田工程测量规范 Code of engineering survey for oil-gas field		大庆油田工程有限公司
486	GB/T 50538-2010	埋地钢质管道防腐保温层技术标准 Technical standard for anti-corrosion and insulation coatings of buried steel pipeline		大庆油田工程有限公司
487	GB/T 50539-2009	油气输送管道工程测量规范 Specifications of survey for oil and gas transportation pipeline engineering		中国石油集团工程设计公司西南分公司
488	GB 50540-2009 （2012年版）	石油天然气站内工艺管道工程施工规范 Code for construction of pipe process in oil and gas transmission pipeline station		中国石油天然气管道局
489	GB 50541-2009	钢铁企业原料场工艺设计规范 Code for design of raw material yard of iron and steel plants		中冶赛迪工程技术股份有限公司
490	GB 50542-2009	石油化工厂区管线综合技术规范 Technical code for pipelines coordination in petrochemical plant		中国石油化工股份有限公司洛阳分公司
491	GB 50543-2009	建筑卫生陶瓷工厂节能设计规范 Code for design of energy conservation of building and sanitary ceramic plant		中国建筑材料工业规划研究院 中国建筑材料集团公司咸阳陶瓷研究设计院
492	GB 50544-2009	有色金属企业总图运输设计规范 Code for design of general layout & transportation for non-ferrous metallurgical enterprises		云南华昆工程技术股份公司（原昆明有色冶金设计研究院）

工程建设国家标准

序号	标准编号	标准名称	被代替标准编号	主编单位
493	GB 50545-2010	110kV～750kV架空输电线路设计规范 Code for design of 110kV～750kV overhead transmission line		中国电力工程顾问集团公司 华东电力设计院
494	GB/T 50546-2009	城市轨道交通线网规划编制标准 Code for compilation of urban railway network planning		中国城市规划设计研究院
495	GB 50547-2010	尾矿堆积坝岩土工程技术规范 Technical code for geotechnical engineering of tailings embankment		中国有色金属工业西安勘察设计研究院
496	GB 50548-2010	330kV～750kV架空输电线路勘测规范 Code for investigation and surveying of 330kV～750kV overhead transmission line		中国电力工程顾问集团中南电力设计院
497	GB/T 50549-2010	电厂标识系统编码标准 Coding standard for power plant identification system		中国电力工程顾问集团公司
498	GB 50550-2010	建筑结构加固工程施工质量验收规范 Code for acceptance of construction quality of strengthening building structures		四川省建筑科学研究院
499	GB 50551-2010	球团机械设备安装工程质量验收规范 Code for quality acceptance of palletizing mechanical equipment installation engineering		中国十七冶集团有限公司
500	GB/T 50552-2010	煤炭工业露天矿工程建设项目设计文件编制标准 Design document preparation standard of open-pit mine construction project in coal industry		中煤国际工程集团沈阳设计研究院
501	GB/T 50553-2010	煤炭工业选煤厂工程建设项目设计文件编制标准 Design document compose standard of coal preparation plant project in coal industry		中煤国际工程集团北京华宇工程有限公司
502	GB/T 50554-2010	煤炭工业矿井工程建设项目设计文件编制标准 Design document preparation standard of mine construction project in coal industry		中煤国际工程集团南京设计研究院
503	GB 50555-2010	民用建筑节水设计标准 Standard for water saving design in civil building		中国建筑设计研究院
504	GB 50556-2010	工业企业电气设备抗震设计规范 Code for aseismic design of electrical facilities in industrial plants		中国石化工程建设公司
505	GB/T 50557-2010	重晶石防辐射混凝土应用技术规范 Technical code for barite concrete against radiation		湖南省第六工程有限公司 湖南省第四工程有限公司
506	GB 50558-2010	水泥工厂环境保护设计规范 Code for design of environmental protection of cement plant		天津水泥工业设计研究院有限公司
507	GB 50559-2010	玻璃工厂环境保护设计规范 Code for design of environmental protection of glass plant		秦皇岛玻璃工业研究设计院 国家建筑材料工业标准定额总站
508	GB 50560-2010	建筑卫生陶瓷工厂设计规范 Code for design of building and sanitary ceramic plant		咸阳陶瓷研究设计院 大连三川建设集团股份有限公司
509	GB/T 50561-2010	建材工业设备安装工程施工及验收规范 Code for equipment installing and acceptance of building material industry		中国建材国际工程有限公司 蚌埠玻璃工业设计研究院

工程建设国家标准

序号	标准编号	标准名称	被代替标准编号	主编单位
510	GB/T 50562-2010	煤炭矿井工程基本术语标准 Code for basic terms of coal mine construction engineering		中国矿业大学 淮南矿业（集团）有限责任公司
511	GB/T 50563-2010	城市园林绿化评价标准 Evaluation standard for urban landscaping and greening		城市建设研究院
512	GB/T 50564-2010	金属非金属矿山采矿制图标准 Mining drawing standard for metal and nonmetal mines		中国恩菲工程技术有限公司
513	GB 50565-2010	纺织工程设计防火规范 Code for design of textile engineering on fire protection and prevention		中国纺织工业设计院
514	GB 50566-2010	冶金除尘设备工程安装与质量验收规范 Code for installation and quality acceptance of metallurgical dedusting equipment		鞍钢建设集团有限公司 北京首钢建设集团有限公司
515	GB 50567-2010	炼铁工艺炉壳体结构技术规范 Technical code for shell structure of ironmaking furnace		中冶赛迪工程技术股份有限公司
516	GB 50568-2010	油气田及管道岩土工程勘察规范 Code for oil and gas field and pipeline investigation of geotechnical engineering		中国石油天然气管道工程有限公司
517	GB 50569-2010	钢铁企业热力设施设计规范 Code of design on thermo for iron & steel metallurgy enterprises		中冶赛迪工程技术股份有限公司
518	GB/T 50571-2010	海上风力发电工程施工规范 The code for construction of offshore wind power project		中国长江三峡集团公司
519	GB/T 50572-2010	核电厂工程地震调查与评价规范 Code for seismic investigation and evaluation of nuclear power plants		电力规划设计总院
520	GB 50573-2010	双曲线冷却塔施工与质量验收规范 Code for construction and quality acceptance of hyperbolic cooling tower engineering		西北电力建设第四工程公司
521	GB 50574-2010	墙体材料应用统一技术规范 Uniform technical code for wall materials used in buildings		中国建筑东北设计研究院有限公司 广厦建设集团有限责任公司
522	GB 50575-2010	1kV及以下配线工程施工与验收规范 Code for construction and acceptance of 1kV and blow wiring work		浙江省工业设备安装集团有限公司 宁波建工股份有限公司
523	GB 50576-2010	铝合金结构工程施工质量验收规范 Code for acceptance of construction quality of aluminium structures		上海市第五建筑有限公司 同济大学
524	GB 50577-2010	水泥工厂职业安全卫生设计规范 Code for design of safety and health of cement plant		中国建筑材料科学研究总院 天津水泥工业设计研究院有限公司
525	GB 50578-2010	城市轨道交通信号工程施工质量验收规范 Code for constructional quality acceptance of urban rail transit signal engineering		中国铁路通信信号上海工程集团有限公司

工程建设国家标准

序号	标准编号	标准名称	被代替标准编号	主编单位
526	GB 50579-2010	航空工业理化测试中心设计规范 Code of design for physical and chemical testing center of aviation industry		中国航空规划建设发展有限公司
527	GB 50580-2010	连铸工程设计规范 Code for design of continuous casting engineering		中冶京诚工程技术有限公司
528	GB 50581-2010	煤炭工业矿井监测监控系统装备配置标准 Standard for allocated mine monitoring and controlling system of coal industry		中煤国际工程集团南京设计研究院 中国煤炭建设协会勘察设计委员会
529	GB 50582-2010	室外作业场地照明设计标准 Standard for lighting design of outdoor work places		中国建筑科学研究院
530	GB 50583-2010	选煤厂建筑结构设计规范 Code for design of building and structure of coal mine of preparation plant		中煤国际工程集团北京华宇工程有限公司 中国煤炭建设协会勘察设计委员会
531	GB 50584-2010	煤气余压发电装置技术规范 Technical code for top gas pressure recovery turbine		中冶南方工程技术有限公司
532	GB 50585-2010	岩土工程勘察安全规范 Occupational safety code for geotechnical investigation		福建省建筑设计研究院 福建省九龙建设集团
533	GB 50586-2010	铝母线焊接工程施工及验收规范 Code for construction and acceptance of aluminium bus-bar welding engineering		七冶建设有限责任公司 （原中国有色金属工业第七冶金建设公司）
534	GB/T 50587-2010	水库调度设计规范 Design code for operation of reservoir		长江水利委员会长江勘测规划设计研究院 水利部水利水电规划设计总院
535	GB 50588-2010	水泥工厂余热发电设计规范 Code for design of waste heat power generation in cement plant		国家建筑材料工业标准定额总站 中材节能发展有限公司
536	GB/T 50589-2010	环氧树脂自流平地面工程技术规范 Technical code of construction for epoxy resins self-leveling flooring		全国化工施工标准化管理中心站
537	GB/T 50590-2010	乙烯基酯树脂防腐蚀工程技术规范 Technical code for anticorrosion engineering of vinyl ester resins		全国化工施工标准化管理中心站
538	GB 50591-2010	洁净室施工及验收规范 Code for construction and acceptance of cleanroom	JGJ 71-90	中国建筑科学研究院
539	GB 50592-2010	煤矿矿井建筑结构设计规范 Code for design of colliery building and structure of coal mine		煤炭工业太原设计研究院 中国煤炭建设协会勘察设计委员会
540	GB/T 50593-2010	煤炭矿井制图标准 Standard for drawings underground mining		煤炭工业合肥设计研究院
541	GB/T 50594-2010	水功能区划分标准 Standard for water function zoning		水利部水利水电规划设计总院 长江流域水资源保护局
542	GB 50595-2010	有色金属矿山节能设计规范 Code for energy conservation design of non-ferrous metal mines		中国恩菲工程技术有限公司（原中国有色工程设计研究总院） 云南华昆工程技术股份有限公司（原昆明有色冶金设计研究院）

工程建设国家标准

序号	标准编号	标准名称	被代替标准编号	主编单位
543	GB/T 50596-2010	雨水集蓄利用工程技术规范 Technical code for rainwater collection, storage and utilization		中国灌溉排水发展中心
544	GB/T 50597-2010	纺织工程常用术语、计量单位及符号标准 Standard for terminology, measure units and symbols in textile engineering		四川省纺织工业设计院
545	GB 50598-2010	水泥原料矿山工程设计规范 Code for design of cement raw material mine engineering		天津水泥工业设计研究院有限公司
546	GB 50599-2010	灌区改造技术规范 Technical code for amelioration of the irrigation and drainage scheme		中国灌溉排水发展中心
547	GB/T 50600-2010	渠道防渗工程技术规范 Technical code for seepage control engineering on canal		中国灌溉排水发展中心
548	GB 50601-2010	建筑物防雷工程施工与质量验收规范 Code for construction and quality acceptance for lightning protection engineering of structures		南通五建建设工程有限公司 江苏顺通建设工程有限公司
549	GB/T 50602-2010	球形储罐γ射线全景曝光现场检测标准 Radiographic testing standard for a single exposure with γ-ray technique of spherical tank installation		中国石油天然气第一建设公司 洛阳中油检测工程有限公司
550	GB 50603-2010	钢铁企业总图运输设计规范 Code for design of general layout and transportation for iron & steel enterprise		中冶南方工程技术有限公司
551	GB/T 50604-2010	民用建筑太阳能热水系统评价标准 Evaluation standard for solar water heating system of civil buildings		中国建筑设计研究院
552	GB/T 50605-2010	住宅区和住宅建筑内通信设施工程设计规范 Code for design of communication engineering in residential districts and residential buildings		中国移动通信集团设计院有限公司
553	GB 50606-2010	智能建筑工程施工规范 Code for installation of intelligent building systems		通州建总集团有限公司 中信建设有限责任公司
554	GB 50607-2010	高炉喷吹煤粉工程设计规范 Design regulations of pulverized coal injection for blast furnace		中冶京诚工程技术有限公司
555	GB 50608-2010	纤维增强复合材料建设工程应用技术规范 Technical code for infrastructure application of FRP composites		中冶建筑研究总院有限公司 国家工业建筑诊断与改造工程技术研究中心
556	GB/T 50609-2010	石油化工工厂信息系统设计规范 Code for design of plant information system in petrochemical engineering		中国石化工程建设公司
557	GB/T 50610-2010	车用乙醇汽油储运设计规范 Code for design of automobile ethanol gasoline storage and transportation		中国石化工程建设公司
558	GB 50611-2010	电子工程防静电设计规范 Code for design of protection of electrostatic discharge in electronic engineering		上海电子工程设计研究院有限公司

工程建设国家标准

序号	标准编号	标准名称	被代替标准编号	主编单位
559	GB 50612-2010	冶金矿山选矿厂工艺设计规范 Code for technological design of metallurgical concentrator		中冶北方工程技术有限公司 中冶长天国际工程有限责任公司
560	GB 50613-2010	城市配电网规划设计规范 Code for planning and design of urban distribution network		中国南方电网有限责任公司 中国国家电网公司
561	GB 50614-2010	跨座式单轨交通施工及验收规范 Code for construction and acceptance of straddle monorail transit		重庆市轨道交通集团有限公司
562	GB 50615-2010	冶金工业水文地质勘察规范 Code for hydrogeological investigation of metallurgical industry		中冶集团武汉勘察设计院有限公司
563	GB 50616-2010	铜冶炼厂工艺设计规范 Code for design of copper smelter processes		中国瑞林工程技术有限公司
564	GB 50617-2010	建筑电器照明装置施工与验收规范 Code for construction and acceptance of electrical lighting installation in building		宁波建工股份有限公司 浙江省工业设备安装集团有限公司
565	GB 50618-2011	房屋建筑和市政基础设施工程质量检测技术管理规范 Testing technology management code for building and municipal infrastructure engineering quality		中国建筑业协会 福建省九龙建设集团
566	GB/T 50619-2010	火力发电厂海水淡化工程设计规范 Code for designing of thermal power plants seawater desalination system		中国电力工程顾问集团华北电力设计院工程有限公司
567	GB 50620-2010	粘胶纤维工厂设计规范 Code for design of viscose fibre plant		江西省纺织工业科研设计院
568	GB/T 50621-2010	钢结构现场检测技术标准 Technical standard for in-site testing of steel structure		中国建筑科学研究院
569	GB/T 50622-2010	用户电话交换系统工程设计规范 Code for design of private telephone switch system engineering		中国移动通信集团设计院有限公司
570	GB/T 50623-2010	用户电话交换系统工程验收规范 Code for acceptance of private telephone switch system engineering		中国移动通信集团设计院有限公司
571	GB/T 50624-2010	住宅区和住宅建筑内通信设施工程设计规范 Code for acceptance of communication engineering in residential districts and residential buildings		中国移动通信集团设计院有限公司
572	GB/T 50625-2010	机井技术规范 Technical code for water wells		中国灌溉排水发展中心
573	GB/T 50626-2010	住房公积金支持保障性住房建设项目贷款业务规范 Code of housing provident fund project loan for affordable housing construction		天津市住房公积金管理中心
574	GB/T 50627-2010	城镇供热系统评价标准 Evaluation standard for district heating system		中国建筑科学研究院 河南省第五建筑安装工程公司
575	GB 50628-2010	钢管混凝土工程施工质量验收规范 Code for quality acceptance of the concrete filled steel tubular engineering		中国工程建设标准化协会 南通华新建工集团有限公司

工程建设国家标准

序号	标准编号	标准名称	被代替标准编号	主编单位
576	GB 50629-2010	板带轧钢工艺设计规范 Code for design of rolling process for plate and strip mill		中冶京诚工程技术有限公司
577	GB 50630-2010	有色金属工程设计防火规范 Code for design on fire prevention of nonferrous metals engineering		中国恩菲工程技术有限公司 （原中国有色工程设计研究总院）
578	GB 50631-2010	住宅信报箱工程技术规范 Technical code for private mail boxes of residential buildings		中国建筑设计研究院 大连九州建设集团有限公司
579	GB 50632-2010	钢铁企业节能设计规范 Code for design of energy saving of iron and steel industry		中冶京诚工程技术有限公司
580	GB 50633-2010	核电厂工程测量技术规范 Technical code for engineering surveying of nuclear power station		广东省电力设计研究院 电力规划设计总院
581	GB 50634-2010 （2015年版）	水泥窑协同处置工业废物设计规范 Code for design of industrial waste composition in cement kiln		天津水泥工业设计研究院有限公司 国家建筑材料工业标准定额总站
582	GB 50635-2010	会议电视会场系统工程设计规范 Code for design of hall system engineering of videoconference		中国电子科技集团公司第三研究所
583	GB 50636-2010	城市轨道交通综合监控系统工程设计规范 Code for design of urban rail transit integrated supervision and control system engineering		北京和利时系统工程有限公司 中国电子科技集团第十四研究所
584	GB/T 50637-2010	弹体毛坯旋压工艺设计规范 Code for design of spinning process for shell body blank		中国兵器工业新技术推广研究所
585	GB/T 50638-2010	麻纺织设备工程安装与质量验收规范 Code of acceptance for installation of textile equipments for bast fibers		浙江金鹰股份有限公司
586	GB 50639-2010	锦纶工厂设计规范 Code for design of polyamide polymer and fiber plant		中国纺织工业设计院
587	GB/T 50640-2010	建筑工程绿色施工评价标准 Evaluation standard for green construction of building		中国建筑股份有限公司 中国建筑第八工程局有限公司
588	GB 50641-2010	有色金属矿山井巷安装工程施工规范 Code for installation engineering construction of non-ferrous metals mine sinking and drifting		十四冶建设集团有限公司
589	GB 50642-2011	无障碍设施施工验收及维护规范 Construction acceptance and maintenance standards of the barrier-free facilities		南京建工集团有限公司 江苏省金陵建工集团有限公司
590	GB 50643-2010	橡胶工厂职业安全与卫生设计规范 Code for design of occupational safety and hygiene of rubber factory		中国石油和化工勘察设计协会 中国石油和化工勘察设计协会橡胶塑料设计专业委员会
591	GB/T 50644-2011	油气管道工程建设项目设计文件编制标准 Standard for compiling the design documents of oil and gas pipeline projects		中国石油天然气管道工程有限公司 中国石油集团工程设计有限责任公司西南分公司

工程建设国家标准

序号	标准编号	标准名称	被代替标准编号	主编单位
592	GB 50645-2011	石油化工绝热工程施工质量验收规范 Code for construction quality acceptance of insulation in petrochemical engineering		中国石化集团第四建设公司
593	GB 50646-2011	特种气体系统工程技术规范 Technical code for speciality gas system engineering		信息产业电子第十一设计研究院有限公司 中国电子系统工程第二建设有限公司
594	GB 50647-2011	城市道路交叉口规划规范 Code for planning of intersections on urban roads		同济大学
595	GB 50648-2011	化学工业循环冷却水系统设计规范 Code for design of recirculating cooling water system in chemical palnt		中国天辰工程有限公司 中国石油和化工勘察设计协会给排水设计专业委员会
596	GB/T 50649-2011	水利水电工程节能设计规范 Code for design of energy saving for water resources and hydropower projects		水利部水利水电规划设计总院
597	GB 50650-2011	石油化工装置防雷设计规范 Code for design protection of petrochemical plant against lighting		中国石化工程建设公司
598	GB/T 50651-2011	煤炭工业矿区总体规划文件编制标准 Preparation standard for general planning on mining area of coal industry		中煤西安设计工程有限责任公司
599	GB 50652-2011	城市轨道交通地下工程建设风险管理规范 Code for risk management of underground works in urban rail transit		中国土木工程学会 同济大学
600	GB 50653-2011	有色金属矿山井巷工程施工规范 Code for construction of non-ferrous metals mine sinking and drifting engineering		十四冶建设集团有限公司
601	GB 50654-2011	有色金属工业安装工程质量验收统一标准 Unified standards for constructional quality acceptance of non-ferrous metals industrial installation engineering		有色金属工业建设工程质量监督总站
602	GB/T 50655-2011	化工厂蒸汽系统设计规范 Code for design of steam system in chemical plant		中国石油和化工勘察设计协会 中国成达工程有限公司
603	GB 50656-2011	施工企业安全生产管理规范 Code for construction company safety manage criterion		上海市建设工程安全质量监督总站 上海城建建设实业（集团）有限公司
604	GB/T 50657-2011	煤炭露天采矿制图标准 Standard of drawing-making for surface mining coal		中煤国际工程集团沈阳设计研究院
605	GB/T 50658-2011	煤炭工业矿区机电设备修理厂工程建设项目设计文件编制标准 Standard for preparing design document of construction project of electromechanical equipment repair plant in mining area of coal industry		中煤国际工程集团南京设计研究院
606	GB/T 50659-2011	煤炭工业矿区水煤浆工程建设项目设计文件编制标准 Engineering design document standard for coal water mixture project in coal industry		中煤国际工程集团北京华宇工程有限公司
607	GB 50660-2011	大中型火力发电厂设计规范 Code for design of fossil fired power plant		中国电力工程顾问集团公司

工程建设国家标准

序号	标准编号	标准名称	被代替标准编号	主编单位
608	GB 50661-2011	钢结构焊接规范 Code for welding of steel structures		中冶建筑研究总院 中国二冶集团有限公司
609	GB/T 50662-2011	水工建筑物抗冰冻设计规范 Code for design of hydraulic structures against ice and freezing action		中水东北勘测设计研究有限责任公司
610	GB/T 50663-2011	核电厂工程水文技术规范 Technical code for engineering hydrology for nuclear power plant		电力规划设计总院
611	GB/T 50664-2011	棉纺织设备工程安装与质量验收规范 Code for installation and quality inspection of cotton textile machinery		中国纺织机械器材工业协会
612	GB 50665-2011	1000kV架空输电线路设计规范 Code for design of 1000kV overhead transmission line		中国电力工程顾问集团公司 国家电网公司
613	GB 50666-2011	混凝土结构工程施工规范 Code for construction of concrete structures		中国建筑科学研究院
614	GB 50667-2011	印染设备工程安装与质量验收规范 Code for installation and quality inspection of dyeing and finishing equipment		中国纺织机械器材工业协会
615	GB/T 50668-2011	节能建筑评价标准 Standard for energy efficient building assessment		中国建筑科学研究院
616	GB 50669-2011	钢筋混凝土筒仓施工与质量验收规范 Code for construction and acceptance of reinforced concrete silos		河北省第四建筑工程公司 河北建工集团有限公司
617	GB/T 50670-2011	机械设备安装工程术语标准 Terminology standard for mechanical equipment installation		中国机械工业建设总公司
618	GB 50671-2011	飞机喷漆机库设计规范 Code for design of aircraft paint hangar		中国航空规划建设发展有限公司 （原中国航空工业规划设计研究院）
619	GB 50672-2011	钢铁企业综合污水处理厂工艺设计规范 Code for process design of comprehensive sewage treatment for iron and steel enterprises		中冶建筑研究总院有限公司
620	GB 50673-2011	有色金属冶炼厂电力设计规范 Code for power design of non-ferrous metals smelters		长沙有色冶金设计研究院
621	GB/T 50674-2013	核电厂工程气象技术规范 Code of meteorology for nuclear power plant		中国电力顾问集团华东电力设计院 电力规划设计总院
622	GB/T 50675-2011	纺织工程制图标准 Standard for textile engineering drawings		中国纺织工业设计院
623	GB/T 50676-2011	铀燃料元件厂混凝土结构厂房可靠性鉴定技术规范 Technical code for appraisal of reliability of concrete structural factory buildings for uranium fuel element plants		中国核电工程有限公司郑州分公司 （原核工业第五研究设计院）
624	GB 50677-2011	空分制氧设备安装工程施工与质量验收规范 Code for construction and acceptance of oxygen plant equipment installation engineering		中冶天工集团有限公司
625	GB 50678-2011	废弃电器电子产品处理工程设计规范 Code for design of the waste electrical and electronic equipment processing engineering		中国电子工程设计院

工程建设国家标准

序号	标准编号	标准名称	被代替标准编号	主编单位
626	GB 50679-2011	炼铁机械设备安装规范 Code for acceptance of mechanical equipment installation of ironmaking system		上海宝冶集团公司
627	GB/T 50680-2012	城镇燃气工程术语标准 Standard for basic terms of city gas engineering		北京市煤气热力工程设计院
628	GB 50681-2011	机械工业厂房建筑设计规范 Code for design of machinery building architecture		机械工业第一设计研究院
629	GB 50682-2011	预制组合立管技术规范 Technical code for pre-fabricated united pipe risers		中建三局第一建设公司 同济大学
630	GB 50683-2011	现场设备、工业管道焊接工程施工质量验收规范 Code for acceptance of field equipment, industrial pipe welding construction quality		中国石油和化工勘察设计协会 中油吉林化建工程股份有限公司
631	GB 50684-2011	化学工业污水处理与回用设计规范 Code for design of wastewater treatment and reuse in chemical industry		中国石油和化工勘察设计协会 东华工程科技股份有限公司
632	GB 50685-2011	电子工业纯水系统设计规范 Code for design of pure water system of electronic industry		信息产业电子第十一设计研究院科技工程股份有限公司
633	GB 50686-2011	传染病医院建筑施工及验收规范 Code for construction and acceptance of infectious diseases hospitals		中国建筑科学研究院
634	GB 50687-2011	食品工业洁净用房建筑技术规范 Architectural and technical code for cleanroom in food industry		中国建筑科学研究院
635	GB 50688-2011	城市道路交通设施设计规范 Code for design of urban road traffic facility		上海市政工程设计研究总院（集团）有限公司
636	GB 50689-2011	通信局（站）防雷与接地工程设计规范 Code for design of lightning protection and earthing engineering for telecommunication bureaus (staions)		中讯邮电咨询设计院
637	GB 50690-2011	石油化工非金属管道工程施工质量验收规范 Code for construction quality acceptance of non-metallic piping engineering in petrochemical engineering		胜利油田胜利石油化工建设有限责任公司
638	GB/T 50691-2011	油气田地面工程建设项目设计文件编制标准 Standard for Compiling the design documents of oil and gas field surface construction projects		中国石油集团工程设计有限责任公司华北分公司
639	GB/T 50692-2011	天然气处理厂工程建设项目设计文件编制标准 Standard for compiling the design documents of natural gas treating plant projects		中国石油集团工程设计有限责任公司西南分公司
640	GB 50693-2011	坡屋面工程技术规范 Technical code for slope roof engineering		中国建筑防水协会
641	GB 50694-2011	酒厂设计防火规范 Code for design of fire protection and prevention of alcoholic beverages factory		四川省公安消防总队
642	GB 50695-2011	涤纶、锦纶、丙纶设备工程安装与质量验收规范 Code for installation of polyester, polyamide, polypropylene fiber-making equipments and quality acceptance		北京中丽制机工程技术有限公司（原北京中丽制机化纤工程技术有限公司）

工程建设国家标准

序号	标准编号	标准名称	被代替标准编号	主编单位
643	GB 50696-2011	钢铁企业冶金设备基础设计规范 Code for design of metallurgical equipment foundation in iron and steel enterprises		中冶赛迪工程技术股份有限公司
644	GB 50697-2011	1000kV 变电站设计规范 Code for design of 1000kV substation		中国电力工程顾问集团公司 国家电网公司
645	GB/T 50698-2011	埋地钢质管道交流干扰防护技术标准 Standard for AC interference mitigation of buried steel pipelines		中国石油集团工程设计有限责任公司西南分公司
646	GB 50699-2011	液压振动台基础技术规范 Technical code for hydraulic vibrator foundation		五洲工程设计研究院（中国五洲工程设计有限公司）
647	GB/T 50700-2011	小型水电站技术改造规范 Technical renovation code for small hydropower station		水利部农村水电及电气化发展局
648	GB 50701-2011	烧结砖瓦工厂设计规范 Code for design of fired brick and tile plant		西安墙体材料研究设计院 中国建筑材料工业规划研究院
649	GB 50702-2011	砌体结构设加固设计规范 Code for design of strengthening masonry structures		四川省建筑科学研究院 中国华西企业有限公司
650	GB/T 50703-2011	电力系统安全自动装置设计规范 Code for design of automaticity equipment for power system security		中国电力工程顾问集团东北电力设计院
651	GB 50704-2011	硅太阳能电池工厂设计规范 Code for design of crystalian silicon solar cell plant		信息产业电子第十一设计研究院科技工程股份有限公司 中国电子系统工程第二建设有限公司
652	GB 50705-2012	服装工厂设计规范 Code for design of garments plant		北京维拓时代建筑设计有限公司
653	GB 50706-2011	水利水电工程劳动安全与工业卫生设计规范 Code for design of occupational safety and health of water resources and hydropower projects		水利部水利水电规划设计总院 长江水利委员会长江勘察规划设计研究院
654	GB 50707-2011	河道整治设计规范 Code for design of river regulation		水利部水利水电规划设计总院 中水淮河规划设计研究有限公司
655	GB/T 50708-2012	胶合木结构技术规范 Technical code of glude laminated timber structures		中国建筑西南设计研究院
656	GB 50709-2011	钢铁企业管道支架设计规范 Code for design of pipe supports in iron and steel enterprises		中冶赛迪工程技术股份有限公司 重庆建工集团股份有限公司
657	GB 50710-2011	电子工程节能设计规范 Code for design of energy conservation of electronic industry		中国电力工程设计院
658	GB 50711-2011	冶炼烟气制酸设备安装工程施工规范 Code for construction of acid-making equipment installtion engineering for metallurgical off-gas		中国十五冶金建设集团公司
659	GB 50712-2011	冶炼烟气制酸设备安装工程质量验收规范 Code for acceptance of construction quality of acid-making equipment installtion enginering for metallurgical off-gas		中国十五冶金建设集团公司
660	GB 50713-2011	板带精整工艺设计规范 Code for design of finishing process of plate and strip steel		中冶赛迪工程技术股份有限公司

工程建设国家标准

序号	标准编号	标准名称	被代替标准编号	主编单位
661	GB 50714-2011	钢管涂层车间工艺设计规范 Code for design of steel pipe coating workshop process		中冶赛迪工程技术股份有限公司
662	GB 50715-2011	地铁工程施工安全评价标准 Standard for construction safety assessment of metro engineering		华中科技大学 武汉市市政建设集团有限公司
663	GB/T 50716-2011	重有色金属冶炼设备安装工程施工规范 Code for extractive metallurgy construction of equipment installation engineering of heavy non-ferrous metals		二十三冶建设集团有限公司
664	GB 50717-2011	重有色金属冶炼设备安装工程质量验收规范 Code for extractive metallurgy construction quality acceptance of mechanical equipment installation engineering of heavy non-ferrous metals		二十三冶建设集团有限公司
665	GB/T 50718-2011	建材工厂工程建设项目设计文件编制标准 Standard for design document editing of construction projects in building materials factories		国家建筑材料工业标准定额总站 中国中材国际工程股份有限公司
666	GB/T 50719-2011	电磁屏蔽室工程技术规范 Technical code for electromagnetic shielded enclosure		北方设计研究院
667	GB 50720-2011	建设工程施工现场消防安全技术规范 Technical code for fire safety of construction site		中国建筑第五工程局有限公司 中国建筑股份有限公司
668	GB 50721-2011	钢铁企业给水排水设计规范 Code for design of water supply & drainage of iron and steel enterprises		中冶赛迪工程技术股份有限公司
669	GB 50722-2011	城市轨道交通建设项目管理规范 Code of project management for urban rail transit construction		中国土木工程学会 北京城建设计研究总院
670	GB 50723-2011	烧结机械设备安装规范 Code for installation of sintering mechanical equipment		中冶天工集团有限公司
671	GB 50724-2011	大宗气体纯化及输送系统工程技术规范 Technical code for bulk gas purification and delivery system engineering		信息产业电子第十一设计研究院 中国电子系统工程第四建设有限公司
672	GB 50725-2011	液晶显示器件生产设备安装工程施工及验收规范 Code for construction and acceptance of liquid crystal display manufacturing equipment installation engineering		中国电子科技集团第二研究所
673	GB 50726-2011	工业设备及管道防腐蚀工程施工规范 Code for anticorrosive engineering construction of industrial euqipment and pipeline		中国石油和化工勘察设计协会 全国化工施工标准化管理中心站
674	GB 50727-2011	工业设备及管道防腐蚀工程施工质量验收规范 Code for acceptance of construction quality of anticorrosive engineering of industrial equipment and pipeline		中国石油和化工勘察设计协会 全国化工施工标准化管理中心站
675	GB 50728-2011	工程结构加固材料安全性鉴定技术规范 Technical code for safety appraisal of engineering structural strengthening materials		四川省建筑科学研究院 中国华西企业股份有限公司

工程建设国家标准

序号	标准编号	标准名称	被代替标准编号	主编单位
676	GB 50729-2012	±800kV 及以下直流换流站土建工程施工质量验收规范 Code for acceptance of construction quality of ±800kV & under HVDC converter substation		国家电网公司直流建设分公司
677	GB 50730-2011	冶金机械液压、润滑和气动设备工程施工规范 Code for construction of metallurgical mechanical hydraulic, lubricating and dynamic equipment engineering		中国一冶集团有限公司
678	GB/T 50731-2011	建材工程术语标准 Standard for terminology of building materials projects		国家建筑材料工业标准定额总站 中国建筑材料工业规划研究院
679	GB/T 50732-2011	城市轨道交通综合监控系统工程施工与质量验收规范 Code for construction and acceptance of urban rail transit integrated supervision and control system engineering		北京和利时系统工程有限公司 中国电子科技集团公司第十四研究所
680	GB/T 50733-2011	预防混凝土碱骨料应用技术规程 Technical code for prevention of alkali-aggregate reaction in concrete		中国建筑科学研究院 浙江舜江建设集团有限公司
681	GB 50734-2012	冶金工业建设钻探技术规范 Technical code for drilling of metallurgical industry construction		中勘冶金勘察设计研究院有限责任公司
682	GB 50735-2011	铁合金工艺及设备设计规范 Code for design of ferroalloy process and equipment		中钢集团工程设计研究院有限公司
683	GB 50736-2012	民用建筑供暖通风与空气调节设计规范 Design code for heating ventilation and air conditioning of civil buildings	GB 50019-2003 相应条文	中国建筑科学研究院
684	GB 50737-2011	石油储备库设计规范 Code for design of petroleum storage depot		中国石化工程建设公司
685	GB 50738-2011	通风与空调工程施工规范 Code for construction of ventilation and air conditioning		中国建筑科学研究院 北京住总集团有限公司
686	GB 50739-2011	复合土钉墙基坑支护技术规范 Technical code for composite soil nailing wall in retaining and protection of excavation		济南大学 江苏第一建筑安装有限公司
687	GB 50741-2012	1000kV 架空输电线路勘测规范 Code for investigation and surveying of 1000kV overhead transmission line		中国电力工程顾问集团公司
688	GB 50742-2012	炼钢机械设备安装规范 Code for installation of mechanical equipment for steel-making		中国一冶集团有限公司
689	GB/T 50743-2012	工程施工废弃物再生利用技术规范 Code for recycling of construction & demolition waste		江苏南通二建集团有限公司 同济大学
690	GB/T 50744-2011	轧机机械设备安装规范 Code for installation of rolling mill mechanical equipment		中国二十冶集团有限公司

工程建设国家标准

序号	标准编号	标准名称	被代替标准编号	主编单位
691	GB 50745-2012	核电厂常规岛设计防火规范 Code for design of fire protection for conventiontional island in nuclear power plants		东北电力设计院
692	GB/T 50746-2012	石油化工循环水场设计规范 Code for design of petrochemical recirculation cooling water unit		中国石化工程建设公司
693	GB 50747-2012	石油化工污水处理设计规范 Code for design of wastewater treatment in petrochemical industry		中国石化集团洛阳石油化工工程公司
694	GB/T 50748-2011	选煤工艺制图标准 Standard for drawing of coal preparation technology		中国煤炭建设协会勘察设计委员会 中煤国际工程集团北京华宇工程公司
695	GB 50749-2012	冶金工业建设岩土工程勘察规范 Code for geotechnical engineering investigation of metallurgical industry construction		中勘冶金勘察设计研究院有限责任公司
696	GB 50750-2012	粘胶纤维设备工程安装与质量验收规范 Code for engineering installment acceptance of viscose fiber machinery		中国纺织机械器材工业协会
697	GB 50751-2012	医用气体工程技术规范 Technical code for medical gases engineering		上海市建筑学会
698	GB 50752-2012	电子辐射工程技术规范 Technical code for electronic radiation engineering		山东高阳建设有限公司 山东蓝孚电子加速器技术有限公司
699	GB 50753-2012	有色金属冶炼厂收尘设计规范 Code for dust collection design of non-ferrous metals plant		中国恩菲工程技术有限公司（原中国有色工程设计研究总院）
700	GB 50754-2012	挤压钢管工程设计规范 Code for design of steel pipe extrusion engineering		中冶京诚工程技术有限公司
701	GB 50755-2012	钢结构工程施工规范 Code for construction of steel structures		中国建筑股份有限公司 中建钢构有限公司
702	GB/T 50756-2012	钢制储罐地基处理技术规范 Technical code for ground treatment of steel tanks		中国石化集团洛阳石油化工公司
703	GB 50757-2012	水泥窑协同处置污泥工程设计规范 Code for design of sludge co-processing in cement kiln		天津水泥工业设计研究院有限公司
704	GB 50758-2012	有色金属加工厂节能设计规范 Code for design of energy conservation in non-ferrous metals processing plants		中色科技股份有限公司 中国有色金属工业工程建设标准规范管理处
705	GB 50759-2012	油品装载系统油气回收设施设计规范 Code for design of vapor recovery facilities of oil products loading system		中国石化集团洛阳石油化工公司
706	GB/T 50760-2012	数字集群通信工程技术规范 Code for engineering technology of digital trunking communication		中讯邮电咨询设计院有限公司
707	GB 50761-2012	石油化工钢制设备抗震设计规范 Code for seismic design of petrochemical steel facilities		中国石化工程建设有限公司
708	GB 50762-2012	秸秆发电厂设计规范 Design code for straw power plant		中国电力工程顾问集团东北电力设计院

工程建设国家标准

序号	标准编号	标准名称	被代替标准编号	主编单位
709	GB 50763-2012	无障碍设计规范 Code for accessibility design	JGJ 50-2001	北京市建筑设计研究院
710	GB 50764-2012	电厂动力管道设计规范 Design code of power piping for power plant		中国电力工程顾问集团东北电力设计院
711	GB 50765-2012	炭素厂工艺设计规范 Design code for carbon plant		贵阳铝镁设计研究院有限公司
712	GB 50766-2012	水电水利工程压力钢管制作安装及验收规范 Code for manufacture installation and acceptance of steel penstocks in hydroelectric and hydraulic engineering		中国水利水电第七工程局
713	GB 50767-2013	火炸药工程设计能耗指标标准 Standard for design of energy consumption index in propellant and explosive engineering		五洲工程设计研究院
714	GB/T 50768-2012	白蚁防治工程基本术语标准 Standard for basic terminology of termite control project		全国白蚁防治中心 贵州建工集团有限公司
715	GB/T 50769-2012	节水灌溉工程验收规范 Code for acceptance of water-saving irrigation engineering		中国灌溉排水发展中心
716	GB/T 50770-2013	石油化工安全仪表系统设计规范 Code for design of safety instrumented system in petrochemical engineering		中国石化工程建设有限公司
717	GB 50771-2012	有色金属采矿设计规范 Code for design of nonferrous metal mining		长沙有色冶金设计研究院有限公司
718	GB/T 50772-2012	木结构工程施工规范 Code for construction of timber structures		哈尔滨工业大学 黑龙江建设集团有限公司
719	GB 50773-2012	蓄滞洪区设计规范 Code for design of flood detention and retarding basin		水利部水利水电规划设计总院 湖南省水利水电勘察设计研究总院
720	GB 50774-2012	±800kV及以下换流站干式平波电抗器施工及验收规范 Code for construction and acceptance of dry-type smoothing reactors in converter stations at ±800kV and below		国家电网公司直流建设分公司
721	GB/T 50775-2012	±800kV及以下换流站换流阀施工及验收规范 Code for construction and acceptance of converter valve in converter station at ±800kV and below		国家电网公司直流建设分公司
722	GB 50776-2012	±800kV及以下换流站换流变压器施工及验收规范 Code for construction and acceptance of converter transformer in converter station at ±800kV and below		国家电网公司直流建设分公司
723	GB 50777-2012	±800kV及以下换流站构支架施工及验收规范 Code for construction and acceptance of frame works in converter station at ±800kV and below		国家电网公司直流建设分公司
724	GB 50778-2012	露天煤矿岩土工程勘察规范 Code for investigation of geotechnical engineering of open pit coal mine		中煤国际工程集团沈阳设计研究院

工程建设国家标准

序号	标准编号	标准名称	被代替标准编号	主编单位
725	GB 50779-2012	石油化工控制室抗爆设计规范 Code for design of blast resistant control building in petrochemical industry		中国石化集团洛阳石油化工公司
726	GB/T 50780-2013	电子工程建设术语标准 Terminology standard of electronics engineering construction		中国电子工程设计院
727	GB 50781-2012	电子工厂化学品系统工程技术规范 Technical code for chemical system of electronic Factory		信息产业电子第十一设计研究院科技工程股份有限公司 信息产业部电子工程标准定额站
728	GB 50782-2012	有色金属选矿厂工艺设计规范 Code for technological design of non-ferrous concentrator		中国恩菲工程技术有限公司 中国有色金属工业工程建设标准规范管理处
729	GB/T 50783-2012	复合地基技术规范 Technical code for composite foundation		浙江大学 浙江中南建设集团有限公司
730	GB/T 50784-2013	混凝土结构现场检测技术标准 Technical standard for in-situ inspection of concrete structure		中国建筑科学研究院 中国新兴建设开发总公司
731	GB/T 50785-2012	民用建筑室内热湿环境评价标准 Evaluation standard for indoor thermal environment in civil buildings		中国建筑科学研究院 重庆大学
732	GB/T 50786-2012	建筑电气制图标准 Standard for building electricity drawings		中国建筑标准设计研究院 中国纺织工业设计院
733	GB 50787-2012	民用建筑太阳能空调工程技术规范 Technical code for solar air conditioning system of civil buildings		中国建筑设计研究院 中国可再生能源学会太阳能建筑专业委员
734	GB 50788-2012	城镇给水排水技术规范 Technical code for water supply and sewerage of urban		住建部标准定额研究所 城市建设研究院
735	GB/T 50789-2012	±800kV 直流换流站设计规范 Code for design of ±800kV DC converter station		中国电力工程顾问集团公司 中国南方电网有限责任公司 中国电力企业联合会
736	GB 50790-2013	±800kV 直流架空输电线路设计规范 Code for designing of ±800kV DC overhead transmission line		中国电力企业联合会 中国电力工程顾问集团公司 中国南方电网有限责任公司
737	GB 50791-2013	地热电站设计规范 Code for design of geothermal power plants		中国电力工程顾问集团西南电力设计院
738	GB 50793-2012	会议电视会场系统工程施工及验收规范 Code for construction and acceptance of hall system engineering of videoconference		中国电力科技集团第三研究所 工业和信息产业部电子工业标准化研究院电子工程标准定额站
739	GB 50794-2012	光伏发电站施工规范 Code for construction of PV power station		协鑫光伏系统有限公司 中国电力企业联合会
740	GB/T 50795-2012	光伏发电工程施工组织设计规范 Code for construction organization planning of photovoltaic power project		华电新能源发展有限公司 中国电力企业联合会
741	GB/T 50796-2012	光伏发电工程验收规范 Code for acceptance of photovoltaic power project		国家能源投资有限公司 中国电力企业联合会
742	GB 50797-2012	光伏发电站设计规范 Code for design of photovoltaic power station		上海电力设计院有限公司 中国电力企业联合会

工程建设国家标准

序号	标准编号	标准名称	被代替标准编号	主编单位
743	GB 50798-2012	石油化工大型设备吊装工程规范 Code for large-size equipment hoisting engineering in petrochemical industry		中石化宁波工程有限公司 中石化第十建设有限公司
744	GB 50799-2012	电子会议系统工程设计规范 Code for design of the electrical conference systems		北京奥特维科技有限公司 深圳市台电实业有限公司 工业和信息化部电子工业标准化研究院电子工程标准定额站
745	GB 50800-2012	消声室和半消声室技术规范 Technical code for anechoic and semi-anechoic rooms		中国电子工程设计院 工业和信息化标准化研究院电子工程标准定额站
746	GB/T 50801-2013	可再生能源建筑应用工程评价标准 Evaluation standard for application of renewable energy in buildings		中国建筑科学研究院 住建部科技发展促进中心
747	GB/T 50805-2012	城市防洪工程设计规范 Code for design of urban flood control project	CJJ 50-92	水利部水利水电规划设计总院 中水北方勘测设计研究有限责任公司
748	GB 50807-2013	铀矿石和铀化合物储存设施安全技术规范 Safety technical code for the storage facilities of uranium ores and uranium compounds		中核第四研究设计工程有限公司
749	GB 50808-2013 （限内部发行）	城市居住区人民防空工程规划规范（内部发行） Code of civil air defence works for urban residential areas planning		中国建筑标准设计研究院
750	GB 50809-2012	硅集成电路芯片工厂设计规范 Code for design of silicon integrated circuits wafer fab		信息产业部第十一设计研究院 信息产业部电子工程标准定额站
751	GB 50810-2012	煤炭工业给水排水设计规范 Code for design of water supply and drainage of coal industry		中国煤炭建设协会勘察设计委员会 中煤西安设计工程有限责任公司
752	GB/T 50811-2012	燃气系统运行安全评价标准 Standard for the operation safety assessment of gas system		中国城市燃气协会
753	GB/T 50812-2013	化工厂蒸汽凝结水系统设计规范 Code for design of steam condensate system in chemical plant		中石油东北炼化工程有限公司吉林设计院 中国石油和化工勘察设计协会
754	GB 50813-2012	石油化工粉体料仓防静电燃爆设计规范 Code for design of static explosion prevention in petrochemical powders silo		中石化南京工程有限公司
755	GB 50814-2013	电子工程环境保护设计规范 Code for design environmental protection of electronic engineering		中国电子工程设计院 工业和信息化部电子工业标准化研究院电子工程标准定额站
756	GB/T 50815-2013	稀硫酸真空浓缩处理技术规范 Technical code of vacuum concentration treatment for weak sulfuric acid		北京北方节能环保公司
757	GB 50816-2012	弹药装药废水处理设计规范 Code for design of wastewater treatment of ammunition loading		北京北方节能环保公司
758	GB/T 50817-2013	农田防护林工程设计规范 Code for design of farmland shelterbelts engineering		河南省林业调查规划院

工程建设国家标准

序号	标准编号	标准名称	被代替标准编号	主编单位
759	GB/T 50818-2013	石油天然气管道工程全自动超声波检测技术规范 Mechanized ultrasonic testing technology specification for oil & gas construction pipeline project		中国石油天然气管道局 中国石油天然气股份有限公司规划总院
760	GB 50819-2013	油气田集输管道施工规范 Code for construction of oil and gas field gathering and transmission pipeline		四川石油天然气建设工程有限责任公司 中国石油天然气股份有限公司规划总院
761	GB/T 50820-2013	建材矿山工程建设项目设计文件编制标准 Standard for design document of construction projects in building materials mines		天津水泥工业设计研究院有限公司 中国建筑材料工业规划研究院
762	GB 50821-2012	煤炭工业环境保护设计规范 Code for design of environmental protection in coal industry		中国煤炭建设协会勘察设计委员会 中煤西安设计工程有限责任公司
763	GB 50822-2012	中密度纤维板工程设计规范 Code for design of medium density fiberboard engineering		国家林业局林产工业规划设计院
764	GB/T 50823-2013	油气田及管道工程计算机控制系统设计规范 Code for computer control system design of oil/gas fields and pipelines		胜利油田胜利勘察设计研究院有限公司
765	GB/T 50824-2013	农村居住建筑节能设计标准 Design standard for energy efficiency of rural residential buildings		中国建筑科学研究院 中国建筑设计研究院
766	GB 50825-2013	钢铁厂加热炉工程质量验收规范 Code for quality acceptance of reheating furnaces in iron and steel works		中冶京诚工程技术有限公司
767	GB 50826-2012	电磁波暗室工程技术规范 Technical code for construction of electromagnetic wave anechoic enclosure		中国电子工程设计院 工业和信息化部电子工业标准化研究院电子工程标准定额站
768	GB 50827-2012	刨花板工程设计规范 Code for design of particleboard engineering		国家林业局林产工业规划设计院
769	GB 50828-2012	防腐木材工程应用技术规范 Technical code for engineering application of preservative treated wood		木材节约发展中心 宁波建工股份有限公司
770	GB 50829-2013	租赁模板脚手架维修保养技术规范 Technical code for maintenance and repair of rentable formwork and scaffold		木材节约发展中心 中国基建物资租赁承包协会
771	GB 50830-2013	冶金矿山采矿设计规范 Code for design of metal mine		中冶北方工程技术有限公司
772	GB/T 50831-2012	城市规划基础资料搜集规范 Code for basic data collection of urban planning		江苏省城市规划设计研究院
773	GB/T 50832-2013	1000kV 系统电气装置安装工程电气设备交接试验标准 Standard for acceptance test of electric equipment of 1000kV system electric equipment installation engineering		国家电网公司
774	GB/T 50833-2012	城市轨道交通工程基本术语标准 Standard for basic terminology of urban rail transit engineering		住建部标准定额研究所 北京交通大学

工程建设国家标准

序号	标准编号	标准名称	被代替标准编号	主编单位
775	GB 50834-2013	1000kV构支架施工及验收规范 Code for construction and acceptance of 1000kV lattice frame and support		中国电力企业联合会 国家电网公司
776	GB 50835-2013	1000kV电力变压器、油浸电抗器、互感器施工及验收规范 Code for construction and acceptance of 1000kV power transformer, oil-immersed reactor and mutual inductor		中国电力企业联合会 国家电网公司
777	GB 50836-2013	1000kV高压电器（GIS、HGIS、隔离开关、避雷器）施工及验收规范 Code for construction and acceptance of 1000kV high voltage electric equipment (GIS, HGIS, disconnector and surge arrester)		中国电力企业联合会 国家电网公司
778	GB/T 50837-2013	有色金属冶炼工程制图标准 Standard for non-ferrous metal refinery drawings		中国恩菲工程技术有限公司
779	GB 50838-2015	城市综合管廊工程技术规范 Technical code for urban utility tunnel engineering	GB 50838-2012	上海市政工程设计研究总院（集团）有限公司 同济大学
780	GB/T 50839-2013	城市轨道交通工程安全控制技术规范 Technical code of urban rail transit engineering safety control		中国城市规划设计研究院
781	GB 50840-2012	矿浆管线施工及验收规范 Code for construction and acceptance of slurry pipeline engineering		中国二冶集团有限公司 中冶建工集团有限公司
782	GB/T 50841-2013	建设工程分类标准 Standard of construction classification		同济大学 中天建设集团
783	GB 50842-2013	建材矿山工程施工与验收规范 Code for construction and acceptance of building materials mine engineering		国家建筑材料工业标准定额总站 中国建筑材料工业规划研究院
784	GB 50843-2013	建筑边坡工程鉴定与加固技术规范 Technical code for appraisal and reinforcement of building slope		重庆一建设集团有限公司 重庆市设计院
785	GB/T 50844-2013	工程建设标准实施评价规范 Evaluation code for implementation of engineering construction standard		住建部标准定额研究所
786	GB/T 50845-2013	小水电电网节能改造工程技术规范 Technical code for energy-saving refurbishment for small hydropower grids engineering		水利部农村水电及电气化发展局 水利部农村电气化研究所
787	GB 50846-2012	住宅区和住宅建筑内光纤到户通信设施工程设计规范 Code for design of communication engineering for fiber to the home in residential districts and residential buildings		中国移动通信集团设计院有限公司
788	GB 50847-2012	住宅区和住宅建筑内光纤到户通信设施工程施工及验收规范 Code for construction and acceptance of communication engineering for fiber to the home in residential districts and residential buildings		中国移动通信集团设计院有限公司

工程建设国家标准

序号	标准编号	标准名称	被代替标准编号	主编单位
789	GB/T 50848-2013	机械工业工程建设项目设计文件编制标准 Documentation standard for engineering construction project design in mechanical industry		中国联合工程公司 上海市机电设计研究院有限公司
790	GB 50849-2014	传染病医院建筑设计规范 Code for design of infectious diseases hospital		中国中元国际工程有限公司
791	GB 50850-2013	铝电解厂工艺设计规范 Code for design of aluminum smelter processes		贵阳铝镁设计研究院有限公司 中国有色金属工业工程建设标准规范管理处
792	GB/T 50851-2013	建筑工程人工材料设备机械数据标准 Construction project data standard of labour, materials, equipments and machines		广东省建设工程造价管理总站 住房和城乡建设部信息中心
793	GB/T 50852-2013	建筑工程咨询分类标准 Classification standard of building engineering consultation		天津大学 内蒙古兴泰建筑有限公司
794	GB/T 50853-2013	城市通信工程规划规范 Code of urban communication engineering planning		中国城市规划设计研究院 深圳市城市规划设计研究院
795	GB 50854-2013	房屋建筑与装饰工程工程量计算规范 Standard method of measurement for building construction and fitting-out works		四川省建设工程造价管理总站 住房和城乡建设部标准定额研究所
796	GB 50855-2013	仿古建筑工程工程量计算规范 Standard method of measurement for pseudo-classic architectural works		江苏省建设工程造价管理总站
797	GB 50856-2013	通用安装工程工程量计算规范 Standard method of measurement for general building services works		住房和城乡建设部标准定额研究所 四川省建设工程造价管理总站
798	GB 50857-2013	市政工程工程量计算规范 Standard method of measurement for public utilities works		浙江省建设工程造价管理总站 住房和城乡建设部标准定额研究所
799	GB 50858-2013	园林绿化工程工程量计算规范 Standard method of measurement for landscape works		江苏省建设工程造价管理总站 住房和城乡建设部标准定额研究所
800	GB 50859-2013	矿山工程工程量计算规范 Standard method of measurement for mining works		中国建设工程造价管理协会 住房和城乡建设部标准定额研究所
801	GB 50860-2013	构筑物工程工程量计算规范 Standard method of measurement for affiliated structure works		中国建设工程造价管理协会 住房和城乡建设部标准定额研究所
802	GB 50861-2013	城市轨道交通工程工程量计算规范 Standard method of measurement for urban transit railway works		住房和城乡建设部标准定额研究所 铁路工程定额所
803	GB 50862-2013	爆破工程工程量计算规范 Standard method of measurement for blasting works		中国工程爆破协会 住房和城乡建设部标准定额研究所
804	GB 50863-2013	尾矿设施设计规范 Code for design of tailings facilities		中国恩菲工程技术有限公司 中国有色金属工业工程建设标准规范管理处
805	GB 50864-2013	尾矿设施施工及验收规范 Code for construction and acceptance of tailings disposal facilities		昆明有色冶金设计研究院股份公司

工程建设国家标准

序号	标准编号	标准名称	被代替标准编号	主编单位
806	GB/T 50865-2013	光伏发电接入配电网设计规范 Code for design of photovoltaic generation connecting to distribution network		国家电网公司 中国电力企业联合会
807	GB/T 50866-2013	光伏发电站接入电力系统设计规范 Design code for photovoltaic power station connecting to power system		国家电网公司 中国电力企业联合会
808	GB 50867-2013	养老设施建筑设计规范 Design code for buildings of elderly facilities		哈尔滨工业大学
809	GB 50868-2013	建筑工程容许振动标准 Standard for allowable vibration of building engineering		中国机械工业集团有限公司
810	GB 50869-2013	生活垃圾卫生填埋处理技术规范 Technical code for municipal solid waste sanitary landfill	CJJ 17-2004	华中科技大学
811	GB 50870-2013	建筑施工安全技术统一规范 Unified code for technique for constructional safety		江苏省建筑工程管理局
812	GB 50872-2014	水电工程设计防火规范 Code for fire protection design of hydropower projects		水利水电规划设计总院
813	GB 50873-2013	化学工业给水排水管道设计规范 Design code for piping of water supply and drainage in chemical industry		中国石油和化工勘察设计协会 华陆工程科技有限责任公司
814	GB 50874-2013	煤炭工业半地下储仓建筑结构设计规范 Code for structure design on semi-underground storage bin in coal industry		中国煤炭建设协会勘察设计委员会 中煤西安设计工程有限责任公司
815	GB/T 50875-2013	工程造价术语标准 Standard glossary of project costs		中国建设工程造价管理协会
816	GB/T 50876-2013	小型水电站安全检测与评价规范 Code for safety detecting and evaluation of small hydropower station		水利部农村水电及电气化发展局 水利部农村电气化研究所
817	GB 50877-2014	防火卷帘、防火门、防火窗施工及验收规范 Code for installation and acceptance of fire resistant shutters, fire resistant doorsets and fire resistant windows		辽宁省公安消防总队 公安部天津消防研究所
818	GB/T 50878-2013	绿色工业建筑评价标准 Evaluation standard for green industrial building		中国建筑科学研究院 机械工业第六设计研究院有限公司
819	GB 50879-2013	人造板生产热能中心工程设计规范 Code for design of thermal energy plant engineering of wood-based panel production		国家林业局林产工业规划设计院
820	GB 50880-2013	冶炼烟气制酸工艺设计规范 Design code for acid-making with smelting off-gas		中国有色金属工业工程建设标准规范管理处 中国恩菲工程技术有限公司
821	GB 50881-2013	疾病预防控制中心建筑技术规范 Architectural and technical code for center for disease control and prevention		中国建筑科学研究院
822	GB 50882-2013	轻金属冶炼机械设备安装工程施工规范 Code for construction specification of light-metal smelting mechanical equipment installation engineering		七冶建设有限责任公司 中国有色金属工业第六冶金建设公司 中国有色金属工业工程建设标准规范管理处

工程建设国家标准

序号	标准编号	标准名称	被代替标准编号	主编单位
823	GB 50883-2013	轻金属冶炼机械设备安装工程质量验收规范 Code for quality acceptance of light-metal smelting mechanical equipment installation engineering		七冶建设有限责任公司 中色十二冶金建设有限公司
824	GB 50884-2013	钢筒仓技术规范 Code for design of steel silo structures		中国瑞林工程技术有限公司
825	GB/T 50885-2013	水源涵养林工程设计规范 Code for design of water resources protection forest engineering		国家林业局调查规划设计院
826	GB/T 50886-2013	林产工业工程术语标准 Glossary standard of forest products industry engineering		国家林业局林产工业规划设计院
827	GB/T 50887-2013	人造板工程环境保护设计规范 Code for environmental protection design of wood based panel engineering		国家林业局林产工业规划设计院
828	GB/T 50888-2013	人造板工程节能设计规范 Code for energy efficiency design of wood based panel engineering		国家林业局林产工业规划设计院
829	GB 50889-2013	人造板工程职业安全卫生设计规范 Code for design of occupational safety and health in wood based panel engineering		国家林业局林产工业规划设计院
830	GB 50890-2013	饰面人造板工程设计规范 Code for design of wood-based panel finishing engineering		国家林业局林产工业规划设计院
831	GB 50891-2013	有色金属冶炼厂自控设计规范 Code for automation design of non-ferrous metals smelters		中国恩菲工程技术有限公司 昆明有色冶金设计研究院有限公司
832	GB/T 50892-2013	油气田及管道工程仪表控制系统设计规范 Code for engineering design of instrument control system for oil/gas fields and pipelines		中国石化集团中原石油勘探局勘察设计研究院
833	GB/T 50893-2013	供热系统节能改造技术规范 Technical code for retrofitting of heating system on energy efficiency		北京城建科技促进会 泛华建设集团公司
834	GB 50894-2013	机械工业环境保护设计规范 Code for design of environment protection for machinery industry		中国新时代国际工程公司
835	GB 50895-2013	烟气脱硫机械设备工程安装及验收规范 Code for installation and quality acceptance of flue gas desulphurization mechanical equipment engeering		中国十七冶集团有限公司 攀钢集团工程技术有限公司
836	GB 50896-2013	压型金属板工程应用技术规范 Technical code for application of profiled metal sheets		中冶建筑研究总院有限公司
837	GB 50897-2013	装饰石材工厂设计规范 Code for design of decorative stone plant		中国建设材料工业规划研究院 中材人工晶体研究院
838	GB 50898-2013	细水雾灭火系统技术规范 Technical code for water mist fire extinguishing system		公安部天津消防研究所
839	GB/T 50899-2013	房地产估价基本术语标准 Standard for basic terminology of real estate appraisal		中国房地产估价师与房地产经纪人学会

工程建设国家标准

序号	标准编号	标准名称	被代替标准编号	主编单位
840	GB/T 50900-2016	村镇住宅结构施工及验收规范		中国建筑科学研究院 福建省抗震防灾技术中心
841	GB 50901-2013	钢-混凝土组合结构施工规范 Code for construction of steel-concrete composite structures		中国建筑第二工程局 中国建筑股份有限公司
842	GB/T 50902-2013	医药工程基本术语标准 Standard for fundamental terms of pharmaceutical engineering		中石化上海工程有限公司
843	GB/T 50903-2013	市政工程施工组织设计规范 Code for construction organization plan of municipal engineering		济南城建集团有限公司 威通建设集团股份有限公司
844	GB/T 50904-2013	非织造布设备工程安装与质量验收规范 Code for installation and quality acceptance of nonwoven equipments engineering		中国纺织工业联合会 恒天重工股份有限公司
845	GB/T 50905-2014	建筑工程绿色施工规范 Code for green construction of building		中国建筑股份有限公司 中国建筑技术集团有限公司
846	GB 50906-2013	机械工业厂房结构设计规范 Code for design of machinery industry workshop structures		中国中元国际工程公司
847	GB 50907-2013	抗爆间室结构设计规范 Code for design of blast resistant chamber structures		中国五洲工程设计集团有限公司
848	GB/T 50908-2013	绿色办公建筑评价标准 Evaluation standard for green office building		住房和城乡建设部科技发展促进中心
849	GB 50909-2014	城市轨道交通结构抗震设计规范 Code for seismic design of urban rail transit structures		同济大学 天津市地下铁道集团有限公司
850	GB 50910-2013	机械工业工程节能设计规范 Code for design of energy conservation of mechanical industrial engineering		中国联合工程公司
851	GB 50911-2013	城市轨道交通工程监测技术规范 code for monitoring measurement of urban rail transit engineering		北京城建勘测设计研究院
852	GB/T 50912-2013	钢铁渣粉混凝土应用技术规范 Technical code for application of ground iron and steel slag concrete		中冶建筑研究总院有限公司 中国京冶工程技术有限公司
853	GB 50913-2013	医药工艺用水系统设计规范 Code for design of pharmaceutical process water system		中国医药集团联合工程有限公司
854	GB 50914-2013	化学工业建（构）筑物抗震设防分类标准 Standard for classification of seismic protection of buildings and structures in chemical industry		中国石油和化工勘察设计协会 中国寰球工程公司
855	GB 50915-2013	有色金属矿山井巷工程设计规范 Code for design of underground opening of non-ferrous metals mine		中国恩菲工程技术公司
856	GB 50917-2013	钢-混凝土组合桥梁设计规范 Code for design of steel and concrete composite bridges		上海城市建设设计研究总院

工程建设国家标准

序号	标准编号	标准名称	被代替标准编号	主编单位
857	GB 50918-2013	城镇建设智能卡系统工程技术规范 Technical code for smart card system in urban construction		住建部信息中心 住建部IC卡应用服务中心
858	GB 50919-2013	有色金属冶炼厂节能设计规范 Code for design of energy conservation of non-ferrous metals smelter		中国瑞林工程技术公司有限公司
859	GB/T 50920-2013	用材竹林工程设计规范 Code for design of culm-producing bamboo plantation engineering		国家林业局调查规划设计院
860	GB/T 50921-2013	速生丰产用材林工程设计规范 Code for design of fast-growing and high-yielding timber plantation engineering		国家林业局调查规划设计院
861	GB 50922-2013	天线工程技术规范 Code for technical antenna engineering		工业和信息化部电子工业标准化研究院电子工程标准定额站 中国电子科技集团公司第三十九研究所
862	GB 50923-2013	钢管混凝土拱桥技术规范 Technical code for concrete-filled steel tube arch bridges		福州大学 中建海峡建设发展有限公司
863	GB 50924-2014	砌体结构工程施工规范 Code for construction of masonry structures engineering		陕西省建筑科学研究院 陕西建工集团第五建筑工程公司
864	GB 50925-2013	城市对外交通规划规范 Code for intercity transportation planning		上海城市规划设计研究院
865	GB 50926-2013	丝绸工厂设计规范 Code for design of silk factory		中国纺织工业联合会 四川省纺织工业设计院
866	GB/T 50927-2013	大中型水电工程建设风险管理规范 Code for risk management of large and medium scale hydropower projects		中国电力企业联合会 中国水力发电工程学会
867	GB 50929-2013	氨纶工厂设计规范 Code for design of spandex plant		江苏省纺织工业设计研究院有限公司 中国纺织工业联合会
868	GB 50930-2013	冷轧带钢工厂设计规范 Code for design of cold rolling strip plant		中冶南方工程技术公司
869	GB/T 50931-2013	医药工程建设项目设计文件编制标准 Standard for formulation of design documents of pharmaceutical construction projects		中石化上海工程有限公司 中国医药集团重庆医药设计院 中国医药集团武汉医药设计院
870	GB/T 50933-2013	石油化工装置设计文件编制标准 Authorized specification for plant design document in petrochemical engineering		中国石化工程建设有限公司
871	GB/T 50934-2013	石油化工工程防渗技术规范 Technical code for seepage prevention in petrochemical engineering		中石化洛阳工程有限公司
872	GB/T 50935-2013	煤矿瓦斯抽采工程设计文件编制标准 Design document preparation standard of the gas drainage engineering in coal mine		中煤科工集团重庆设计研究院有限公司
873	GB 50936-2014	钢管混凝土结构技术规范 Technical code for concrete filled steel tubular structures		哈尔滨工业大学 中国建筑科学研究院
874	GB 50937-2013	选煤厂管道安装工程施工与验收规范 Code for construction and acceptance of piping installation in coal preparation plant		中煤建筑安装工程集团公司 河南锦源建设有限公司

工程建设国家标准

序号	标准编号	标准名称	被代替标准编号	主编单位
875	GB/T 50938-2013	石油化工钢制低温储罐技术规范 Technical code for low temperature steel storage tanks in petrochemical engineering		中石化洛阳工程有限公司
876	GB/T 50939-2013	急救中心建筑设计规范 Design code for first-aid station		上海建筑设计研究院
877	GB/T 50941-2014	建筑地基基础术语标准 Standard for terms used in building foundation		中国建筑科学研究院 浙江宝业建设集团公司
878	GB/T 50942-2014	盐渍土地区建筑技术规范 Technical code for building in saline soil regions		合肥工业大学 中建三局第三建筑工程有限责任公司
879	GB/T 50943-2015	海岸软土地基堤坝工程技术规范 Technical code for levee engineering on soft ground in coast area		浙江科技学院 红阳建设集团有限公司
880	GB 50944-2013	防静电工程施工与质量验收规范 Code for construction and quality acceptance of antistatic engineering		工业和信息化部电子工业标准化研究院电子工程标准定额站
881	GB 50945-2013	光纤厂工程技术规范 Technical code for optical fiber plant engineering		工业和信息化部电子工业标准化研究院电子工程标准定额站 信息产业电子第十一设计研究院科技工程股份有限公司
882	GB 50946-2013	煤矿设备安装工程质量验收规范 Code for acceptance of construction quality of coal mine equipment installation		平煤神马建工集团公司有限公司
883	GB/T 50947-2014	建筑日照计算参数标准 Standard for assessment parameters of sunlight on building		中国城市规划设计研究院 北京市城市规划设计研究院
884	GB/T 50948-2013	体育场建筑声学技术规范 Technical code for architectural acoustics of stadium		工业和信息化部电子工业标准化研究院电子工程标准定额站 中国电子科技集团公司第三研究所
885	GB 50949-2013	扩声系统工程施工规范 Construction code for sound reinforcement system		工业和信息化部电子工业标准化研究院电子工程标准定额站 中国电子科技集团公司第三研究所
886	GB 50950-2013	光缆厂生产设备安装工程施工及质量验收规范 Code for construction and quality acceptance for installation engineering of manufacturing equipments on optical cable plant		工业和信息化部电子工业标准化研究院电子工程标准定额站 中国电子科技集团公司第八研究所
887	GB/T 50951-2013	有色金属矿山工程建设项目设计文件编制标准 Design document preparation standard of construction project for nonferrous mine		中国恩菲工程技术有限公司 中国有色金属工业工程建设标准规范管理处
888	GB 50952-2013	农村民居雷电防护工程技术规范 Technical code for lightning protection engineering for rural residential areas		湖南省防雷中心 河北省防雷中心
889	GB 50953-2014	网络互联调度系统工程技术规范 Technical code for engineering of network interconnection and dispatch system		工业和信息化部电子工业标准化研究院电子工程标准定额站 安徽四创电子股份有限公司
890	GB 50954-2014	水泥窑协同处置垃圾工程设计规范 Code for design of municipal solid waste co-processing in cement kiln		中材国际环境工程（北京）有限公司

工程建设国家标准

序号	标准编号	标准名称	被代替标准编号	主编单位
891	GB 50955-2013	石灰石矿山工程勘察技术规范 Technical code for investigation of limestone mine engineering		西安建材地质工程勘察院 中材地质工程勘查研究院
892	GB 50956-2013	化纤工厂验收规范 Code for acceptance of chemical fiber plant		中国纺织工业设计院 中国纺织工业联合会
893	GB 50957-2013	生物液体燃料工厂设计规范 Code for design of liquid biofuel plant		中国海诚工程科技股份有限公司
894	GB/T 50958-2013	核电厂常规岛设计规范 Code for design of conventional island for nuclear power plant		中国能源建设集团广东省电力设计研究院 中国电力工程顾问集团华东电力设计院 中国电力企业联合会
895	GB 50959-2013	有色金属工程结构荷载规范 Load code of nonferrous metal engineering structures		中国恩菲工程技术有限公司
896	GB/T 50960-2014	小水电电网安全运行技术规范 Technical code of safe operating for small hydropower grid		水利部农村水电及电气化发展局 水利部农村电气化研究所
897	GB 50961-2014	有色金属矿山井巷安装工程质量验收规范 Code for underground opening installation acceptance of non-ferrous metal mines		中国有色金属工业第十四冶金建设公司 十一冶建设集团有限责任公司
898	GB 50962-2014	铜加工厂工艺设计规范 Code for design of copper processing plant		中色科技股份有限公司 中国有色工程有限公司
899	GB 50963-2014	硫酸、磷肥生产污水处理设计规范 Code for design of wastewater treatment in sulfuric acid and phosphate fertilizer production		中国石油和化工勘察设计协会 中国石化集团南京工程有限公司
900	GB/T 50964-2014	小型水电站运行维护技术规范 Code for operation and maintenance of small hydropower stations		水利部农村水电及电气化发展局 水利部农村电气化研究所
901	GB 50965-2014	冶金烧结球团烟气氨法脱硫设计规范 Code for design of sinter and pellet flue gas ammonia desulfurization		中钢集团工程设计研究院有限公司
902	GB 50966-2014	电动汽车充电站设计规范 Code for design of electric vehicle charging station		国家电网公司 中国电力企业联合会
903	GB 50967-2014	焦化机械设备安装规范 Code for installation of coking and chemical mechanical equipment		中国五冶集团有限公司
904	GB 50968-2014	露天煤矿工程施工规范 Code for construction of surface coal mine engineering		中煤平朔集团有限公司
905	GB 50970-2014	装饰石材矿山露天开采工程设计规范 Code for design of open-pit mining engineering in decoration stone mine		中国建筑材料工业规划研究院 浙江中材工程勘测设计有限公司
906	GB 50971-2014	钢铁企业余能发电机械设备工程安装与质量验收规范 Code for installation and quality acceptance of power generating by remain energy mechanical equipment in iron industry		北京首钢建设集团有限公司 中国十九冶集团有限公司

工程建设国家标准

序号	标准编号	标准名称	被代替标准编号	主编单位
907	GB 50972-2014	循环流化床锅炉施工及质量验收规范 Code for construction and quality acceptance of circulating fluidized-bed boiler		中国电力企业联合会 中国电力建设企业协会
908	GB 50973-2014	联合循环机组燃气轮机施工及质量验收规范 Code for construction and quality acceptance of gas turbine of combined cycle unit		中国电力企业联合会 中国电力建设企业协会
909	GB 50974-2014	消防给水及消火栓系统技术规范 Technical code for fire protection water supply and hydrant systems		中国中元国际工程公司
910	GB/T 50975-2014	铀浓缩工厂工艺水管道工程施工及验收规范 Code for construction and acceptance of process water piping in uranium enrichment plant		中核新能源工业工程有限责任公司
911	GB/T 50976-2014	继电保护及二次回路安装及验收规范 Code for installation and acceptance of protection equipment and secondary circuit in infrastructure project		中国电力企业联合会 重庆市电力公司
912	GB/T 50977-2014	化学工程节水设计规范 Code for design of water saving in chemical engineering		中国石油和化工勘察设计协会 华陆工程科技有限责任公司
913	GB/T 50978-2014	电子工业工程建设项目设计文件编制标准 Documentary standard for engineering construction project design in electronics industry		工业和信息化部电子工业标准化研究院电子工程标准定额站 中国电子工程设计院 信息产业电子第十一设计研究院科技工程股份有限公司
914	GB/T 50979-2014	橡胶坝工程技术规范 Technical code for rubber dam engineering		中国水利水电科学研究院
915	GB/T 50980-2014	电力调度通信中心工程设计规范 Code for design of power dispatching and communication center		中国电力企业联合会 中国能源建设集团广东省电力设计院
916	GB 50981-2014	建筑机电工程抗震设计规范 Code for seismic design of mechanical and electrical equipment		中国建筑设计院有限公司
917	GB 50982-2014	建筑与桥梁结构监测技术规范 Technical code for monitoring of building and bridge structures		中国建筑科学研究院 海南建设工程股份有限公司
918	GB/T 50983-2014	低、中水平放射性废物处置场岩土工程勘察规范 Code for site investigation of geotechnical engineering for low and intermediate level radioactive wastedi sposal		中国电力企业联合会 中国能源建设集团广东省电力设计研究院
919	GB 50984-2014	石油化工工厂布置设计规范 Code for design of petrochemical plant layout		中国石化工程建设有限公司 中石化洛阳工程有限公司
920	GB 50985-2014	铅锌冶炼厂工艺设计规范 Code for process design of lead and zinc smelters		中国恩菲工程技术有限公司 中国有色金属工业工程建设标准规范管理处
921	GB 50986-2014	干法赤泥堆场设计规范 Code for design of dry red mud stack		中国有色工程有限公司 贵阳铝镁设计研究院有限公司
922	GB 50987-2014	水利工程设计防火规范 Code for fire protection design of hydraulic engineering		中水珠江规划勘测设计有限公司

工程建设国家标准

序号	标准编号	标准名称	被代替标准编号	主编单位
923	GB 50988-2014	有色金属工业环境保护工程设计规范 Code for design of environment protection engineering of nonferrous metals industry		中国瑞林工程技术有限公司
924	GB/T 50989-2014	大型螺旋塑料管道输水灌溉工程技术规范 Technical code for buried twisting plastic pipeline engineering		中国水利水电科学研究院 水利部兰州勘测设计研究院
925	GB 50990-2014	加气混凝土工厂设计规范 Code for design of autoclaved aerated concrete plant		武汉建筑材料工业设计研究院有限公司 国家建筑材料工业标准定额总站
926	GB 50991-2014	埋地钢质管道直流干扰防护技术标准 Technical standard for DC interference mitigation of buried steel pipeline		中国石油管道公司沈阳调度中心
927	GB 50992-2014	石油化工工程地震破坏鉴定标准 Authentication standard for earthquake destruction in petrochemical engineering		中国石化工程建设有限公司
928	GB 50993-2014	1000kV输变电工程竣工验收规范 Code for completion acceptance of 1000kV transmission and transformation engineering		中国电力企业联合会 国家电网公司
929	GB 50994-2014	工业企业电气设备抗震鉴定标准 Standard for aseismatic appraisal of electrical facilities in industrial plants		中国石化工程建设有限公司
930	GB 50995-2014	冶金工程测量规范 Code for surveying of metallurgical engineering		中冶集团武汉勘察研究院有限公司
931	GB 50996-2014	地下水封石洞油库施工及验收规范 Code for construction and acceptance of underground oil storage in rock caverns		中铁隧道集团有限公司
932	GB/T 50997-2014	冷轧电工钢工程设计规范 Code for engineering of cold-rolled electrical steel strip plant		中冶南方工程技术有限公司
933	GB 50998-2014	乳制品厂设计规范 Design code for dairy products plant		中国轻工业武汉设计工程有限责任公司
934	GB/T 51003-2014	矿物掺合料应用技术规范 Technical code for application of mineral admixture		中国建筑科学研究院
935	GB 51004-2015	建筑地基基础工程施工规范 Code for construction of building foundation engineering		上海建工集团股份有限公司 上海市基础工程集团有限公司
936	GB 51005-2014	水泥工厂余热发电工程施工与质量验收规范 Code for construction and acceptance of waste heat recovery power generation in cement plant		中材节能股份有限公司 国家建筑材料工业标准定额总站
937	GB 51006-2014	石油化工建（构）筑物结构荷载规范 Load code for design of buildings and special structures in petrochemical industry		中国石化工程建设有限公司
938	GB/T 51007-2014	石油化工用机泵工程设计规范 Design code for rotary machines of petrochemical industries		中石化上海工程有限公司
939	GB 51008-2016	高耸与复杂钢结构检测与鉴定标准 Standard for inspection and appraisal of high-rise and complicated steel structures		同济大学 中天建设集团有限公司

工程建设国家标准

序号	标准编号	标准名称	被代替标准编号	主编单位
940	GB 51009-2014	火炸药生产厂房设计规范 Code for design of propellant and explosive work architecture		中国兵器工业标准化研究所 中国五洲工程设计集团有限公司
941	GB 51010-2014	铝电解系列不停电停开槽设计规范 Code for design of shutting down/restarting aluminum reduction cells without power interruption		中国有色工程有限公司 贵阳铝镁设计研究院有限公司
942	GB 51011-2014	煤矿选煤设备安装工程施工与验收规范 Code for construction and acceptance of coal mine preparation equipment installation engineering		中煤矿山建设集团有限责任公司 平煤神马建工集团有限公司
943	GB/T 51012-2014	铀浓缩工厂工艺气体管道工程施工及验收规范 Code for construction and acceptance of process gas piping in uranium enrichment plant		中核新能源工业工程有限责任公司
944	GB/T 51013-2014	铀转化设施设计规范 Code for design of uranium conversion facility		中核新能源工业工程有限责任公司
945	GB 51014-2014	水泥工厂岩土工程勘察规范 Code for geotechnical engineering investigation of cement plant		建材广州地质工程勘察院 建材桂林地质工程勘察院
946	GB/T 51015-2014	海堤工程设计规范 Code for design of sea dike project		水利部水利水电规划设计总院 广东省水利水电科学研究院
947	GB 51016-2014	非煤露天矿边坡工程技术规范 Technical code for non-coal open-pit mine slope engineering		中勘冶金勘察设计研究院有限责任公司
948	GB 51017-2014	古建筑防雷工程技术规范 Technical code for lightning protection engineering of ancient buildings		中国建筑科学研究院
949	GB 51018-2014	水土保持工程设计规范 Code for design of soil and water conservation engineering		水利部水利水电规划设计总院 黄河勘测规划设计有限公司
950	GB 51019-2014	化工工程管架、管墩设计规范 Code for design of pipe racks and pipe sleepers in chemical industry		中国石油和化工勘察设计协会 中国石化工程建设有限公司
951	GB 51020-2014	铝电解厂通风除尘与烟气净化设计规范 Code for design of ventilation dedusting and fume scrubbing of aluminum smelter		中国有色工程有限公司 沈阳铝镁设计研究院有限公司
952	GB/T 51021-2014	轻金属冶炼工程术语标准 Standard for terminology of light metal smelting engineering		中国有色工程有限公司 沈阳铝镁设计研究院有限公司
953	GB 51022-2015	门式刚架轻型房屋钢结构技术规范 Technical code for steel structure of light-weight building with gabled frames		中国建筑标准设计研究院
954	GB/T 51023-2014	有色金属冶炼工程建设项目设计文件编制标准 Design document preparation standard of construction project for non-ferrous metallurgical engineering		中国恩菲工程技术有限公司 中国有色金属工业工程建设标准规范管理处
955	GB 51024-2014	煤矿安全生产智能监控系统设计规范 Code for design of intelligent monitoring and control system of coal mine safety production		中国煤炭建设协会勘察设计委员会 中国煤炭科工集团南京设计研究院有限公司
956	GB/T 51025-2016	超大面积混凝土地面无缝施工技术规范 Technical code for jointless construction of super-large area slabs-on-ground		中建五局第三建设有限公司 重庆大学

工程建设国家标准

序号	标准编号	标准名称	被代替标准编号	主编单位
957	GB/T 51026-2014	石油库设计文件编制标准 Standard for compilation of the design document of oil depot		中石化洛阳工程有限公司
958	GB/T 51027-2014	石油化工企业总图制图标准 Standard for general layout drawings of petrochemical enterprises		中石化洛阳工程有限公司
959	GB/T 51028-2015	大体积混凝土温度测控技术规范 Technical code for temperature measurement and control of mass concrete		陕西省建筑科学研究院 广州富利建筑安装工程有限公司
960	GB 51029-2014	火炬工程施工及验收规范 Code for construction and acceptance of flare project		中国石油和化工勘察设计协会 陕西化建工程有限责任公司
961	GB 51030-2014	再生铜冶炼厂工艺设计规范 Code for design of secondary copper smelter processes		中国有色工程有限公司 中国瑞林工程技术有限公司
962	GB/T 51031-2014	火力发电厂岩土工程勘察规范 Code for geotechnical investigation of fossil fuel power plant		中国电力企业联合会 中国电力工程顾问集团华北电力设计院有限公司
963	GB 51032-2014	铁尾矿砂混凝土应用技术规范 Technical code for application of iron tailings sand concrete		中国十七冶集团有限公司 中冶建工集团有限公司
964	GB/T 51033-2014	水利泵站施工及验收规范 Code for construction and acceptance of pumping station		中国灌溉排水发展中心
965	GB 51034-2014	多晶硅工厂设计规范 Code for design of polysilicon plant		中国有色工程有限公司 中国恩菲工程技术有限公司
966	GB 51035-2014	电子工业纯水系统安装与验收规范 Code for installation and acceptance of pure water system of electronic industry		工业和信息化部电子工业标准化研究院电子工程标准定额站 中国电子系统工程第二建设有限公司
967	GB 51036-2014	有色金属矿山井巷工程质量验收规范 Code for underground opening acceptance of non-ferrous metal mines		中国有色金属工业第十四冶金建设公司 金诚信矿业管理股份有限公司
968	GB 51037-2014	微组装生产线工艺设备安装工程施工及验收规范 Code for construction and acceptance of micro-assembling production line process equipment installation engineering		工业和信息化部电子工业标准化研究院电子工程标准定额站 中国电子科技集团第二研究所
969	GB 51038-2015	城市道路交通标志和标线设置规范 Code for layout of urban rode traffic signs and markings		上海市政工程设计研究总院（集团）有限公司 公安部交通管理科学研究所
970	GB 51039-2014	综合医院建筑设计规范 Code for design of general hospital	JGJ 49-88	国家卫生和计划生育委员会规划与信息司 中国医院协会医院建筑系统研究分会
971	GB/T 51040-2014	地下水监测工程技术规范 Technical code for groundwater monitoring		水利部水文局
972	GB 51041-2014	核电厂岩土工程勘察规范 Code for geotechnical investigation of nuclear power plants		中国核电工程有限公司 电力规划设计总院
973	GB 51042-2014	医药工业废弃物处理设施工程技术规范 Techical code for waste treatment facilities of pharmaceutical industry		中国医药集团重庆医药设计院

工程建设国家标准

序号	标准编号	标准名称	被代替标准编号	主编单位
974	GB 51043-2014	电子会议系统工程施工与质量验收规范 Code for construction and quality acceptance of the electronical conference systems		工业和信息化部电子工业标准化研究院电子工程标准定额站 北京奥特维科技有限公司
975	GB 51044-2014	煤矿采空区岩土工程勘察规范 Code for investigation of geotechnical engineering in the coal mine goaf		中煤科工集团武汉设计研究院有限公司
976	GB 51045-2014	水泥工厂脱硝工程技术规范 Technical code for denitration project of cement plant		中国中材国际环境工程（北京）有限公司 天津水泥工业设计研究院有限公司
977	GB/T 51046-2014	国家森林公园设计规范 Code for design of national forest park		国家林业局调查规划设计院
978	GB 51047-2014	医药工业总图运输设计规范 Code for design of general plot plan and transportation of pharmaceutical industry		中国医药集团联合工程有限公司
979	GB 51048-2014	电化学储能电站设计规范 Design code for electrochemical energy storage station		中国电力企业联合会 中国南方电网有限责任公司调峰调频发电公司
980	GB 51049-2014	电气装置安装工程串联电容器补偿装置施工及验收规范 Code for construction and acceptance of series capacitor installation electrical equipment installation engineering		中国电力企业联合会 中国电力科学研究院
981	GB/T 51050-2014	钢铁企业能源计量和监测工程技术规范 Technical code for energy measuring and monitoring engineering in the iron and steel enterprises		中冶京诚工程技术有限公司 中冶京诚鼎宇管理系统有限公司
982	GB/T 51051-2014	水资源规划规范 Code for water resources planning		水利部水利水电规划设计总院
983	GB 51052-2014	毛纺织工厂设计规范 Code for design of wool textile factory		中国纺织工业联合会 江苏省纺织工业设计研究院有限公司
984	GB 51053-2014	煤炭工业矿井节能设计规范 Code for mine energy conservation design of coal industry		煤炭工业合肥设计研究院
985	GB 51054-2014	城市消防站设计规范 Design code for fire station		公安部上海消防研究所
986	GB 51055-2014	有色金属工业厂房结构设计规范 Code for design of non-ferrous industrial plant structures		中国有色金属工业工程建设标准规范管理处 中国瑞林工程技术有限公司 中国有色工程有限公司
987	GB 51056-2014	烟囱可靠性鉴定标准 Standard for appraisal of reliability of chimneys		中冶建筑研究总院有限公司
988	GB/T 51057-2015	种植塑料大棚工程技术规范 Technical code for horticultural plastic tunnel engineering		农业部规划设计研究院
989	GB 51058-2014	精神专科医院建筑设计规范 Code for design of psychiatric hospital		中国中元国际工程公司
990	GB 51059-2014	有色金属加工机械安装工程施工与质量验收规范 Code for construction and acceptance of nonferrous metals processing mechanical installation engineering		中国有色工程有限公司 中国十五冶金建设集团有限公司

工程建设国家标准

序号	标准编号	标准名称	被代替标准编号	主编单位
991	GB 51060-2014	有色金属矿山水文地质勘探规范 Code for exploration of hydrogeology in nonferrous metal mines		中国有色工程有限公司 华北有色工程勘察院有限公司
992	GB/T 51061-2014	电网工程标识系统编码规范 Code for grid identification system		中国电力企业联合会 中国电力工程顾问集团华北电力设计院有限公司
993	GB 51062-2014	煤矿设备安装工程施工规范 Code for construction of coal mine equipment installation engineering		中煤矿山建设集团有限责任公司
994	GB/T 51063-2014	大中型沼气工程技术规范 Technical code for large and medium-scale biogas engineering		北京市公用事业科学研究所
995	GB/T 51064-2015	吹填土地基处理技术规范 Technical code for ground treatment of hydraulic fill		河海大学 上海港湾基础建设（集团）有限公司
996	GB/T 51065-2014	煤矿提升系统工程设计规范 Code for design of coal mine hoisting system engineering		中国煤炭建设协会勘察设计委员会 中煤科工集团沈阳设计研究院有限公司
997	GB 51066-2014	工业企业干式煤气柜安全技术规范 Technical code for safety of waterless gasholder in industrial enterprise		中冶赛迪工程技术股份有限公司
998	GB 51067-2014	光缆生产厂工艺设计规范 Code for design of process of optical cable plant		工业和信息化部电子工业标准化研究院电子工程标准定额站 中国电子科技集团第八研究所
999	GB/T 51068-2014	煤炭工业露天矿机电设备修理设施设计规范 Code for open pit mine mechanical and electrical equipments maintenance facilities design of coal industry		中煤科工集团沈阳设计研究院有限公司
1000	GB 51069-2014	中药药品生产厂工程技术规范 Technical code for traditional Chiese medicine production plant engineering		中国医药集团重庆医药设计院
1001	GB 51070-2014	煤炭矿井防治水设计规范 Code for design of water prevention and control of coal mine		中煤科工集团武汉设计研究院有限公司
1002	GB/T 51071-2014	330kV～750kV智能变电站设计规范 Code for design of 330kV～750kV smart substation		国家电网公司
1003	GB/T 51072-2014	110（66）kV～220 kV智能变电站设计规范 Code for design of 110（66）kV～220kV smart substation		国家电网公司
1004	GB 51073-2014	医药工业仓储工程设计规范 Code for design of warehousing project of pharmaceutical industry		中国医药集团联合工程有限公司
1005	GB/T 51074-2015	城市供热规划规范 Code for urban heating supply planning		北京市城市规划设计研究院
1006	GB/T 51075-2015	选矿机械设备工程安装规范 Code for installation of mineral processing equipment engineering		中国三冶集团有限公司

工程建设国家标准

序号	标准编号	标准名称	被代替标准编号	主编单位
1007	GB 51076-2015	电子工业防微振工程技术规范 Technical code for anti-microvibration engineering of electronics industry		工业和信息化部电子工业标准化研究院电子工程标准定额站 中国电子工程设计院
1008	GB/T 51077-2015	电动汽车电池更换站设计规范 Code for design of electric vehicle battery-swap station		中国电力企业联合会 国家电网公司
1009	GB 51078-2015	煤炭矿井设计防火规范 Code for design of prevention of mine fire in coal mines		中煤科工集团重庆设计研究院有限公司
1010	GB 51079-2016	城市防洪规划规范 Code for urban planning on flood control		湖北省城市规划设计研究院
1011	GB 51080-2015	城市消防规划规范 Code for planning of urban fire control		重庆市规划设计研究院
1012	GB 51081-2015	低温环境混凝土应用技术规范 Technical code for application of concrete under cryogenic circumstance		中国石油和化工勘察设计协会 中国寰球工程公司
1013	GB/T 51082-2015	工业建筑涂装设计规范 Code for coating design of industrial construction		中国石油和化工勘察设计协会 中国寰球工程公司
1014	GB/T 51083-2015	城市节水评价标准 Standard for urban water conservation evaluation		北京建筑大学 中国城镇供水排水协会
1015	GB 51084-2015	有色金属工程设备基础技术规范 Technical code for equipment foundation of nonferrous metals engineering		中国有色金属工业工程建设标准规范管理处 中国恩菲工程技术有限公司
1016	GB/T 51085-2015	防风固沙林工程设计规范 Code for design of windbreak and sand fixation forest		国家林业局调查规划设计院
1017	GB/T 51086-2015	医药实验工程术语标准 Standard for terminology of pharmaceutical laboratory engineering		中石化上海工程有限公司
1018	GB/T 51087-2015	船厂既有水工构筑物结构改造和加固设计规范 Design code for renovation and strengthening of existing shipyard marine structure		中船第九设计研究院工程有限公司
1019	GB/T 51088-2015	丝绸设备工程安装与质量验收规范 Code for installation and quality acceptance of silk equipment engineering		中国纺织工业联合会 浙江金鹰股份有限公司
1020	GB/T 51089-2015	针织设备工程安装与质量验收规范 Code for installation and quality acceptance of knitting equipment engineering		中国纺织工业联合会 国家纺织机械质量监督检验中心
1021	GB/T 51090-2015	色织设备工程安装与质量验收规范 Code for installation and quality acceptance of yarn-dyed woven equipment engineering		中国纺织工业联合会 恒天重工股份有限公司
1022	GB/T 51091-2015	试听室工程技术规范 Technical code for listening test room engineering		工业和信息化部电子工业标准化研究院电子工程标准定额站 北京奥特维科技有限公司
1023	GB 51092-2015	制浆造纸厂设计规范 Code for design of pulp and papermaking plant		中国海诚工程科技股份有限公司
1024	GB 51093-2015	钢铁企业喷雾焙烧法盐酸废液再生工程技术规范 Technical code for hydrochloric acid regeneration engineering according to spraying and roasting technology of iron and steel enterprises		中冶南方工程技术有限公司

工程建设国家标准

序号	标准编号	标准名称	被代替标准编号	主编单位
1025	GB/T 51094-2015	工业企业湿式气柜技术规范 Technical code for fluid seal gas holder of industrial enterprises		中国石油和化工勘察设计协会 东华工程科技股份有限公司
1026	GB/T 51095-2015	建设工程造价咨询规范 Code for construction cost consultation		中国建设工程造价管理协会
1027	GB 51096-2015	风力发电场设计规范 Code for design of wind farm		中国电力企业联合会 中国电力工程顾问集团华北电力设计院工程有限公司
1028	GB/T 51097-2015	水土保持林工程设计规范 Code for design of soil and water conservation forest engineering		国家林业局调查规划设计院
1029	GB/T 51098-2015	城镇燃气规划规范 Code for planning of city gas		北京市煤气热力工程设计院有限公司
1030	GB 51099-2015	有色金属工业岩土工程勘察规范 Code for investigation of geotechnical engineering for nonferrous metals industry		中国有色工程有限公司 中国有色金属工业西安勘察设计研究院
1031	GB/T 51100-2015	绿色商店建筑评价标准 Assessment standard for green store building		中国建筑科学研究院
1032	GB 51101-2016	太阳能发电站支架基础技术规范 Technical code for supporting bracket foundation of solar power station		中国电力企业联合会 诺斯曼能源科技（北京）有限公司
1033	GB 51102-2016	压缩天然气供应站设计规范		中国市政工程华北设计研究院
1034	GB/T 51103-2015	电磁屏蔽室工程施工及质量验收规范 code for construction and acceptance of electromagnetic shielding enclosure		工业和信息化部电子工业标准化研究院电子工程标准定额站 常州雷宁电磁屏蔽设备有限公司
1035	GB 51104-2015	取向硅钢生产线设备安装与验收规范 code for installation and acceptance for engineering equipment of grain-oriented silicon steel production line		中国五冶集团有限公司
1036	GB 51105-2015	挤压钢管工程设备安装与验收规范 code for installation and acceptance for engineering equipment of extrusion steel pipe		中国五冶集团有限公司
1037	GB/T 51106-2015	火力发电厂节能设计规范 Energy conservation code for design of fossil fired power plants		中国电力企业联合会 中国电力工程顾问集团有限公司
1038	GB 51107-2015	纤维增强硅酸钙板工厂设计规范 Code for design of fibre reinforced calcium silicate board plant		武汉建筑材料工业设计研究院有限公司 国家建筑材料工业标准定额总站
1039	GB 51108-2015	尾矿库在线安全监测系统工程技术规范 Technical coad for online safety monitoring system of tailing pond		中国有色金属工业工程建设标准规范管理处 中国有色金属长沙勘察设计院有限公司
1040	GB/T 51109-2015	氨纶设备工程安装与质量验收规范 Code for installation and quality acceptance of spandex equipment		中国纺织工业联合会 江阴中绿化纤工艺技术有限公司
1041	GB 51110-2015	洁净厂房施工及质量验收规范 Code of construction and quality acceptance of industrial cleanroom		中国电子工程设计院 工业和信息化部电子工业标准化研究院电子工程标准定额站

工程建设国家标准

序号	标准编号	标准名称	被代替标准编号	主编单位
1042	GB/T 51111-2015	露天金属矿施工组织设计规范 Code of construction organization design of open-pit metal mine		八冶建设集团有限公司 九冶建设有限公司
1043	GB 51112-2015	针织工厂设计规范 Code for design of knitting factory		中国纺织工业联合会 河北中纺工程设计有限公司
1044	GB 51113-2015	光伏压延玻璃工厂设计规范 Code for design of photovoltaic rolled glass plant		蚌埠玻璃工业设计研究院
1045	GB 51114-2015	露天煤矿施工组织设计规范 Code of construction organization plan of surface coal mine		中国煤炭建设协会勘察设计委员会 中煤科工集团沈阳设计研究院有限公司
1046	GB 51115-2015	固相缩聚工厂设计规范 Code for design of solid-state polycondensation plant		上海纺织建筑设计研究院
1047	GB/T 51116-2016	医药工程安全风险评估技术标准 Pharmaceutical engineering technical standard for safety risk assessment		中国医药集团重庆医药设计院
1048	GB/T 51117-2015	数字同步网工程技术规范 Technical code for engineering of digital synchronization network		中讯邮电咨询设计院有限公司
1049	GB 51118-2015	尾矿堆积坝排渗加固工程技术规范 Technical code for drainage-consolidation of tailings embankment		中冶集团武汉勘察研究院有限公司
1050	GB 51119-2015	冶金矿山排土场设计规范 Code for waste dump design of metal mine		中冶北方工程技术有限公司
1051	GB 51120-2015	通信局（站）防雷与接地工程验收规范 Acceptance code for lightning protection and earthing engineering of telecommunication bureaus (stations)		中国通信建设集团有限公司
1052	GB/T 51121-2015	风力发电工程施工与验收规范 Code for construction and acceptance of wind power project		中国电力企业联合会 中国长江三峡集团公司
1053	GB 51122-2015	集成电路封装测试厂设计规范 Code for design of integrated circuit assembly and test factory		工业和信息化部电子工业标准化研究院电子工程标准定额站 信息产业电子第十一设计研究院科技工程股份有限公司
1054	GB 51123-2015	光纤器件生产厂工艺设计规范 Code for design of process of optical fiber component plant		工业和信息化部电子工业标准化研究院 中国电子科技集团公司第八研究所
1055	GB/T 51124-2015	马铃薯贮藏设施设计规范 Design code for potato storage facilities		农业部规划设计研究院
1056	GB/T 51125-2015	通信局站共建共享技术规范 Code for joint construction and sharing of telecommunications stations		中讯邮电咨询设计院有限公司
1057	GB/T 51126-2015	波分复用（WDM）光纤传输系统工程验收规范 Code for acceptance of wavelength division multiplexing (WDM) optical fiber transmission system engineering		中国通信建设集团有限公司
1058	GB 51127-2015	印制电路板工厂设计规范 Code for design of printed circuit board plant		工业和信息化部电子工业标准化研究院电子工程标准定额站 中国电子工程设计院

工程建设国家标准

序号	标准编号	标准名称	被代替标准编号	主编单位
1059	GB 51128-2015	钢铁企业煤气储存和输配系统设计规范 Code for design of gas storage & transportation and distribution system for iron steel enterprises		中冶华天工程技术有限公司
1060	GB/T 51129-2015	工业化建筑评价标准 Standard for assessment of industrialized building		住房和城乡建设部住宅产业化促进中心 中国建筑科学研究院
1061	GB/T 51130-2016	沉井与气压沉箱施工规范 Code for construction of open caisson and pneumatic caisson		上海市基础工程有限公司 舜杰建设（集团）有限公司
1062	GB 51132-2015	工业有色金属管道工程施工及质量验收规范 Code for construction and quality acceptance of industrial non-ferrous metallic piping engineering		中十冶集团有限公司 二十一冶建设有限公司
1063	GB 51133-2015	医药工业环境保护设计规范 Code for design of environmental protection of pharmaceutical industry		中石化上海工程有限公司
1064	GB 51134-2015	煤矿瓦斯发电工程设计规范 Design code for coal mine gas power project		中国煤炭建设协会勘察设计委员会 煤炭工业合肥设计研究院
1065	GB 51135-2015	转炉煤气净化及回收工程技术规范 Technical code for converter gas cleaning and recovery system engineering		中冶南方工程技术有限公司
1066	GB 51136-2015	薄膜晶体管液晶显示器工厂设计规范 Code for design of thin film transistor liquid crystal display plant		工业和信息化部电子工业标准化研究院电子工程标准定额站 中国电子工程设计院
1067	GB 51137-2015	电子工业废水废气处理工程施工及验收规范 Code for construction and acceptance of waste water and exhaust treatment engineering of the electronics industry		工业和信息化部电子工业标准化研究院电子工程标准定额站 中国电子系统工程第二建设有限公司
1068	GB 51138-2015	尿素造粒塔工程施工及质量验收规范 Code for construction and quality acceptance of urea prilling tower engineering		中国石油和化工勘察设计协会 中化二建集团有限公司
1069	GB 51139-2015	纤维素纤维用浆粕工厂设计规范 Code for design of pulp plant for cellulose fibre		中国纺织工业联合会 新乡白鹭化纤集团设计研究所
1070	GB/T 51140-2015	建筑节能基本术语标准 Standard for basic terminology of building energy-saving		住房和城乡建设部科技发展促进中心
1071	GB/T 51141-2015	既有建筑绿色改造评价标准 Assessment standard for green retrofitting of existing building		中国建筑科学研究院 住房和城乡建设部科技发展促进中心
1072	GB 51142-2015	液化石油气供应工程设计规范 Code for design liquefied petroleum gas (LPG) supply engineering		中国市政工程华北设计研究总院有限公司
1073	GB 51143-2015	防灾避难场所设计规范 Code for design of disasters mitigation emergency congregate shelter		河北省地震工程研究中心 北京工业大学北京城市与工程安全减灾中心
1074	GB 51144-2015	煤炭工业矿井建设岩土工程勘察规范 Code for investigation of geotechnical engineering of mine construction in coal industry		中煤邯郸设计工程有限责任公司
1075	GB 51145-2015	煤矿电气设备安装工程施工与验收规范 Code for construction and acceptance of coal mine electrical equipment installation engineering		中煤矿山建设集团有限责任公司 江苏省矿业工程集团有限公司

工程建设国家标准

序号	标准编号	标准名称	被代替标准编号	主编单位
1076	GB/T 51146-2015	硝化甘油生产废水处理设施技术规范 Technical code for wastewater treatment facilities from nitroglycerine production		中国兵器工业标准化研究所
1077	GB/T 51147-2015	硝胺类废水处理设施技术规范 Technical code for wastewater treatment facilities from nitroamine explosives		中国兵器工业标准化研究所
1078	GB/T 51148-2016	绿色博览建筑评价标准 Assessment standard for green museum and exhibition building		中国建筑科学研究院
1079	GB/T 51149-2016	城市停车规划规范 Code for urban parking plan		北京市城市规划设计研究院
1080	GB/T 51150-2016	城市轨道交通客流预测规范		北京交通发展研究中心
1081	GB 51151-2016	城市轨道交通公共安全防范系统工程技术规范 Technical code for engineering of public security and protection system of urban rail transit		公安部第三研究所
1082	GB/T 51152-2015	波分复用（WDM）光纤传输系统工程设计规范 Code for engineering design of wavelength division multiplexing (WDM) optical fiber transmission systems		中国移动通信集团设计院有限公司
1083	GB/T 51153-2015	绿色医院建筑评价标准 Evaluation standard for green hospital building		中国建筑科学研究院 住房和城乡建设部科技与产业化发展中心
1084	GB/T 51154-2015	海底光缆工程设计规范 Code for engineering design of optical fiber submarine cable systems		中国移动通信集团设计院有限公司
1085	GB 51155-2016	机械工程建设项目职业安全卫生设计规范 Code for design of occupational safety and health in machinery industry		中国新时代国际工程公司
1086	GB 51156-2015	液化天然气接收站工程设计规范 Code for design of liquefied natural gas receiving terminal		中国寰球工程公司 中石化洛阳工程有限公司
1087	GB 51157-2016	物流建筑设计规范 Code for design of logistics building		中国中元国际工程公司
1088	GB 51158-2015	通信线路工程设计规范 Code for design of telecommunication cable line engineering		中讯邮电咨询设计院有限公司
1089	GB 51159-2016	色织和牛仔布工厂设计规范 Code for design of yarn dyed and denim fabric plant		中国纺织工业协会 上海纺织建筑设计研究院
1090	GB 51160-2016	纤维增强塑料设备和管道工程技术规范 Technical code for fibre reinforced plastics equipment and piping engineering		中国石油和化工勘察设计协会 上海富晨化工有限公司
1091	GB/T 51161-2016	民用建筑能耗标准		住房和城乡建设部标准定额研究所
1092	GB 51162-2016	重型结构和设备整体提升技术规范 Technical code for integral lifting of heavy structure and equipment		上海建工股份有限公司 同济大学
1093	GB/T 51163-2016	城市绿线划定技术规范 Technical code for delimitation of urban green line		城市建设研究院

工程建设国家标准

序号	标准编号	标准名称	被代替标准编号	主编单位
1094	GB 51164-2016	钢铁企业煤气储存和输配系统施工及质量验收规范 Code for construction and quality acceptance of gas storage & transportation and distribution system for iron steel enterprises		天津二十冶建设有限公司 中国二十冶集团有限公司
1095	GB/T 51165-2016	绿色饭店建筑评价标准 Assessment standard for green hotel building		住房和城乡建设部科技发展促进中心 中国饭店协会
1096	GB/T 51167-2016	海底光缆工程验收规范 Code for acceptance of optical fiber submarine cable systems engineering code for acceptance of optical fiber submarine cable systems engineering		中国移动通信集团设计院有限公司
1097	GB/T 51168-2016	城市古树名木养护和复壮工程技术规范		中国城市建设研究院有限公司
1098	GB/T 51169-2016	煤炭工业矿井采掘设备配备标准 Stand for mining machine coal mine		北京圆之翰煤炭工程设计有限公司
1099	GB 51170-2016	航空工业工程设计规范 Code for design of aviation industry engineering		中国航空规划设计研究总院有限公司
1100	GB 51171-2016	通信线路工程验收规范 Code for acceptance of telecommunication cable line engineering		中国通信建设集团有限公司
1101	GB/T 51172-2016	在役油气管道工程检测技术规范 Inapection code for in-service oil & gas pipeline		中国石油天然气股份有限公司管道分公司管道科技研究中心 中石化第十建设有限公司
1102	GB 51173-2016	煤炭工业露天矿疏干排水设计规范 Code for design of dewatering and draining in open pit mine of coal industry		大地工程开发（集团）有限公司
1103	GB/T 51175-2016	炼油装置火焰加热炉工程技术规范 The technical specification of refining fired heater		中国石化工程建设有限公司
1104	GB 51176-2016	干混砂浆生产线设计规范 Code for design of dry-mixed mortar production line		北京建筑材料科学研究总院有限公司 常州市建筑科学研究院股份有限公司
1105	GB 51177-2016	升船机设计规范 Design code of shiplift		水利部水利水电规划设计总院 长江勘测规划设计研究院
1106	GB/T 51178-2016	建材矿山工程测量技术规范 Technical code for engineering surveying of building materials mine		建材成都地质工程勘察院 沈阳建材地质工程勘察院
1107	GB 51179-2016	煤矿井下煤炭运输设计规范 Code for design of underground coal haulage system of coal mine		煤炭工业济南设计研究院有限公司
1108	GB 51180-2016	煤矿采空区建（构）筑物地基处理技术规范 Technical code for ground treatment of buildings in coal mine goaf		煤炭工业太原设计研究院
1109	GB 51181-2016	煤炭洗选工程节能设计规范 Design Code for energy efficiency of coal cleaning engineering		中煤科工集团北京华宇工程有限公司
1110	GB 51182-2016	火炸药及其制品工厂建筑结构设计规范 Code for design of Architecture and structure for the factory of explosives and their products		中国五洲工程设计集团有限公司

工程建设国家标准

序号	标准编号	标准名称	被代替标准编号	主编单位
1111	GB/T 51183-2016	农业温室结构荷载规范 Code for the design load of horticultural greenhouse structures		农业部规划设计研究院
1112	GB 51184-2016	矿山提升井塔设计规范 Code for design of mine winding tower		中国煤炭建设协会勘察设计委员会 煤炭工业合肥设计研究院
1113	GB 51185-2016	煤炭工业矿井抗震设计规范 Code for mine seismic design of coal industry		中国煤炭建设协会勘察设计委员会 中煤邯郸设计工程有限责任公司
1114	GB 51186-2016	机制砂石骨料工厂设计规范 Code for design of machine-made gravel aggregate plant		苏州中材非金属矿工业设计研究院有限公司 中国建筑材料工业规划研究院
1115	GB/T 51187-2016	城市排水防涝设施数据采集与维护技术规范 Technical code for data acquisition and maintenance of urban drainage and local flooding prevention and control facilities		清华大学 中国城市规划设计研究院
1116	GB/T 51188-2016	建筑与工业给水排水系统安全评价标准		中国中元国际工程有限公司
1117	GB/T 51189-2016	火力发电厂海水淡化工程调试及验收规范 Specification of the commissioning and acceptance for seawater desalination project in thermal power plants		中国电力企业联合会 西安热工研究院有限公司
1118	GB/T 51190-2016	海底电力电缆输电工程设计规范 Code for design of submarine power cables project		中国电力企业联合会 国网浙江省电力公司
1119	GB/T 51191-2016	海底电力电缆输电工程施工及验收规范 Code for construction and acceptance of submarine power cable project		中国电力企业联合会 国网浙江省电力公司
1120	GB 51192-2016	公园设计规范 Code for the design of public park		北京市园林绿化局
1121	GB/T 51193-2016	聚酯及固相缩聚设备工程安装与质量验收规范 Code for installation and quality acceptance of PET and SSP equipments engineering		中国纺织工业联合会 中国昆仑工程公司
1122	GB 51194-2016	通信电源设备安装工程设计规范 Code for design of engineering for telecommunication power supply equipment installation		中讯邮电咨询设计院有限公司
1123	GB 51195-2016	互联网数据中心工程技术规范 Technical code for Iternet data center engineering		中国移动通信集团设计院有限公司
1124	GB/T 51196-2016	有色金属矿山工程测控设计规范 Code for design of measurement & control non-ferrous metals mines		中国有色工程有限公司 中国恩菲工程技术有限公司
1125	GB 51197-2016	煤炭工业露天矿节能设计规范 Code for energy conservation design on open pit mine of coal industry		中煤西安设计工程有限责任公司
1126	GB/T 51198-2016	微组装生产线工艺设计规范 Code for micro-assembling production line process design		工业和信息化部电子工业标准化研究院电子工程标准定额站 中国电子科技集团公司第二研究所
1127	GB 51199-2016	通信电源设备安装工程验收规范 Code for acceptance of construction engineering of telecommunication power supply facility		中国通信建设集团有限公司

工程建设国家标准

序号	标准编号	标准名称	被代替标准编号	主编单位
1128	GB/T 51200-2016	高压直流换流站设计规范 Code for design of HVDC converter station		中国电力工程顾问集团中南电力设计院有限公司
1129	GB 51201-2016	沉管法隧道施工与质量验收规范 Code for construction and quality acceptance of immersed tunnel		广州市市政集团有限公司 交通运输部广州打捞局
1130	GB 51202-2016	冰雪景观建筑技术标准	JGJ 247-2011	哈尔滨市建筑设计院 哈尔滨五建工程有限责任公司
1131	GB 51203-2016	高耸结构工程施工质量验收规范		同济大学 青岛东方铁塔股份有限公司
1132	GB 51204-2016	建筑电气工程电磁兼容技术规范		上海建筑设计研究院有限公司 上海现代建筑设计(集团)有限公司
1133	GB 51205-2016	精对苯二甲酸工厂设计规范		中国纺织工业联合会 中国昆仑工程公司
1134	GB 51206-2016	太阳能电池生产设备安装工程施工及质量验收规范		工业和信息化部电子工业标准化研究院电子工程标准定额站 中国电子系统工程第四建设有限公司
1135	GB/T 51207-2016	钢铁工程设计文件编制标准		中冶南方工程技术有限公司 中冶长天工程技术有限公司
1136	GB 51208-2016	人工制气厂站设计规范		中冶焦耐工程技术有限公司
1137	GB 51209-2016	发光二极管工厂设计规范		工业和信息化部电子工业标准化研究院电子工程标准定额站 中国电子工程设计院
1138	GB/T 51210-2016	建筑施工脚手架安全技术统一标准		中国建筑业协会 内蒙古兴泰建筑有限责任公司
1139	GB/T 51211-2016	城市轨道交通无线局域网宽带工程技术规范		上海申通轨道交通研究咨询有限公司
1140	GB/T 51212-2016	建筑信息模型应用统一标准		中国建筑科学研究院
1141	GB/T 51231-2016	装配式混凝土建筑技术标准		中国建筑标准设计研究院有限公司
1142	GB/T 51232-2016	装配式钢结构建筑技术标准		中国建筑标准设计研究院有限公司
1143	GB/T 51233-2016	装配式木结构建筑技术标准		中国建筑西南设计院有限公司

二、工程建设标准局部修订公告

序号	公告编号	标准编号	标准名称	公告时间
1	公告第1号	GBJ 11-89	建筑抗震设计规范	1993年
2	公告第2号	GBJ 10-90	混凝土结构设计规范	1993年
3	公告第3号	GBJ 74-84	石油库设计规范	1995年
4	公告第4号	GBJ 16-87	建筑设计防火规范	1995年
5	公告第5号	GBJ 10-89	混凝土结构设计规范	1997年
6	公告第6号	GBJ 15-88	建筑给水排水设计规范	1997年
7	公告第7号	GBJ 16-87	建筑设计防火规范	1997年
8	公告第8号	GB 50045-95	高层民用建筑设计防火规范	1997年
9	公告第9号	GBJ 98-87	人民防空工程设计防火规范	1997年
10	公告第10号	GBJ 140-90	建筑灭火器配置设计规范	1997年
11	公告第11号	GBJ 13-86	室外给水设计规范	1998年
12	公告第12号	GBJ 14-87	室外排水设计规范	1998年
13	公告第13号	GB 50028-93	城镇燃气设计规范	1998年
14	公告第14号	GB 50181-93	蓄滞洪区建筑工程技术规范	1998年
15	公告第15号	GBJ 108-87	地下工程防水技术规范	1998年
16	公告第16号	JGJ 79-91	建筑地基处理技术规范	1998年
17	公告第17号	CJJ 37-90	城市道路设计规范	1998年
18	公告第18号	CJJ 69-95	城市人行天桥于人行地道技术规范	1998年
19	公告第19号	CJJ 17-88	城市生活垃圾卫生填埋技术标准	1998年
20	公告第20号	GB 50045-95	高层民用建筑设计防火规范	1999年
21	公告第21号	GB 50160-92	石油化工企业设计防火规范	1999年
22	公告第22号	GB 50222-95	建筑内部装修设计防火规范	1999年
23	公告第23号	GB 50193-93	二氧化碳灭火系统设计规范	2000年
24	公告第24号	GB 50057-94	建筑物防雷设计规范	2000年
25	公告第25号	GB 50151-92	低倍数泡沫灭火系统设计规范	2000年
26	公告第26号	GBJ 19-87	采暖通风与空气调节设计规范	2001年
27	公告第27号	GBJ 16-87	建筑设计防火规范	2001年
28	公告第28号	GB 50045-95	高层民用建筑设计防火规范	2001年
29	公告第29号	GB 50222-95	建筑内部装修设计防火规范	2001年
30	公告第30号	GB 50095-98	人民防空工程设计防火规范	2001年
31	公告第31号	GB 50180-93	城市居住区规划设计规范	2002年
32	公告第32号	GB 50196-93	高倍数、中倍数泡沫灭火系统设计规范	2002年
33	公告第51号	GB 50028-93	城镇燃气设计规范	2002年
34	公告第66号	JGJ 130-2001	建筑施工扣件试钢管脚手架安全技术规范	2003年
35	公告第67号	GB 50003-2001	砌体结构设计规范	2002年
36	公告第69号	JGJ 130-2001	多孔砖砌体结构设计规范	2003年
37	公告第135号	GB 50261-2001	自动喷水灭火系统施工及验收规范	2003年
38	公告第142号	GB 50096-1999	住宅设计规范	2003年
39	公告第153号	GB 50038-94	人民防空地下室设计规范	2003年
40	公告第187号	GB 50299-1999	地下铁道工程施工及验收规范	2003年
41	公告第190号	GJJ 69-95	城市人行天桥与地道技术规范	2003年
42	公告第313号	GB 50500-2003	建筑工程工程量清单计价规范	2005年
43	公告第360号	GB 50084-2001	自动喷水灭火系统设计规范	2005年
44	公告第361号	GB 50045-95（2001年版）	高层民用建筑设计防火规范	2005年

工程建设标准局部修订公告

序号	公告编号	标准编号	标准名称	公告时间
45	公告第 375 号	GB 50005-2003	木结构设计规范	2005 年
46	公告第 396 号	GB 50156-2002	汽车加油加气站设计及施工规范	2006 年
47	公告第 428 号	GB 50325-2001	民用建筑工程室内环境污染控制规范	2006 年
48	公告第 453 号	JGJ 127-2000	看守所建筑设计规范（内部发行）	2006 年
49	公告第 458 号	GB 50009-2001	建筑结构荷载规范	2006 年
50	公告第 462 号	GB 50253-2003	输油管道工程设计规范	2006 年
51	公告第 796 号	GB 50316-2000	业金属管道设计规范	2008 年
52	公告第 71 号	GB 50011-2001	建筑抗震设计规范	2008 年
53	公告第 234 号	GB 50366-2005	地源热泵系统工程技术规范	2009 年
54	公告第 314 号	GB 50021-2001	岩土工程勘察规范	2009 年
55	公告第 329 号	GB 50111-2006	铁路工程抗震设计规范	2009 年
56	公告第 409 号	GB 50015-2003	建筑给水排水设计规范	2009 年
57	公告第 559 号	GB 50193-93（1999 版）	二氧化碳灭火系统设计规范	2010 年
58	公告第 849 号	GB 50204-2002	混凝土结构工程施工质量验收规范	2011 年
59	公告第 1114 号	GB 50014-2006	室外排水设计规范	2011 年
60	公告第 1146 号	GB 50226-2007	铁路旅客车站建筑设计规范	2011 年
61	公告第 1562 号	GB 50540-2009	石油天然气站内工艺管道工程施工规范	2012 年
62	公告第 64 号	GB 50325-2010	民用建筑工程室内环境污染控制规范	2013 年
63	公告第 311 号	GB 50014-2006（2011 年版）	室外排水设计规范	2014 年
64	公告第 498 号	GB 50156-2012	汽车加油加气站设计与施工规范	2014 年
65	公告第 847 号	GB 50634-2010	水泥窑协同处置工业废物设计规范	2015 年
66	公告第 909 号	GB 50010-2010	混凝土结构设计规范	2015 年
67	公告第 1190 号	GB 50180-93（2002 年版）	城市居住区规划设计规范	2016 年
68	公告第 1191 号	GB 50014-2006（2014 年版）	室外排水设计规范	2016 年
69	公告第 1192 号	GB 50420-2007	城市绿地设计规范	2016 年
70	公告第 1193 号	CJJ 37-2012	城市道路工程设计规范	2016 年
71	公告第 1199 号	GB 50011-2010	建筑抗震设计规范	2016 年
72	公告第 1282 号	GB 50513-2009	城市水系规划规范	2016 年

三、工程建设行业标准

建筑工程

序号	标准编号	标准名称	被代替标准编号	备案号	主编单位
1	JGJ 1-2014	装配式混凝土结构技术规程 Technical specification for precast concrete structures	JGJ 1-91	J1736-2014	中国建筑标准设计研究院 中国建筑科学研究院
2	JGJ 3-2010	高层建筑混凝土结构技术规程 Technical specification for concrete structures of tall building	JGJ 3-2002	J183-2010	中国建筑科学研究院
3	JGJ 6-2011	高层建筑筏形与箱形基础技术规范 Technical code for tall building raft foundations and box foundations	JGJ 6-99	J1160-2011	中国建筑科学研究院
4	JGJ 7-2010	空间网格结构技术规程 Technical specification for space frame structures	JGJ 61-2003	J1072-2010	中国建筑科学研究院
5	JGJ 8-2016	建筑变形测量规范 Code for deformation measurement of building and structure	JGJ/T 8-97 JGJ 8-2007	J719-2016	建设综合勘察研究设计院
6	JGJ/T 10-2011	混凝土泵送施工技术规程 Technical specification for construction of concrete pumping	JGJ/T 10-95	J1223-2011	中国建筑科学研究院 浙江省二建建设集团有限公司
7	JGJ 12-2006	轻骨料混凝土结构技术规程 Technical Specification for lightweight aggregate concrete structures	JGJ 12-99	J515-2006	中国建筑科学研究院
8	JGJ 13-2014	约束砌体与配筋砌体结构技术规程 Technical specification for constrained or reinforced masonry buildings	JGJ/T 13-94	J1802-2014	中国建筑科学研究院 巨匠建设集团有限公司
9	JGJ/T 14-2011	混凝土小型空心砌块建筑技术规程 Technical specification for concrete small-sized hollow block masonry building	JGJ/T 14-2004	J361-2011	四川省建筑科学研究院 广西建工集团第五建筑工程有限责任公司
10	JGJ/T 15-2008	早期推定混凝土强度试验方法标准 Standard for test method of early estimating compressive strength of concrete	JGJ 15-83	J784-2008	中国建筑科学研究院
11	JGJ 16-2008	民用建筑电气设计规范 Code for electrical design of civil buildings	JGJ/T 16-92	J778-2008	中国建筑东北设计研究院
12	JGJ/T 17-2008	蒸压加气混凝土建筑应用技术规程 Technical specification for application of autoclaved aerocrete	JGJ 17-84	J824-2008	北京市建筑设计研究院 哈尔滨市建筑设计院
13	JGJ 18-2012	钢筋焊接及验收规程 Specification for welding and acceptance of reinforcing steel bars	JGJ 18-2003	J253-2012	陕西省建筑科学研究院
14	JGJ 19-2010	冷拔低碳钢丝应用技术规程 Technical specification for application or cold-drawn low-carbon wires	JGJ 19-92	J992-2010	中国建筑科学研究院 江西省建工集团公司
15	JGJ 22-2012	钢筋混凝土薄壳结构设计规程 Specification for design of reinforced concrete shell structures	JGJ/T 22-98	J1400-2012	中国建筑科学研究院

建筑工程

序号	标准编号	标准名称	被代替标准编号	备案号	主编单位
16	JGJ/T 23-2011	回弹法检测混凝土抗压强度技术规程 Technical specification for inspection of concrete compressive strength by rebound method	JGJ/T 23-2001 JGJ/T 23-92	J115-2011	陕西省建筑科学研究院 浙江海天建设集团有限公司
17	JGJ 25-2010	档案馆建筑设计规范 Code for design of archives buildings	JGJ 25-2000	J1079-2010	国家档案局档案科学技术研究所
18	JGJ 26-2010	严寒和寒冷地区居住建筑节能设计标准 Design standard for energy efficiency of residential buildings in severe cold and cold zones	JGJ 26-95	J997-2010	中国建筑科学研究院
19	JGJ/T 27-2014	钢筋焊接接头试验方法标准 Standard for test methods of welded joint of reinforcing steel bars	JGJ/T 27-2001	J1829-2014	陕西省建筑科学研究院 陕西建工集团总公司
20	JGJ/T 29-2015	建筑涂饰工程施工及验收规程 Specification for construction and acceptance of building painting operation	JGJ/T 29-2003	J250-2003	住房和城乡建设部住宅产业化促进中心 山东宁建建设集团有限公司
21	JGJ/T 30-2015	房地产业基本术语标准 Standard for basic terminology of real estate industry	JGJ/T 30-2003	J251-2015	北京市国土资源和房屋管理局
22	JGJ 31-2003	体育建筑设计规范 Design code for sports building		J265-2003	北京市建筑设计研究院
23	JGJ 33-2012	建筑机械使用安全技术规程 Technical specification for safety operation of constructional machinery	JGJ 33-2001	J119-2012	江苏省华建建设股份有限公司 江苏邗建集团有限公司
24	JGJ 35-87	建筑气象参数标准 Standard for parameter of construction atmosphere			中南地区建筑标准设计协作组办公室
25	JGJ 36-2016	宿舍建筑设计规范 Code for design of dormitory buildings	JGJ 36-2005 JGJ 36-87	J480-2016	中国建筑标准设计研究院有限公司
26	JGJ 38-2015	图书馆建筑设计规范 Code for design of library buildings	JGJ 38-99	J2081-2015	中国建筑西北设计研究院有限公司
27	JGJ 39-2016	托儿所、幼儿园建筑设计规范 Code for design of nursery and infant school buildings	JGJ 39-87	J2176-2016	黑龙江省建筑设计研究院
28	JGJ 40-87	疗养院建筑设计规范 Code for design of sanatorium buildings			福建省建筑设计院
29	JGJ/T 41-2014	文化馆建筑设计规范 Code for design of the public culture center	JGJ 41-87	J1912-2014	吉林省建苑设计集团有限公司 中国艺术科技研究所
30	JGJ 46-2005	施工现场临时用电安全技术规范 Technical code for safety of temporary electrification on construction Site	JGJ 46-88	J405-2005	沈阳建筑大学
31	JGJ 48-2014	商店建筑设计规范 Code for design of store buildings	JGJ 48-88	J1839-2014	中南建筑设计院股份有限公司
32	JGJ 51-2002	轻骨料混凝土技术规程 Technical specification of lightweight aggregate concrete	JGJ 51-90	J215-2002	中国建筑科学研究院
33	JGJ 52-2006	普通混凝土用砂、石质量及检验方法标准 Standard for technical requirements and test methods of sand and crushed stone (or gravel) for ordinary concrete	JGJ 52-92 JGJ 53-92	J628-2006	中国建筑科学研究院

建筑工程

序号	标准编号	标准名称	被代替标准编号	备案号	主编单位
34	JGJ/T 53-2011	房屋渗漏修缮技术规程 Technical specification for repairing water seepage of building	CJJ 62-95	J1158-2011	河南国基建设集团有限公司 新蒲建设集团有限公司
35	JGJ 55-2011	普通混凝土配合比设计规程 specification for mix proportion design of ordinary concrete	JGJ 55-2000	J64-2011	中国建筑科学研究院
36	JGJ 57-2016	剧场建筑设计规范 Design code for theater	JGJ 57-88 JGJ 57-2000	J67-2016	中国建筑西南设计研究院
37	JGJ 58-2008	电影院建筑设计规范 Code for architectural design of cinema	JGJ 58-88		中广电广播电影电视设计研究院 中国电影科学技术研究所
38	JGJ 59-2011	建筑施工安全检查标准 Standard for construction safety inspection	JGJ 59-99	J1334-2011	天津市建工工程总承包有限公司 中启胶建集团有限公司
39	JGJ/T 60-2012	交通客运站建筑设计规范 Code for design of passenger transportation building	JGJ 60-99 JGJ 86-92	J1473-2012	大连市建筑设计研究院有限公司 甘肃省建筑设计研究院
40	JGJ 62-2014	旅馆建筑设计规范 Code for design of hotel building	JGJ 62-90	J1895-2014	中国建筑设计院有限公司
41	JGJ 63-2006	混凝土用水标准 Standard of Water for Concrete	JGJ 63-89	J531-2006	中国建筑科学研究院
42	JGJ 64-89	饮食建筑设计规范 Code for design of dietetic buildings			中国建筑东北设计院 辽宁省食品卫生监督检验所
43	JGJ 65-2013	液压滑动模板施工安全技术规程 Technical specification for safety of the hydraulic slipform in construction	JGJ 65-89	J1600-2013	中冶建筑研究总院有限公司 江苏江都建设集团有限公司
44	JGJ 66-2015	博物馆建筑设计规范 Code for design of museum building	JGJ 66-91	J2055-2015	华东建筑设计研究院有限公司
45	JGJ 67-2006	办公建筑设计规范 Design code for office building	JGJ 67-89	J556-2006	浙江省建筑设计研究院
46	JGJ 69-90	PY型预钻式旁压试验规程 Specification for side-pressure test of type TY predrill			常州市建筑设计院
47	JGJ/T 70-2009	建筑砂浆基本性能试验方法标准 Standard for test method of basic properties of construction mortar	JGJ 70-90	J856-2009	陕西省建筑科学研究院 山河建设集团有限公司
48	JGJ 72-2004	高层建筑岩土工程勘察规程 Specification for geotechnical investigation of tall buildings	JGJ 72-90	J366-2004	机械工业勘察设计研究院
49	JGJ 74-2003	建筑工程大模板技术规程 Technical specification for large-area formwork building construction		J270-2003	中国建筑科学研究院建筑机械化研究分院
50	JGJ 75-2012	夏热冬暖地区居住建筑节能设计标准 Design standard for energy efficiency of residential buildings in hot summer and warm winter zone	JGJ 75-2003	J1482-2012	中国建筑科学研究院 广东省建筑科学研究院
51	JGJ 76-2003	特殊教育学校建筑设计规范 Code for design of special education schools		J282-2003	西安建筑科技大学
52	JGJ/T 77-2010	施工企业安全生产评价标准 Standard for the work safety assesment of construction company	JGJ/T 77-2003	J278-2010	上海市建设工程安全质量监督总站

建筑工程

序号	标准编号	标准名称	被代替标准编号	备案号	主编单位
53	JGJ 79-2012	建筑地基处理技术规范 Technical code for ground treatment of buildings	JGJ 79-2002	J220-2012	中国建筑科学研究院
54	JGJ 80-2016	建筑施工高处作业安全技术规范 Technical Code for safety of working at height of building construction	JGJ 80-91	J2227-2016	上海市建筑施工技术研究所
55	JGJ 82-2011	钢结构高强度螺栓连接技术规程 Technical specification for high strength bolt connections of steel structures	JGJ 82-91	J1141-2011	中冶建筑研究总院有限公司
56	JGJ 83-2011	软土地区岩土工程勘察规程 Specification for geotechnical investigation in soft clay area	JGJ 83-91	J1186-2011	中国建筑科学研究院
57	JGJ/T 84-2015	岩土工程勘察术语标准 Standard for terminology of geotechnical investigation	JGJ 84-92	J1981-2015	建设综合勘察研究设计院
58	JGJ 85-2010	预应力筋用锚具、夹具和连接器应用技术规程 Technical specification for application of anchorage, grip and coupler for prestressing tendons	JGJ 85-2002	J1006-2010	中国建筑科学研究院 歌山建设集团有限公司
59	JGJ/T 87-2012	建筑工程地质勘探与取样技术规程 Technical specification for engineering gelogical prospecting and sampling of constructions	JGJ 87-92 JGJ 89-92	J1343-2012	中南勘察设计院有限公司
60	JGJ 88-2010	龙门架及井架物料提升机安全技术规范 Technical code for safety of gantry frame and headframe hoisters	JGJ 88-92	J1083-2010	天津市建工集团（控股）有限公司 天津市建工工程总承包有限公司
61	JGJ 91-83	科学实验建筑设计规范 Design code for scientific experiment buildings			中国科学院北京建筑设计研究院
62	JGJ 92-2016	无粘结预应力混凝土结构技术规程 Technical specification for concrete strutures prestressed with unbounded tendons	JGJ/T 92-93 JGJ 92-2004	J409-2005	中国建筑科学研究院
63	JGJ 94-2008	建筑桩基技术规范 Technical code for building pile foundations	JGJ 94-94	J793-2008	中国建筑科学研究院
64	JGJ 95-2011	冷轧带肋钢筋混凝土结构技术规程 Technical specification for concrete structures with cold-rolled ribbed steel wires and bars	JGJ 95-2003	J254-2011	中国建筑科学研究院 中鑫建设集团有限公司
65	JGJ 96-2011	钢框胶合板模板技术规程 Technical specification for plywood form with steel frame	JGJ 96-95	J1140-2011	中国建筑科学研究院 温州中城建设集团有限公司
66	JGJ/T 97-2011	工程抗震术语标准 Standard for terminology in earthquake engineering	JGJ/T 97-95	J1159-2011	中国建筑科学研究院
67	JGJ/T 98-2010	砌筑砂浆配合比设计规程 Specification for mix proportion design of masonry mortar 备案号 J 65-2010	JGJ 98-2000	J65-2010	陕西省建筑科学研究院 浙江八达建设集团有限公司

建筑工程

序号	标准编号	标准名称	被代替标准编号	备案号	主编单位
68	JGJ 99-2015	高层民用建筑钢结构技术规程 Technical specification for steel structure of tall buildings	JGJ 99-98	J2105-2015	中国建筑技术研究院标准设计研究所
69	JGJ 100-2015	车库建筑设计规范 Code for design of parking garage building	JGJ 100-98	J1996-2015	北京建筑工程学院
70	JGJ/T 101-2015	建筑抗震试验规程 Specification for seismic test of buildings	JGJ 101-96	J1988-2015	中国建筑科学研究院
71	JGJ 102-2003	玻璃幕墙工程技术规范 Technical code for glass curtain wall engineering	JGJ 102-96	J280-2003	中国建筑科学研究院
72	JGJ 103-2008	塑料门窗工程技术规程 Technical specification for PVC-U doors and windows engineering	JGJ 103-96	J811-2008	中国建筑科学研究院
73	JGJ/T 104-2011	建筑工程冬期施工规程 Specification for winter construction of building engineering	JGJ/T 104-97	J1189-2011	黑龙江省寒地建筑科学研究院 天元建设集团有限公司
74	JGJ/T 105-2011	机械喷涂抹灰施工规程 Specification for construction of plastering by mortar spraying	JGJ/T 105-96	J1232-2011	中国建筑科学研究院 华丰建设股份有限公司
75	JGJ 106-2014	建筑基桩检测技术规范 Technical code for testing of building foundation piles	JGJ 106-2003	J256-2014	中国建筑科学研究院
76	JGJ 107-2016	钢筋机械连接技术规程 Technical specification for mechanical splicing of steel reinforcing bars	JGJ 107-2003 JGJ 108-96 JGJ 109-96 JGJ 107-2010	J986-2010	中国建筑科学研究院
77	JGJ 110-2008	建筑工程饰面砖粘结强度检验标准 Testing standard for adhesive strength of tapestry brick of construction engineering	JGJ 110-97	J787-2008	中国建筑科学研究院
78	JGJ 111-2016	建筑与市政工程地下水控制技术规范 Technical code for groundwater control in building and municipal engineering	JGJ/T 111-98	J2772-2016	建设部综合勘察研究设计院
79	JGJ 113-2015	建筑玻璃应用技术规程 Technical specification for application of architectural glass	JGJ 113-2009	J255-2009	中国建筑材料科学研究总院 云南省第三建筑工程公司
80	JGJ 114-2014	钢筋焊接网混凝土结构技术规程 Technical specification for concrete structures reinforced with welded steel fabric	JGJ 114-2003	J1739-2014	中国建筑科学研究院 浙江鸿翔建设集团有限公司
81	JGJ 115-2006	冷轧扭钢筋混凝土构件技术规程 Technical specification for concrete structural element with cold-rolled and twisted bars	JGJ 115-97	J530-2006	北京市建筑设计研究院
82	JGJ 116-2009	建筑抗震加固技术规程 Technical specification for seismic strengthening of buildings	JGJ 116-98	J886-2009	中国建筑科学研究院
83	JGJ 117-98	民用建筑修缮工程查勘与设计规程 Specification for engineering examination and design of repairing civil architecture			上海市房屋土地管理局
84	JGJ 118-2011	冻土地区建筑地基基础设计规范 Code for design of soil and foundation of buildings in frozen soil region	JGJ 118-98	J1231-2011	黑龙江省寒地建筑科学研究院 大连阿尔滨集团有限公司

建筑工程

序号	标准编号	标准名称	被代替标准编号	备案号	主编单位
85	JGJ/T 119-2008	建筑照明术语标准 Standard for terminology of architectural lighting	JGJ/T 119-98	J827-2008	中国建筑科学研究院
86	JGJ 120-2012	建筑基坑支护技术规程 Technical specification for retaining and protection of building foundation excavations	JGJ 120-99	J1412-2012	中国建筑科学研究院
87	JGJ/T 121-2015	工程网络计划技术规程 Technical specification for engineering network planning and scheduling	JGJ/T 121-99	J1992-2015	江苏中南建筑产业集团有限责任公司 东南大学
88	JGJ 123-2012	既有建筑地基基础加固技术规范 Technical code for improvement of soil and foundation of existing buildings	JGJ 123-2000	J1447-2012	中国建筑科学研究院
89	JGJ 124-99	殡仪馆建筑设计规范 Code for design of funeral parlor's buildings			民政部101研究所
90	JGJ 125-2016	危险房屋鉴定标准 Standard for dangerous building appraisal	CJ 13-86 JGJ 125-99 JGJ 125-99 （2004年版）	J2228-2016	重庆市土地房屋管理局
91	JGJ 126-2015	外墙饰面砖工程施工及验收规程 Specification for construction and acceptance of tapestry brick work for exterior wall	JGJ 126-2000	J23-2015	中国建筑科学研究院 浙江环宇建设集团有限公司
92	JGJ 127-2000 （2006年版）	看守所建筑设计规范（内部发行） Code for design of detention houses	GA9-91	J37-2000	中华人民共和国公安部监所管理局
93	JGJ 128-2010	建筑施工门式钢管脚手架安全技术规范 Technical code for safety of frame scaffoldings with steel tubules in construction	JGJ 128-2000	J43-2010	哈尔滨工业大学 浙江宝业建设集团有限公司
94	JGJ/T 129-2012	既有居住建筑节能改造技术规程 Technical specification for energy efficiency retrofitting of existing residential buildings	JGJ 129-2000	J1468-2012	中国建筑科学研究院
95	JGJ 130-2011	建筑施工扣件式钢管脚手架安全技术规范 Technical code for safety of steel tubular scaffold with couplers in construction	JGJ 130-2001	J84-2011	中国建筑科学研究院 江苏南通二建集团有限公司
96	JGJ/T 131-2012	体育场馆声学设计及测量规程 Specification for acoustical design and measurement of gymnasium and stadium	JGJ/T 131-2000	J42-2012	中国建筑科学研究院
97	JGJ/T 132-2009	居住建筑节能检测标准 Standard for energy efficiency test of residential buildings	JGJ 132-2001	J85-2009	中国建筑科学研究院
98	JGJ 133-2001	金属与石材幕墙工程技术规范 Technical code for metal and stone curtain walls engineering		J113-2001	中国建筑科学研究院
99	JGJ 134-2010	夏热冬冷地区居住建筑节能设计标准 Design standard for energy efficiency of residential buildings in hot summer and cold winter zone	JGJ 134-2001	J995-2010	中国建筑科学研究院
100	JGJ 135-2007	载体桩设计规程 Specification for design of ram-compaction piles with composite bearing base	JGJ/T 135-2001	J121-2007	北京波森特岩土工程有限公司

建筑工程

序号	标准编号	标准名称	被代替标准编号	备案号	主编单位
101	JGJ/T 136-2001	贯入法检测砌筑砂浆抗压强度技术规程 Technical specification for testing compressive strength of masonry mortar by penetration resistance method		J131-2001	中国建筑科学研究院
102	JGJ 138-2016	组合结构设计规范 Code for design of composite structures	JGJ 138-2001	J130-2016	中国建筑科学研究院
103	JGJ/T 139-2001	玻璃幕墙工程质量检验标准 Standard for testing of engineering quality of glass curtain walls		J139-2001	国家建筑工程质量监督检验中心
104	JGJ 140-2004	预应力混凝土结构抗震设计规程 Specification for seismic design of prestressed concrete structures		J301-2004	中国建筑科学研究院
105	JGJ 141-2004	通风管道技术规程 Technical specification of air duct		J363-2004	中国安装协会
106	JGJ 142-2012	辐射供暖供冷技术规程 Technical specification for radiant heating and cooling	JGJ 142-2004	J1445-2012	中国建筑科学研究院
107	JGJ/T 143-2004	多道瞬态面波勘察技术规程 Technical specification for multi-channel transient surface wave investigation		J370-2004	北京市水电物探研究所
108	JGJ 144-2004	外墙外保温工程技术规程 Technical specification for external thermal insulation on walls		J408-2005	建设部科技发展促进中心
109	JGJ 145-2013	混凝土结构后锚固技术规程 Technical specification for post-installed fastenings in concrete structures	JGJ 145-2004	J1595-2013	中国建筑科学研究院 科达集团股份有限公司
110	JGJ 146-2013	建设工程施工现场环境与卫生标准 Standard for environment and sanitation of construction site	JGJ 146-2004	J375-2013	北京建工一建工程建设有限公司 北京市第三建筑工程有限公司
111	JGJ 147-2016	建筑拆除工程安全技术规范 Technical code for safety of buildings demolition engineering	JGJ 147-2004	J376-2004	北京建工集团有限责任公司
112	JGJ 149-2006	混凝土异形柱结构技术规程 Technical specification for concrete structures with specially shaped columns		J514-2006	天津大学
113	JGJ 150-2008	擦窗机安装工程质量验收规程 Specification for construction and acceptance of installation of permanently installed suspended access equipment		J779-2008	中国建筑科学研究院
114	JGJ/T 151-2008	建筑门窗玻璃幕墙热工计算规程 Calculation specification for thermal performance of windows, doors and glass curtain-walls		J828-2008	广东省建筑科学研究院 广东省建筑工程集团有限公司
115	JGJ/T 152-2008	混凝土中钢筋检测技术规程 Technical specification for test of reinforcing steel bar in concrete		J794-2008	中国建筑科学研究院
116	JGJ 153-2016	体育场馆照明设计及检测标准 Standard for lighting design and test of sports venues	JGJ 153-2007	J684-2016	中国建筑科学研究院 深圳市建安(集团)股份有限公司

建筑工程

序号	标准编号	标准名称	被代替标准编号	备案号	主编单位
117	JGJ/T 154-2007	民用建筑能耗数据采集标准 Standard for energy consumption survey of civil buildings		J685-2007	深圳市建筑科学研究院
118	JGJ 155-2013	种植屋面工程技术规程 Technical specification for green roof	JGJ 155-2007	J683-2013	中国建筑防水协会 天津天一建设集团有限公司
119	JGJ 156-2008	镇（乡）村文化中心建筑设计规范 Code for design of cultural centen buildings in towns and villages		J797-2008	中国建筑设计研究院
120	JGJ/T 157-2014	建筑轻质条板隔墙技术规程 Technical specification for lightweight panel partition walls of buildings	JGJ/T 157-2008	J786-2014	国家住宅与居住环境工程技术研究中心 江苏九鼎环球建设科技集团有限公司
121	JGJ 158-2008	蓄冷空调工程技术规程 Technical specification for cool storage air-conditioning system		J812-2008	中国建筑科学研究院
122	JGJ 159-2008	古建筑修建工程施工与质量验收规范 Code for construction and acceptance of ancient chinese architecture engineering		J813-2008	苏州市房产管理局
123	JGJ 160-2016	施工现场机械设备检查技术规范 Technical specification for inspection of machinery and equipment on construction site	JGJ 160-2008	J817-2016	中国建筑业协会机械管理与租赁分会
124	JGJ 161-2008	镇（乡）村建筑抗震技术规程 Seismic technical specification for building construction in town and village		J798-2008	中国建筑科学研究院
125	JGJ 162-2008	建筑施工模板安全技术规范 Technical code for safety of forms in construtuon		J814-2008	沈阳建筑大学
126	JGJ/T 163-2008	城市夜景照明设计规范 Code for lighting design of urban nightscape		J822-2008	中国建筑科学研究院
127	JGJ 164-2008	建筑施工木脚手架安全技术规范 Technical code for safety of wooden scaffold in construction		J815-2008	沈阳建筑大学 浙江八达建设集团有限公司
128	JGJ 165-2010	地下建筑工程逆作法技术规程 Technical specification for top-down construction method of underground buildings		J1138-2010	黑龙江省建工集团有限责任公司 哈尔滨市城乡建设委员会
129	JGJ 166-2016	建筑施工碗扣式钢管脚手架安全技术规范 Technical code for safety of cuplok steel tubular scaffolding in construction	JGJ 166-2008	J823-2008	河北建设集团有限公司 中天建设集团有限公司
130	JGJ 167-2009	湿陷性黄土地区建筑基坑工程安全技术规程 Technical specifications for safe retaining and protection of building foundation excavation engineering in collapsible loess regions		J859-2009	陕西省建设工程质量安全监督总站
131	JGJ 168-2009	建筑外墙清洗维护技术规程 Technical specification for cleaning maintenance of building external wall		J857-2009	上海市房地产科学研究院
132	JGJ 169-2009	清水混凝土应用技术规程 Technical specification for fair-faced concrete construction		J858-2009	中国建筑股份有限公司 中建三局建设工程股份有限公司

建筑工程

序号	标准编号	标准名称	被代替标准编号	备案号	主编单位
133	JGJ/T 170-2009	城市轨道交通引起建筑物振动与二次辐射噪声限值及其测量方法标准 Standard for limit and measuring method of building vibration and secondary noise caused by rail transit		J855-2009	建设部科技发展促进中心 深圳市地铁有限公司
134	JGJ 171-2009	三岔双向挤扩灌注桩设计规程 Design specification for cast-in-place piles with expanded branches and bells by 3-way extruding arms		J864-2009	北京中阔地基基础技术有限公司
135	JGJ/T 172-2012	建筑陶瓷薄板应用技术规程 Technical specification for application of building ceramic sheet board	JGJ/T 172-2009	J861-2012	北京新型材料建筑设计研究院有限公司 广东蒙娜丽莎新型材料集团有限公司
136	JGJ 173-2009	供热计量技术规程 Technical specification for heat metering of district heating system		J860-2009	中国建筑科学研究院
137	JGJ 174-2010	多联机空调系统工程技术规程 Technical specification for multi-connected split air condition system		J1001-2010	中国建筑科学研究院
138	JGJ/T 175-2009	自流平地面工程技术规程 Technical specification of self-leveling flooring construction		J884-2009	中国建筑材料检验认证中心
139	JGJ 176-2009	公共建筑节能改造技术规范 Technical code for the retrofitting of public building on energy efficiency		J885-2009	中国建筑科学研究院
140	JGJ/T 177-2009	公共建筑节能检测标准 Standard for energy efficiency test of public buildings		J970-2009	中国建筑科学研究院
141	JGJ/T 178-2009	补偿收缩混凝土应用技术规程 Technical specification for application of shrinkage-compensating concrete		J887-2009	中国建筑材料科学研究总院 长业建设集团有限公司
142	JGJ/T 179-2009	体育建筑智能化系统工程技术规程 Technical specification for intelligent system engineering of sports building		J889-2009	中国建筑标准设计研究院 国家体育总局体育设施建设和标准办公室
143	JGJ 180-2009	建筑施工土石方工程安全技术规范 Technical code for safety in earthwork of building construction		J888-2009	中国建筑技术集团有限公司 江苏省华建建设股份有限公司
144	JGJ/T 181-2009	房屋建筑与市政基础设施工程检测分类标准 Classification standard of test for building and municipal engineering		J968-2009	广州市建筑科学研究院有限公司 国家建筑工程质量监督检验中心
145	JGJ/T 182-2009	锚杆锚固质量无损检测技术规程 Technical specification for nondestructive testing of rock bolt system		J955-2009	长江大学
146	JGJ 183-2009	液压升降整体脚手架安全技术规程 Technical specification for safety of hydraulic lifting integral scaffold		J892-2009	南通四建集团有限公司 苏州二建筑集团有限公司

建筑工程

序号	标准编号	标准名称	被代替标准编号	备案号	主编单位
147	JGJ 184-2009	建筑施工作业劳动防护用品配备及使用标准 Standard for outfit and used of labour protection articles on construction site		J964-2009	北京建工集团有限责任公司 北京六建集团公司
148	JGJ/T 185-2009	建筑工程资料管理规程 Specification for building engineering document management		J951-2009	中建一局集团建设发展有限公司 苏州第一建筑集团有限公司
149	JGJ/T 186-2009	逆作复合桩基技术规程 Technical specification for composite pile foundation with top-down method		J952-2009	江苏南通六建建设集团有限公司 江苏江中集团有限公司
150	JGJ/T 187-2009	塔式起重机混凝土基础工程技术规程 Technical specification for concrete foundation engineering of tower cranes		J953-2009	华丰建设股份有限公司 中国建筑科学研究院
151	JGJ/T 188-2009	施工现场临时建筑物技术规范 Technical code of temporary building of construction site		J954-2009	福建建科建筑设计院 中国建筑第七工程局有限公司
152	JGJ/T 189-2009	建筑起重机械安全评估技术规程 Technical specification for safety assessment of building crane on construction site		J969-2009	上海市建工设计研究院有限公司 龙元建设集团股份有限公司
153	JGJ 190-2010	建筑工程检测试验技术管理规范 Code for technical management of building engineering inspection and testing		J973-2010	中国建筑一局（集团）有限公司 浙江勤业建工集团有限公司
154	JGJ/T 191-2009	建筑材料术语标准 Standard for terminology of building materials		J996-2009	中国建筑科学研究院 安徽建工集团有限公司
155	JGJ/T 192-2009	钢筋阻锈剂应用技术规程 Technical specification for application of corrosion inhibitor for steel bar		J956-2009	中国建筑科学研究院 浙江中成建工集团有限公司
156	JGJ/T 193-2009	混凝土耐久性检验评定标准 Standard for inspection and assessment of concrete durability		J957-2009	中国建筑科学研究院 中设建工集团有限公司
157	JGJ/T 194-2009	钢管满堂支架预压技术规程 Technical specification for preloading in full scaffold construction		J958-2009	宏润建设集团股份有限公司
158	JGJ 195-2010	液压爬升模板工程技术规程 Technical specification for the hydraulic climbing formwork engineering		J987-2010	江苏江都建设工程有限公司
159	JGJ 196-2010	建筑施工塔式起重机安装、使用、拆卸安全技术规程 Technical specification for safety installation operation and dismantlement of tower crane in construction		J974-2010	上海市建工设计研究院有限公司 上海市第四建筑有限公司
160	JGJ/T 197-2010	混凝土预制拼装塔机基础技术规程 Technical specification for prefabricated concrete block assembled base of tower crane		J1086-2010	江苏省苏中建设集团股份有限公司
161	JGJ/T 198-2010	施工企业工程建设技术标准化管理规范 Code for management of technical standardization of project construction of construction enterprises		J993-2010	中国工程建设标准化协会建筑施工专业委员会 中天建设集团有限公司

建筑工程

序号	标准编号	标准名称	被代替标准编号	备案号	主编单位
162	JGJ/T 199-2010	型钢水泥土搅拌墙技术规程 Technical specification for soil mixed wall		J994-2010	上海现代建筑设计（集团）有限公司 浙江环宇建设集团有限公司
163	JGJ/T 200-2010	喷涂聚脲防水工程技术规程 Technical specification for spray polyurea waterproofing		J988-2010	中国建筑科学研究院 浙江昆仑建设集团股份有限公司
164	JGJ/T 201-2010	石膏砌块砌体技术规程 Technical specification for gypsum bloch masonry		J1003-2010	南通建筑工程总承包有限公司 龙信建设集团有限公司
165	JGJ 202-2010	建筑施工工具式脚手架安全技术规范 Technical code for safety of implementation scaffold practice in construction		J999-2010	中国建筑业协会建筑安全分会
166	JGJ 203-2010	民用建筑太阳能光伏系统应用技术规范 Technical code for application of solar photovoltaic system of civil buildings		J996-2010	中国建筑设计研究院 中国可再生能源学会太阳能建筑专业委员会
167	JGJ/T 204-2010	建筑施工企业管理基础数据标准 Standard for basic data of construction enterprise management		J976-2010	中国建筑第七工程局有限公司
168	JGJ/T 205-2010	建筑门窗工程检测技术规程 Technical specification for inspection of building doors and windows		J991-2010	中国建筑科学研究院 浙扛省建工集团有限责任公司
169	JGJ 206-2010	海砂混凝土应用技术规范 Technical code for application of sea sand concrete		J1038-2010	中国建筑科学研究院 浙江中联建设集团有限公司
170	JGJ/T 207-2010	装配箱混凝土空心楼盖结构技术规程 Technical specification for assembly box concrete hollow floor structure		J1008-2010	山东天齐置业集团股份有限公司 南通建工集团股份有限公司
171	JGJ/T 208-2010	后锚固法检测混凝土抗压强度技术规程 Technical specification for inspection of concrete compressive strength by post-installed adhesive anchorage method		J1007-2010	山东省建筑科学研究院 江苏盐城二建集团有限公司
172	JGJ 209-2010	轻型钢结构住宅技术规程 Technical specification for lightweight residential buildings of steel structure		J1009-2010	中国建筑科学研究院
173	JGJ/T 210-2010	刚-柔性桩复合地基技术规程 Technical specification for rigid-flexible pile composite foundation		J1005-2010	温州东瓯建设集团有限公司
174	JGJ/T 211-2010	建筑工程水泥-水玻璃双液注浆技术规程 Technical specification for cement-silicate grouting in building engineering		J1004-2010	湖南省建筑工程集团总公司 湖南省第六工程有限公司
175	JGJ/T 212-2010	地下工程渗漏治理技术规程 Technical specification for remedial waterproofing of the underground works		J1080-2010	中国建筑科学研究院 浙江国泰建设集团有限公司
176	JGJ/T 213-2010	现浇混凝土大直径管桩复合地基技术规程 Technical specification for composite foundation of cast-in-place concrete large-diameter pipe pile		J1073-2010	河海大学 江苏弘盛建设工程集团有限公司
177	JGJ 214-2010	铝合金门窗工程技术规范 Technical code for aluminum alloy window and door engineering		J1071-2010	中国建筑金属结构协会

建筑工程

序号	标准编号	标准名称	被代替标准编号	备案号	主编单位
178	JGJ 215-2010	建筑施工升降机安装、使用、拆卸安全技术规程 Technical specification for safety of installation, use and disassembly of building hoist in construction		J1067-2010	浙江展诚建设集团股份有限公司 浙江大学
179	JGJ/T 216-2010	铝合金结构工程施工规程 Specification for construction of aluminium structures		J1068-2010	上海市第二建筑有限公司 浙江中南建设集团有限公司
180	JGJ 217-2010	纤维石膏空心大板复合墙体结构技术规程 Technical specification for composite wall structures with glass fiber reinforced gypsum panels		J1092-2010	山东省建设建工（集团）有限责任公司 山东建筑大学
181	JGJ 218-2010	展览建筑设计规范 Design code for exhibition building		J1081-2010	同济大学建筑设计研究院（集团）有限公司
182	JGJ/T 219-2010	混凝土结构用钢筋间隔件应用技术规程 Technical specification for application of reinforcement spacings used in concrete structures		J1139-2010	江苏南通六建建设集团有限公司 同济大学
183	JGJ/T 220-2010	抹灰砂浆技术规程 Technical specification for plastering mortar		J1077-2010	陕西省建筑科学研究院 正太集团有限公司
184	JGJ/T 221-2010	纤维混凝土应用技术规程 Technical specification for application of fiber reinforced concrete		J1076-2010	中国建筑科学研究院 大连悦泰建设工程有限公司
185	JGJ/T 222-2011	建筑工程可持续性评价标准 Standard for sustainability assessment of building project		J1217-2011	清华大学 河南红旗渠建设集团有限公司
186	JGJ/T 223-2010	预拌砂浆应用技术规程 Technical specification for application of ready-mixed mortar		J1082-2010	中国建筑科学研究院 广州市建筑集团有限公司
187	JGJ 224-2010	预制预应力混凝土装配整体式框架结构技术规程 Technical specification for framed structures comprised of precast prestressed concrete components		J1127-2010	南京大地建设集团有限责任公司 启东建筑集团有限公司
188	JGJ/T 225-2010	大直径扩底灌注桩技术规程 Technical specification for large-diameter belled cast-in-place pile foundation		J1122-2010	合肥工业大学 浙江省东阳第三建筑工程有限公司
189	JGJ/T 226-2011	低张拉控制应力拉索技术规程 Technical specification for tension cable of low control stress for tensioning		J1195-2011	浙江省二建建设集团 浙江省一建建设集团
190	JGJ 227-2011	低层冷弯薄壁型钢房屋建筑技术规程 Technical specification for low-rise cold-formed thin-walled steel buildings		J1162-2011	中国建筑标准设计研究院
191	JGJ/T 228-2010	植物纤维工业灰渣混凝土砌块建筑技术规程 Technical specification for plant fiber-industrial waste slag concrete block masonry buildings		J1126-2010	中国建筑设计研究院 中博建设集团有限公司
192	JGJ/T 229-2010	民用建筑绿色设计规范 Code for green design of civil buildings		J1125-2010	中国建筑科学研究院 深圳市建筑科学研究院有限公司

建筑工程

序号	标准编号	标准名称	被代替标准编号	备案号	主编单位
193	JGJ 230-2010	倒置式屋面工程技术规程 Technical specification for inversion type roof		J1124-2010	中达建设集团股份有限公司 广东金辉华集团有限公司
194	JGJ 231-2010	建筑施工承插型盘扣式钢管支架安全技术规程 Technical specification for safety of disk lock steel tubular scaffold in construction		J1128-2010	南通新华建筑集团有限公司 无锡市锡山三建实业有限公司
195	JGJ 232-2011	矿物绝缘电缆敷设技术规程 Technical specification for mineral insulated cable laying		J1142-2011	中国新兴建设开发总公司
196	JGJ/T 233-2011	水泥土配合比设计规程 Specification for mix proportion design of cement soil		J1144-2011	福建省建筑科学研究院 福建建工集团总公司
197	JGJ/T 234-2011	择压法检测砌筑砂浆抗压强度技术规程 Technical specification for compressive strength of masonry mortar bed testing by selective pressing method		J1157-2011	江苏省金陵建工集团有限公司 江苏南通三建集团有限公司
198	JGJ/T 235-2011	建筑外墙防水工程技术规程 Technical specification for waterproofing of exterior wall of building		J1166-2011	中国建筑科学研究院 方远建设集团股份有限公司
199	JGJ/T 236-2011	建筑产品信息系统基础数据规范 Code for basic data of construction products information system		J1146-2011	中国建筑标准设计研究院 华升建设集团有限公司
200	JGJ 237-2011	建筑遮阳工程技术规范 Technical code for solar shading engineering of buildings		J1161-2011	北京中建建筑科学研究院有限公司 中国建筑业协会建筑节能分会
201	JGJ/T 238-2011	混凝土基层喷浆处理技术规程 Technical specification for interface guniting on concrete base		J1156-2011	云南工程建设总承包公司 云南建工集团有限公司
202	JGJ/T 239-2011	建(构)筑物移位工程技术规程 Technical specification for moving engineering of buildings		J1183-2011	山东建筑大学 烟建集团有限公司
203	JGJ/T 240-2011	再生骨料应用技术规程 Technical specification for application of recycled aggregate		J1187-2011	中国建筑科学研究院 青建集团股份公司
204	JGJ/T 241-2011	人工砂混凝土应用技术规程 Technical specification for application of manufactured sand concrete		J1188-2011	重庆大学 中建五局第三建设有限公司
205	JGJ 242-2011	住宅建筑电气设计规范 Code for electrical design of residential buildings		J1193-2011	中国建筑标准设计研究院
206	JGJ 243-2011	交通建筑电气设计规范 Code for electrical design of transportation buildings		J1227-2011	现代设计集团华东建筑设计研究院有限公司
207	JGJ/T 244-2011	房屋建筑室内装饰装修制图标准 Drawing standard for interior decoration and renovation of building		J1216-2011	东南大学建筑学院 江苏广宇建设集团
208	JGJ/T 245-2011	房屋白蚁预防技术规程 Technical specification for termite prevention in buildings		J1228-2011	全国白蚁防治中心

建筑工程

序号	标准编号	标准名称	被代替标准编号	备案号	主编单位
209	JGJ/T 246-2012	房屋代码编码标准 Standard for house coding		J1395-2012	北京市建设信息中心 中国建筑科学研究院
210	JGJ 248-2012	底部框架-抗震墙砌体房屋抗震技术规程 Technical specification for earthquake-resistant of masonry buildings with frame and seismic-wall in the lower stories		J1397-2012	中国建筑科学研究院
211	JGJ/T 249-2011	拱形钢结构技术规程 Technical specification for steel arch structure		J1215-2011	清华大学 五洋建设集团股份有限公司
212	JGJ/T 250-2011	建筑与市政工程施工现场专业人员职业标准 Occupational standards for construction site technician or building and municipal engineering		J1225-2011	中国建设教育协会 苏州二建建筑集团有限公司
213	JGJ/T 251-2011	建筑钢结构防腐蚀技术规程 Technical specification for anticorrosion of building steel structure		J1218-2011	河南省第一建筑工程集团 林州建总建筑工程公司
214	JGJ/T 252-2011	房地产市场基础信息数据标准 Standard for data of real estate market's basic information		J1326-2011	上海市房屋土地资源信息中心
215	JGJ 253-2011	无机轻集料砂浆保温系统技术规程 Technical specification for thermal insulating systems of inorganic lightweight aggregate mortar		J1329-2011	广厦建设集团有限责任公司 宁波荣山新型材料有限公司
216	JGJ 254-2011	建筑施工竹脚手架安全技术规范 Technical code for safety of bamboo scaffold in construction		J1336-2011	深圳市建设（集团）有限公司 湖南长大建设集团股份有限公司
217	JGJ 255-2012	采光顶与金属屋面技术规程 Technical specification for skylight and metal roof		J1410-2012	中国建筑科学研究院 中国新兴建设开发总公司
218	JGJ 256-2011	钢筋锚固板应用技术规程 Technical specification for application of headed bars		J1230-2011	中国建筑科学研究院 北京韩建集团有限公司
219	JGJ 257-2012	索结构技术规程 Technical specification for cable structures		J1402-2012	中国建筑科学研究院
220	JGJ/T 258-2011	预制带肋底板混凝土叠合楼板技术规程 Technical specification for concrete composite slab with precast ribbed panel		J1233-2011	湖南高岭建设集团股份有限公司
221	JGJ/T 259-2012	混凝土结构耐久性修复与防护技术规程 Technical specification for rehabilitation and protection of concrete structures durability		J1403-2012	中冶建筑研究总院有限公司
222	JGJ/T 260-2011	采暖通风与空气调节工程检测技术规程 Technical specification for test of heating & ventilating and air-conditioning engineering		J1229-2011	中国建筑科学研究院 湖南望新建设集团股份有限公司
223	JGJ/T 261-2011	外墙内保温工程技术规程 Technical specification for interior thermal insulation on external walls		J1332-2011	中国建筑标准设计研究院 武汉建工股份有限公司
224	JGJ/T 262-2012	住宅厨房模数协调标准 Standard for modular coordination of residential kitchen		J1350-2012	国家住宅与居住环境工程技术研究中心

建筑工程

序号	标准编号	标准名称	被代替标准编号	备案号	主编单位
225	JGJ/T 263-2012	住宅卫生间模数协调标准 Standard for module coordination of residential bathroom		J1351-2012	国家住宅与居住环境工程技术研究中心
226	JGJ/T 264-2012	光伏建筑一体化系统运行与维护规范 Code for operation and maintenance of building mounted photovoltaic system		J1344-2012	无锡尚德太阳能电力有限公司 华仁建设集团有限公司
227	JGJ/T 265-2012	轻型木桁架技术规范 Technical code for light wood trusses		J1398-2012	中国建筑西南设计研究院有限公司
228	JGJ 266-2011	市政架桥机安全使用技术规程 Technical specification for safe use of municipal bridge erecting machine		J1333-2011	鹏达建设集团有限公司 上海市第七建筑有限公司
229	JGJ/T 267-2012	被动式太阳能建筑技术规范 Technical code for passive solar buildings		J1345-2012	中国建筑设计研究院 山东建筑大学
230	JGJ/T 268-2012	现浇混凝土空心楼盖技术规程 Technical specification for cast-in-situ concrete hollow floor structure		J1399-2012	中冶建筑研究总院有限公司
231	JGJ/T 269-2012	轻型钢丝网架聚苯板混凝土构件应用技术规程 Technical specification for the application of concrete elements reinforced with light steel mesh framed expanded polystyrene panel		J1338-2011	上海沪标工程建设咨询有限公司 新八建设集团有限公司
232	JGJ 270-2012	建筑物倾斜纠偏技术规程 Technical specification for incline-rectifying of buildings		J1446-2012	中国建筑第六工程局有限公司 中国建筑第四工程局有限公司
233	JGJ/T 271-2012	混凝土结构工程无机材料后锚固技术规程 Technical specification for post-anchoring used in concrete structure with inorganic anchoring material		J1393-2012	济南四建（集团）有限责任公司 潍坊昌大建设集团有限责任公司
234	JGJ/T 272-2012	建筑施工企业信息化评价标准 Standard for evaluating the informatization of construction enterprises		J1341-2012	中国建筑业协会 中建国际建设有限公司
235	JGJ/T 273-2012	钢丝网架混凝土复合板结构技术规程 Technical specification for wire grids concrete composite slab structure		J1411-2012	华声（天津）国际企业有限公司 天津市建筑设计院
236	JGJ/T 274-2012	装饰多孔砖夹心复合墙技术规程 Technical pecification for cavity wall filled with insulation and decorative perforated brick		J1409-2012	西安墙体材料研究设计院，西安建筑科技大学
237	JGJ/T 275-2013	密肋复合板结构技术规程 Technical specification for multi-ribbed composite panel structures		J1701-2013	北京交通大学 华北建设集团有限公司
238	JGJ 276-2012	建筑施工起重吊装工程安全技术规范 Technical code for safety of lifting in construction		J1354-2012	沈阳建筑大学 东北金城建设股份有限公司
239	JGJ/T 277-2012	红外热像法检测建筑外墙饰面粘结质量技术规程 Technical specification for inspecting the defects of exterior walls cement coating of building with infrared thermography method		J1346-2012	甘肃省建设投资（控股）集团总公司 中国建筑科学研究院

建筑工程					
序号	标准编号	标准名称	被代替标准编号	备案号	主编单位
240	JGJ 278-2012	房地产登记技术规程 Technical specification of real estate registration		J1396-2012	中国房地产研究会房地产产权产籍和测量委员会
241	JGJ/T 279-2012	建筑结构体外预应力加固技术规程 Technical specification for strengthening building structures with external prestressing tendons		J1339-2012	中国京冶工程技术有限公司 浙江舜杰建筑集团股份有限公司
242	JGJ/T 280-2012	中小学校体育设施技术规程 Technical specification for sports facilities of primary and middle school		J1371-2012	中国建筑标准设计研究院 河南国安建设集团有限公司
243	JGJ/T 281-2012	高强混凝土应用技术规程 Technical specification for application of high strength concrete		J1414-2012	中国建筑科学研究院 浙江大东吴集团有限公司
244	JGJ/T 282-2012	高压喷射扩大头锚杆技术规程 Technical specification for underreamed anchor by jet grouting		J1432-2012	深圳钜联锚杆技术有限公司 标力建设集团有限公司
245	JGJ/T 283-2012	自密实混凝土应用技术规程 Technical specification for application of self-compacting concrete		J1404-2012	厦门市建筑科学研究院集团股份有限公司 福建六建集团有限公司
246	JGJ 284-2012	金融建筑电气设计规范 Code for electrical design of financial buildings		J1442-2012	上海建筑设计研究院有限公司
247	JGJ/T 285-2014	公共建筑能耗远程监测系统技术规程 Technical specification for the remote monitoring system of public building energy consumption		J1918-2014	深圳市建筑科学研究院股份有限公司
248	JGJ 286-2013	城市居住区热环境设计标准 Design standard for thermal environment of urban residential areas		J1644-2013	华南理工大学
249	JGJ/T 287-2014	建筑反射隔热涂料节能检测标准 Standard for energy efficiency test of solar heat reflecting insulation coatings of buildings		J1921-2014	华南理工大学
250	JGJ/T 288-2012	建筑能效标识技术标准 Standard for building energy performance certification		J1474-2012	中国建筑科学研究院 住房和城乡建设部科技发展促进中心
251	JGJ 289-2012	建筑外墙外保温防火隔离带技术规程 Technical specification for fire barrier zone of external thermal insulation composite system on walls		J1478-2012	中国建筑科学研究院 江苏省建筑科学研究院有限公司
252	JGJ/T 290-2012	组合锤法地基处理技术规程 Technical specification for ground treatment of combination hammer		J1466-2012	江西中恒建设集团有限公司 江西中煤建设集团有限公司
253	JGJ/T 291-2012	现浇塑性混凝土防渗芯墙施工技术规程 Technical specification for construction of plastic concrete core wall		J1494-2013	云南建工水利水电建设有限公司 云南建工第四建设有限公司
254	JGJ/T 292-2012	建筑工程施工现场视频监控技术规范 Technical code for video surveillance on construction site		J1469-2012	南通建筑工程总承包有限公司

建筑工程

序号	标准编号	标准名称	被代替标准编号	备案号	主编单位
255	JGJ/T 293-2013	淤泥多孔砖应用技术规程 Technical specification for application of silt perforated bricks		J1587-2013	中国建筑标准设计研究院 山东德建集团有限公司
256	JGJ/T 294-2013	高强混凝土强度检测技术规程 Technical specification for strength testing of high strength concrete		J1592-2013	中国建筑科学研究院
257	JGJ/T 295-2013	建筑采光追逐镜施工技术规程 Technical specification for construction of heliostat automatic tracking system of building daylight		J1654-2013	中建三局第二建设工程有限责任公司 重庆大学
258	JGJ/T 296-2013	高抛免振捣混凝土应用技术规程 Technical specification for application of high dropping non vibration concrete		J1519-2013	重庆建工集团股份有限公司 重庆建工住宅建设有限公司
259	JGJ 297-2013	建筑消能减震技术规程 Technical specification for seismic energy dissipation of buildings		J1596-2013	广州大学
260	JGJ 298-2013	住宅室内防水工程技术规范 Technical code for interior waterproof of residential buildings		J1589-2013	中国建筑标准设计研究院 北京韩建集团有限公司
261	JGJ/T 299-2013	建筑防水工程现场检测技术规范 Technical code for in-site testing of building waterproof engineering		J1593-2013	中国建筑科学研究院 四川省晟茂建设有限公司
262	JGJ 300-2013	建筑施工临时支撑结构技术规范 Technical code for temporary support structures in construction		J1599-2013	中国建筑一局（集团）有限公司 中国建筑股份有限公司
263	JGJ/T 301-2013	大型塔式起重机混凝土基础工程技术规程 Technical specification for concrete foundation engineering of large tower cranes		J1597-2013	北京九鼎同方技术发展有限公司 国强建设集团有限公司
264	JGJ/T 302-2013	建筑工程施工过程结构分析与监测技术规范 Technical code for construction process analyzing and monitoring of building engineering		J1598-2013	中国建筑股份有限公司 中建八局第一建设有限公司
265	JGJ/T 303-2013	渠式切割水泥土连续墙技术规程 Technical specification for trench cutting re-mixing deep wall		J1637-2013	浙江省建筑设计研究院 东通岩土科技（杭州）有限公司
266	JGJ/T 304-2013	住宅室内装饰装修工程质量验收规范 Code for construction quality acceptance of housing interior decoration		J1594-2013	住房和城乡建设部住宅产业化促进中心 龙信建设集团有限公司
267	JGJ 305-2013	建筑施工升降设备设施检验标准 Standard for testing of lifting equipments and facilities in construction		J1601-2013	鹏达建设集团有限公司 舜元建设（集团）有限公司
268	JGJ/T 306-2016	建筑工程安装职业技能标准 Standard for occupational skill of construction and installation engineering		J2208-2016	
269	JGJ/T 307-2013	城市照明节能评价标准 Evaluation standard for urban green lighting energy efficiency		J1652-2013	中国城市科学研究会 通广建工集团有限公司
270	JGJ/T 308-2013	磷渣混凝土应用技术规程 Technical specification for application of phosphorous slag powder concrete		J1636-2013	云南省建筑科学研究院 云南建工第五建设有限公司

建筑工程

序号	标准编号	标准名称	被代替标准编号	备案号	主编单位
271	JGJ/T 309-2013	建筑通风效果测试与评价标准 The standard of the measurement and evaluation for efficiency of building ventilation		J1638-2013	中国建筑科学研究院 天津三建建筑工程有限公司
272	JGJ 310-2013	教育建筑电气设计规范 Code for electrical design of education buildings		J1648-2013	清华大学建筑设计研究院有限公司 同济大学
273	JGJ 311-2013	建筑深基坑工程施工安全技术规范 Technical code for construction safety of deep building foundation excavations		J1650-2013	上海星宇建设集团有限公司 郑州大学
274	JGJ 312-2013	医疗建筑电气设计规范 Code for electrical design of medical buildings		J1651-2013	中国建筑设计研究院
275	JGJ/T 313-2013	建设领域信息技术应用基本术语标准 Standard for basic terms of information technology applications in construction field		J1647-2013	建设综合勘察研究设计院有限公司 住房和城乡建设部信息中心
276	JGJ/T 314-2016	建筑工程施工职业技能标准 Standard for occupational skill of construction engineering		J2210-016	上海市城乡建设和管理委员会 人才服务考核评价中心
277	JGJ/T 315-2016	建筑装饰装修职业技能标准 Standard for occupational skill of building interior finish and cladding construction		J2209-2016	上海市城乡建设和管理委员会 人才服务考核评价中心
278	JGJ/T 316-2013	单层防水卷材屋面工程技术规程 Technical specification for single-ply roofing		J1671-2013	中国建筑防水协会 中国江苏国际经济技术合作集团有限公司
279	JGJ/T 317-2014	建筑工程裂缝防治技术规程 Technical specification for prevention and treatment of crack on building engineering		J1740-2014	中国建筑科学研究院 中铁建设集团有限公司
280	JGJ/T 318-2014	石灰石粉在混凝土中应用技术规程 Technical specification for application of ground limestone in concrete		J1737-2014	中国建筑科学研究院 温州建设集团有限公司
281	JGJ 319-2013	低温辐射电热膜供暖系统应用技术规程 Technical specification for heating systems of low temperature electric radiant heating film		J1670-2013	哈尔滨工业大学 黑龙江中惠地热股份有限公司
282	JGJ/T 320-2014	住房公积金基础数据标准 Standard for basic data of housing provident fund		J1758-2014	住房和城乡建设部信息中心 北京住房公积金管理中心
283	JGJ 321-2014	点挂外墙板装饰工程技术规程 Technical apecification for dot-hanging exterior wall panel engineering		J1842-2014	中国建筑科学研究院 中国新兴建设开发总公司
284	JGJ/T 322-2013	混凝土中氯离子含量检测技术规程 Technical specification for test of chloride ion content in concrete		J1702-2013	中国建筑科学研究院 江西昌南建设集团有限公司
285	JGJ/T 323-2014	自保温混凝土复合砌块墙体应用技术规程 Technical specification for application of self-insulation concrete compound block walls		J1738-2014	福建省建筑科学研究院 江苏省建筑科学研究院有限公司
286	JGJ/T 324-2014	建筑幕墙工程检测方法标准 Standard for test method of building curtain wall engineering		J1743-2014	中国建筑科学研究院 广西建工集团第一建筑工程有限责任公司

建筑工程

序号	标准编号	标准名称	被代替标准编号	备案号	主编单位
287	JGJ/T 325-2014	预应力高强钢丝绳加固混凝土结构技术规程 Technical specification for strengthening concrete structures with prestressed high strength steel wire ropes		J1744-2014	东南大学 北京特希达科技有限公司
288	JGJ/T 326-2014	机械式停车库工程技术规范 Technical code for mechanical parking garage engineering		J1742-2014	中国建筑第七工程局有限公司 福建省第五建筑工程公司
289	JGJ/T 327-2014	劲性复合桩技术规程 Technical specification for strength composite piles		J1741-2014	万通建设集团有限公司 昆明二建建设（集团）有限公司
290	JGJ/T 328-2014	预拌混凝土绿色生产及管理技术规程 Technical specification for green production and management of ready-mixed concrete		J1805-2014	中国建筑科学研究院 博坤建设集团公司
291	JGJ/T 329-2015	交错桁架钢结构设计规程 Specification for design of staggered steel truss framing systems		J1997-2015	中国建筑标准设计研究院 福建四海建设有限公司
292	JGJ/T 330-2014	水泥土复合管桩基础技术规程 Technical specification for pile foundation of pipe pile embedded in cement soil		J1806-2014	山东省建筑科学研究院 中建八局第一建设有限公司
293	JGJ/T 331-2014	建筑地面工程防滑技术规程 Technical specification for slip resistance of building floor		J1843-2014	北京城建科技促进会 山东兴华建设集团有限公司
294	JGJ 332-2014	建筑塔式起重机安全监控系统应用技术规程 Technical specification for safety monitoring system of tower cranes		J1888-2014	河南省建筑科学研究院有限公司 新乡克瑞重型机械科技股份有限公司
295	JGJ 333-2014	会展建筑电气设计规范 Code for electrical design of conference and exhibition buildings		J1838-2014	北京市建筑设计研究院有限公司
296	JGJ/T 334-2014	建筑设备监控系统工程技术规范 Technical code for engineering of building automation system		J1840-2014	同方股份有限公司 中国建筑业协会智能建筑分会
297	JGJ/T 335-2014	城市地下空间利用基本术语标准 Standard for basic terminology of urban underground space utilization		J1892-2014	上海市政工程设计研究总院（集团）有限公司
298	JGJ 336-2016	人造板材幕墙工程技术规范 Technical code for artificial panel curtain wall engineering		J2224-2016	中国建筑标准设计研究院 广东建筑科学研究院
299	JGJ 337-2015	钢绞线网片聚合物砂浆加固技术规程 Technical specification for strengthening of building structures with steel stranded wire mesh and polymer mortar		J2085-2015	中国建筑科学研究院 中达建设集团股份有限公司
300	JGJ/T 338-2014	建筑工程风洞试验方法标准 Standard for wind tunnel test of buildings and structures		J1967-2015	中国建筑科学研究院 广东省建筑科学研究院
301	JGJ 339-2015	非结构构件抗震设计规范 Code for seismic design of non-structural components		J1990-2015	中国建筑科学研究院

建筑工程

序号	标准编号	标准名称	被代替标准编号	备案号	主编单位
302	JGJ 340-2015	建筑地基检测技术规范 Technical code for testing of building foundation soils		J1998-2015	福建省建筑科学研究院 福州建工（集团）总公司
303	JGJ/T 341-2014	泡沫混凝土应用技术规程 Technical specification for application of foamed concrete		J1965-2015	中国建筑科学研究院 温州建设集团有限公司
304	JGJ 342-2014	蒸发冷却制冷系统工程技术规程 Technical specification for evaporative cooling refrigeration system		J1894-2014	中国建筑科学研究院
305	JGJ 343-2014	变风量空调系统工程技术规程 Technical specification for VAV air conditioning system		J1887-2014	中国建筑科学研究院 浙江鸿翔建设集团有限公司
306	JGJ/T 344-2014	随钻跟管桩技术规程 Technical specification for drilling with prestressed high strength concrete pipe cased pile		J1970-2015	广州市建筑科学研究院有限公司 广州建筑股份有限公司
307	JGJ 345-2014	公共建筑吊顶工程技术规程 Technical specification for ceiling engineering of public building		J1968-2015	中国建筑标准设计研究院有限公司 山东寿光第一建筑有限公司
308	JGJ/T 346-2014	建筑节能气象参数标准 Standard for weather data of building energy efficiency		J1923-2014	中国建筑科学研究院
309	JGJ/T 347-2014	建筑热环境测试方法标准 Standard of test methods for thermal environment of building		J1891-2014	华南理工大学
310	JGJ 348-2014	建筑工程施工现场标志设置技术规程 Technical specification for signs layout of construction site		J1919-2014	杭州天和建设集团有限公司 重庆建工第三建设有限责任公司
311	JGJ/T 349-2015	民用建筑氡防治技术规程 Technical specification for radon control of civil building		J1989-2015	深圳市建筑科学研究院股份有限公司 泰宏建设发展有限公司
312	JGJ/T 350-2015	保温防火复合板应用技术规程 Technical specification for application of thermal insulated fireproof composite panels		J2052-2015	中国建筑科学研究院 江西建工第一建筑有限责任公司
313	JGJ/T 351-2015	建筑玻璃膜应用技术规程 Technical specification for application of building glass film and coating		J2050-2015	中国建筑科学研究院 天津住宅集团建设工程总承包有限公司
314	JGJ/T 352-2014	建筑塑料复合模板工程技术规程 Technical specification for engineering of plastic composite formwork		J1920-2014	中阳建设集团有限公司 发达控股集团有限公司
315	JGJ 354-2014	体育建筑电气设计规范 Code for electrical design of sports buildings		J1917-2014	悉地国际设计顾问（深圳）有限公司
316	JGJ 355-2015	钢筋套筒灌浆连接应用技术规程 Technical specification for grout sleeve splicing of rebars		J1983-2015	中国建筑科学研究院 云南省第二建筑工程公司
317	JGJ/T 357-2015	围护结构传热系数现场检测技术规程 Technical specification for in-situ measurement of thermal transmittance of building envelope		J1986-2015	广州市建筑科学研究院有限公司

建筑工程

序号	标准编号	标准名称	被代替标准编号	备案号	主编单位
318	JGJ/T 358-2015	农村火炕系统通用技术规范 Technical specification for rural kang system		J1995-2015	沈阳建筑大学 哈尔滨工业大学
319	JGJ/T 359-2015	建筑反射隔热涂料应用技术规程 Technical specification for application of architectural reflective thermal insulation coating		J2054-2015	福建省建筑科学研究院 恒亿集团有限公司
320	JGJ 360-2015	建筑隔震工程施工及验收规范 Code for construction and acceptance of building isolation engineering		J2051-2015	中国建筑标准设计研究院 山西太行建设开发有限公司
321	JGJ 361-2014	人工碎卵石复合砂应用技术规程 Technical specification for the application of mixed sand with crushed gravel		J1972-2015	重庆建工第三建设有限责任公司 中国土木工程学会
322	JGJ 362-2016	塑料门窗设计及组装技术规程 Technical specification for design and fabricating of plastic doors and windows		J2225-2016	中国建筑金属结构协会塑料门窗委员会 维卡塑料（上海）有限公司
323	JGJ/T 363-2014	农村住房危险性鉴定标准 Standard for fatalness evaluation of rural area building		J1971-2015	同济大学 上海建工一建集团有限公司
324	JGJ/T 364-2016	地下工程盖挖法施工规程 Specification for construction of underground structure engineering by covered excavation		J2213-2016	中建交通建设集团有限公司 中建海峡建设发展有限公司
325	JGJ/T 365-2015	太阳能光伏玻璃幕墙电气设计规范 Code for electrical design of solar photovoltaic glass curtain wall		J1993-2015	深圳市创益科技发展有限公司
326	JGJ 366-2015	混凝土结构成型钢筋应用技术规程 Technical specification for application of fabricated steel bars of concrete structure		J2012-2015	中国建筑科学研究院 广州市恒盛建设工程有限公司
327	JGJ 367-2015	住宅室内装饰装修设计规范 Code for design of the residential interior decoration		J2053-2015	东南大学 永升建设集团有限公司
328	JGJ/T 368-2015	钻芯法检测砌体抗剪强度及砌筑砂浆强度技术规程 Technical specification for testing shear strength of masonry and strength of mortar with drilled core		J2083-2015	山东省建筑科学研究院 江西建工第二建筑有限责任公司
329	JGJ 369-2016	预应力混凝土结构设计规范 Code for design of prestressed concrete structures		J1268-2016	同济大学 上海市第七建筑有限公司
330	JGJ/T 370-2015	悬挂式竖井施工规程 Specification for construction of suspended shaft		J2089-2015	郑州市市政工程总公司 合肥建工集团有限公司
331	JGJ/T 371-2016	非烧结砖砌体现场检测技术规程 Technical specification for in-site testing of non fired block masonry		J2155-2016	四川省建筑科学研究院 成都市第六建筑工程公司
332	JGJ/T 372-2016	喷射混凝土应用技术规程 Technical specification for application of sprayed concrete		J2156-2016	厦门市建筑科学研究院集团股份有限公司 厦门特房建设工程集团有限公司

建筑工程

序号	标准编号	标准名称	被代替标准编号	备案号	主编单位
333	JGJ/T 373-2016	白蚁防治工职业技能标准 Standard for occupational skills of termite control operators		J2177-2016	全国白蚁防治中心
334	JGJ/T 374-2015	导光管采光系统技术规程 Technical specification for tubular daylighting system		J2084-2015	中国建筑科学研究院 湖南省沙坪建筑有限公司
335	JGJ/T 375-2016	管幕预筑法施工技术规范 Technical code for construction of pipe-roof pre-construction method		J2215-2016	中建市政建设有限公司，山东三箭建设工程股份有限公司
336	JGJ 376-2015	建筑外墙外保温系统修缮标准 Standard for building external thermal insulation system repair		J2104-2015	上海市房地产科学研究院 青岛海川建设集团有限公司
337	JGJ/T 377-2016	木丝水泥板应用技术规程 Technical specification for application of wood wool cement boards		J2157-2016	中国建筑科学研究院 无锡泛亚环保科技有限公司
338	JGJ/T 378-2016	拉脱法检测混凝土抗压强度技术规程 Technical specification for inspection of concrete compressive strength by pull off method		J2271-2016	建研科技股份有限公司 天颂建设集团有限公司
339	JGJ/T 379-2016	螺纹桩技术规程 Technical specification for screw concrete pile		J2212-2016	重庆建工住宅建有限公司 北京波森特岩土工程有限公司
340	JGJ/T 380-2015	钢板剪力墙技术规程 Technical specification for steel plate shear walls		J2101-2015	哈尔滨工业大学 中建四局第六建筑工程有限公司
341	JGJ/T 381-2016	纤维片材加固砌体结构技术规范 Technical code for strengthening masonry structures with fiber reinforced polymer laminates		J2202-2016	武汉大学土木建筑工程学院 武汉建工股份有限公司
342	JGJ 383-2016	轻钢轻混凝土结构技术规程 Technical specification of lightweight steel and lightweight concrete structures		J2154-2016	中国建筑技术集团有限公司
343	JGJ/T 384-2016	钻芯法检测混凝土强度技术规程 Technical specification for testing concrete strength with drilled core method		J2211-2016	中国建筑科学研究院 江苏兴邦建工集团有限公司
344	JGJ/T 385-2015	高性能混凝土评价标准 Standard for assessment of high performance concrete		J2087-2015	中国建筑科学研究院
345	JGJ 386-2016	组合铝合金模板工程技术规程 Technial specification for combinel aluminum alloy form work engineering		J2201-2016	广东省建筑科学研究院 广东合迪科技有限公司
346	JGJ/T 388-2016	住房公积金信息系统技术规范 Technical code for information system of housing provident fund		J2178-2016	苏州市住房公积金管理中心
347	JGJ/T 389-2016	组装式桁架模板支撑应用技术规程 Techinical specification for application of assembled truss form work-support		J2320-2016	山东德建集团有限公司 江苏永泰建造工程有限公司
348	JGJ/T 390-2016	既有住宅建筑功能改造技术规范 Techinical code for the retrofitting of existing residential building on using function		J2174-2016	上海维固工程实业有限公司 上海建筑设计研究院

建筑工程

序号	标准编号	标准名称	被代替标准编号	备案号	主编单位
349	JGJ/T 391-2016	绿色建筑运行维护技术规范 Techinical specification for operation and maintenance of green building		J2319-2016	中国建筑科学研究院
350	JGJ 392-2016	商店建筑电气设计规范 Code for electrical design of the store buildings		J2274-2016	合肥工业大学建筑设计研究院
351	JGJ/T 397-2016	公墓和骨灰寄存建筑设计规范 Code for design of public cemetery and ashes depositing buildng		J2288-2016	哈尔滨工业大学 西安建筑科技大学建筑设计研究院
352	JGJ/T 399-2016	城市雕塑工程技术规程 Techinical specification for public sculpture project construction		J2317-2016	中国建筑文化中心 华仁建设集团有限公司

城镇建设工程

序号	标准编号	标准名称	被代替标准编号	备案号	主编单位
1	CJJ 1-2008	城镇道路工程施工与质量验收规范 Code for construction and quality acceptance of road works in city and town	CJJ 1-90		北京市政建设集团有限责任公司 中国市政工程协会
2	CJJ 2-2008	城市桥梁工程施工与质量验收规范 Code for construction and quality acceptance of bridge works in city	CJJ 2-90	J820-2008	北京市政建设集团有限责任公司 北京市公路桥梁建设集团有限公司
3	CJJ 6-2009	城镇排水管道维护安全技术规程 Technical sepcification for safety of urban sewer maintenance	CJJ 6-85	J950-2009	天津市排水管理处
4	CJJ 7-2007	城市工程地球物理探测规范 Code for engineering geophysical prospecting and testing in city	CJJ 7-85	J720-2007	山东正元地理信息工程有限责任公司
5	CJJ/T 8-2011	城市测量规范 Code for urban survey	CJJ 8-99	J1330-2011	北京市测绘设计研究院
6	CJJ 11-2011	城市桥梁设计规范 Code for design of the municipal bridge	CJJ 11-93	J1190-2011	上海市政工程设计研究总院
7	CJJ 12-2013	家用燃气燃烧器具安装及验收规程 Specification for installation and acceptance of domestic gas burning appliances	CJJ 12-99	J1640-2013	中国市政工程华北设计研究总院
8	CJJ/T 13-2013	供水水文地质钻探与管井施工操作规程 Specification for hydro geologic drilling and well construction of water-supply	CJJ 13-87	J1588-2013	中国市政工程中南设计研究总院有限公司
9	CJJ 14-2016	城市公共厕所设计标准 Standard for design of urban public toilets	CJJ 14-87 CJJ 14-2005	J476-2016	北京市环境卫生设计科学研究所
10	CJJ/T 15-2011	城市道路公共交通站、场、厂工程设计规范 Code for design of urban road public transportation stop, terminus and depot engineering	CJJ 15-87	J1331-2011	武汉市交通科学研究所
11	CJJ 27-2012	环境卫生设施设置标准 Standard for setting of environmental sanitation facilities	CJJ 27-2005	J1497-2013	上海市环境工程设计科学研究院有限公司
12	CJJ 28-2014	城镇供热管网工程施工及验收规范 Code for construction and acceptance of city heating pipelines	CJJ 28-2004	J372-2014	北京市热力集团有限责任公司
13	CJJ/T 29-2010	建筑排水塑料管道工程技术规程 Technical specification for plastic pipeline for building drainage	CJJ/T 29-98	J1136-2010	上海现代建筑设计（集团）有限公司
14	CJJ 30-2009	粪便处理厂运行维护及其安全技术规程 Technical specification for operation maintenance and safety of night soil treatment plants	CJJ/T 30-99	J894-2009	华中科技大学
15	CJJ 32-2011	含藻水给水处理设计规范 Code for design of algae water treatment	CJJ 32-89	J1184-2011	中国市政工程中南设计研究总院
16	CJJ 33-2005	城镇燃气输配工程施工及验收规范 Code for construction and acceptance of city and town gas distribution works	CJJ 33-89	J404-2005	城市建设研究院

城镇建设工程

序号	标准编号	标准名称	被代替标准编号	备案号	主编单位
17	CJJ 34-2010	城镇供热管网设计规范 Design code for city heating network	CJJ 34-2002	J1074-2010	北京市煤气热力工程设计院有限公司
18	CJJ 36-2016	城镇道路养护技术规范 Technical code of maintenance for urban road	CJJ 36-90 CJJ 36-2006	J528-2016	北京市政路桥管理养护集团有限公司 南通英雄建设集团有限公司
19	CJJ 37-2012 （2016年版）	城市道路工程设计规范 Code for design of urban road engineering	CJJ 37-90 CJJ 37-2012	J1353-2012	北京市市政工程设计研究总院
20	CJJ 39-91	古建筑修建工程质量检验评定标准（北方地区） Standard for quality test and estimation of ancient building repairing engineerings（in northern area）			北京市房地产管理局
21	CJJ 40-2011	高浊度水给水设计规范 Code for design of water supply engineering using high-turbidity raw water	CJJ 40-91	J1185-2011	中国市政工程西北设计研究院有限公司
22	CJJ/T 43-2014	城镇道路沥青路面再生利用技术规程 Technical specification for asphalt pavement recycling of urban road	CJJ 43-91	J1916-2014	上海市市政规划设计研究院 广东电白建设集团有限公司
23	CJJ 45-2015	城市道路照明设计标准 Standard for lighting design of urban road	CJJ 45-91 CJJ 45-2006	J627-2006	中国建筑科学研究院
24	CJJ 47-2016	生活垃圾转运站技术规范 Technical code for transfer station of municipal solid waste	CJJ 47-91 CJJ 47-2006	J511-2016	华中科技大学
25	CJJ 49-92	地铁杂散电流腐蚀防护技术规程			北京市地下铁道科学技术研究所
26	CJJ 51-2016	城镇燃气设施运行、维护和抢修安全技术规程 Safety technical specification for operation, maintenance and rush-repair of city gas facilities	CJJ 51-2001 CJJ 51-2006	J112-2016	中国城市燃气协会
27	CJJ 52-2014	生活垃圾堆肥处理技术规范 Technical code for the composting of municipal solid waste	CJJ/T 52-93	J1969-2015	同济大学 中国城市建设研究院有限公司
28	CJJ/T 53-93	民用房屋修缮工程施工规程 Code for repairing construction of civil buildings			天津市房产住宅科学研究所
29	CJJ/T 54-93	污水稳定塘设计规范 Code for design of wastewater stabilization ponds			哈尔滨建筑工程学院（全国氧化塘协作组）
30	CJJ/T 55-2011	供热术语标准 Standard for terminology of heating	CJJ 55-93	J1220-2011	哈尔滨工业大学
31	CJJ 56-2012	市政工程勘察规范 Code for geotechnical investigation of municipal engineering projects	CJJ 56-94	J1493-2013	北京市勘察设计研究院有限公司
32	CJJ 57-2012	城乡规划工程地质勘察规范 Code for geo-engineering site investigation and evaluation of urban and rural planning	CJJ 57-94	J1472-2012	北京市勘察设计研究院有限公司

城镇建设工程

序号	标准编号	标准名称	被代替标准编号	备案号	主编单位
33	CJJ 58-2009	城镇供水厂运行、维护及安全技术规程 Technical specification for operation, maintenance and safety of city and town waterworks	CJJ 58-94	J967-2009	中国城镇供水排水协会 北京市自来水集团有限责任公司
34	CJJ 60-2011	城镇污水处理厂运行、维护及安全技术规程 Technical specification for operation, maintenance and safety of municipal wastewater treatment plant	CJJ 60-94	J1182-2011	中国城镇供水排水协会 天津创业环保集团股份有限公司
35	CJJ 61-2003	城市地下管线探测技术规程 Technical specification for detecting and surveying of under-ground pipelines and cables in city	CJJ 61-94	J271-2003	北京市测绘设计研究院
36	CJJ 63-2008	聚乙烯燃气管道工程技术规程 Technical specification for polyethylene (PE) gas pipeline engineerings	CJJ 63-95	J780-2008	建设部科技发展促进中心
37	CJJ 64-2009	粪便处理厂设计规范 Code for design of night soil treatment plant	CJJ 64-95	J893-2009	华中科技大学
38	CJJ/T 65-2004	市容环境卫生术语标准 Standard for terminology of environmental sanitation	CJJ 65-95	J374-2004	海市环境工程设计科学研究院
39	CJJ/T 66-2011	路面稀浆罩面技术规程 Technical specification for slurry surfacing of pavements	CJJ 66-95	J1221-2011	北京市政路桥建材集团有限公司 深圳市政工程总公司
40	CJJ/T 67-2015	风景园林制图标准 Standard for drawing of landscape architecture	CJJ 67-95	J1982-2015	中国城市建设研究院有限公司 同济大学
41	CJJ 68-2016	城镇排水管渠与泵站运行、维护及安全技术规程 Technical specification for maintenance of sewers & channels and pumping station in city	CJJ/T 68-96 CJJ 68-2007	J659-2016	上海市排水管理处
42	CJJ 69-95 (2003年版)	城市人行天桥与人行地道技术规范 Technical specification of urban pedestrian overcrossing and underpass			北京市市政工程研究院
43	CJJ 70-96	古建筑修建工程质量检验评定标准（南方地区） Standard for quality test and estimation of ancient building repairing engineerings (in southern area)			苏州市房地产管理局
44	CJJ/T 71-2011	机动车清洗站技术规范 Technical code for automotive vehicle washing station	CJJ 71-2000	J24-2011	天津市环境卫生工程设计院
45	CJJ/T 72-2015	无轨电车牵引供电网工程技术规范 Technical code for trolleybus supply network engineering	CJJ 72-97	J2107-2015	北京市公交总公司电车公司供电所
46	CJJ/T 73-2010	卫星定位城市测量技术规范 Technical code for urban surveying using satellite positioning system	CJJ 73-97	J990-2010	北京市测绘设计研究院

城镇建设工程

序号	标准编号	标准名称	被代替标准编号	备案号	主编单位
47	CJJ 74-99	城镇地道桥顶进施工及验收规程 Specification for construction and acceptance of underpass bridges in town by jacking method			石家庄市市政建设总公司
48	CJJ 75-97	城市道路绿化规划与设计规范 Code for planting planning and design on urban roads			中国城市规划设计研究院
49	CJJ 76-2012	城市地下水动态观测规程 Specification for dynamic observation of groundwater in urban area	CJJ/T 76-98	J1342-2012	建设综合勘察研究设计院有限公司
50	CJJ/T 78-2010	供热工程制图标准 Drawing standard for heating engineering	CJJ/T 78-97	J1137-2010	哈尔滨工业大学
51	CJJ/T 81-2013	城镇供热直埋热水管道技术规程 Technical specification for directly buried hot-water heating pipeline in city	CJJ/T 81-1998	J1639-2013	城市建设研究院 北京市煤气热力工程设计院有限公司
52	CJJ 82-2012	园林绿化工程施工及验收规范 Code for construction and acceptance of landscaping engineering	CJJ/T 82-1999	J1496-2013	天津市市容和园林管理委员会
53	CJJ 83-2016	城乡用地竖向规划规范 Code for vertical planning on urban field	CJJ 83-99	J2207-2016	四川省城乡规划设计研究院
54	CJJ/T 85-2002	城市绿地分类标准 Standard for classification of urban green space		J185-2002	北京北林地景园林规划设计院有限责任公司
55	CJJ 86-2014	生活垃圾堆肥处理厂运行维护技术规程 Technical specification for operation and maintenance of municipal solid waste composting plant	CJJ/T 86-2000	J1964-2015	华中科技大学
56	CJJ/T 87-2000	乡镇集贸市场规划设计标准 Standard for market planning of town and township			中国建筑技术研究院村镇规划设计研究所
57	CJJ 88-2014	城镇供热系统运行维护技术规程 Technical specification for operation and maintenance of city heating system	CJJ 88-2000	J25-2014	沈阳惠天热电股份有限公司
58	CJJ 89-2012	城市道路照明工程施工及验收规程 Specification for construction and inspection of urban road lighting engineering	CJJ 89-2001	CJJ 89-2012	北京市城市照明管理中心 中国市政工程协会城市道路照明专业委员会
59	CJJ 90-2009	生活垃圾焚烧处理工程技术规范 Technical code for projects of municipal solid waste incineration	CJJ 90-2002	CJJ 90-2002	城市建设研究院 五洲工程设计研究院
60	CJJ/T 91-2002	园林基本术语标准 Standard for basic terminology of landscape architecture		J217-2002	建设部城市建设研究院
61	CJJ 92-2016	城市供水管网漏损控制及评定标准 Standard for leakage control and assessment of urban water supply distribution system	CJJ 92-2002	J187-2016	中国城镇供水协会
62	CJJ 93-2011	生活垃圾卫生填埋场运行维护技术规程 Technical specification for operation and maintenance of municipal solid waste sanitary landfill	CJJ 93-2003	J252-2011	华中科技大学 杭州市固体废弃物处理有限公司

城镇建设工程

序号	标准编号	标准名称	被代替标准编号	备案号	主编单位
63	CJJ 94-2009	城镇燃气室内工程施工与质量验收规范 Code for construction and quality acceptance of city indoor gas engineering	CJJ 94-2003	J264-2003	北京市煤气热力工程设计院有限公司
64	CJJ 95-2013	城镇燃气埋地钢质管道腐蚀控制技术规程 Technical specification for external corrosion control of buried steel pipeline for city gas	CJJ 95-2003	J273-2013	北京市燃气集团有限责任公司
65	CJJ 96-2003	地铁限界标准 Standard of metro gauges		J274-2003	同济大学铁道与城市轨道交通研究院 北京城建设计研究总院
66	CJJ/T 97-2003	城市规划制图标准 Standard for drawing in urban planning			浙江省建设厅
67	CJJ/T 98-2014	建筑给水塑料管道工程技术规程 Technical specification for plastic pipeline engineering of building water supply	CJJ/T 98-2003 GB/T 50349-2005	J279-2014	上海现代建筑设计集团 上海建筑设计研究院有限公司
68	CJJ 99-2003	城市桥梁养护技术规范 Technical code maintenance for city bridge			北京市市政工程管理处
69	CJJ 100-2004	城市基础地理信息系统技术规范 Technical specification for urban fundametal geographic information system		J298-2004	北京市测绘设计研究院 重庆市勘测院 上海城市发展信息研究中心
70	CJJ 101-2016	埋地塑料给水管道工程技术规程 Technical specification for buried polyethylene pipeline of water supply engineering ?	CJJ 101-2004	J362-2016	北京中环工程设计监理有限责任公司
71	CJJ/T 102-2004	城市生活垃圾分类及其评价标准 Classification and evaluation standard of municipal solid waste		J373-2004	广州市市容环境卫生局
72	CJJ/T 103-2013	城市地理空间框架数据标准 Standard for urban geospatial framework data	CJJ 103-2004	J367-2013	建设综合勘察研究设计院有限公司
73	CJJ/T 104-2014	城镇供热直埋蒸汽管道技术规程 Technical specification for directly buried steam heating pipeline in city	CJJ 104-2005	J456-2014	中国市政工程华北设计研究总院
74	CJJ 105-2005	城镇供热管网结构设计规范 Code for structural design of heating pipelines in city and town		J457-2005	北京市煤气热力工程设计院有限公司
75	CJJ/T 106-2010	城市市政综合监管信息系统技术规范 Technical code for urban municipal supervision and management information system	CJJ/T 106-2005	J455-2010	北京市东城区城市管理监督中心
76	CJJ/T 107-2005	生活垃圾填埋场无害化评价标准 Standard of assessment on municipal solid wastelandfill		J477-2005	中国城市环境卫生协会
77	CJJ/T 108-2006	城市道路除雪作业技术规程 Techincal specification of snow remove operation for city road		J495-2006	北京市环境卫生设计科学研究所
78	CJJ 109-2006	生活垃圾转运站运行维护技术规程 Techincal specification for operation and maintenance of municipal solid waste transfer station		J512-2006	城市建设研究院
79	CJJ 110-2006	管道直饮水系统技术规程 Technical specification of pipe system for fine drinking water		J479-2006	中国建筑设计研究院 深圳市水务集团深水海纳水务有限公司等

城镇建设工程

序号	标准编号	标准名称	被代替标准编号	备案号	主编单位
80	CJJ/T 111-2006	预应力混凝土桥梁预制节段逐跨拼装施工技术规程 Techincal specification for construction of span by span method of precast segment in prestressed concrete bridge		J533-2006	上海市第一市政工程有限公司
81	CJJ 112-2007	生活垃圾卫生填埋场封场技术规程 Techincal code for municipal solid waste sanitary landfill closure		J657-2007	深圳市环境卫生管理处
82	CJJ 113-2007	生活垃圾卫生填埋场防渗系统工程技术规范 Techincal code for liner system of municipal solid waste landfill		J658-2007	城市建设研究院
83	CJJ/T 114-2007	城市公共交通分类标准 Standard for classification of urban public transportation		J682-2007	城市建设研究院
84	CJJ/T 115-2007	房地产市场信息系统技术规范 Techincal code for real estate market information system		J662-2007	上海市房地产交易中心
85	CJJ/T 116-2014	建设领域应用软件测评工作通用规范 General code for the evaluation work of application software in the field of construction	CJJ/T 116-2008	J782-2014	住房和城乡建设部信息中心 黑龙江省建工集团有限责任公司
86	CJJ/T 117-2007	建设电子文件与电子档案管理规范 Code for management of electronic construction records and archives		J725-2007	广州市城建档案馆 建设部城建档案馆工作办公室
87	CJJ/T 119-2008	城市公共交通工程术语标准 Terminology standard for urban public transport engineering		J782-2008	中国城市公共交通协会
88	CJJ 120-2008	城镇排水系统电气与自动化工程技术规程 Technical specification of electrical & automation engineering for city drainage system		J783-2008	J783-2008
89	CJJ/T 121-2008	风景名胜区分类标准 Standard for scenic and historic areas classification		J816-2008	城市建设研究院
90	CJJ 122-2008	游泳池给水排水工程技术规程 Technical specification for water supply and drainage engineering of swimming pool		J821-2008	中国建筑设计研究院
91	CJJ 123-2008	镇（乡）村给水工程技术规程 Technical specification of water supply engineering for town and village		J799-2008	上海市政工程设计研究总院
92	CJJ 124-2008	镇（乡）村排水工程技术规程 Technical specification of wastewater engineering for town and village		J800-2008	上海市政工程设计研究总院
93	CJJ/T 125-2008	环境卫生图形符号标准 Standard for figure symbols of environmental sanitation	CJ/T 13-1999 CJ/T 14-1999 CJ/T 15-1999	J825-2008	华中科技大学
94	CJJ/T 126-2008	城市道路清扫保洁质量与评价标准 Standard for quality and assessment of city road sweeping and cleaning		J826-2008	北京市环境卫生设计科学研究所
95	CJJ 127-2009	建筑排水金属管道工程技术规程 Technical specification of metal pipe work for building drainage		J865-2009	中国金属学会铸铁管分会

城镇建设工程

序号	标准编号	标准名称	被代替标准编号	备案号	主编单位
96	CJJ 128-2009	生活垃圾焚烧厂运行维护与安全技术规程 Technical specification for operation maintenance and safety of municipal solid waste incineration plant		J854-2009	深圳市市政环卫综合处理厂
97	CJJ 129-2009	城市快速路设计规程 Specification for design of urban expressway		J863-2009	北京市市政工程设计研究总院
98	CJJ/T 130-2009	燃气工程制图标准 standard of gas engineering drawing		J890-2009	中国市政工程华北设计研究总院
99	CJJ 131-2009	城镇污水处理厂污泥处理技术规程 technical specification for sludge treatment of municipl wastewater treatment plant		J891-2009	北京城市排水集团有限责任公司
100	CJJ 132-2009	城乡用地评定标准 standard for urban and rural loud evaluation			陕西省城乡规划设计研究院
101	CJJ 133-2009	生活垃圾填埋场填埋气体收集处理及利用工程技术规范 Technical code for projects of landfill gas collection treatment and utilization		J959-2009	城市建设研究院 广州工程总承包集团有限公司
102	CJJ 134-2009	建筑垃圾处理技术规范 Technical code for construction and demolition waste treatment		J960-2009	上海市环境工程设计科学研究院有限公司 江苏中兴建设有限公司
103	CJJ/T 135-2009	透水水泥混凝土路面技术规程 Technical specification for pervious cement concrete pavement		J965-2009	江苏省建工集团有限公司 河南省第一建筑工程集团有限责任公司
104	CJJ 136-2010	快速公共汽车交通系统设计规范 Code for design of bus rapid transit system		J1000-2010	北京市市政工程设计研究总院
105	CJJ/T 137-2010	生活垃圾焚烧厂评价标准 Standard for assessment on municipal solid waste incineration plant		J1002-2010	城市建设研究院
106	CJJ 138-2010	城镇地热供热工程技术规程 Technical specification for geothermal space heating engineering		J1010-2010	天津大学
107	CJJ 139-2010	城市桥梁桥面防水工程技术规程 Technical specification for waterproofing of city bridge decks		J975-2010	北京市市政工程设计研究总院
108	CJJ 140-2010	二次供水工程技术规程 Technical specification for secondary water supply engineering		J1011-2010	天津市供水管理处
109	CJJ/T 141-2010	建设项目交通影响评价技术标准 Technical standards of traffic impact analysis of construction projects		J998-2010	中国城市规划设计研究院
110	CJJ 142-2014	建筑屋面雨水排水系统技术规程 Technical specification for raindrainage system of building roof		J1757-2014	中国建筑设计研究院 深圳市建工集团股份有限公司
111	CJJ 143-2010	埋地塑料排水管道工程技术规程 Technical specification of buried plastic pipeline of sewer engineering		J1037-2010	住房和城乡建设部科技发展促进中心 汕头市达濠市政建设有限公司
112	CJJ/T 144-2010	城市地理空间信息共享与服务元数据标准 Standard of metadata for urban geospatial information sharing and services		J1039-2010	建设综合勘察研究设计院有限公司

城镇建设工程

序号	标准编号	标准名称	被代替标准编号	备案号	主编单位
113	CJJ 145-2010	燃气冷热电三联供工程技术规程 Technical specification for gas-fired combined cooling, heating and power engineering		J1085-2010	城市建设研究院 北京市煤气热力工程设计院有限公司
114	CJJ/T 146-2011	城镇燃气报警控制系统技术规程 Technical specification for gas alarm and control system		J1165-2011	中国城市燃气协会
115	CJJ/T 147-2010	城镇燃气管道非开挖修复更新工程技术规程 Technical specification for trenchless rehabilitation and replacement engineering of city gas pipe		J1075-2010	北京市燃气集团有限责任公司
116	CJJ/T 148-2010	城镇燃气加臭技术规程 Technical specification for city gas odorization		J1091-2010	中国市政工程华北设计研究总院
117	CJJ 149-2010	城市户外广告设施技术规范 Technical code for urban outdoor advertising facilities		J1069-2010	上海环境卫生工程设计院 江苏省苏中建设集团股份有限公司
118	CJJ 150-2010	生活垃圾渗沥液处理技术规范 Technical code for leachate treatment of municipal solid waste		J1070-2010	城市建设研究院 上海环境卫生工程设计院
119	CJJ/T 151-2010	城市遥感信息应用技术规范 Code for urban remote sensing information applications		J1130-2010	建设综合勘察研究设计院有限公司
120	CJJ 152-2010	城市道路交叉口设计规程 Specification for design of intersections on urban roads		J1084-2010	华中科技大学
121	CJJ/T 153-2010	城镇燃气标志标准 Standard for city gas signs		J1093-2010	北京市燃气集团有限责任公司
122	CJJ/T 154-2011	建筑给水金属管道工程技术规程 Technical specification for metallic pipeline engineering of building water supply		J1164-2011	中国建筑金属结构协会 中太建设集团股份有限公司
123	CJJ/T 155-2011	建筑给水复合管道工程技术规程 Technical specification for composite pipeline engineering of building water supply		J1163-2011	中国建筑金属结构协会 浙江宝业建设集团有限公司
124	CJJ/T 156-2010	生活垃圾转运站评价标准 Standard for assessment on municipal solid waste transfer station		J1121-2010	华中科技大学
125	CJJ/T 157-2010	城市三维建模技术规范 Technical code for three dimensional city modeling		J1129-2010	武汉市国土资源和规划局
126	CJJ/T 158-2011	城建档案业务管理规范 Code for management of urban construction archives		J1143-2011	住房和城乡建设部城建档案工作办公室 北京市城建档案馆
127	CJJ 159-2011	城镇供水管网漏水探测技术规程 Technical specification for leak detection of water supply pipe nets in cities and towns		J1145-2011	城市建设研究院
128	CJJ 160-2011	公共浴场给水排水工程技术规程 Technical specification for public SPA pool water supply and drainage engineering		J1222-2011	中国建筑设计研究院 杭州萧宏建设集团有限公司
129	CJJ 161-2011	污水处理卵形消化池工程技术规程 Technical specification for egg-shaped digester of sewage treatment		J1181-2011	中建八局第三建设有限公司 中国建筑股份有限公司

序号	标准编号	标准名称	被代替标准编号	备案号	主编单位
		城镇建设工程			
130	CJJ/T 162-2011	城市轨道交通自动售检票系统检测技术规程 Technical specification for test technology of urban rail transit automatic fare collection system		J1191-2011	广州市地下铁道总公司
131	CJJ/T 163-2011	村庄污水处理设施技术规程 Technical specification of wastewater treatment facilities for village		J1219-2011	中国科学院生态环境研究中心
132	CJJ/T 164-2011	盾构隧道管片质量检测技术标准 Standard for quality inspection of shield tunnel segment		J1194-2011	广东省建筑科学研究院 中铁二十五局集团有限公司
133	CJJ/T 165-2011	建筑排水复合管道工程技术规程 Technical specification for composite pipeline engineering of building drainage		J1328-2011	中国建筑金属结构协会 上海城建建设实业（集团）有限公司
134	CJJ 166-2011	城市桥梁抗震设计规范 Code for seismic design of urban bridges		J1224-2011	同济大学
135	CJJ 167-2012	城市轨道交通直线电机牵引系统设计规范 Code for design of urban rail transit by linear motor		J1394-2012	广州地铁设计研究院有限公司
136	CJJ/T 168-2011	镇（乡）村绿地分类标准 Standard for green land classification of town and village		J1327-2011	湖南城市学院
137	CJJ 169-2012	城镇道路路面设计规范 Code for pavement design of urban road		J1340-2012	上海市政工程设计研究总院（集团）有限公司
138	CJJ/T 170-2011	地铁与轻轨系统运营管理规范 Code for operation management of Metro and LRT Systems		J1235-2011	住房和城乡建设部科技发展促进中心 广州市地下铁道总公司
139	CJJ/T 171-2012	风景园林标志标准 Standard for marks of landscape architecture		J1392-2012	天津师范大学
140	CJJ/T 172-2011	生活垃圾堆肥厂评价标准 Standard for assessment on municipal solid waste compost plant		J1337-2011	城市建设研究院
141	CJJ/T 173-2012	风景名胜区游览解说系统标准 Standard for interpretation system in scenic and historic areas		J1349-2012	城市建设研究院
142	CJJ/T 174-2013	城市水域保洁作业及质量标准 Standard for work and quality of city water area cleaning		J1649-2013	上海环境实业有限公司
143	CJJ 175-2012	生活垃圾卫生填埋气体收集处理及利用工程运行维护技术规程 Technical specification for operation and maintenance of landfill gas collection treatment and utilization projects		J1348-2012	中国科学院武汉岩土力学研究所 杭州市环境集团有限公司
144	CJJ 176-2012	生活垃圾卫生填埋场岩土工程技术规范 Technical code for geotechnical engineering of municipal solid waste sanitary landfill		J1347-2012	浙江大学
145	CJJ/T 177-2012	气泡混合轻质土填筑工程技术规程 Technical specification for foamed mixture lightweight soil filling engineering		J1352-2012	广东冠生土木工程技术有限公司 深圳市市政工程总公司

城镇建设工程

序号	标准编号	标准名称	被代替标准编号	备案号	主编单位
146	CJJ/T 178-2012	公共汽电车行车监控及集中调度系统技术规程 Technical specification for driving surveillance and centralized dispatch system of bus and trolleybus		J1415-2012	中国城市公共交通协会
147	CJJ 179-2012	生活垃圾收集站技术规程 Technical specification for municipal solid waste collecting station		J1430-2012	青岛市环境卫生科研所
148	CJJ/T 180-2012	城市轨道交通工程档案整理标准 Standard for archives arrangement of urban railtransit project		J1465-2012	天津市地下铁道集团有限公司 天津市城市建设档案馆
149	CJJ 181-2012	城镇排水管道检测与评估技术规程 Technical specification for inspection and evaluation of urban sewer		J1441-2012	广州市市政集团有限公司
150	CJJ/T 182-2014	城镇供水与污水处理化验室技术规范 Technical code for the laboratory of urban waterworks and wastewater treatment plant		J1890-2014	天津市供水管理处 中国城镇供水排水协会
151	CJJ 183-2012	城市轨道交通站台屏蔽门系统技术规范 Technical code for platform screen door system of urban railway transit		J1444-2012	广州市地下铁道总公司
152	CJJ 184-2012	餐厨垃圾处理技术规范 Technical code for food waste treatment		J1495-2013	城市建设研究院
153	CJJ/T 185-2012	城镇供热系统节能技术规范 Technical code for energy efficiency of city heating system		J1480-2012	北京市煤气热力工程设计院有限公司
154	CJJ/T 186-2012	城市地理编码技术规范 Technical code for city geocoding		J1481-2012	建设综合勘察研究设计院有限公司
155	CJJ/T 187-2012	建设电子档案元数据标准 Standard for electronic construction recordkeeping metadata		J1479-2012	住房和城乡建设部城建档案工作办公室 珠海市城市建设档案馆
156	CJJ/T 188-2012	透水砖路面技术规程 Technical specification for pavement of water permeable brick		J1475-2012	大连九洲建设集团有限公司 北京城乡建设集团有限责任公司
157	CJJ/T 189-2014	镇（乡）村仓储用地规划规范 Code for warehouse land plan in town and village		J1734-2014	吉林省城乡规划设计研究院
158	CJJ/T 190-2012	透水沥青路面技术规程 Technical specification for permeable asphalt pavement		J1443-2012	长安大学
159	CJJ/T 191-2012	浮置板轨道技术规范 Technical code for floating slab track		J1477-2012	深圳市地铁集团有限公司
160	CJJ/T 192-2012	盾构可切削混凝土配筋技术规程 Technical specification for shield-cuttable concrete reinforcement		J1471-2012	深圳市海川实业股份有限公司 中铁二院工程集团有限责任公司
161	CJJ 193-2012	城市道路路线设计规范 Code for design of urban road alignment		J1470-2012	上海市政工程设计研究总院（集团）有限公司
162	CJJ 194-2013	城市道路路基设计规范 Code for design of urban road subgrades		J1590-2013	同济大学

城镇建设工程

序号	标准编号	标准名称	被代替标准编号	备案号	主编单位
163	CJJ/T 195-2013	风景名胜区监督管理信息系统技术规范 Technical code for supervision and management information system of scenic and historic areas		J1504-2013	北京建设数字科技股份有限公司 中国风景名胜区协会
164	CJJ/T 196-2012	住房保障信息系统技术规范 Technical code for housing security information system		J1476-2012	哈尔滨工业大学 住房和城乡建设部信息中心
165	CJJ/T 197-2012	住房保障基础信息数据标准 Standard for data of housing security's basic information		J1483-2012	江苏省住房和城乡建设厅住宅与房地产业促进中心 住房和城乡建设部信息中心
166	CJJ/T 198-2013	城市轨道交通接触轨供电系统技术规范 Technical code for contact rail power supply system of urban rail transit		J1643-2013	广州市地下铁道总公司
167	CJJ/T 199-2013	城市规划数据标准 Standard for urban planning data		J1655-2013	中国城市规划设计研究院
168	CJJ 200-2014	城市供热管网暗挖工程技术规程 Technical specification for city heat-supplying pipe net constructed by mining method		J1893-2014	北京市热力集团有限责任公司
169	CJJ 201-2013	直线电机轨道交通施工及验收规范 Code for construction and acceptance of linear motor for urban rail transit		J1646-2013	广州市地下铁道总公司
170	CJJ/T 202-2013	城市轨道交通结构安全保护技术规范 Technical code for protection structures of urban rail transit		J1645-2013	广州地铁设计研究院有限公司
171	CJJ 203-2013	城镇供热系统抢修技术规程 Technical specification for emergency repair of district heating system		J1653-2013	北京市热力集团有限责任公司
172	CJJ/T 204-2013	生活垃圾土土工试验技术规程 Technical specification for soil test of landfilled municipal solid waste		J1642-2013	中国科学院武汉岩土力学研究所
173	CJJ 205-2013	生活垃圾收集运输技术规程 Technical specification for collection and transportation of municipal solid waste		J1672-2013	华中科技大学 城市建设研究院
174	CJJ/T 206-2013	城市道路低吸热路面技术规范 Technical code for lower-heat-absorbing pavement of urban road		J1688-2013	长安大学
175	CJJ 207-2013	城镇供水管网运行、维护及安全技术规程 Technical specification for operation, maintenance and safety of urban water supply pipe-networks		J1669-2013	中国城镇供水排水协会
176	CJJ/T 208-2014	磷矿尾矿砂道路基（垫）层施工及质量验收规范 Code for construction and quality acceptance of phosphate tailing ore of road base (pad) layer		J1830-2014	锦宸集团有限公司 江苏省盐阜建设集团有限公司
177	CJJ/T 209-2013	塑料排水检查井应用技术规程 Technical specification for application of plastics manholes and inspection chambers for sewerage		J1703-2013	昆明普尔顿环保科技股份有限公司 云南巨和建设集团有限公司

城镇建设工程

序号	标准编号	标准名称	被代替标准编号	备案号	主编单位
178	CJJ/T 210-2014	城镇排水管道非开挖修复更新工程技术规程 Technical specification for trenchless rehabilitation and renewal of urban sewer pipeline		J1735-2014	中国地质大学（武汉）
179	CJJ/T 211-2014	粪便处理厂评价标准 Standard for assessment of night soil treatment plant		J1841-2014	华中科技大学
180	CJJ/T 212-2015	生活垃圾焚烧厂运行监管标准 Standard for supervision on operation of municipal solid waste incineration plants		J1991-2015	中国城市建设研究院有限公司
181	CJJ/T 213-2016	生活垃圾卫生填埋场运行监管标准 Standard for superrision on operation of municipal solid waste landfill		J2226-2016	上海市环境工程设计科学研究院有限公司 中国城市建设研究院有限公司
182	CJJ/T 214-2016	生活垃圾填埋场防渗土工膜渗漏破损探测技术规程 Technical specification for leak location surveys of geomembrane in municipal solid waste landfill		J1269-2016	武汉市环境卫生科学研究设计院 中国科学院武汉岩土力学研究所
183	CJJ /215-2014	城镇燃气管网泄漏检测技术规程 Technical specification for leak detection of city gas piping system		J1755-2014	北京市燃气集团有限责任公司
184	CJJ/T 216-2014	燃气热泵空调系统工程技术规程 Technical specification for gas engine heat pump air-conditioning system		J1756-2014	北京市公用事业科学研究所
185	CJJ 217-2014	盾构法开仓及气压作业技术规范 Technical code for operation in excavation chamber of shield tunneling machine at atmospheric or compressed air		J1837-2014	广州地铁设计研究院有限公司 中铁隧道股份有限公司
186	CJJ/T 218-2014	城市道路彩色沥青混凝土路面技术规程 Technical specification for colored asphalt concrete pavement of urban road		J1914-2014	河南省公路工程局集团有限公司 安阳建工（集团）有限责任公司
187	CJJ/T 220-2014	城镇供热系统标志标准 Standard for city heating system signs		J1826-2014	唐山市热力总公司
188	CJJ 221-2015	城市地下道路工程设计规范 Code for design of urban underground road engineering		J1994-2015	上海市政工程设计研究总院
189	CJJ/T 222-2015	喷泉水景工程技术规程 Technical specification for artificial fountain and water scenery engineering		J1987-2015	中国建筑金属结构协会 浙江鸿翔建设集团有限公司
190	CJJ/T 223-2014	供热计量系统运行技术规程 Technical specification for heat metering system operation		J1896-2014	哈尔滨工业大学
191	CJJ 224-2014	城镇给水预应力钢筒混凝土管管道工程技术规程 Technical specification for prestressed concrete cylinder pipeline of city water supply engineering		J1922-2014	中国市政工程华北设计研究总院
192	CJJ/T 225-2016	城镇供水行业职业技能标准 Standard for occupational skills of cities and towns water supply industry		J2172-2016	南京市自来水总公司

城镇建设工程

序号	标准编号	标准名称	被代替标准编号	备案号	主编单位
193	CJJ/T 226-2014	城镇供水管网抢修技术规程 Technical specification for rush-repair of water supply network in city and town		J1889-2014	绍兴市水联建设工程有限责任公司 无锡市给排水工程有限责任公司
194	CJJ/T 227-2014	城市照明自动控制系统技术规范 Technical code for automatic control system of urban lighting		J1913-2014	中国城市科学研究会
195	CJJ/T 228-2014	城镇污水处理厂运营质量评价标准 Standard for operation and maintenance quality assessment of municipal wastewater treatment plant		J1966-2015	北京城市排水集团有限责任公司 中国人民大学
196	CJJ/T 229-2015	城镇给水微污染水预处理技术规程 Technical specification for micro-polluted water pretreatment of municipal water supply		J1985-2015	上海市政工程设计研究总院（集团）有限公司
197	CJJ/T 230-2015	排水工程混凝土模块砌体结构技术规程 Technical specification for concrete small hollow block masonry structures of drainage engineering		J1984-2015	北京市市政工程研究院 云南官房建筑集团股份有限公司
198	CJJ 231-2015	生活垃圾焚烧厂检修规程 spealiction for maintenance of municipal solid waste incineration plant		J2086-2015	深圳能源集团股份有限公司 深圳市能源环保有限公司
199	CJJ 232-2016	建筑同层排水工程技术规程 Technical specification for same-floor drainage engineering in buildings		J2203-2016	中国建筑金属结构协会 华东建筑设计研究院有限公司
200	CJJ/T 233-2015	城市桥梁检测与评定技术规范 Technical code for test and evaluation of city bridges		J2091-2015	中国建筑科学研究院 广东省建筑科学研究院
201	CJJ/T 234-2015	国家重点公园评价标准 Standard for national key parks evaluation		J2056-2015	北京市公园管理中心
202	CJJ/T 235-2015	城镇桥梁钢结构防腐蚀涂装工程技术规程 Technical specification for protective coating engineering of bridge steel structure in city and town		J2090-2015	华恒建设集团有限公司 浙江新东阳建设集团有限公司
203	CJJ/T 236-2015	垂直绿化工程技术规程 Technical specification for vertical planting		J2082-2015	城市建设研究院 江苏中兴建设有限公司
204	CJJ/T 237-2016	园林行业职业技能标准 Standard for occupational skills of garden and park industry		J2173-2016	江苏省常州建设高等职业技术学校
205	CJJ/T 238-2016	抗车辙沥青混合料应用技术规程 Technical specification for oppliction of anti-rutting asphalt mixture		J2273-2016	济南城建集团有限公司 浙江欣捷建设有限公司
206	CJJ/T 239-2016	城市桥梁结构加固技术规程 Technical specifications for structural strengthening of urban bridges		J2286-2016	广州市市政集团有限公司 杭州市市政工程集团有限公司
207	CJJ/T 240-2015	动物园术语标准 Standard for terminology of zoo		J2106-2015	北京动物园
208	CJJ/T 241-2016	城镇供热监测与调控系统技术规程 Technical specification for monitoring and controlling system of urban heating		J2285-2016	北京市热力集团有限责任公司 北京市热力工程设计有限责任公司

城镇建设工程

序号	标准编号	标准名称	被代替标准编号	备案号	主编单位
209	CJJ 242-2016	城市道路与轨道交通合建桥梁设计规范 Code for design of bridge combined with urban road and rail transit		J2275-2016	上海市政工程设计研究总院（集团）有限公司
210	CJJ/T 243-2016	城镇污水处理厂臭气处理技术规程 Technical specification for odor control of municipal wastewater treatment plant		J1270-2016	上海市政工程设计研究总院（集团）有限公司
211	CJJ/T 244-2016	城镇给水管道非开挖修复更新工程技术规程 Technical specification for trenchless rehabilitation and renewal of urban water supply pipelines			中国城镇供水排水协会
212	CJJ/T 245-2016	住宅生活排水系统立管排水能力测试标准 Standard for capacity test of vertical pipe of the domestic residential drainage system			中国建筑设计研究院 大元建业集团股份有限公司
213	CJJ/T 246-2016	镇（乡）村给水工程规划规范 Code for planning of town and village water supply engineering		J2204-2016	上海市政工程设计研究总院
214	CJJ/T 247-2016	供热站房噪声与振动控制技术规程 Technical specification for noise and vibration control of heating station		J2269-2016	北京市热力工程设计有限责任公司
215	CJJ 248-2016	城市梁桥拆除工程安全技术规范 Technical code for safety of demolishing and removing of urban beam bridge		J2175-2016	宁波市政工程建设集团股份有限公司 宏润建设集团股份有限公司
216	CJJ/T 249-2016	市政公用设施运行管理人员职业标准 Occupation standard for operation management technician of urban municipal facility		J2214-2016	中国建设教育协会
217	CJJ/T 250-2016	城镇燃气管道穿跨越工程技术规程 Technical specification for crossing and aerial crossing engineering of city gas pipeline		J2205-2016	建设部沈阳煤气热力研究设计院
218	CJJ 252-2016	城镇污水再生利用设施运行、维护及安全技术规程 Technical specificatin for operation、maintenance and safety of municipal wastewater reclamation facilities		J2283-2016	北京城市排水集团有限责任公司 天津中水有限公司
219	CJJ/T 253-2016	再生骨料透水混凝土应用技术规程 Technical specification for application of pervious recycled aggregate concrete		J2200-2016	湖南顺天建设集团有限公司 安徽三建工程有限公司
220	CJJ/T 254-2016	城镇供热直埋热水管道泄漏监测系统技术规程 Technical specification for leakage surveillance system of airectly buried heating pipe		J2270-2016	北京豪特耐管道设备有限公司
221	CJJ/T 256-2016	中低速磁浮交通供电技术规范 Technical code for Power supply of medium and low speed maglev transportation		J2284-2016	北京控股磁悬浮技术发展有限公司 铁道第三勘察设计院集团有限公司
222	CJJ/T 259-2016	城镇燃气自动化系统技术规范 Technical code for automatic system of city gas		J2287-2016	中国城市燃气协会
223	CJJ/T 260-2016	道路深层病害非开挖处治技术规程 Technical specification for trenchless treatment of road deep diseases		J2318-2016	郑州安源工程技术有限公司 中国建筑第七工程局有限公司

电力工程

序号	标准编号	标准名称	被代替标准编号	备案号	主编单位
1	DL/T 866-2015	电流互感器和电压互感器选择及计算规程 Code for selection and calculation of current transformer and voltage transformer		J2030-2015	中国电力工程顾问集团华北电力设计院工程有限公司
2	DL/T 5001-2014	火力发电厂工程测量技术规程 Technical code for engineering survey of fossil-fired power plant		J1958-2015	中国电力工程顾问集团东北电力设计院
3	DL/T 5006-2007	水电水利工程岩体观测规程 Code for rock mass observations of hydroelectric and water consercancy engineering		J686-2007	中国水电顾问集团成都勘测设计院
4	DL 5009.1-2014	电力建设安全工作规程 第1部分：火力发电 Code of safety operation in power engineering construction Part 1: steam power		J1925-2014	中国电力建设企业协会
5	DL 5009.2-2013	电力建设安全工作规程 第2部分：电力线路 Code of safety operation in power engineering construction		J1677-2013	国家电网公司安全监察质量部 中国电力科学研究院
6	DL 5009.3-2013	电力建设安全工作规程 第3部分：变电站 Code of safety operation in power engineering construction		J1676-2013	国家电网公司安全监察质量部 中国电力科学研究院
7	DL/T 5017-2007	水电水利工程压力钢管制造安装及验收规范 Specifications for manufacture installation and acceptance of steel penstocks in hydroelectric and hydraulic engineering		J687-2007	水电第七工程局 郑州机械设计研究院
8	DL/T 5020-2007	水电工程可行性研究报告编制规程 Code for preparation of hydroelectric project feasibility study report		J688-2007	水电水利规划设计总院
9	DL 5022-2012	火力发电厂土建结构设计技术规程 Technical code for the design of civil structure of fossil fuel power plant		J1355-2012	中国电力工程顾问集团西北电力设计院
10	DL 5027-2015	电力设备典型消防规程 Specification of typical extinguishing and protection for electrical equipment		J2006-2015	国网上海市电力公司 上海电力股份有限公司 华东电力设计院 等
11	DL/T 5028.1-2015	电力工程制图标准 第1部分 一般规则部分 Standard for drawing of electric power engineering Part 1: general regulation		J2057-2015	中国电力工程顾问集团华北电力设计院工程有限公司
12	DL/T 5028.2-2015	电力工程制图标准 第2部分：机械部分 Standard for drawing of electric power engineering Part 2: mechanical		J2058-2015	中国电力工程顾问集团华北电力设计院工程有限公司
13	DL/T 5028.3-2015	电力工程制图标准 第3部分：电气、仪表与控制部分 Standard for drawing of electric power engineering Part 3: electric, instrumentation and control		J2059-2015	中国电力工程顾问集团华北电力设计院工程有限公司
14	DL/T 5028.4-2015	电力工程制图标准 第4部分：土建部分 Standard for drawing of electric power engineering Part 4: civil work		J2060-2015	中国电力工程顾问集团华北电力设计院工程有限公司

电力工程

序号	标准编号	标准名称	被代替标准编号	备案号	主编单位
15	DL/T 5029-2012	火力发电厂建筑装修设计标准		J1463-2012	中国电力工程顾问集团西南电力设计院
16	DL/T 5038-2012	灯泡贯流式水轮发电机组安装工艺规程 Technological guide for installation of bulb turbine generator units		J1391-2012	中国水利水电建设集团公司 中国水利水电闽江工程局
17	DL/T 5040-2006	输电线路对无线电台影响防护设计规程 The design rules of radio stations against effects from power transmission lines		J540-2006	电力规划设计协会
18	DL/T 5041-2012	火力发电厂厂内通信设计技术规定 Technical code for inner communication design of thermal power plants		J1361-2012	中国电力工程顾问集团西南电力设计院
19	DL/T 5044-2014	电力工程直流电源系统设计技术规程 Technical code for design of DC auxiliary power supply system for power projects		J1959-2015	中国电力工程顾问集团华北电力设计院工程有限公司
20	DL/T 5045-2006	梯级水电厂集中控制工程设计规范 Design specification for centralized monitoring and hydropower plants		J547-2006	水电水利规划设计总院
21	DL/T 5046-2006	火力发电厂废水治理设计技术规程 Technical code for the design of waste water treatment of fossil fuel power plants		J541-2006	电力规划设计协会
22	DL/T 5052-2016	火力发电厂辅助及附属建筑物建筑面积标准 Standard for auxiliary and ancillary building's area of fossil-fired power plant		J2165-2016	电力规划设计总院
23	DL 5053-2012	火力发电厂职业安全设计规程 Design code of occupational safety for fossil fuel power plants		J1356-2012	中国电力工程顾问集团东北电力设计院
24	DL/T 5054-2016	火力发电厂汽水管道设计规范 Design code for stream/water piping of fossil fired power plant		J2166-2016	中国电力工程顾问集团东北电力设计院有限公司
25	DL/T 5055-2007	水工混凝土掺用粉煤灰技术规范 Technical specification of fly ash for use in hydraulic concrete		J689-2007	长江科学研究院
26	DL/T 5056-2007	变电站总布置设计技术规程 Technical code of general plan design for substation		J761-2007	西北电力设计院
27	DL/T 5064-2007	水电工程建设征地移民安置规划设计规范 Design specifications for land requisition resident relocation planning of hydroelectric project		J690-2007	水电水利规划设计总院
28	DL 5068-2014	发电厂化学设计规范 Code for designing chemistry of power plant		J1960-2015	中国电力工程顾问集团西北电力设计院
29	DL/T 5070-2012	水轮机金属蜗壳现场制造安装及焊接工艺导则 Technological guide for manufacturing assembling and welding of metal spiral cas		J1390-2012	中国水利水电建设股份有限公司
30	DL/T 5071-2012	混流式水轮机转轮现场制造工艺导则 Technological guide for the field manufacturing of turbine runner		J1389-2012	中国水利水电建设股份有限公司

电力工程

序号	标准编号	标准名称	被代替标准编号	备案号	主编单位
31	DL/T 5072-2007	火力发电厂保温油漆设计规程 Code for designing insulation and painting of fossil fuel power plant		J691-2007	西南电力设计院
32	DL/T 5076-2008	220kV 及以下架空送电线路勘测技术规程 Technical code of exploration and surveying for 220kv and lower level overhead transmission line		J801-2008	中国电力工程顾问集团公司东北电力设计院等
33	DL/T 5083-2010	水电水利工程预应力锚索施工技术规范 Specification of prestressing tendon onstruction for hydropower and water resources project			中国葛洲坝集团股份有限公司
34	DL/T 5084-2012	电力工程水文技术规程 Technical code of hydrology for electrical power projects		J1462-2012	中国电力工程顾问集团华东电力设计院
35	DL/T 5093-2016	电力岩土工程勘测资料整编技术规程 Technical code of data processing for investigation of geotechnical engineering for electric power projects		J2158-2016	中国电力工程顾问集团中南电力设计院有限公司
36	DL/T 5094-2012	火力发电厂建筑设计规程 Code for designing building of fossil fuel power plant		J1359-2012	中国电力工程顾问集团西南电力设计院
37	DL/T 5095-2013	火电厂和核电厂常规岛主厂房荷载设计技术规程 Technical code for design load of mainbuilding in fossil-fired power plant and the conventional island		J1709-2014	中国电力工程顾问集团西南电力设计院
38	DL/T 5096-2008	电力工程钻探技术规程 Drilling technical regulations of electrical power engineering		J802-2008	中国电力工程顾问集团公司东北电力设计院
39	DL/T 5097-2014	火力发电厂贮灰场岩土工程勘测技术规程 Technical code for investigation of geotechnical engineering of ash yard of fossil fuel power plant		J1956-2015	中国电力工程顾问集团西南电力设计院
40	DL/T 5100-2014	水工混凝土外加剂技术规程 Technical specification for chemical admixtures for hydraulic concrete		J1754-2014	南京水利科学研究院
41	DL/T 5102-2013	土工离心模型试验技术规程 Specification for geotechnical centrifuge model test techniques		J1700-2013	中国水利水电科学研究院
42	DL/T 5103-2012	35kV~220kV 无人值班变电站设计技术规程 Technical code for design of 35kV~220kV unattended substation		J1362-2012	江苏省电力设计院
43	DL/T 5104-2016	电力工程工程地质测绘技术规程 Technical code for engineering geological survey and mapping of electric power engineering		J2159-2016	中国电力工程顾问集团西南电力设计院有限公司
44	DL/T 5111-2012	水电水利工程施工监理规范 Specific ations for construction supervision of hydroelectric and water resource		J1454-2012	中国建设监理协会水电建设监理分会

电力工程

序号	标准编号	标准名称	被代替标准编号	备案号	主编单位
45	DL/T 5113.3-2012	水电水利基本建设工程 单元工程质量等级评定标准 第3部分：水轮发电机组安装工程 Quality degree evaluate standard of unit engineering for hydropower and water conservancy construction engineering Part 3: turbine and generator units installation engineering		J1388-2012	中国水利水电建设集团公司 中国水利水电第四工程局
46	DL/T 5113.4-2012	水电水利基本建设工程 单元工程质量等级评定标准 第4部分：水力机械辅助设备安装工程 Quality degree evaluate standard of unit engineering for hydropower and water conservancy construction engineering Part 4: auxiliary equipment of hydraulic machinery installation engineering		J1387-2012	中国葛洲坝集团股份有限公司
47	DL/T 5113.5-2012	水电水利基本建设工程 单元工程质量等级评定标准 第5部分：发电电气设备安装工程 Quality degree evaluate standard of unit engineering for hydropower and water conservancy construction engineering Part 5: electric generation equipment installation engineering		J1386-2012	中国葛洲坝集团股份有限公司
48	DL/T 5113.6-2012	水电水利基本建设工程 单元工程质量等级评定标准 第6部分：升压变电电气设备安装工程 Quality degree evaluate standard of unit engineering for hydropower and water conservancy construction engineering Part 6: step-up substation electrical equipment installation engineering		J1385-2012	中国水利水电第八工程局有限公司
49	DL/T 5113.7-2015	水电水利基本建设工程单元工程质量等级评定标准 第7部分：碾压式土石坝工程 Quality grade evaluation standard of unit engineering for hydropower and water conservancy construction engineering Part 7: rolled earth-rock dam		J2003-2015	中国水利水电第七工程局有限公司 中国水利水电第五工程局有限公司
50	DL/T 5113.10-2012	水电水利基本建设工程 单元工程质量等级评定标准 第10部分：沥青混凝土工程 Quality degree evaluate standard of unit engineering forhydropower and water conservancy construction engineering Part 10: bituminous concrete engineering		J1384-2012	中国葛洲坝集团股份有限公司
51	DL/T 5115-2008	混凝土面板堆石坝接缝止水技术规范 Technical specifications for joint seal of concrete face rockfill dam		J803-2008	中国水电顾问集团公司华东勘测设计院
52	DL/T 5129-2013	碾压式土石坝施工规范 Specifications for rolled earth-rock fill dam construction		J1699-2013	中国水利水电第五工程局有限公司 中国水利水电第十五工程局有限公司
53	DL/T 5131-2015	农网建设与改造技术导则 The rural electric network construct and reconstruct technical guide		J2004-2015	中国电力科学研究院 国家电网公司农电工作部 国网河北省电力公司

电力工程

序号	标准编号	标准名称	被代替标准编号	备案号	主编单位
54	DL/T 5135-2013	水电水利工程爆破施工技术规范 Specifications of excavation blasting for hydropower and water resources projects		J1698-2013	中国葛洲坝集团股份有限公司
55	DL/T 5136-2012	火力发电厂、变电站二次接线设计技术规程 Technical code for designing of electrical secondary wiring in fossil fuel power plants and substations		J142-2012	中国电力工程顾问集团西北电力设计院
56	DL/T 5138-2014	电力工程数字摄影测量规程 Code for digital photogrammetry in electric power engineering		J1951-2015	中国电力工程顾问集团中南电力设计院
57	DL/T 5142-2012	火力发电厂除灰设计技术规程 Technical code for the design of ash handling system of fossil-fired power plant		J1461-2012	中国电力工程顾问集团西北电力设计院
58	DL/T 5144-2015	水工混凝土施工规范 Specifications for hydraulic concrete construction		J148-2015	中国长江三峡集团公司
59	DL/T 5145-2012	火力发电厂制粉系统设计计算技术规定		J1453-2012	西安热工研究院有限公司
60	DL/T 5148-2012	水工建筑物水泥灌浆施工技术规范		J1383-2012	中国水电基础局有限公司
61	DL/T 5151-2014	水工混凝土砂石骨料试验规程 Specification for testing aggregates of hydraulic concrete		J1753-2014	南京水利科学研究院
62	DL/T 5153-2014	火力发电厂厂用电设计技术规程 Technical code for the design of auxiliary power system of fossil-fired power plant		J1954-2015	中国电力工程顾问集团华东电力设计院
63	DL/T 5154-2012	架空输电线路杆塔结构设计技术规定 Technical code for the design of tower and pole structures of overhead transmiss		J172-2012	中国电力工程顾问集团西南电力设计院、中国电力工程顾问集团公司
64	DL/T 5156.1-2015	电力工程勘测制图标准 第1部分：测量 Standard of electric power engineering surveying drawing Part 1：surveying		J2016-2015	中国电力工程顾问集团华北电力设计院工程有限公司
65	DL/T 5156.2-2015	电力工程勘测制图标准 第2部分：岩土工程 Standard of electric power engineering surveying drawing Part 2：geotechnical engineering		J2017-2015	中国电力工程顾问集团华北电力设计院工程有限公司
66	DL/T 5156.3-2015	电力工程勘测制图标准 第3部分：水文气象 Standard of electric power engineering surveying drawing Part 3：hydrometeorology		J2018-2015	中国电力工程顾问集团华北电力设计院工程有限公司
67	DL/T 5156.4-2015	电力工程勘测制图标准 第4部分：水文地质 Standard of electric power engineering surveying drawing Part 4：hydrogeology		J2019-2015	中国电力工程顾问集团华北电力设计院工程有限公司
68	DL/T 5156.5-2015	电力工程勘测制图标准 第5部分：物探 Standard of electric power engineering surveying drawing Part 5：geophysical		J2020-2015	中国电力工程顾问集团华北电力设计院工程有限公司
69	DL/T 5157-2012	电力系统调度通信交换网设计技术规程		J179-2012	中国电力工程顾问集团华北电力设计院工程有限公司

电力工程

序号	标准编号	标准名称	被代替标准编号	备案号	主编单位
70	DL/T 5158-2012	电力工程气象勘测技术规程 Technical code of meteorological surveying for electrical power projects		J1360-2012	中国电力工程顾问集团西南电力设计院
71	DL/T 5159-2012	电力工程物探技术规程 Technical code for geophysical exploration of electric power engineering		J1460-2012	中国电力工程顾问集团西北电力设计院
72	DL/T 5160-2015	电力工程岩土描述技术规程 Technical code of rock and soil description for geotechnical investigation of electric power engineering		J2021-2015	中国电力工程顾问集团西北电力设计院有限公司
73	DL/T 5161.1-2002	电气装置安装工程 质量检验及评定规程 第1部分：通则 Specification for construction quality checkout and evaluation of electric equipment installation Part 1: general rules		J189-2002	国家电力公司电力建设研究所
74	DL/T 5161.2-2002	电气装置安装工程 质量检验及评定规程 第2部分：高压电器施工质量检验 Specification for construction quality checkout and evaluation of electric equipment installation Part 2: high-voltage electric power equipments		J190-2002	国家电力公司电力建设研究所
75	DL/T 5161.3-2002	电气装置安装工程 质量检验及评定规程 第3部分：电力变压器、油浸电抗器、互感器施工质量检验 Specification for construction quality checkout and evaluation of electric equipment installation Part 3: power transformers, oil-immersed type reactors and instrument transformers		J191-2002	国家电力公司电力建设研究所
76	DL/T 5161.4-2002	电气装置安装工程 质量检验及评定规程 第4部分：母线装置施工质量检验		J192-2002	国家电力公司电力建设研究所
77	DL/T 5161.5-2002	电气装置安装工程 质量检验及评定规程 第5部分：电缆线路施工质量检验 Specification for construction quality checkout and evaluation of electric equipment installation Part 5: power cable lines		J193-2002	国家电力公司电力建设研究所
78	DL/T 5161.6-2002	电气装置安装工程 质量检验及评定规程 第6部分：接地装置施工质量检验 Specification for construction quality checkout and evaluation of electric equipment installation Part 6: grounding devices		J194-2002	国家电力公司电力建设研究所
79	DL/T 5161.7-2002	电气装置安装工程 质量检验及评定规程 第7部分：旋转电机施工质量检验 Specification for construction quality checkout and evaluation of electric equipment installation Part 7: rotating electrical machines		J195-2002	国家电力公司电力建设研究所
80	DL/T 5161.8-2002	电气装置安装工程 质量检验及评定规程 第8部分：盘、柜及二次回路结线施工质量检验 Specification for construction quality checkout and evaluation of electric equipment installation Part 8: panels, cabinets and secondary circuit wirings		J196-2002	国家电力公司电力建设研究所

电力工程

序号	标准编号	标准名称	被代替标准编号	备案号	主编单位
81	DL/T 5161.9-2002	电气装置安装工程 质量检验及评定规程 第9部分：蓄电池施工质量检验 Specification for construction quality checkout and evaluation of electric equipment installation Part 9：batteries		J197-2002	国家电力公司电力建设研究所
82	DL/T 5161.10-2002	电气装置安装工程 质量检验及评定规程 第10部分：35kV及以下架空电力线路施工质量检验 Specification for construction quality checkout and evaluation of electric equipment installation Part 10：35kV and below overhead power line installation		J198-2002	国家电力公司电力建设研究所
83	DL/T 5161.11-2002	电气装置安装工程 质量检验及评定规程 第11部分：电梯电气装置施工质量检验 Specification for construction quality checkout and evaluation of electric equipment installation Part 11：elevator		J199-2002	国家电力公司电力建设研究所
84	DL/T 5161.12-2002	电气装置安装工程 质量检验及评定规程 第12部分：低压电器施工质量检验 Specification for construction quality checkout and evaluation of electric equipment installation Part 12：low-voltage apparatus		J200-2002	国家电力公司电力建设研究所
85	DL/T 5161.13-2002	电气装置安装工程 质量检验及评定规程 第13部分：电力交流设备施工质量检验 Specification for construction quality checkout and evaluation of electric equipment installation Part 13：power converter equipments		J201-2002	国家电力公司电力建设研究所
86	DL/T 5161.14-2002	电气装置安装工程 质量检验及评定规程 第14部分：起重机电气装置施工质量检验 Specification for construction quality checkout and evaluation of electric equipment installation Part 14：crane		J202-2002	国家电力公司电力建设研究所
87	DL/T 5161.15-2002	电气装置安装工程 质量检验及评定规程 第15部分：爆炸及火灾危险环境电气装置施工质量检验 Specification for construction quality checkout and evaluation of electric equipment installation Part 15：electric equipments in explosive or inflammable hazardous environment		J203-2002	国家电力公司电力建设研究所
88	DL/T 5161.16-2002	电气装置安装工程 质量检验及评定规程 第16部分：1kV及以下配线工程施工质量检验 Specification for construction quality checkout and evaluation of electric equipment installation Part 16：1kV and under feeder cable engineering		J204-2002	国家电力公司电力建设研究所
89	DL/T 5161.17-2002	电气装置安装工程 质量检验及评定规程 第17部分：电气照明装置施工质量检验 Specification for construction quality checkout and evaluation of electric equipment installation Part 17：electric lighting devices		J205-2002	国家电力公司电力建设研究所

电力工程

序号	标准编号	标准名称	被代替标准编号	备案号	主编单位
90	DL 5162-2013	水电水利工程施工安全防护设施技术规范 Technical code of safety and protection facility for hydropower and water conservancy construction		J1675-2013	中国水利水电建设股份有限公司
91	DL/T 5168-2016	110kV～750kV架空输电线路施工质量检验及评定规程 Code for construction quality inspection and evaluation of 110kV～750kV overhead transmission line		J2143-2016	国家电网公司交流建设分公司 中国电力科学研究院等
92	DL/T 5170-2015	变电站岩土工程勘测技术规程 Technical code for investigation of geotechnical engineering of substation		J2022-2015	中国电力工程顾问集团华北电力设计院工程有限公司
93	DL/T 5173-2012	水电水利工程施工测量规范 Technical specification of construction surveyin hydroelectric		J1382-2012	中国葛洲坝集团股份有限公司
94	DL/T 5178-2016	混凝土坝安全监测技术规范 Technical specification for concrete dam safety monitoring		J2144-2016	国家能源局大坝安全监察中心
95	DL/T 5187.3-2012	火力发电厂运煤设计技术规程 第3部分：运煤自动化		J1363-2012	中国电力工程顾问集团华北电力设计院工程有限公司
96	DL 5190.2-2012	电力建设施工技术规范 第2部分：锅炉机组 The code for erection of electric power construction Part 2: boiler unit		J1377-2012	广东火电工程总公司
97	DL5190.3-2012	电力建设施工技术规范 第3部分：汽轮发电机组 Technical specification for thermal power erection and construction Part 3: steam turbine generator unit		J1376-2012	江苏省电力建设第三工程公司
98	DL5190.4-2012	电力建设施工技术规范 第4部分：热工仪表及控制装置 Technical specification for thermal power erection and construction Part 4: instrumentation and control		J1375-2012	东北电业管理局第一工程公司
99	DL5190.5-2012	电力建设施工技术规范 第5部分：管道及系统 Technical specification for thermal power erection and construction Part 5: piping and system		J1374-2012	西北电力建设第一工程公司
100	DL5190.6-2012	电力建设施工技术规范 第6部分：水处理及制氢设备和系统 Technical specification for thermal power erection and construction Part 6: water treatment and hydrogen generation equipment and system		J1373-2012	江苏省电力建设第一工程公司
101	DL5190.8-2012	电力建设施工技术规范 第8部分：加工配制 Technical specification for thermal power erection and construction Part 8: processing and prefabrication		J1372-2012	安徽电力建设第一工程公司

电力工程

序号	标准编号	标准名称	被代替标准编号	备案号	主编单位
102	DL/T 5196-2016	火力发电厂石灰石-石膏湿法烟气脱硫系统设计规程 Code for design of limestone/gypsum wet flue gas desulfurization system of fossil fuel power plant		J2160-2016	中国电力工程顾问集团西南电力设计院有限公司
103	DL/T 5204-2016	发电厂油气管道设计规程 Code for oil/gas piping design of power plant		J2161-2016	中国电力工程顾问集团西南电力设计院
104	DL/T 5207-2005	水工建筑物抗冲磨防空蚀技术规范 Technical specification for abrasion and cavitation resistance of concrete in hydraulic structures		J417-2005	南京水利科学研究院
105	DL/T 5210.1-2012	电力建设施工质量验收及评价规程 第1部分：土建工程 Specification for construction quality acceptance and evaluation of electric power construct Part 1：civil construction engineering		J1417-2012	上海电力建筑工程公司 浙江省电力建设质量监督中心站
106	DL/T 5210.2-2009	电力建设施工质量验收及评价规程 第2部分：锅炉机组 Code for construction quality acceptance and evaluation of electric power construct Part 2：boiler unit		J901-2009	天津电力建设公司 山东电力建设第二工程公司
107	DL/T 5210.3-2009	电力建设施工质量验收及评价规程 第3部分：汽轮发电机组 Code for construction quality acceptance and evaluation of electric power construct Part 3：turbine generator unit		J902-2009	浙江省火电建设公司 江苏省电力建设第三工程公司
108	DL/T 5210.4-2009	电力建设施工质量验收及评价规程 第4部分：热工仪表及控制装置 Code for construction quality acceptance and evaluation of electric power construct Part 4：instrumentation and controlcontrol		J903-2009	广东火电工程总公司 江苏省电力建设第一工程公司
109	DL/T 5210.5-2009	电力建设施工质量验收及评价规程 第5部分：管道及系统 Code for construction quality acceptance and evaluation of electric power construction Part 5：piping & system		J904-009	湖南省火电建设公司
110	DL/T 5210.6-2009	电力建设施工质量验收及评价规程 第6部分：水处理剂制 氢设备和系统 Code for construction quality acceptance and evaluation of electric power construct Part 6：the equipment and the system with the water treatment and hydrogen generation		J905-2009	河北省电力建设第一工程公司
111	DL/T 5210.7-2010	电力建设施工质量验收及评定规程 第7部分：焊接 Code for construction quality acceptance and evaluation of electric power construct Part 7：welding		J1066-2010	国网北京电力建设研究院

电力工程

序号	标准编号	标准名称	被代替标准编号	备案号	主编单位
112	DL/T 5210.8-2009	电力建设施工质量验收及评价规程 第8部分：加工配制 Code for construction quality acceptance and evaluation of electric power construct Part 8: processing preparation		J906-2009	四川电力建设三公司
113	DL/T 5217-2013	220kV～500kV 紧凑型架空输电线路设计技术规程 Technical code for design of 220kV～500kV compact overhead transmission line		J1710-2014	华北电力设计院工程有限公司
114	DL/T 5218-2012	220kV～750kV 变电站设计技术规程 Technical code for the design of 220kV～750kV substation		J1459-2012	中国电力工程顾问集团华东电力设计院
115	DL/T 5219-2014	架空输电线路基础设计技术规程 Technical code for design of foundation of overhead transmission line		J1957-2015	中国电力工程顾问集团中南电力设计院
116	DL/T 5224-2014	高压直流输电大地返回系统设计技术规程 Technical code for design of HVDC earth return system		J1853-2014	中国电力工程顾问集团中南电力设计院
117	DL/T 5226-2013	发电厂电力网络计算机监控系统设计技术规程 Technical code for the design of networkcomputerized monitoring and control system inpower plant		J1711-2014	中国电力工程顾问集团中南电力设计院
118	DL/T 5229-2016	电力工程竣工图文件编制规定 Stipulations for drafting as-built-drawing documents of electric power engineering		J2162-2016	中国电力工程顾问集团华东电力设计院有限公司
119	DL/T 5231-2010	±800kV 及以下直流输电接地极施工及验收规程 Specification for construction and acceptance of earthing pole in ±800kV & under DC transmission		J1065-2010	国网直流工程建设有限公司 南网超高压公司
120	DL/T 5232-2010	±800kV 及以下直流换流站电气装置安装工程施工及验收规程 Specification for construction and acceptance of electric equipments in ±800kV & under converter station		J1064-2010	国网直流工程建设有限公司 南网超高压公司
121	DL/T 5233-2010	±800kV 及以下直流换流站电气装置施工质量检验及评定规程 Specification for construction quality checkout and acceptance of electric equipments in ±800kV & under DC converter station		J1063-2010	国网直流工程建设有限公司 南网超高压公司
122	DL/T 5234-2010	±800kV 及以下直流输电工程启动及竣工验收规程 Specification for starting and acceptance of ±800kV & under DC transmission project		J1062-2010	国网直流工程建设有限公司
123	DL/T 5235-2010	±800kV 及以下直流架空输电线路工程施工及验收规程 Code for construction and acceptance of ±800kV and under DC overhead transmission line		J1061-2010	国网直流工程建设有限公司

电力工程

序号	标准编号	标准名称	被代替标准编号	备案号	主编单位
124	DL/T 5236-2010	±800kV 及以下直流架空输电线路工程施工质量检验及评定规程 Specification for construction quality checkout and evaluation of ±800kV and under DC overhead transmission line		J1060-2010	国网直流工程建设有限公司
125	DL/T 5237-2010	灌浆记录仪技术导则 Technical guide for grouting recorder		J1059-2010	中国水电基础局有限公司
126	DL/T 5238-2010	土坝灌浆技术规范 Technical specification of grouting for soil dam		J1058-2010	中国水电基础局有限公司
127	DL/T 5240-2010	火力发电厂燃烧系统设计计算技术规程 Technical code for design and calculation		J1057-2010	西北电力设计院
128	DL/T 5241-2010	水工混凝土耐久性技术规范 Technical specifications for durability of hydraulic concrete		J1056-2010	南京水利科学研究院
129	DL/T 5242-2010	35kV～220kV 变电站无功补偿装置设计技术规定 Echnical rules for designing of reactive power compensation equipment in 35kV～220kV substations		J1055-2010	上海电力设计院有限公司 济南迪生电子电气有限公司
130	DL/T 5243-2010	水电水利工程场内施工道路技术规范 Technical specification for construction road in site of hydropower and water resources project		J1054-2010	中国水利水电第三工程局有限公司
131	DL/T 5244-2010	水电水利工程常规水工模型试验规程 Code for normal hydraulics model investigation for hydropower & water resources		J1053-2010	水资源与水电工程科学国家重点实验室（武汉大学） 长江水利委员会长江科学院
132	DL/T 5245-2010	水电水利工程掺气减蚀模型试验规 Code for hydraulic investigation on aeration-cavitation resistance for hydropower & water resources		J1052-2010	长江水利委员会长江科学院 中国水电顾问集团中南勘测设计研究院
133	DL/T 5246-2010	水电水利工程滑坡涌浪模拟技术规程 Code for landslide-generated waves simulation for hydropower & water resources		J1051-2010	长江水利委员会长江科学院 资源与水工程科学国家重点实验室（武汉大学）
134	DL/T 5247-2010	水电站有压输水系统水工模型试验规程 Code for hydraulic model investigation on pressure water-delivery system for hydropower & water resources		J1050-2010	长江水利委员会长江科学院 资源与水工程科学国家重点实验室（武汉大学）
135	DL/T 5248-2010	履带起重机安全操作规程 Safety operation code for crawler crane		J1049-2010	中国水利水电第三工程局有限公司
136	DL/T 5249-2010	门座起重机安全操作规程 Safety operation code for portal crane		J1048-2010	中国水利水电第三工程局有限公司 中国水电建设集团十五工程局有限公司
137	DL/T 5250-2010	汽车起重机安全操作规程 Safety operation code for truck crane		J1047-2010	中国水利水电第三工程局有限公司 中国水利水电第二工程局公司
138	DL/T 5251-2010	水工混凝土建筑物缺陷检测和评估技术规程 Technical code for detection and evaluation		J1046-2010	北京中水科海利工程技术公司

电力工程

序号	标准编号	标准名称	被代替标准编号	备案号	主编单位
139	DL/T 5252-2010	火力发电厂环境影响评价气象测试技术规定 Meteorological testing specification for environment impact assessment of thermal powerplant		J1045-2010	西北电力设计院 中南电力设计院 新疆电力设计院
140	DL/T 5257-2010	火电厂烟气脱硝工程施工验收技术规程 Specification for construction acceptance of thermal power plant flue gas denitration project		J1177-2011	中电投远达环保工程有限公司
141	DL/T 5262-2010	水电水利工程施工机械安全操作规程 推土机 Operation code of safety for construction equipment of hydroelectric and hydraulic engineering bulldozer		J1169-2011	中国水利水电第四工程局有限公司
142	DL/T 5263-2010	水电水利工程施工机械安全操作规程 装载机 Operation code of safety for construction equipment of hydroelectric and hydraulic engineering loader		J1168-2011	中国水利水电第四工程局有限公司
143	DL/T 5267-2012	水电水利工程覆盖层灌浆技术规范 Specification of overburden grouting for hydropower and water resources projects		J1381-2012	中国水电基础局有限公司
144	DL/T 5268-2012	混凝土面板堆石坝翻模固坡施工技术规程 Construction technique code on technology of turning-over formwork and fixing slope for CFRD		J1380-2012	中国水利水电第一工程局有限公司
145	DL/T 5269-2012	水电水利工程砾石土心墙堆石坝施工规范 Specification on rockfill dam built with gravel and soil core wall in hydroelectric and hydraulic engineering		J1379-2012	中国水利水电第七工程局有限公司
146	DL5270-2012	核子法密度及含水量测试规程 Test code for density and moisture by nuclear methods		J1378-2012	中国葛洲坝集团股份有限公司
147	DL/T 5280-2012	水电水利工程施工机械安全操作规程 凿岩台车 Safety operation code for construction equipment of hydropower and water conservancy engineering rock drilling jumbo		J1452-2012	中国水利水电第六工程局有限公司
148	DL/T 5281-2012	水电水利工程施工机械安全操作规程 平地机 Operation code of safety for construction epuipment of hydropower and water resources grader		J1451-2012	中国水利水电第二工程局有限公司
149	DL/T 5282-2012	水电水利工程施工机械安全操作规程 塔式起重机 Operation code of safety for construction epuipment of hydropower and water resources tower crane		J1450-2012	中国水利水电第二工程局有限公司
150	DL/T 5283-2012	水电水利工程施工机械安全操作规程 混凝土泵车 Operation code of safety for construction equipment of hydropower and water resources concrete pump truck-mounted		J1449-2012	中国水利水电第二工程局有限公司

电力工程

序号	标准编号	标准名称	被代替标准编号	备案号	主编单位
151	DL/T 5284-2012	碳纤维复合芯铝绞线施工工艺及验收导则 Guide for construction and acceptance of carbon fibre complex core aluminium stranded wire		J1448-2012	中国电力科学研究院
152	DL/T 5293-2013	电气装置安装工程 电气设备交接试验报告统一格式 Standardized report form of hand-over test for electric equipment		J1697-2013	中国电力科学研究院
153	DL/T 5294-2013	火力发电建设工程机组调试技术规范 The unit commissioning technical code for fossil power construction project		J1696-2013	华北电力科学研究院有限责任公司
154	DL/T 5295-2013	火力发电建设工程机组调试质量验收及评价规程 Code for the unit commissioning quality acceptance and evaliation of fossil power construction projects		J1695-2013	浙江省电力公司电力科学研究院
155	DL/T 5296-2013	水工混凝土掺用氧化镁技术规范 Technical specification of magnesium oxide expansive for use		J1694-2013	中国长江三峡集团公司
156	DL/T 5297-2013	混凝土面板堆石坝挤压边墙技术规范 Technology specification for extruded-crub in the concrete-faced rockfill dam		J1693-2013	中国水电建设集团十五工程局有限公司
157	DL/T 5298-2013	水工混凝土抑制碱-骨料反应技术规范 Technical specifications for inhibiting alkali-aggregate reaction		J1692-2013	长江水利委员会长江科学院
158	DL/T 5299-2013	大坝混凝土声波检测技术规程 Technical code for inspection concrete of dam by sonic method		J1691-2013	浙江华东工程安全技术有限公司
159	DL/T 5300-2013	1000kV架空输电线路工程施工质量检验及评定规程 Specification for construction quality checkout and evaluation of 1000kV overhead transmission line		J1690-2013	国家电网公司
160	DL/T 5301-2013	架空输电线路无跨越架不停电跨越架线施工工艺导则 Construction technology guide for tension stringing of overhead transmission line when over-crossing live lines without cross frame		J1689-2013	中国水利水电第四工程局有限公司 中国水利水电第十一工程局有限公司
161	DL/T 5302-2013	水电水利工程施工机械安全操作规程 专用汽车 Safety operation specification of construction machinery for hydroelectric and hydraulic engineering		J1688-2013	中国水利水电第四工程局有限公司 中国水利水电第十一工程局有限公司
162	DL/T 5303-2013	水工塑性混凝土试验规程 Test code for hydraulic plastic concrete		J1687-2013	中国葛洲坝集团股份有限公司
163	DL/T 5304-2013	水工混凝土掺用石灰石粉技术规范 Technical specification of limestone powder		J1686-2013	长江水利委员会长江科学院

电力工程

序号	标准编号	标准名称	被代替标准编号	备案号	主编单位
164	DL/T 5305-2013	水电水利工程施工机械安全操作规程 运输类车辆 Safety operation specification of construction machinery for hydroelectric and hydraulic engineering transport vehicle		J1685-2013	中国水利水电第四工程局有限公司 中国水利水电第十一工程局有限公司
165	DL/T 5306-2013	水电水利工程清水混凝土施工规范 Specifications for fair-faced concrete construction of hydroelectric and hydraulic engineering		J1684-2013	中国水利水电建设股份公司
166	DL/T 5307-2013	水电水利工程施工度汛风险评估规程 Risk assessment code for flood control on construction of hydropower and water resources project		J1683-2013	中国水利水电第三工程局有限公司
167	DL/T 5308-2013	水电水利工程施工安全监测技术规范 Technical specification for construction safety monitoring of hydropower and water resources project		J1682-2013	中国水利水电第三工程局有限公司
168	DL/T 5309-2013	水电水利工程水下混凝土施工规范 Specifications for underwater concrete construction of hydroelectric		J1681-2013	中国水利水电建设股份公司
169	DL/T 5310-2013	沥青混凝土面板堆石坝及库盆施工规范 Specifications for asphalt concrete face rockfill dam and reservoir basin construction		J1680-2013	中国水利水电第五工程局有限公司
170	DL/T 5311-2013	水电水利工程砂石料开采及加工系统运行规范 Specification for aggregate quarrying and processing system of the hydroelectric and water conservancy projects		J1679-2013	中国葛洲坝集团股份有限公司
171	DL/T 5312-2013	1000kV变电站电气装置安装工程施工质量检验及评定规程 Specification for construction quality checkout and evaluation of electric equipment installation for 1000kV substation		J1678-2013	国家电网公司
172	DL/T 5313-2014	水电站大坝运行安全评价导则 Guide for safety assessment of large dams for hydropower station in operation		J1752-2014	国家能源局大坝安全监察中心
173	DL/T 5314-2014	水电水利工程施工安全生产应急能力评估导则 Guide of emergency capability assessment for safety construction in hydropower and water resources projects		J1751-2014	中国水利水电建设股份有限公司 中国水电建设集团十五工程局有限公司 陕西省安全生产科学技术中心 陕西省安全生产科学技术有限公司
174	DL/T 5315-2014	水工混凝土建筑物修补加固技术规程 Technical code for repair and reinforcement of hydraulic concrete structures		J1750-2014	中国水利水电科学研究院 中国葛洲坝集团股份有限公司
175	DL/T 5316-2014	水电水利工程软土地基施工监测技术规范 Technical specification for soft clay foundation construction safety monitoring of hydropower project		J1749-2014	水利部交通运输部国家能源局南京水利科学研究院

电力工程

序号	标准编号	标准名称	被代替标准编号	备案号	主编单位
176	DL/T 5317-2014	水电水利工程聚脲涂层施工技术规程 Technical specification for construction of polyurea coating in hydropower and water resources engineering		J1748-2014	中国水利水电科学研究院
177	DL/T 5318-2014	架空输电线路扩径导线架线施工工艺导则 Construction technology guide for the tension stringing of expanded diameter conductors for overhead transmission Line		J1747-2014	中国电力科学研究院
178	DL 5319-2014	架空输电线路大跨越工程施工及验收规范 Code for construction and acceptance of large crossing overhead transmission line		J1745-2014	国家电网公司交流建设分公司 江苏省送变电公司
179	DL/T 5320-2014	架空输电线路大跨越工程架线施工工艺导则 Construction technology guidance for the tension stringing of large crossing overhead transmission line		J1746-2014	国家电网公司交流建设分公司 安徽送变电工程公司
180	DL/T 5330-2015	水工混凝土配合比设计规程 Code for mix design of hydraulic concrete		J490-2015	长江水利委员会长江科学院
181	DL/T 5340-2015	直流架空输电线路对电信线路危险和干扰影响防护设计技术规程 Technical code for protective designing of DC overhead transmission lines to occur danger and interference effects to telecommunication lines		J522-2015	中国电力工程顾问集团中南电力设计院
182	DL/T 5341-2006	电力建设工程量清单计价规范 变电工程 Code of valuating with bill quantity of electric power construction works transformer substation works		J543-2006	电力建设定额站
183	DL/T 5342-2006	750kV架空送电线路铁塔组立施工工艺导则 Construction technology guidance for the assembling and erection of the steel towers of 750kV overhead transmission line		J544-2006	国电电力建设研究院
184	DL/T 5343-2006	750kV架空送电线路张力架线施工工艺导则 Construction technology guidance for tension stringing of 750kV overhead transmission line		J545-2006	国电电力建设研究院
185	DL/T 5344-2006	电力光纤通信工程验收规范 Acceptance specification for optical fiber telecommunication engineering of electric power		J546-2006	国电通信中心
186	DL/T 5346-2006	混凝土拱坝设计规范 Drawing standard for concrete arch dams		J548-2006	水电水利规划设计总院
187	DL/T 5349-2006	水电水利工程水利机械制图标准 Drawing standard for hydraulic machinery of hydropower and water conservancy project		J551-2006	水电水利规划设计总院
188	DL/T 5350-2006	水电水利工程电气制图标准 Drawing standard of electrical engineering for hydropower and water resources project		J552-2006	水电水利规划设计总院

电力工程

序号	标准编号	标准名称	被代替标准编号	备案号	主编单位
189	DL/T 5351-2006	水电水利工程地质制图标准 Drawing standard of geologic for hydropower and water resources project		J553-2006	水电水利规划设计总院
190	DL/T 5352-2006	高压配电装置设计技术规程 Technical code for designing high voltage electrical switchgear		J554-2006	电力规划设计协会
191	DL/T 5353-2006	水电水利工程边坡设计规范 Drawing standard for slope of hydropower and water conservancy project		J555-2006	水电水利规划设计总院
192	DL/T 5354-2006	水电水利工程钻孔土工试验规程 Code for soil tests in borehole of hydroelectric and water conservancy engineering		J629-2007	中国水电顾问集团公司贵阳勘测设计院
193	DL/T 5355-2006	水电水利工程土工试验规程 Code for soil tests for hydropower and water conservancy engineering		J630-2007	中国水电顾问集团公司成都勘测设计院
194	DL/T 5356-2006	水电水利工程粗粒土试验规程 Code for coarse-graied soid tests for hydropower and water conservancy engineering		J631-2007	中国水电顾问集团公司成都勘测设计院
195	DL/T 5357-2006	水电水利工程岩土化学分析试验规程 Code for chemical analysis tests of rock and soil for hydropower and water conservancy engineeing		J632-2007	中国水电顾问集团公司成都勘测设计院
196	DL/T 5358-2006	水电水利工程金属结构设备防腐蚀技术规程 Technical code for anticorrosion of metal structures in hydroelectric and hydraulic engineering		J633-2007	南京水科院
197	DL/T 5359-2006	水电水利工程水流空化模型试验规程 Code for model test of cavitation for hydropower & hydraulic engineering		J634-2007	长江水利委员会长江科学院 武汉大学水资源与水电工程科学国家重点实验室
198	DL/T 5360-2006	水电水利工程溃坝洪水模拟技术规程 Code for simulation of dam-break flow for hydropower & hydraulic engineering		J635-2007	武汉大学水资源与水电工程科学国家重点实验室 长江水利委员会长江科学院
199	DL/T 5361-2006	水电水利工程施工导截流模型试验规程 Code for model test of construction diversion and river closure for hydropower & hydraulic engineering		J636-2007	武汉大学水资源与水电工程科学国家重点实验室 长江水利委员会长江科学院
200	DL/T 5363-2006	水工碾压式沥青混凝土施工规范 Specifications for construction of hydraulic roller compacter bituminous concrete		J638-2007	中国葛洲坝水利水电工程集团有限公司
201	DL/T 5364-2015	电力调度数据网络工程初步设计内容深度规定 Regulations of content depth for electric power dispatching data network project preliminary design		J639-2015	中国电力工程顾问集团华东电力设计院
202	DL/T 5365-2006	电力数据通信网络工程初步设计内容深度规定 Rules of content depth for electric power data communication network project preliminary design		J640-2007	电力规划设计标委会

电力工程

序号	标准编号	标准名称	被代替标准编号	备案号	主编单位
203	DL/T 5366-2014	发电厂汽水管道应力计算技术规程 Technical code for stress calculating of steam/water piping in power plant		J1852-2014	中国电力工程顾问集团华东电力设计院
204	DL/T 5367-2007	水电水利工程岩体应力测试规程 Code for rock mass stress measurements of hydroelectric and water conservancy engineering		J693-2007	中国水电顾问集团成都勘测设计院
205	DL/T 5368-2007	水电水利工程岩石试验规程 Code for rock tests of hydroelectric and water conservancy engineering		J694-2007	中国水电顾问集团成都勘测设计院
206	DL/T 5369-2016	电力建设工程量清单计价规范 火力发电厂工程 Code of valuation with bill quantity of electric power construction works fossil fuel power plants works		J281-2017	东北电力设计院 西北电力设计院
207	DL/T 5370-2007	水电水利工程施工通用安全技术规程 General technical specification for safety of hydroelectric and hydraulic engineering construction		J696-2007	中国水利水电建设集团公司
208	DL/T 5371-2007	水电水利工程土建施工安全技术规程 Technical code of safety for hydroelectric and hydraulic civil construction engineering		J697-2007	中国水利水电建设集团公司
209	DL/T 5372-2007	水电水利工程金属结构与机电设备安装安全技术规程 Technical specification for safety of installation of metal structure and mechanical & electrical equipment of hydroelectric and hydraulic engineering		J698-2007	中国水利水电建设集团公司
210	DL/T 5373-2007	水电水利工程施工作业人员安全技术操作规程 Technical operation code of safety for workmen of hydroelectric and hydraulic construction engineering		J699-2007	中国水利水电建设集团公司
211	DL/T 5374-2007	火力发电厂初步可行性研究报告内容深度规定 Regulation for content and depth of pre-feasibility study report of fossil fuel power plants		J804-2008	电力规划设计总院等
212	DL/T 5375-2007	火力发电厂可行性研究报告内容深度规定 Regulation for content and depth of feasibility study report of fossil fuel power plants		J805-2008	电力规划设计总院
213	DL/T 5376-2007	水电工程建设征地处理范围界定规范 Specifications for drawing-up of boundary line of land requisition and treatment of hydroelectric project		J700-207	水电水利规划设计总院 中国水电顾问集团成都勘测设计院
214	DL/T 5377-2007	水电工程建设征地实物指标调查规范 Specifications for investigation of land requisition material index of hydroelectric project		J701-2007	水电水利规划设计总院 中国水电顾问集团昆明勘测设计院

电力工程

序号	标准编号	标准名称	被代替标准编号	备案号	主编单位
215	DL/T 5378-2007	水电工程农村移民安置规划设计规范 Design specifications for rural resident relocation planning of hydroelectric project		J702-2007	水电水利规划设计总院 中国水电顾问集团成都勘测设计院
216	DL/T 5379-2007	水电工程移民专业项目规划设计规范 Design specifications for professional works for resident relocation planning of hydroelectric project		J703-2007	水电水利规划设计总院 中国水电顾问集团中南勘测设计院
217	DL/T 5380-2007	水电工程移民安置城镇迁建规划设计规范 Design specifications for resident relocation and township construction planning of hydroelectric project		J704-2007	水电水利规划设计总院 中国水电顾问集团中南勘测设计院
218	DL/T 5381-2007	水电工程水库库底清理设计规范 Design specifications for cleaning of reservoir zone of hydroelectric project		J705-2007	水电水利规划设计总院 中国水电顾问集团西北勘测设计院
219	DL/T 5382-2007	水电工程建设征地移民安置补偿费用概（估）算编制规范 Specifications for preparation of cost estimation (or evaluation) for land requisition and resident relocation of hydroelectric project		J706-2007	水电水利规划设计总院 中国水电顾问集团华东勘测设计院
220	DL/T 5383-2007	风力发电场设计技术规范 Technical specification of wind power plant design		J707-2007	新疆风电工程设计咨询有限公司
221	DL/T 5384-2007	风力发电工程施工组织设计规范 Specification for construction management and design of wind power project		J708-2007	新疆风力发电厂、新疆风电设计研究院
222	DL/T 5385-2007	大坝安全监测系统施工监理规范 Specification for construction supervision of dam safety monitoring system		J709-2007	中国水电顾问集团成都勘测设计院
223	DL/T 5386-2007	水电水利工程混凝土预冷系统设计导则 Design guide for concrete precooling system of hydropower and water conservancy project		J710-2007	中国水电顾问集团中南勘测设计院
224	DL/T 5387-2007	水工混凝土掺用磷渣粉技术规范 Technical specification for phosphorous slag powder use in hydraulic concrete		J689-2007	长江水利委员会长江科学院
225	DL/T 5388-2007	水电水利工程天然建筑材料勘察规程 Code of natural building material investigation for hydropower and water resources project		J712-2007	水电水利规划设计总院
226	DL/T 5389-2007	水工建筑物岩石基础开挖工程施工技术规范 Constructing technical specifications on rock-foundation excavating engineering of hydraulic structures		J713-2007	长江水利委员会长江科学院
227	DL/T 5390-2014	发电厂和变电站照明设计技术规定 Technical code for designing of lighting in power plants and substations		J1953-2015	中国电力工程顾问集团西北电力设计院
228	DL/T 5391-2007	电力系统通信设计技术规定 Design technical code of dispatching commmnication of electric power system		J715-2007	西北电力设计院

电力工程

序号	标准编号	标准名称	被代替标准编号	备案号	主编单位
229	DL/T 5392-2007	电力系统数字同步网工程设计规范 Specifications of engineering design for digital synchronization network of electric power system		J716-2007	中南电力设计院
230	DL/T 5393-2007	高压直流换流站接入系统设计内容深度规定 Design regulations of content and profundity for high voltage direct current converter station connecting to the system		J723-2007	中南电力设计院
231	DL/T 5394-2007	电力工程地下金属构筑物防腐技术导则 Code for anticorrosion of underground steel structure in power project		J724-2007	中国电力工程顾问集团中南电力设计院等
232	DL/T 5395-2007	碾压式土石坝设计规范 Design specification for rolled earth-rock fill dams		J763-2007	水电规划设计总院
233	DL/T 5396-2007	水力发电厂高压电气设备选择及布置设计规范 High voltage electric equipment selection and arrangement design code for Hydro-power station		J764-2007	水电规划设计总院
234	DL/T 5397-2007	水电工程施工组织设计规范 Specification for construction planning of hydropower engineering		J765-2007	水电规划设计总院
235	DL/T 5398-2007	水电站进水口设计规范 Design specification for intake of hydropower station		J766-2007	水电规划设计总院
236	DL/T 5399-2007	水电水利工程垂直升船机设计导则 Design guide for vertical shiplift in hydropower and water resources projects		J767-2007	水电规划设计总院
237	DL/T 5400-2016	水工建筑物滑动模板施工技术规范 Construction technique specifications for hydraulic structure sliding formwork		J2153-2016	中国水利水电第三工程局有限公司
238	DL/T 5401-2007	水力发电厂电气试验设备配置导则 Instrument allocation guide for electrical test of hydropower plant		J769-2007	水电规划设计总院
239	DL/T 5402-2007	水电水利工程环境保护设计规范 Specification for environmental protection design of water conservancy and hydropower project		J770-2007	中国水电顾问集团成都勘测设计院
240	DL/T 5404-2007	电力系统同步数字系列（SDH）光缆通信工程设计技术规定 Technical code of electric power system engineering design for SDH optical fiber cable communication project		J772-2007	电力规划设计标委会
241	DL/T 5405-2008	城市电力电缆线路初步设计内容深度规程 Code for content depth on preliminary design of urban power cables		J806-2008	北京电力设计院等
242	DL/T 5406-2010	水工建筑物化学灌浆施工规范 Specification for chemical grouting construction of hydraulic structures		J1044-2010	中国葛洲坝集团公司 中国水利水电基础工程局

电力工程

序号	标准编号	标准名称	被代替标准编号	备案号	主编单位
243	DL/T 5409.1-2009	核电厂工程勘测技术规程 第1部分：地震地质 Technical code for engineering investigation of nuclear power plants Part 1: seismic hazard		J1043-2010	中国地震局地球物理研究所 广东省电力设计研究院 华东电力设计院
244	DL/T 5409.2-2010	核电厂工程勘测技术规程 第2部分：岩土工程 Technical code for engineering investigation of nuclear power plants Part 2: geotechnical engineering		J1042-2010	广东省电力设计研究院 华东电力设计院
245	DL/T 5409.3-2010	核电厂工程勘测技术规程 第3部分：水文气象 Technical code for engineering investigation of Nuclear power plants Part 3: hydrological and meteorological survey		J1041-2010	广东省电力设计研究院 华东电力设计院
246	DL/T 5409.4-2010	核电厂工程勘测技术规程 第4部分：测量 Technical code for engineering investigation of nuclear power plants Part 4: surveying		J1040-2010	广东省电力设计研究院 华东电力设计院
247	DL/T 5417-2009	火电厂烟气脱硫工程施工质量验收及评定规程 Specification for construction quality inspection and assessment of thermal power plant flue gas desulphurization		J916-2009	中国华电工程（集团）有限公司
248	DL/T 5418-2009	火电厂烟气脱硫吸收塔施工及验收规程 Code for construction and acceptance of flue gas desulphurization absorber of thermal power plant		J917-2009	苏源环保工程股份有限公司 中国华电工程（集团）有限公司
249	DL/T 5426-2009	±800kV高压直流输电系统成套设计规程 Specification for system design of ±800kV UHVDC		J924-2009	国家电网公司直流工程建设有限公司 南方电网技术研究中心
250	DL/T 5430-2009	无人值班变电站远方监控中心设计技术规程 Technical code for designing of remote monitoring and control center about unattended substation		J928-2009	江苏省电力设计院
251	DL/T 5434-2009	电力建设工程监理规范 The code of power construction project management		J932-2009	中国电力建设企业协会
252	DL/T 5437-2009	火力发电建设工程启动试运及验收规程 Code for fossil power construction project from the unit commissioning to completed acceptance		J935-2009	华北电力科学研究院有限责任公司 中国电力建设企业协会
253	DL/T 5446-2012	电力系统调度自动化工程可行性研究报告内容深度规定 Regulation for content and depth of feasibility study report of power system dispatch automation engineering		J1370-2012	中国电力工程顾问集团公司
254	DL/T 5447-2012	电力系统通信系统设计内容深度规定 Code of profundity for communication system design of power system		J1369-2012	中国电力工程顾问集团西南电力设计院

电力工程

序号	标准编号	标准名称	被代替标准编号	备案号	主编单位
255	DL/T 5448-2012	输变电工程可行性研究内容深度规定 Regulation for content of feasibility study of transmission and substation		J1368-2012	电力规划设计总院
256	DL 5449-2012	20kV配电设计技术规定 Technical rule for design of 20kV electrical installation		J1358-2012	中国南方电网有限责任公司
257	DL/T 5450-2012	20kV配电设备选型技术规定 Technical rule for selecting of 20kV distribution electrical equipment		J1367-2012	中国南方电网有限责任公司
258	DL/T 5451-2012	架空输电线路工程初步设计内容深度规定 Code of content profundity for preliminary design for overhead transmission line		J1366-2012	中国电力工程顾问集团公司
259	DL/T 5452-2012	变电工程初步设计内容深度规定 Code of content profundity for preliminary design of substation		J1365-2012	中国电力工程顾问集团公司
260	DL/T 5453-2012	串补站设计技术规程 Technical code for designing series compensator station		J1364-2012	中国电力工程顾问集团中南电力设计院
261	DL 5454-2012	火力发电厂职业卫生设计规程 Design code of occupational health for fossil fuel power plants		J1357-2012	中国电力工程顾问集团东北电力设计院
262	DL/T 5455-2012	火力发电厂热工电源及气源系统设计技术规程 Technical code for the design of power supply andair supply system of instrument		J1458-2012	中国电力工程顾问集团华北电力设计院工程有限公司
263	DL/T 5456-2012	火力发电厂信息系统设计技术规定 Technical code for information system design of fossil fired power plant		J1457-2012	中国电力工程顾问集团公司
264	DL/T 5457-2012	变电站建筑结构设计技术规程 Technical code for the design of substation buildings and structures		J1456-2012	广东省电力设计研究院
265	DL/T 5458-2012	变电工程施工图设计内容深度规定 Code for content profundity of detail designingsubstation project		J1492-2012	中国电力工程顾问集团华东电力设计院
266	DL/T 5459-2012	换流站建筑结构设计技术规程 Technical code of architectural & structural design for converter station		J1491-2012	中国电力工程顾问集团中南电力设计院
267	DL/T 5460-2012	换流站站用电设计技术规定 Technical code for the design of auxiliary power system of converter station		J1490-2012	中国电力工程顾问集团中南电力设计院
268	DL/T 5461.1-2012	火力发电厂施工图设计文件内容深度规定 第1部分：总的部分 Regulations for content and depth of detailed design ducuments of fossil-fired power plant Part 1：General		J1489-2012	中国电力工程顾问集团中南电力设计院
269	DL/T 5461.2-2013	火力发电厂施工图设计文件内容深度规定 第2部分：总图运输 Regulations for content and depth of detailed design ducuments of fossil-fired power plant Part 2：general plan and transportation		J1712-2014	中国电力工程顾问集团中南电力设计院

电力工程

序号	标准编号	标准名称	被代替标准编号	备案号	主编单位
270	DL/T 5461.3-2013	火力发电厂施工图设计文件内容深度规定 第3部分：热机 Regulations for content and depth of detailed design ducuments of fossil-fired power plant Part 3：turbin/boiler and auxiliaries		J1713-2014	中国电力工程顾问集团中南电力设计院
271	DL/T 5461.4-2013	火力发电厂施工图设计文件内容深度规定 第4部分：运煤 Regulations for content and depth of detailed design ducuments of fossil-fired power plant Part 4：coal handling		J1714-2014	中国电力工程顾问集团中南电力设计院
272	DL/T 5461.5-2013	火力发电厂施工图设计文件内容深度规定 第5部分：除灰渣 Regulations for content and depth of detailed design ducuments of fossil-fired power plant Part 5：ash handling		J1715-2014	中国电力工程顾问集团中南电力设计院
273	DL/T 5461.6-2013	火力发电厂施工图设计文件内容深度规定 第6部分：电厂化学 Regulations for content and depth of detailed design ducuments of fossil-fired power plant Part 6：chemical		J1716-2014	中国电力工程顾问集团中南电力设计院
274	DL/T 5461.7-2013	火力发电厂施工图设计文件内容深度规定 第7部分：烟气脱硫 Regulations for content and depth of detailed design ducuments of fossil-fired power plant Part 7：flue gas desulpfhurization		J1717-2014	中国电力工程顾问集团中南电力设计院
275	DL/T 5461.8-2013	火力发电厂施工图设计文件内容深度规定 第8部分：电气 Regulations for content and depth of detailed design ducuments of fossil-fired power plant Part 8：electrical		J1718-2014	中国电力工程顾问集团中南电力设计院
276	DL/T 5461.9-2013	火力发电厂施工图设计文件内容深度规定 第9部分：仪表与控制 Regulations for content and depth of detailed design ducuments of fossil-fired power plant Part 9：instrumentation and control		J1719-2014	中国电力工程顾问集团中南电力设计院
277	DL/T 5461.10-2013	火力发电厂施工图设计文件内容深度规定 第10部分：建筑 Regulations for content and depth of detailed design ducuments of fossil-fired power plant Part 10：architecture		J1720-2014	中国电力工程顾问集团中南电力设计院
278	DL/T 5461.11-2013	火力发电厂施工图设计文件内容深度规定 第11部分：土建结构 Regulations for content and depth of detailed design ducuments of fossil-fired power plant Part 11：Civil structure		J1721-2014	中国电力工程顾问集团中南电力设计院
279	DL/T 5461.12-2013	火力发电厂施工图设计文件内容深度规定 第12部分：采暖通风及空气调节 Regulations for content and depth of detailed design ducuments of fossil-fired power plant Part 12：HVAC		J1722-2014	中国电力工程顾问集团中南电力设计院

电力工程

序号	标准编号	标准名称	被代替标准编号	备案号	主编单位
280	DL/T 5461.13-2013	火力发电厂施工图设计文件内容深度规定 第13部分：水工工艺 Regulations for content and depth of detailed design ducuments of fossil-fired power plant Part 13：hydraulic processes		J1723-2014	中国电力工程顾问集团中南电力设计院
281	DL/T 5461.14-2013	火力发电厂施工图设计文件内容深度规定 第14部分：水工结构 Regulations for content and depth of detailed design ducuments of fossil-fired power plant Part 14：hydraulic structure		J1724-2014	中国电力工程顾问集团中南电力设计院
282	DL/T 5461.15-2013	火力发电厂施工图设计文件内容深度规定 第15部分：通信 Regulations for content and depth of detailed design ducuments of fossil-fired power plant Part 15：communicarion		J1725-2014	中国电力工程顾问集团中南电力设计院
283	DL/T 5461.16-2013	火力发电厂施工图设计文件内容深度规定 第16部分：信息系统 Regulations for content and depth of detailed design ducuments of fossil-fired power plant Part 16：information system		J1726-2014	中国电力工程顾问集团中南电力设计院
284	DL/T 5462-2012	架空输电线路覆冰观测技术规定 Technical regulation of icing measurement for overhead transmission line		J1488-2012	中国电力工程顾问集团西南电力设计院
285	DL/T 5463-2012	110kV～750kV架空输电线路施工图设计内容深度规定 Regulations for content and depth of the working drawing design of 110kV～750kV		J1487-2012	中国电力工程顾问集团西南电力设计院
286	DL/T 5464-2013	火力发电工程初步设计概算编制导则 Guidelines for primary design budgetary estimate of fossil fuel power project		J1617-2013	中国电力工程顾问集团公司电力规划设计总院
287	DL/T 5465-2013	火力发电工程施工图预算编制导则 Guidelines for construction detail design estimate of fossil fuel power project		J1616-2013	中国电力工程顾问集团公司电力规划设计总院
288	DL/T 5466-2013	火力发电工程可行性研究投资估算编制导则 Guidelines for feasibility study cost estimation of fossil fuel power project		J1615-2013	中国电力工程顾问集团公司电力规划设计总院
289	DL/T 5467-2013	输变电工程初步设计概算编制导则 Guidelines for primary design budgetary estimate of transmission and transformer project		J1614-2013	中国电力工程顾问集团公司电力规划设计总院
290	DL/T 5468-2013	输变电工程施工图预算编制导则 Guidelines for construction detail design estimate of transmission and transformer project		J1613-2013	中国电力工程顾问集团公司电力规划设计总院
291	DL/T 5469-2013	输变电工程可行性研究投资估算编制导则 Guidelines for feasibility study cost estimation of transmission and transformer project		J1612-2013	中国电力工程顾问集团公司电力规划设计总院
292	DL/T 5470-2013	燃煤发电工程建设预算项目划分导则 Guidelines for item segregation of construction budget for thermal power generation project		J1611-2013	中国电力工程顾问集团华东电力设计院

电力工程

序号	标准编号	标准名称	被代替标准编号	备案号	主编单位
293	DL/T 5471-2013	变电站、开关站、换流站工程建设预算项目划分导则 Guidelines for item segregation of construction budget for substation, switch station and convertor		J1610-2013	中国电力工程顾问集团西北电力设计院
294	DL/T 5472-2013	架空输电线路工程建设预算项目划分导则 Guidelines for item segregation of construction budget for overhead transmission line projects		J1609-2013	中国电力工程顾问集团中南电力设计院
295	DL/T 5473-2013	燃气-蒸汽联合循环发电工程建设预算项目划分导则 Guidelines for item segregation of construction budget for gas-steam combined cycles power plant project		J1608-2013	中国电力工程顾问集团西北电力设计院
296	DL/T 5474-2013	生物质发电工程建设预算项目划分导则 Guidelines for item segregation of construction budget for biomass power generation project		J1607-2013	中国电力工程顾问集团东北电力设计院
297	DL/T 5475-2013	垃圾发电工程建设预算项目划分导则 Guidelines for item segregation of construction budget for garbage power generation project		J1606-2013	中国电力工程顾问集团中南电力设计院
298	DL/T 5476-2013	电缆输电线路工程建设预算项目划分导则 Guidelines for item segregation of construction budget for cable transmission line project		J1605-2013	国家电网公司
299	DL/T 5477-2013	串联补偿站及静止无功补偿工程建设预算项目划分导则 Guidelines for item segregation of construction budget for series compensator station and static var compensator projects		J1604-2013	中国电力工程顾问集团公司华北电力设计院工程有限公司
300	DL/T 5478-2013	20kV 及以下配电网工程建设预算项目划分导则 Guidelines for item segregation of construction budget for 20kV and under electricity distribution network projects		J1603-2013	中国南方电网有限责任公司
301	DL/T 5479-2013	通信工程建设预算项目划分导则 Guidelines for item segregation of construction budget for communication engineering		J1602-2013	国家电网公司
302	DL/T 5480-2013	火力发电厂烟气脱硝设计技术规程 Technical code for the design of flue gas denitration of fossil fired power plant		J1727-2014	中国电力工程顾问集团华东电力设计院
303	DL/T 5481-2013	电力岩土工程监理规程 Code for project management of geotechnical engineering of electrical power engineering		J1728-2014	中国电力工程顾问集团华北电力设计院工程有限公司
304	DL/T 5482-2013	整体煤气化联合循环技术及设备名词术语 Terms of integrated gasification combined cycle technology and equipment		J1729-2014	中国电力工程顾问集团华北电力设计院工程有限公司
305	DL/T 5483-2013	火力发电厂再生水深度处理设计规范 Code for design of advanced treatment for reclaimed water of fossil-fired power plants		J1730-2014	中国电力工程顾问集团西北电力设计院

电力工程

序号	标准编号	标准名称	被代替标准编号	备案号	主编单位
306	DL/T 5484-2013	电力电缆隧道设计规程 Code for design of power cables tunnel		J1731-2014	北京电力经济技术研究院
307	DL/T 5485-2013	110kV～750kV 架空输电线路大跨越设计技术规程 Technical code for design of long span crossing of 110kV～750kV overhead transmission line		J1732-2014	中国电力工程顾问集团华东电力设计院
308	DL/T 5486-2013	特高压架空输电线路杆塔结构设计技术规程 Technical code for design of tower structures of UHV overhead transmission line		J1733-2014	中国电力工程顾问集团西南电力设计院
309	DL/T 5487-2014	IGCC 发电工程估算编制及项目划分导则 Guidelines for item segregation and estimation of feasibility study for IGCC power plant projects		J1851-2014	中国电力工程顾问集团华北电力设计院工程有限公司
310	DL/T 5488-2014	火力发电厂干式贮灰场设计规程 Code for design of dry ash disposal area of fossil-fired power plant		J1850-2014	中国电力工程顾问集团东北电力设计院
311	DL/T 5489-2014	火力发电厂循环水泵房进水流道设计规范 Technical code for design of inlet flow passage of circulating water pump house of fossil-fired power plant		J1849-2014	中国电力工程顾问集团中南电力设计院
312	DL/T 5490-2014	500kV 交流海底电缆线路设计技术规程 Technical code for design of 500kV AC submarine power cables		J1848-2014	中国电力工程顾问集团中南电力设计院
313	DL/T 5491-2014	电力工程交流不间断电源系统设计技术规程 Technical code for the design of AC uninterruptible power system in power projects		J1950-2015	中国电力工程顾问集团中南电力设计院
314	DL/T 5492-2014	电力工程遥感调查技术规程 Technical code for remote sensing survey of electric power		J1952-2015	中国电力工程顾问集团西北电力设计院
315	DL/T 5493-2014	电力工程基桩检测技术规程 Technical code for testing of electric power engineering foundation piles		J1955-2015	中国电力工程顾问集团华东电力设计院
316	DL/T 5494-2014	电力工程场地地震安全性评价规程 Code for seismic safety evaluation of power engineering		J1949-2015	中国电力工程顾问集团华东电力设计院
317	DL/T 5495-2015	35kV～110kV 户内变电站设计规程 Technical code for design of 35kV～110kV indoor substation		J2023-2015	北京电力经济技术研究院
318	DL/T 5496-2015	220kV～500kV 户内变电站设计规程 Technical code for design of 220kV～500kV indoor substation		J2024-2015	北京电力经济技术研究院
319	DL/T 5497-2015	高压直流架空输电线路设计技术规程 Technical code for design of HVDC overhead transmission line		J2025-2015	中国电力工程顾问集团西北电力设计院有限公司

电力工程

序号	标准编号	标准名称	被代替标准编号	备案号	主编单位
320	DL/T 5498-2015	330kV~500kV 无人值班变电站设计技术规程 Technical code for design of 330kV~500kV unattended substation		J2026-2015	中国电力工程顾问集团中南电力设计院有限公司
321	DL/T 5499-2015	换流站二次系统设计技术规程 Technical code for design of electrical secondary system of converter station		J2027-2015	中国电力工程顾问集团中南电力设计院
322	DL/T 5500-2015	配电自动化系统信息采集及分类技术规范 Technical code for data acquisition and catalog of distribution automation system		J2028-2015	湖南省电力勘测设计院
323	DL/T 5501-2015	冻土地区架空输电线路基础设计技术规程 Technical code for foundation design of overhead transmission line in frozen soil region		J2029-2015	中国电力工程顾问集团西北电力设计院有限公司
324	DL/T 5502-2015	串补站初步设计文件内容深度规定 Regulations for content and depth of preliminary design documents of series compensator station		J2062-2015	中国电力工程顾问集团中南电力设计院
325	DL/T 5503-2015	直流换流站施工图设计内容深度规定 Regulations for content and depth of detail design of HVDC converter station		J2063-2015	中国电力工程顾问集团中南电力设计院
326	DL/T 5504-2015	特高压架空输电线路大跨越设计技术规程 Technical code for design of long span crossing of UHV overhead transmission line		J2064-2015	中国电力工程顾问集团华东电力设计院
327	DL/T 5505-2015	电力应急通信设计技术规程 Technical code for designing electric power emergency communication		J2065-2015	中国能源建设集团广东省电力设计研究院
328	DL/T 5506-2015	电力系统继电保护设计技术规范 Technical code for design of electric power system relay protection		J2066-2015	中国电力工程顾问集团东北电力设计院
329	DL/T 5507-2015	火力发电厂水工设计基础资料及其深度规定 Regulation for basic data and depth of the hydraulic design for fossil-fired power plant		J2061-2015	中国电力工程顾问集团中南电力设计院
330	DL/T 5508-2015	燃气分布式供能站设计规范 Code for design of gas-fired distributed energy station		J2067-2015	上海电力设计院有限公司
331	DL/T 5509-2015	架空输电线路覆冰勘测规程 Code of icing survey for overhead transmission line		J2068-2015	中国电力工程顾问集团西南电力设计院
332	DL/T 5510-2016	智能变电站设计技术规定 Technical code for design of smart substation		J2163-2016	国家电网公司
333	DL/T 5511-2016	直流融冰系统设计技术规程 Technical code for the design of DC de-icing system		J2164-2016	中国电力工程顾问集团中南电力设计院
334	DL/T 5700-2014	城市居住区供配电设施建设规范 Construction code of power supply and distribution establishment for urban residential district		J1938-2014	国网湖北省电力公司

电力工程

序号	标准编号	标准名称	被代替标准编号	备案号	主编单位
335	DL/T 5701-2014	水电水利工程施工机械安全操作规程 反井钻机 Safety operation code for construction equipment of hydropower and water conservancy engineering		J1935-2014	中国水利水电第七工程局有限公司
336	DL/T 5702-2014	水电水利工程沉井施工技术规程 Technology code for construction of open caisson of hydropower and water conservancy engineering		J1963-2015	中国水利水电第七工程局有限公司 陕西建工第一建设集团有限公司
337	DL/T 5703-2014	水电水利工程预应力锚杆用水泥锚固剂技术规程 Technical specification for cement anchoring agent for prestressed anchor of hydropower and water conservancy		J1936-2014	中国水利水电第七工程局有限公司
338	DL/T 5704-2014	火力发电厂热力设备及管道保温防腐施工质量验收规程 Specification for acceptance of construction quality of insulation and anticorrosion of thermal equipment and pipeline		J1937-2014	山东电力建设第一工程公司 重庆电力建设总公司
339	DL/T 5705-2014	循环流化床锅炉砌筑工艺导则 Technical guidance of CFB boiler refractory construction		J1939-2014	河南第二火电建设公司 四川电力建设三公司
340	DL/T 5706-2014	火力发电工程施工组织设计导则 Guide for construction organization design of thermal power engineering		J1940-2014	中国电力建设企业协会
341	DL/T 5707-2014	电力工程电缆防火封堵施工工艺导则 Construction workmanship guide for cable fireproof sealing of electrical engineering		J1941-2014	中国华电工程（集团）有限公司 河南第二火电建设公司
342	DL/T 5708-2014	架空输电线路戈壁碎石土地基掏挖基础设计与施工技术导则 Technical guide for digged foundation of overhead transmission line in gobi gravel soil		J1945-2015	中国电力科学研究院
343	DL/T 5709-2014	配电自动化规划设计导则 Guidelines for distribution automation planning		J1933-2014	中国南方电网有限责任公司
344	DL/T 5710-2014	电力建设土建工程施工技术检验规范 Technical inspection specifications for construction of electric power civil engineering		J1947-2015	浙江电力建设土建工程质量检测中心有限公司
345	DL/T 5711-2014	水电水利工程施工机械安全操作规程 带式输送机 Safety operation code for construction equipment of hydroelectric and water conservancy engineering belt conveyer		J1948-2015	中国水利水电第七工程局有限公司
346	DL/T 5712-2014	水电水利工程接缝灌浆施工技术规范 Construction technology specifications for joint grouting		J1946-2015	中国水利水电第四工程局有限公司
347	DL/T 5713-2014	火力发电厂热力设备及管道保温施工工艺导则 Technical guide for insulation construction of thermal equipments and piping in thermal power plant		J1961-2015	中国电力建设集团河南第一火电建设公司

电力工程

序号	标准编号	标准名称	被代替标准编号	备案号	主编单位
348	DL/T 5714-2014	火力发电厂热力设备及管道保温防腐施工技术规范 Technical code for constructing anticorrosion and insulation of thermal equipment and pipeline of fossil fuel power plant		J1962-2015	中国能源建设集团安徽电力建设第一工程公司
349	DL/T 5715-2015	电力光纤到户组网技术规程 Technical code for power fiber to the home network		J2005-2015	中国能源建设集团广东省电力设计研究院 北京国电通网络技术有限公司
350	DL/T 5716-2015	电力光纤到户施工及验收规范 Code for construction and acceptance of power fiber to the home		J2007-2015	中国能源建设集团广东省电力设计研究院 北京国电通网络技术有限公司
351	DL/T 5717-2015	农村住宅电气工程技术规范 Technical code for electrical engineering of rural residential buildings		J2008-2015	中国电力科学研究院 国家电网公司农电工作部 国网甘肃省电力公司
352	DL/T 5718-2015	单三相混合配电方式设计规范 Code for design of monophase and triphase hybrid distribution mode		J2009-2015	中国电力科学研究院 国网安徽省电力公司
353	DL/T 5719-2015	水电水利工程施工基坑排水技术规范 Technical specification of foundation pit drainage for hydropower and water conservancy engineering construction		J2010-2015	葛洲坝集团第一工程有限公司 中国葛洲坝集团第三工程有限公司 中国葛洲坝集团股份有限公司
354	DL/T 5720-2015	水工自密实混凝土技术规程 Technical code of hydraulic self-compacting concrete		J2011-2015	长江水利委员会长江科学院 中国长江三峡集团公司 三峡地区地质灾害与生态环境湖北省协同创新中心
355	DL/T 5721-2015	水工喷射混凝土试验规程 Test code for hydraulic shotcrete		J2012-2015	长江水利委员会长江科学院 中国长江三峡工程开发总公司 三峡地区地质灾害与生态环境湖北省协同创新中心
356	DL/T 5722-2015	水电水利工程施工机械安全操作规程 塔带机 The regulations for safe operation of the construction equipment for hydroelectric and water conservancy engineering construction tower-belt-crane		J2013-2015	葛洲坝集团第一工程有限公司 中国葛洲坝集团三峡建设工程有限公司 中国葛洲坝集团股份有限公司
357	DL/T 5723-2015	水电水利工程施工机械安全操作规程 履带式布料机 Safety operation code for construction equipment of hydropower and water conservancy engineering crawler spreader		J2014-2015	中国水利水电第八工程局有限公司
358	DL/T 5724-2015	水电工程砂石系统废水处理技术规范 Technical specifications for wastewater treating about aggregate system of hydropower project		J2015-2015	中国葛洲坝集团股份有限公司 葛洲坝集团第五工程有限公司
359	DL/T 5725-2015	35kV 及以下电力用户变电所建设规范 The construction standard of client substation up to 35kV		J2070-2015	江苏省电力公司

电力工程

序号	标准编号	标准名称	被代替标准编号	备案号	主编单位
360	DL/T 5726-2015	1000kV 串联电容器补偿装置施工工艺导则 Technic guidance for construction of 1000kV series capacitor compensation device installation		J2071-2015	国家电网公司 国家电网公司交流建设分公司
361	DL/T 5727-2016	绝缘子用常温固化硅橡胶防污闪涂料现场施工技术规范 Technical specification of room temperature vulcanized silicon rubber anti-pollution coating field execution for insulators		J2145-2016	国网冀北电力有限公司
362	DL/T 5728-2016	水电水利工程控制性灌浆施工规范 Specification of controlled grouting for hydropower and water resources projects		J2146-2016	中国水利水电第三工程局有限公司 中国水利水电第十三工程局有限公司等
363	DL/T 5729-2016	配电网规划设计技术导则 The guide for planning and design of distribution network		J2147-2016	中国电力科学研究院
364	DL/T 5730-2016	水电水利工程施工机械安全操作规程 振捣机械 Safety operation code for construction epuipment of hydropower and water conservancy engineering vibrating machine		J2148-2016	中国水利水电第二工程局有限公司
365	DL/T 5731-2016	水电水利工程施工机械安全操作规程 振动碾 Safety operation code for construction epuipment of hydropower and water conservancy engineering vibrator roller		J2149-2016	中国水利水电第二工程局有限公司
366	DL/T 5732-2016	架空输电线路大跨越工程施工质量检验及评定规程 Specification for construction quality inspection and evaluation of large crossing overhead transmission line		J2150-2016	国家电网公司交流建设分公司、中国电力科学研究院等
367	DL/T 5733-2016	架空输电线路接地模块施工工艺导则 Construction technology guide for the grounding module of overhead transmission line		J2150-2016	国家电网公司交流建设分公司 安徽送变电工程公司等
368	DL/T 5734-2016	电力通信超长站距光传输工程设计技术规程 Design technical specification of ultra-long haul optical transmission engineering of electric power communication		J2152-2016	国家电网公司信息通信分公司等
369	NB/T 25046-2015	核电厂水工设计规范 Code for hydraulic design of nuclear power plants		J2069-2015	东北电力设计院
370	NB/T 31003-2011	大型风电场并网设计技术规范 Design regulations for large-scale wind power connecting to the system		J1271-2011	中国电力工程顾问集团公司
371	NB/T 31073-2015	风电场工程劳动安全与工业卫生验收规程 Occupational safety and health acceptance specification for wind power projects		J2040-2015	水电水利规划设计总院

电力工程

序号	标准编号	标准名称	被代替标准编号	备案号	主编单位
372	NB/T 31074-2015	高海拔风力发电机组技术导则 Technical guide for usage of high attitude wind turbine generator systems		J2041-2015	中国电建集团西北勘测设计研究院有限公司
373	NB/T 31085-2016	风电场项目经济评价规范 Code for economic evaluation of wind farm		J2182-2016	河北省电力勘测设计研究院
374	NB/T 31086-2016	风电场工程水土保持方案编制技术规范 Technical code on preparation of soil and water conservation for wind farm projects		J2183-2016	中国电建集团中南勘测设计研究院有限公司
375	NB/T 31087-2016	风电场项目环境影响评价技术规范 Technical code for environmental impact assessment of wind farm projects		J2184-2016	中国电建集团中南勘测设计研究院有限公司
376	NB/T 31088-2016	风电场安全标识设置设计规范 Code for design of wind farm safety signs arrangement		J2194-2016	中国电建集团中南勘测设计研究院有限公司 水电水利规划设计总院
377	NB 31089-2016	风电场设计防火规范 Code for design of fire protection for wind farms		J2181-2016	中国电建集团北京勘测设计研究院有限公司
378	NB/T 31098-2016	风电场工程规划报告编制规程 Preparation specification for planning report of wind farmprojects		J2185-2016	水电水利规划设计总院
379	NB/T 32027-2016	光伏发电工程设计概算编制规定及费用标准 Preparation regulation for cost estimation of photovoltaic power projects		J2186-2016	水电水利规划设计总院（可再生能源定额站）
380	NB/T 32028-2016	光热发电工程安全验收评价规程 Specification for safety assessment upon completion of concentrating solar power projects		J2187-2016	水电水利规划设计总院
381	NB/T 32029-2016	光热发电工程安全预评价规程 Specification for safety pre-assessment of concentrating solar power projects		J2188-2016	水电水利规划设计总院
382	NB/T 32030-2016	光伏发电工程勘察设计费计算标准 Basis of calculation for investigation and design of photovoltaic power projects		J2189-2016	水电水利规划设计总院（可再生能源定额站）
383	NB/T 33022-2015	电动汽车充电站初步设计内容深度规定 Regulations of content and depth for preliminary design of electric vehicle charging station		J2038-2015	中国能源建设集团广东省电力设计研究院
384	NB/T 33023-2015	电动汽车充换电设施规划导则 Guide for electric vehicle charging/battery swap infrastructure planning		J2039-2015	国家电网公司
385	NB/T 35021-2014	水电站调压室设计规范 Design code for surge chamber of hydropower stations		J1908-2014	中国电力建设集团华东勘测设计研究院有限公司
386	NB/T 35023-2014	水闸设计规范 Design code for sluice		J1911-2014	中国电力建设集团成都勘测设计研究院有限公司
387	NB/T 35024-2014	水工建筑物抗冰冻设计规范 Design code for hydraulic structures		J1910-2014	中国电力建设集团西北勘测设计研究院有限公司

电力工程

序号	标准编号	标准名称	被代替标准编号	备案号	主编单位
388	NB/T 35025-2014	水电工程劳动安全与工业卫生验收规程 Occupational safety and health acceptance specification for hydropower projects		J1909-2014	水电水利规划设计总院
389	NB/T 35026-2014	混凝土重力坝设计规范 Design code for concrete gravity dams		J18-2014	中国电力建设集团华东勘测设计研究院有限公司
390	NB/T 35028-2014	水电工程勘探验收规程 Specification for acceptance of hydropower engineering exploration		J1907-2014	中国电力建设集团昆明勘测设计研究院有限公司
391	NB/T 35029-2014	水电工程测量规范		J1906-2014	中国电力建设集团北京勘测设计研究院有限公司
392	NB/T 35030-2014	水电工程投资匡算编制规定 Preparation regulation of rough estimation for investment		J1905-2014	水电水利规划设计总院（可再生能源定额站）
393	NB/T 35031-2014	水电工程安全监测系统专项投资编制细则 Preparation regulation of special investment on safety monitoring system of hydropower project		J1904-2014	水电水利规划设计总院（可再生能源定额站）
394	NB/T 35032-2014	水电工程调整概算编制规定 Compiling provisions of budgetary estimate adjustment of hydropower project		J1903-2014	水电水利规划设计总院（可再生能源定额站）
395	NB/T 35033-2014	水电工程环境保护专项投资编制细则 Preparation regulation for special investment on environmental protection design of hydropower project		J1902-2014	水电水利规划设计总院（可再生能源定额站）
396	NB/T 35034-2014	水电工程投资估算编制规定 Preparation regulation on investment estimation for pre-feasibility		J1901-2014	水电水利规划设计总院（可再生能源定额站）
397	NB/T 35035-2014	水力发电厂水力机械辅助设备系统设计技术规定 Design rule of hydraulic mechanical auxiliary equipment		J1900-2014	中国电力建设集团北京勘测设计研究院有限公司
398	NB/T 35036-2014	水电工程固定卷扬式启闭机通用技术条件		J1899-2014	中国水电建设集团夹江水工机械有限公司
399	NB/T 35037-2014	水电工程鱼类增殖放流站设计规范 Code for the design of fish restocking station of		J1898-2014	中国电力建设集团成都勘测设计研究院有限公司
400	NB/T 35038-2014	水电工程建设征地移民安置综合监理规范 Code of the comprehensive supervision for hydropower projects land requisition and resettlement		J1897-2014	水电水利规划设计总院
401	NB/T 35039-2014	水电工程地质观测规程 Specification for geological observation of hydropower projects		J1926-2014	中国电力建设集团华东勘测设计研究院有限公司
402	NB/T 35040-2014	水力发电厂供暖通风与空气调节设计规范 Design code for heating ventilation and air conditioning of hydropower plants		J1927-2014	中国电力建设集团西北勘测设计研究院有限公司
403	NB/T 35041-2014	水电工程施工导流设计规范 Design code of construction diversion for hydropower engineering		J1928-2014	中国电力建设集团北京勘测设计研究院有限公司

电力工程

序号	标准编号	标准名称	被代替标准编号	备案号	主编单位
404	NB/T 35042-2014	水力发电厂通信设计规范 Design code of communication for hydroelectric power plants	DL/T 5080-1997	J1929-2014	中国电力建设集团西北勘测设计研究院有限公司
405	NB/T 35043-2014	水电工程三相交流系统短路电流计算导则 Guide for short-circuit current calculation in three-phase AC systems of hydropower projects	DL/T 5163-2002	J1930-2014	中国电力建设集团北京勘测设计研究院有限公司
406	NB/T 35044-2014	水力发电厂厂用电设计规程 Specification for designing service power system for hydropower station	DL/T 5164-2002	J1931-2014	中国电力建设集团华东勘测设计研究院
407	NB/T 35045-2014	水电工程钢闸门制造安装及验收规范 Code for manufacture installation and acceptance of steel gates in hydropower engineering	DL/T 5018-2004	J1934-2014	中国葛洲坝集团公司
408	NB/T 35046-2014	水电工程设计洪水计算规范 Code for calculating design flood of hydropower projects	SL 44-93	J1932-2014	中国电力建设集团西北勘测设计研究院有限公司
409	NB 35047-2015	水电工程水工建筑物抗震设计规范 Code for seismic design of hydraulic structures of hydropower project		J2042-2015	水电水利规划设计总院
410	NB/T 35048-2015	水电工程验收规程 Specification for acceptance of hydropower projects		J2043-2015	水电水利规划设计总院
411	NB/T 35049-2015	水电工程泥沙设计规范 Code for sediment design of hydropower projects		J2044-2015	中国电建集团成都勘测设计研究院有限公司
412	NB/T 35050-2015	水力发电厂接地设计技术导则 Earthing design guide for hydropower station		J2045-2015	中国电建集团成都勘测设计研究院有限公司
413	NB/T 35051-2015	水电工程启闭机制造安装及验收规范 Code for manufacture erection and acceptance of gate hoists in hydropower projects		J2046-2015	水电水利规划设计总院
414	NB/T 35052-2015	水电工程地质勘察水质分析规程 Specification of water quality analysis for hydropowerengineering geological investigation		J2047-2015	中国电建集团华东勘测设计研究院有限公司
415	NB/T 35053-2015	水电站分层取水进水口设计规范 Design code for layered water intake of hydropower station		J2048-2015	中国电建集团华东勘测设计研究院有限公司
416	NB/T 35054-2015	水电工程过鱼设施设计规范 Design code for fish passage facilities in hydropower projects		J2049-2015	中国电建集团华东勘测设计研究院有限公司
417	NB 35055-2015	水电工程钢闸门设计规范 Design code for steel gate in hydropower projects		J2113-2016	能源行业水电金属结构及启闭机标准化技术委员会
418	NB/T 35056-2015	水电站压力钢管设计规范 Design code for steel penstocks of hydroelectric stations		J2114-2016	能源行业水电金属结构及启闭机标准化技术委员会

电力工程

序号	标准编号	标准名称	被代替标准编号	备案号	主编单位
419	NB 35057-2015	水电工程防震抗震设计规范 Code for seismic design of hydropower projects		J2115-2016	水电水利规划设计总院 中国水电工程顾问集团有限公司
420	NB/T 35058-2015	水电工程岩体质量检测技术规程 Technical specification for inspection of rock mass quality of hydropower projects		J2116-2016	中国电建集团贵阳勘测设计研究院有限公司
421	NB/T 35059-2015	河流水电开发环境影响后评价规范 Code for post assessment of environmental impacts of river hydropower development		J2117-2016	水电水利规划设计总院 中国电建集团贵阳勘测设计研究院有限公司
422	NB/T 35060-2015	水电工程移民安置环境保护设计规范 Code for environmental protection design of hydropower projects resettlement		J2118-2016	水电水利规划设计总院、中国电建集团贵阳勘测设计研究院有限公司
423	NB/T 35061-2015	水电工程动能设计规范 Code for energy economy design of hydropower project		J2119-2016	中国电建集团中南勘测设计研究院有限公司
424	NB/T 35062-2015	碾压式土石坝施工组织设计规范 Design code for construction planning of rolled earth-rock dam		J2120-2016	中国电建集团昆明勘测设计研究院有限公司
425	NB/T 35063-2015	水电工程环境监理规范 Code for environmental supervision of hydropower project		J2121-2016	水电水利规划设计总院 中国电建集团成都勘测设计研究院有限公司
426	NB/T 35064-2015	水电工程安全鉴定规程 Specification for safety appraisal of hydropower engineering		J2122-2016	水电水利规划设计总院
427	NB/T 35065-2015	水电工程地震勘探技术规程 Technical specification for seismic exploration of hydropower projects		J2123-2016	中国电建集团成都勘测设计研究院有限公司 四川中水成勘院工程勘察有限责任公司
428	NB/T 35066-2015	水电工程覆盖层钻探技术规程 Specification of overburden drilling techniques for hydropower projects		J2124-2016	中国电建集团成都勘测设计研究院有限公司 成都水利水电建设有限责任公司
429	NB/T 35067-2015	水力发电厂过电压保护和绝缘配合设计技术导则 Overvoltage protection and insulation coordination design guide for hydropower station		J2125-2016	中国电建集团成都勘测设计研究院有限公司
430	NB/T 35068-2015	河流水电规划环境影响评价规范 Code for environmental impact assessment of river hydropower planning		J2126-2016	水电水利规划设计总院 中国电建集团成都勘测设计研究院有限公司
431	NB/T 35069-2015	水电工程建设征地移民安置规划大纲编制规程 Preparation specification for planning outline of land acquisition and resettlementof hydropower projects		J2127-2016	水电水利规划设计总院 中国电建集团西北勘测设计研究院有限公司
432	NB/T 35070-2015	水电工程建设征地移民安置规划报告编制规程 Preparation specification for planning report of land acquisition and resettlement of hydropower projects		J2128-2016	水电水利规划设计总院 中国电建集团西北勘测设计研究院有限公司

电力工程

序号	标准编号	标准名称	被代替标准编号	备案号	主编单位
433	NB/T 35071-2015	抽水蓄能电站水能规划设计规范 Code for hydropower planning of pumped storage power station		J2129-2016	中国电建集团华东勘测设计研究院有限公司
434	NB/T 35072-2015	水电工程水土保持专项投资编制细则 Preparation regulation for special investment on soil and water conservation of hydropower projects		J2130-2016	水电水利规划设计总院（可再生能源定额站）
435	NB/T 35073-2015	水电工程水文测报和泥沙监测专项投资编制细则 Preparation regulation for special investment on hydrologic surveying and forecasting and sediment monitoring of hydropower projects		J2131-2016	水电水利规划设计总院（可再生能源定额站）
436	NB 35074-2015	水电工程劳动安全与工业卫生设计规范 Code for design of occupational safety and health of hydropower projects		J2132-2016	水电水利规划设计总院
437	NB/T 35075-2015	水电工程项目编号及产品文件管理规定 Rule of record number and management for product documents of hydropower projects		J2133-2016	中国电建集团中南勘测设计研究院有限公司 中国水电工程顾问集团有限公司 中国水利水电建设工程咨询有限公司
438	NB/T 35076-2016	水力发电厂二次接线设计规范 Design code for secondary circuit of hydropower plants		J2190-2016	中国电建集团成都勘测设计研究院有限公司
439	NB/T 35077-2016	水电工程数字流域基础地理信息系统技术规范 Technical code for fundamental geographic information system of digital basin based on hydropower projects		J2191-2016	中国电建集团成都勘测设计研究院有限公司
440	NB/T 35079-2016	地下厂房岩壁吊车梁设计规范 Design code for rock-bolted crane girders in underground powerhouses		J2192-2016	中国电建集团中南勘测设计研究院有限公司
441	NB/T 35080-2016	水电站气垫式调压室设计规范 Design code for air cushion surge chamber of hydropower stations		J2193-2016	中国电建集团成都勘测设计研究院有限公司

石油天然气工程

序号	标准编号	标准名称	被代替标准编号	备案号	主编单位
1	SY/T 0003-2012	石油天然气工程制图标准 Specification for petroleum and natural gas engineering drawings	SY/T 0003-2003	35082-2012	大庆油田工程有限公司
2	SY/T 0009-2012	石油地面工程设计文件编制规程 Specification for compiling design documents of petroleum surface facilities	SY/T 0009-2004	37457-2012	中国石化集团江汉石油管理局勘察设计研究院
3	SY/T 0011-2007	天然气净化厂设计规范 Design specification for natural gas conditioning plants	SY/T 0011-1996	21999-2007	中国石油集团工程设计有限责任公司西南分公司
4	SY/T 0017-2006	埋地钢质管道直流排流保护技术标准 Specifications for D.C. drainage protection for buried steel pipelines	SY/T 0017-1996	19093-2006	中国石油管道公司沈阳调度中心
5	SY/T 0021-2008	油气田和管道工程建筑设计规范 Specifications for the designing of field facility and pipeline construction projects	SYJ 21-1990	24269-2008	新疆时代是有工程有限公司
6	SY/T 0026-1999	水腐蚀性测试方法 Test method for corrosivity of water	SYJ 26-1987	3060-1999	江汉石油管理局勘察设计研究院
7	SY/T 0027-2014	稠油注汽系统设计规范 Code for design of steam injection system for viscous crude	SY/T 0027-2007	48143-2015	中油辽河工程有限公司
8	SY/T 0029-2012	埋地钢质检查片应用技术规范 Specification of application for underground steel coupons	SY/T 0029-1998	37458-2012	中国石油集团工程设计有限责任公司西南分公司
9	SY/T 0030-2008	油气田及管道腐蚀与防护工程基本词汇 Glossary of corrosion protection-engineering of oil and gas field and pipeline	SYJ 30-1987	24270-2008	中国石油天然气管道工程有限公司
10	SY 0031-2012	石油工业用加热炉安全规程 Safety regulations for heaters used in petroleum industry	SY 0031-2004	37442-2012	大庆油田工程有限公司（原大庆油田建设设计研究院）
11	SY/T 0033-2009	油气田变配电设计规范 Code for design of oil-gas field electrical power substation and distribution	SYJ 33-1988	27430-2010	华北石油勘察设计研究院
12	SY/T 0037-2012	管道防腐层阴极剥离试验方法 Test methods for cathodic disbonding of pipeline coatings	SY/T 0037-1997	35083-2012	中国石油管道学院
13	SY/T 0038-2013	管道防腐层特定可弯曲性试验方法 Test method for specific bendability of pipeline coatings	SY/T 0038-1997	43160-2014	中国石油管道学院
14	SY/T 0039-2013	管道防腐层化学稳定性试验方法 Test method for chemical resistance of pipeline coatings	SY/T 0039-1997	43161-2014	中国石油管道学院
15	SY/T 0040-2013	管道防腐层抗冲击性试验方法（落锤试验法） Test method for impact resistance of pipeline coatings	SY/T 0040-1997	43162-2014	中国石油管道学院
16	SY/T 0041-2012	防腐涂料与金属黏结的剪切强度试验方法 Standard test method for apparent shear strength of coatings to metal	SY/T 0041-1997	37459-2012	中国石油管道学院

石油天然气工程

序号	标准编号	标准名称	被代替标准编号	备案号	主编单位
17	SY/T 0042-2002	防腐蚀工程经济计算方法 Calculation of economic appraisals for corrosion control projects	SYJ 42-1989	10410-2002	中国石油天然气股份有限公司规划总院
18	SY/T 0043-2006	油气田地面管线和设备涂色规范 Code for painting color of pipelines and equipment in oil-gas fields	SY/T 0043-1996	18026-2006	中国石油天然气股份有限公司石油规划总院
19	SY/T 0045-2008	原油电脱水设计规范 Design specification of crude oil electric dehydration	SY/T 0045-1999	24271-2008	胜利油田胜利工程设计咨询有限责任公司
20	SY/T 0047-2012	油气处理容器内壁牺牲阳极阴极保护技术规范 Technical specification for sacrificial anode cathodic protection of internal surfaces of oil & gas-treating vessels	SY/T 0047-1999	37460-2012	中国石化集团江汉石油管理局勘察设计研究院
21	SY/T 0048-2009	石油天然气工程总图设计规范 Code for layout of geotechnical investigation report	SY/T 0048-2000	27431-2010	中国石油天然气股份有限公司规划总院
22	SY/T 0049-2006	油田地面工程建设规划设计规范 Planning code for oil-field surface production facilities construction	SY/T 0049-1994	18027-2006	中国石油天然气股份有限公司石油规划总院
23	SY/T 0051-2012	岩土工程勘察报告格式规范 Code for layout of geotechnical investigation report	SY/T 0051-2003	37461-2012	中国石化集团江汉石油管理局勘察设计研究院
24	SY/T 0060-2010	油气田防静电接地设计规范 Specifications for designing the electrostatic grounding of oilfield facilities	SY/T 0060-1992	29404-2010	吉林油田石油工程有限责任公司
25	SY/T 0061-2004	埋地钢质管道外壁有机防腐层技术规范 Technical specification of external protective organic coating for buried steel pipelines	SY/T 0061-1992	14013-2004	大庆油田有限公司油田建设设计研究院
26	SY/T 0062-2000	管道防腐层针入度试验方法（钝杆法） Test method for penetration resistance of pipeline coatings (blunt rod)	SY/T 0062-1992	8175-2001	管理局职工学院
27	SY/T 0063-1999	管道防腐层检漏试验方法 Test method for holiday detection in pipeline coatings	SY/T 0063-1992	3065-1999	中国石油天然气管道局职工学院
28	SY/T 0064-2000	管道防腐层水渗透性试验方法 Test method for water penetration into pipeline coatings	SY/T 0064-1992	8176-2001	管道局职工学院
29	SY/T 0065-2000	管道防腐层耐磨性能试验方法（滚筒法） Test method for abrasion resistance of pipeline coatings (drum)	SY/T 0065-1992	8177-2001	管道局职工学院
30	SY/T 0066-1999	钢管防腐层厚度的无损测量方法（磁性法） Test method for nondestructive measurement of film thickness of pipeline coatings on steel	SY/T 0066-1992	3066-1999	中国石油天然气管道局职工学院
31	SY/T 0067-1999	管道防腐层耐冲击性试验方法（石灰石落下法） Test method for impact resistance of pipeline coatings (limestone drop test)	SY/T 0067-1992	3067-1999	中国石油天然气集团公司管道局职工学院

石油天然气工程

序号	标准编号	标准名称	被代替标准编号	备案号	主编单位
32	SY/T 0069-2008	原油稳定设计规范 Design code for crude oil stabilization unit	SY/T 0069-2000	24272-2008	河南石油勘探局勘察设计研究院
33	SY/T 0071-2010	油气集输管道组成件选用标准 Selection of pipe and accessories in oil-gas gathering and transportation engineering	SY/T 0071-1993	29405-2010	大庆油田工程有限公司
34	SY/T 0072-2012	管道防腐层高温阴极剥离试验方法 Test method for cathodic disbonding of pipeline coatings subjected to elevated temperatures	SY/T 0072-1993	35084-2012	中国石油管道学院
35	SY/T 0073-2012	管道防腐层补伤材料评价试验方法 Standard test method for evaluating pipeline coating patch materials	SY/T 0073-1993	37462-2012	中国石油管道学院
36	SY/T 0076-2008	天然气脱水设计规范 Design specification of natural gas dehydration	SY/T 0076-2003	24273-2008	中国石油天然气管道工程有限公司天津分公司
37	SY/T 0077-2008	天然气凝液回收设计规范 Design specification of natural gas liquid recovery	SY/T 0077-2003	24274-2008	胜利石油胜利工程设计咨询有限责任公司
38	SY/T 0080-2008	油气田柴油机发电站设计规范 The design specification of power station with diesel engine in oil and gas field	SY/T 0080-1993	24275-2008	新疆时代石油工程有限公司
39	SY/T 0081-2010	原油热化学沉降脱水设计规范 Design specification of thermo-chemical setting dehydration for crude oil	SY/T 0081-1993	29406-2010	中国石化集团江汉石油管理局勘察设计研究院
40	SY/T 0083-2008	除油罐设计规范 Design specifications of oil removal tank	SY/T 0083-1994	24276-2008	中国石油天然气管道工程有限公司天津分公司
41	SY/T 0084-2012	管道防腐层环状弯曲性能试验方法 Test method for ring bendability of pipeline coating	SY/T 0084-1994	35085-2012	中国石油管道学院
42	SY/T 0085-2012	管道防腐层自然气候暴露试验方法 Standard test method for outdoor weathering exposure of pipeline coatings	SY/T 0085-1994	37463-2012	中国石油管道学院
43	SY/T 0086-2012	阴极保护管道的电绝缘标准 The electrical isolation of cathodical protected pipelines	SY/T 0086-2003	37464-2012	中国石油天然气管道工程有限公司
44	SY/T 0087.1-2006	钢制管道及储罐腐蚀评价标准 埋地钢质管道外腐蚀直接评价 Standard of steel pipeline and tank corrosion assessment-steel pipeline external corrosion direct assessment	SY/T 0087-1995	19094-2006	中国石油天然气股份公司规划总院
45	SY/T 0087.2-2012	钢制管道及储罐腐蚀评价标准 埋地钢质管道内腐蚀直接评价 Standard of steel pipeline and tank corrosion assessment-steel pipeline internal corrosion direct assessment	SY/T 0087-1995	35086-2012	中国石油天然气股份有限公司规划总院
46	SY/T 0087.3-2010	钢制管道及储罐腐蚀评价标准 钢制储罐腐蚀直接评价 Standard of steel pipeline and tank corrosion assessment-corrosion direct assessment of steel tank		29820-2010	中国石油大学（北京）

石油天然气工程

序号	标准编号	标准名称	被代替标准编号	备案号	主编单位
47	SY/T 0088-2016	钢质储罐罐底外壁阴极保护技术标准 Specification of external surface cathodic protection for the bottom of steel storage tank	SY/T 0088-2006	53363-2016	西安长庆科技工程有限责任公司
48	SY/T 0089-2006	油气厂、站、库给水排水设计规范 Code for design of water and wastewater for oil/gas plants, stations and depots	SY/T 0089-1996	18031-2006	中国石油集团工程设计有限责任公司西南分公司
49	SY/T 0094-1999	管道防腐层阴极剥离试验方法（黏结电解槽法） Test method for cathodic disbondment test of pipeline coatings (attached cell method)		3068-1999	中国石油天然气集团公司管道局职工学院
50	SY/T 0095-2000	埋地镁牺牲阳极试样试验室评价的试验方法 Test method for laboratory evaluation of magnesium sacrificial anode test specimens for underground applications		8178-2001	中国石油天然气管道工程有限公司
51	SY/T 0096-2013	强制电流深阳极地床技术规范 Specification of Impressed current deep anode beds	SY/T 0096-2000	43163-2014	中国石油天然气管道工程有限公司天津设计院
52	SY/T 0097-2016	油田采出水用于注汽锅炉给水处理设计规范 Code for design of oilfield produced water used as steam generator	SY/T 0097-2000	53364-2016	中石化石油工程设计有限公司
53	SY/T 0315-2013	钢质管道熔结环氧粉末外涂层技术规范 Technological specification of external fusion bonded epoxy coating for steel pipeline	SY/T 0315-2005	43164-2014	中国石油天然气管道科学研究院
54	SY/T 0317-2012	盐渍土地区建筑规范 Building code for saline soil region	SY/T 0317-1997	37466-2012	中国石油集团工程设计有限公司华北分公司
55	SY/T 0319-2012	钢质储罐液体涂料内防腐层技术标准 Standard of liquid internal coatings for steel tank	SY/T 0319-1998	37467-2012	中国石油集团工程技术研究院
56	SY/T 0320-2010	钢质储罐外防腐层技术标准 Standard of external anticorrosive coating for steel tank	SY/T 0320-1998	29821-2010	中国石油集团工程技术研究院
57	SY/T 0321-2000	钢质管道水泥沙浆衬里技术标准 Technical specification of cement mortar lining of steel pipelines		6962-2000	胜利石油管理局工程建设一公司
58	SY/T 0324-2014	直埋高温钢质管道保温技术规范 Technical specification for thermal insulation of directly buried high temperature steel pipeline	SY/T 0324-2001	48144-2015	中国石油集团工程技术研究院
59	SY/T 0326-2012	钢质储罐内衬环氧玻璃钢技术标准 Technical specification of the lining of glass fibre reinforced plastic of epoxy resin for steel tanks	SY/T 0326-2002	37468-2012	中原石油勘探局勘察设计研究院
60	SY/T 0328-2004	管道焊口内喷焊防腐技术标准 Specification for the internal girth weld coating with the metal powder jetting		14014-2004	石油大学（华东）
61	SY/T 0329-2004	大型油罐基础检测方法 Test methods for large oil tank foundations		14015-2004	中国石油天然气管道工程有限公司

石油天然气工程

序号	标准编号	标准名称	被代替标准编号	备案号	主编单位
62	SY/T 0330-2004	现役管道的不停输移动推荐作法 Recommend practice for movement of in-sevice pipelines		14016-2004	中国石油集团工程技术研究院
63	SY/T 0379-2013	埋地钢质管道煤焦油瓷漆防腐层技术规范 Standard of coal tar enamel external protective coating for buried steel pipeline	SY/T 0379-1998	43165-2014	中国石油天然气管道科学研究院
64	SY/T 0403-2014	输油泵组安装技术规范 Specification for construction of oil pump set	SY/T 0403-1998	48145-2015	中国石油天然气管道第二工程公司
65	SY/T 0404-1998	加热炉工程施工及验收规范 Heating furnace construction and acceptance specifications	SYJ 4004-1990	2672-1999	中国石油天然气管道第二工程公司
66	SY/T 0407-2012	涂装前钢材表面处理规范 Specification of steel surface preparation before application of paint	SY/T 0407-1997	37469-2012	中国石油集团工程技术研究院
67	SY/T 0414-2007	钢质管道聚乙烯胶黏带防腐层技术标准 Technical specification of polyethene tape coating for steel pipelines	SY/T 0414-1998	22001-2007	中国石油集团工程技术研究院
68	SY/T 0415-1996	埋地钢质管道硬质聚氨酯泡沫塑料防腐保温层技术标准 Specification of polyurethane foamed coating for buried steel pipelines	SYJ 18-1986 SYJ 4015-1987 SYJ 4016-1987	0076-1996	大庆石油管理局油田建设设计研究院
69	SY/T 0420-1997	埋地钢质管道石油沥青防腐层技术标准 Technical specification of petroleum asphalt coating for buried steel pipelines	SYJ 4020-1988 SYJ 8-1984	1099-1998	中国石油天然气管道第三工程公司
70	SY/T 0439-2012	石油天然气工程建设基本术语 Fundamental terminology of oil and gas engineering	SYJ 4039-1989	37996-2013	中国石油规划总院
71	SY/T 0440-2010	工业燃气轮机安装技术规范 Specifications for installation of industrial gas turbines	SYJ 4040-1989	29409-2010	中国石油天然气第八建设有限公司
72	SY/T 0441-2010	油田注汽锅炉制造安装技术规范 Technical specification for manufacture and installation of steam generators used in oil-fields	SY/T 4041-2001	29745-2010	中国石油集团工程技术研究院
73	SY/T 0442-2010	钢质管道熔结环氧粉末内防腐层技术标准 Technical standard for internal fusion bonded epoxy coating of steel pipe	SY/T 0442-1997	29746-2010	中国石油集团工程技术研究院
74	SY/T 0447-2014	埋地钢质管道环氧煤沥青防腐层技术标准 Standard of coal tar epoxy coating for buried steel pipeline	SY/T 0447-1996	48146-2015	中国石油天然气管道科学研究院
75	SY/T 0448-2008	油气田油气处理用钢制容器施工技术规范 Technical specification of construction steel vessel for oil and gas treatment in oil and gas field	SY/T 0448-1997 SY/T 0449-1997	24277-2008	天津大港油田集团工程建设有限责任公司
76	SY/T 0452-2012	石油天然气金属管道焊接工艺评定 Specification for weld procedure qualification of oil and gas metal pipelines	SY/T 0452-2002	37470-2012	中国石油天然气管道局第五工程公司
77	SY/T 0457-2010	钢质管道液体环氧涂料内防腐层技术标准 Technical standard of liquid epoxy Internal coating for steel pipeline	SY/T 0457-2000	29747-2010	四川石油天然气建设工程有限责任公司

石油天然气工程

序号	标准编号	标准名称	被代替标准编号	备案号	主编单位
78	SY/T 0460-2010	天然气净化装置设备与管道安装工程施工技术规范 Technical code for construction of equipment and piping installation engineering for natural gas conditioning purification facilities	SY/T 0460-2000	29748-2010	四川石油天然气建设工程有限公司
79	SY/T 0510-2010	钢制对焊管件规范 Code for steel butt-welding fittings	SY/T 0510-1998 SY/T 0518-2002	29749-2010	中国石油天然气管道工程有限公司
80	SY/T 0511.1-2010	石油储罐附件 第1部分：呼吸阀 Oil tank appurtenances Part 1: breather valve	SY/T 0511-1996	30777-2011	中国石油天然气管道工程有限公司
81	SY/T 0511.2-2010	石油储罐附件 第2部分：液压安全阀 Oil tank appurtenances Part 2: hydraulic safety valve	SY/T 0525.1-1993	30778-2011	中国石油天然气管道工程有限公司
82	SY/T 0511.3-2010	石油储罐附件 第3部分：自动通气阀 Oil tank appurtenances Part 3: automatic bleeder vent		30779-2011	中国石油天然气管道工程有限公司
83	SY/T 0511.4-2010	石油储罐附件 第4部分：泡沫塑料一次密封装置 Oil tank appurtenances Part 4: foam-filled primary seal		30780-2011	安徽美祥实业有限公司
84	SY/T 0511.5-2010	石油储罐附件 第5部分：二次密封装置 Oil tank appurtenances Part 5: secondary seal		30781-2011	安徽美祥实业有限公司
85	SY/T 0511.6-2010	石油储罐附件 第6部分：浮顶排水管系统 Oil tank appurtenances Part 6: primary roof drains		30782-2011	中国石油天然气管道工程有限公司
86	SY/T 0511.7-2010	石油储罐附件 第7部分：重锤式刮蜡装置 Oil tank appurtenances Part 7: wax scrappers with weights		30783-2011	中国石油天然气管道工程有限公司
87	SY/T 0511.8-2010	石油储罐附件 第8部分：钢制孔类附件 Oil tank appurtenances Part 8: steel holes	SY/T 0525.4-1993	30784-2011	中国石油天然气管道工程有限公司
88	SY/T 0511.9-2010	石油储罐附件 第9部分：量油孔 Oil tank appurtenances Part 9: dip hatch	SY/T 0525.5-1993	30785-2011	中国石油天然气管道工程有限公司
89	SY/T 0515-2014	油气分离器规范 Specification for oil and gas separators	SY/T 0515-2007	48147-2015	大庆油田工程有限公司（原大庆油田建设设计院）
90	SY/T 0516-2008	绝缘接头与绝缘法兰技术规范 Technical code for insulating joint and insulating flange	SY/T 0516-1997	24278-2008	中国石油集团工程设计有限责任公司西南分工司
91	SY/T 0523-2008	油田水处理过滤器 Oilfield water treatment filter	SY/T 0523-1993	24282-2008	中国石油集团钻井工程技术研究院江汉机械研究所
92	SY/T 0524-2008	导热油加热炉系统规范 Code for heater system with heat-transfer oil	SY/T 0524-1993	24283-2008	中国石油天然气管道工程有限公司
93	SY/T 0525.2-1993	石油储罐回转接头 Rotary joints for oil storage tanks		0016-1994	抚顺石油机械厂
94	SY/T 0526.1-1993	煤焦油瓷漆覆盖层 底漆 干提取物灰分测定 Tar enamel coating Primer Determination of ash content in dry extract		0103-1993	中国石油天然气管道科学研究院

石油天然气工程

序号	标准编号	标准名称	被代替标准编号	备案号	主编单位
95	SY/T 0526.2-1993	煤焦油瓷漆覆盖层 瓷漆 试样准备 Tar enamel coating Enamel Sample preparation		0103-1993	中国石油天然气管道科学研究院
96	SY/T 0526.3-1993	煤焦油瓷漆覆盖层 瓷漆 针入度测定 Tar enamel coating Enamel Penetration determination		0103-1993	中国石油天然气管道科学研究院
97	SY/T 0526.4-1993	煤焦油瓷漆覆盖层 瓷漆 冷弯试验 Tar enamel coating Enamel Cold bending test		0103-1993	中国石油天然气管道科学研究院
98	SY/T 0526.5-1993	煤焦油瓷漆覆盖层 瓷漆 压痕测定 Tar enamel coating Enamel Indentation determination		0103-1993	中国石油天然气管道科学研究院
99	SY/T 0526.6-1993	煤焦油瓷漆覆盖层 瓷漆 流淌性测定（ISO 法） Tar enamel coating Enamel Mobility determination (ISO method)		0103-1993	中国石油天然气管道科学研究院
100	SY/T 0526.7-1993	煤焦油瓷漆覆盖层 瓷漆 流淌性测定（BS 法） Tar enamel coating Enamel Mobility determination (BS method)		0103-1993	中国石油天然气管道科学研究院
101	SY/T 0526.8-1993	煤焦油瓷漆覆盖层 瓷漆 黏结相容性试验 Tar enamel coating Enamel Bonding compatibility test		0103-1993	中国石油天然气管道科学研究院
102	SY/T 0526.9-1993	煤焦油瓷漆覆盖层 瓷漆 加热变化试验 Tar enamel coating Enamel Heating change test		0103-1993	中国石油天然气管道科学研究院
103	SY/T 0526.10-1993	煤焦油瓷漆覆盖层 瓷漆 吸水率测定 Tar enamel coating Enamel Water absorption determination		0103-1993	中国石油天然气管道科学研究院
104	SY/T 0526.11-1993	煤焦油瓷漆覆盖层 瓷漆 灰分测定 Tar enamel coating Enamel Ash determination		0103-1993	中国石油天然气管道科学研究院
105	SY/T 0526.12-1993	煤焦油瓷漆覆盖层 瓷漆 低温脆裂和剥离试验 Tar enamel coating Enamel Cryogenic brittlement and peeling test		0103-1993	中国石油天然气管道科学研究院
106	SY/T 0526.13-1993	煤焦油瓷漆覆盖层 瓷漆 冲击试验 Tar enamel coating Enamel Impact test		0103-1993	中国石油天然气管道科学研究院
107	SY/T 0526.14-1993	煤焦油瓷漆覆盖层 瓷漆 阴极剥离试验 Tar enamel coating Enamel Cathode peeling test		0103-1993	中国石油天然气管道科学研究院
108	SY/T 0526.15-1993	煤焦油瓷漆覆盖层 缠绕带 拉伸强度测定 Tar enamel coating Winding tape Tensile strength determination		0103-1993	中国石油天然气管道科学研究院
109	SY/T 0526.16-1993	煤焦油瓷漆覆盖层 缠绕带 耐水性试验 Tar enamel coating Winding tape Water resistance test		0103-1993	中国石油天然气管道科学研究院

石油天然气工程

序号	标准编号	标准名称	被代替标准编号	备案号	主编单位
110	SY/T 0526.17-1993	煤焦油瓷漆覆盖层 缠绕带 热稳定性试验 Tar enamel coating Winding tape Thermal stability test		0103-1993	中国石油天然气管道科学研究院
111	SY/T 0526.18-1993	煤焦油瓷漆覆盖层 缠绕带 柔韧性试验 Tar enamel coating Winding tape Flexibility test		0103-1993	中国石油天然气管道科学研究院
112	SY/T 0526.19-1993	煤焦油瓷漆覆盖层 缠绕带 单位面积质量测定 Tar enamel coating Winding tape Mass per unit area determination		0103-1993	中国石油天然气管道科学研究院
113	SY/T 0526.20-1993	煤焦油瓷漆覆盖层 缠绕带 厚度测定 Tar enamel coating Winding tape Thickness determination		0103-1993	中国石油天然气管道科学研究院
114	SY/T 0526.21-1993	煤焦油瓷漆覆盖层 热烤缠带 瓷漆取样 Tar enamel coating Thermal winding tape Enamel sampling		0103-1993	中国石油天然气管道科学研究院
115	SY/T 0526.22-1993	煤焦油瓷漆覆盖层 热烤缠带 黏结性试验 Tar enamel coating Thermal winding tape Bonding test		0103-1993	中国石油天然气管道科学研究院
116	SY/T 0538-2012	管式加热炉规范 Code for tubular heaters	SY/T 0538-2004	37471-2012	中国石油天然气管道工程有限公司
117	SY/T 0540-2013	石油工业用加热炉型式与基本参数 Type and basal parameters for heater used in petroleum industry	SY/T 0540-2006	18034-2006	中油辽河工程有限公司
118	SY/T 0546-1996	腐蚀产物的采集与鉴定 Collection and identification of corrosion products		0076-1996	中原石油勘探局勘察设计研究院
119	SY/T 0556-2010	快速开关盲板技术规范 Technical code for quick opening closure	SY/T 0556-1999	29751-2010	中国石油集团工程设计有限责任公司西南分公司
120	SY/T 0599-2006	天然气地面设施抗硫化物应力开裂和抗应力腐蚀开裂的金属材料要求 Metallic material requirements-resistance to sulfide stress cracking and stress corrosion cracking for gas surface equipment	SY/T 0599-1997	19096-2006	中国石油集团工程设计有限责任公司西南分公司
121	SY/T 0602-2005	甘醇型天然气脱水装置规范 Specification for glycol-type gas dehydration units		16430-2005	天津大港油田集团石油工程有限公司
122	SY/T 0603-2005	玻璃纤维增强塑料储罐规范 Specification for fiberglass reinforced plastic tanks		16430-2005	中国石化集团江汉石油管理局勘察设计研究院
123	SY/T 0604-2005	工厂焊接液体储罐规范 Specification for shop welded tanks for storage of production liquids		16432-2005	大庆油田工程有限公司
124	SY/T 0605-2016	凝析气田地面工程设计规范 Code for surface engineering design of gas condensate field	SY/T 0605-2008	53368-2016	中油辽河工程有限公司

石油天然气工程

序号	标准编号	标准名称	被代替标准编号	备案号	主编单位
125	SY/T 0606-2014	现场焊接液体储罐规范 Specification for field welded tanks for storage of production liquids	SY/T 0606-2006	48148-2015	胜利油田胜利工程设计咨询有限责任公司
126	SY/T 0607-2006	转运油库和储罐设施的设计、施工、操作、维护与检验 Design, construction, operation, maintenance, and inspection of terminal & tank facilities		18036-2006	中油辽河工程有限公司
127	SY/T 0608-2014	大型焊接低压储罐的设计与建造 Design and construction of large, welded, low-pressure storage tanks	SY/T 0608-2006	48149-2015	中国石油天然气管道工程有限公司天津分公司
128	SY/T 0609-2016	优质钢制对焊管件规范 Specification for high-test wrought butt-welding fittings	SY/T 0609-2006	53369-2016	中国石油天然气管道工程有限公司廊坊分公司
129	SY/T 0610-2008	地下水封洞库岩土工程勘察规范 Standard for underground water enclosed cavern investigation of geotechnical engineering		24289-2008	中国石油天然气管道工程有限公司
130	SY/T 0611-2008	高含硫化氢气田集输管道系统内腐蚀控制要求 Requirements of controlling internal corrosion in gathering pipeline system for highly hydrogen sulphide gas field		24290-2008	中国石油集团工程设计有限责任公司西南分公司
131	SY/T 0612-2014	高含硫化氢气田地面集输系统设计规范 Code for design of gathering & transmission systems in highly hydrogen sulfide gas field	SY/T 0612-2008	48150-2015	中国石油集团工程设计有限责任公司西南分工司
132	SY 4063-1993	电气设施抗震鉴定技术标准 Specification for earthquake resistance evaluation of electric facilities		0039-1993	大庆石油管理局油田建设设计研究院
133	SY/T 4073-1994	储罐抗震用金属软管和波纹补偿器选用标准 Specifications for selecting hose assembly and bellows expansion joints used for tank anti-earthquake		0056-1994	中国石油天然气总公司基建工程局
134	SY/T 4074-1995	钢制管道水泥砂浆衬里涂敷机涂敷工艺 Process of cement mortar lining with coating machine on the inside surface of steel pipes		0045-1996	胜利石油管理局油建二公司
135	SY/T 4075-1995	钢质管道粉煤灰水泥砂浆衬里离心成型施工工艺 Technological process of coal ash cement mortar lining of steel pipelines with centrifugal lining sets		0045-1996	胜利石油管理局油建二公司
136	SY/T 4076-1995	钢质管道液体涂料内涂层风送挤涂工艺 Technological process of the pneumatically-pig-squeezed liquid inner coating for steel pipelines		0045-1996	胜利石油管理局油建一公司
137	SY/T 4077-1995	钢质管道水泥砂浆衬里风送挤涂工艺 Technological process of the pneumatically-pig-squeezed cement-mortar lining for steel pipelines		0045-1996	胜利石油管理局油建一公司

石油天然气工程

序号	标准编号	标准名称	被代替标准编号	备案号	主编单位
138	SY/T 4078-2014	钢质管道内涂层液体涂料补口机补口工艺规范 Procedure specification of inner coating for welded joint with liquid paint by special machine for steel pipeline	SY/T 4078-1995	48151-2015	中石化胜利油建工程有限公司
139	SY 4081-1995	钢制球型储罐抗震鉴定技术标准 Specification for earthquake resistance evaluation of steel spherical tanks		0045-1996	中国石油天然气总公司工程技术研究院
140	SY/T 4083-2012	电热法消除管道焊接残余应力热处理工艺规范 Technical specification for electric heat method of eliminating pipeline welding residual stress	SY/T 4083-1995	37474-2012	四川石油天然气建设工程有限责任公司
141	SY/T 4102-2013	阀门检验与安装规范 Specification for inspection and installation of valve	SY/T 4102-1995	43169-2014	中国石油天然气第一建设公司
142	SY/T 4103-2006	钢质管道焊接及验收 Welding and acceptance standard for steel pipings and pipelines	SY/T 4103-1995	18039-2006	中国石油天然气管道局职业教育培训中心焊接培训中心
143	SY/T 4105-2005	钢制储罐无溶剂聚氨酯涂料内防腐层技术规范 Technical specification of non-solvent polyurethane internal coating for steel containers		16434-2005	天津大港油田蒲海化工有限公司
144	SY/T 4106-2005	管道无溶剂聚氨酯涂料内外防腐层技术规范 Technical specification of non-solvent polyurethane internal and external coating for pipelines		16435-2005	天津大港油田蒲海化工有限公司
145	SY/T 4107-2005	复合防腐涂层各层厚度破坏性测量方法 Test methods for measurement of dry film thickness of multi-coating system by destructive means		16436-2005	中国石油管道科技中心
146	SY/T 4108-2012	输油（气）管道同沟敷设光缆（硅芯管）设计及施工规范 Code of design and construction acceptance of FOC（HDPE tube）laying in the same trench with oil（gas）pipelines	SY/T 4108-2005	35092-2012	中国石油天然气管道工程有限公司
147	SY/T 4109-2013	石油天然气钢质管道无损检测 Nondestructive testing standard of oil and gas steel pipeline	SY/T 4109-2005	43170-2014	徐州东方工程检测有限公司
148	SY/T 4110-2007	采用聚乙烯内衬修复现存管道施工技术规范 Code for construction technical of rehabilitation exist pipeline using PE liner		22003-2007	中国石油集团工程技术研究院
149	SY/T 4111-2007	天然气压缩机（组）安装工程施工技术规范 Code for construction specification of natural gas compressor set installation		22004-2007	四川石油天然气建设工程有限责任公司
150	SY/T 4112-2007	石油天然气钢质管道对接环焊缝全自动超声波检测试块 Mechanized ultrasonic examination block of girth butt welds in oil and gas steel pipeline		22005-2007	中国石油天然气管道局职业教育培训中心

石油天然气工程

序号	标准编号	标准名称	被代替标准编号	备案号	主编单位
151	SY/T 4113-2007	防腐涂层的耐划伤试验方法 Standard test method for measurement of gouge resistance of coating system		22006-2007	中国石油管道科技公司
152	SY/T 4114-2008	天然气输送管道干燥施工技术规范 Technical specification for drying construction of gas transmission pipeline		24292-2008	中国石油天然气管道局第四工程分公司
153	SY/T 4115-2008	油气输送管道工程建设施工组织设计编制规则 Code for compile construction scheme of oil & gas transmission pipelining		24293-2008	中国石油天然气管道局
154	SY 4116-2008	石油天然气管道工程建设监理规范 Construction supervision standard for petroleum & natural gas pipeline		24250-2008	廊坊中油朗威监理有限责任公司
155	SY/T 4117-2010	高含硫化氢气田集输管道焊接技术规范 Technical code for welding of gathering and transmission pipeline in highly hydrogen sulfide gas field		29752-2010	四川石油天然气建设工程有限责任公司
156	SY 4118-2010	高含硫化氢气田集输场站工程施工技术规范 Code for construction of gathering & transmission distributing station projects in highly hydrogen sulfide gas field		29742-2010	四川石油天然气建设工程有限责任公司
157	SY/T 4119-2010	高含硫化氢气田集输管道工程施工技术规范 Technical code for construction of gathering and transmission pipeline in highly hydrogen sulfide gas field		29753-2010	四川石油天然气建设工程有限责任公司
158	SY/T 4120-2012	高含硫化氢气田钢质管道环焊缝射线检测 Radiographic inspection for circumferential fusion welding joints of steel pipeline in highly hydrogen sulfide gas field		35093-2012	中国石油天然气股份有限公司西南油气田分公司
159	SY/T 4121-2012	光纤管道安全预警系统设计及施工规范 Code for design and construction of optical fiber pipeline security forewarning system project		35094-2012	中国石油天然气管道局
160	SY/T 4122-2012	油田注水工程施工技术规范 Code for construction of oilfield water injection		35095-2012	大庆油田建设集团有限责任公司
161	SY/T 4124-2013	油气输送管道工程竣工验收规范 Asbuilt acception specification for the oil & natural gas pipeline project		43171-2014	中国石油天然气管道局
162	SY/T 4125-2013	钢制管道焊接规程 Welding specification for steel pipeline		43172-2014	中国石油天然气管道科学研究院
163	SY/T 4126-2013	油气输送管道线路工程水工保护施工规范 Code for construction of oil and gas transportation pipeline hydraulic protection		43173-2014	中国石油天然气管道工程有限公司
164	SY/T 4127-2014	钢质管道冷弯管制作及验收规范 Specifications for fabrication and acceptance of steel cold bending pipe		48152-2015	中国石油天然气管道局

石油天然气工程

序号	标准编号	标准名称	被代替标准编号	备案号	主编单位
165	SY/T 4128-2014	大型设备内热法现场整体焊后热处理工艺规程 PWHT process for internal heating method of large & integral equipment in construction site		48153-2015	中国石油天然气第七建设公司
166	SY/T 4129-2014	输油输气管道自动化仪表工程施工技术规范 Technical specification for construction of automation instrumentation engineering of the oil and gas transportation pipeline		48154-2015	中国石油天然气第一建设公司
167	SY/T 4130-2016	玻璃纤维增强热固性树脂现场缠绕立式储罐施工规范 Specification for construction of glass fiber reinforced thermosetting resin on-site winding vertical storage tanks		53370-2016	胜利油田新大管业科技发展有限责任公司
168	SY 4200-2007	石油天然气建设工程施工质量验收规范 通则 Code for quality acceptance of oil and gas construction engineering-general rule		21980-2007	中国石油集团工程技术研究院
169	SY 4201.1-2016	石油天然气建设工程施工质量验收规范 设备安装工程 第1部分：机泵类 Code for quality acceptance of oil and gas construction engineering equipment installation engineering Part 1: mechanism and pump	SY 4201.1-2007	53342-2016	大庆油田建设集团有限责任公司
170	SY 4201.2-2016	石油天然气建设工程施工质量验收规范 设备安装工程 第2部分：塔类 Code for quality acceptance of oil and gas construction engineering equipment installation engineering Part 2: column	SY 4201.2-2007	53343-2016	大庆油田建设集团有限责任公司
171	SY 4201.3-2016	石油天然气建设工程施工质量验收规范 设备安装工程 第3部分：容器类 Code for quality acceptance of oil and gas construction engineering equipment installation engineering Part3: container	SY 4201.3-2007	53344-2016	大庆油田建设集团有限责任公司
172	SY 4201.4-2016	石油天然气建设工程施工质量验收规范 设备安装工程 第4部分：炉类 Code for quality acceptance of oil and gas construction engineering equipment installation engineering Pant 4: furance	SY 4201.4-2007	53345-2016	大庆油田建设集团有限责任公司
173	SY 4202-2016	石油天然气建设工程施工质量验收规范 储罐工程 Code for quality acceptance of oil and gas construction engineering storage tank project	SY 4202-2007	53346-2016	中国石油天然气第一建设公司
174	SY 4203-2016	石油天然气建设工程施工质量验收规范 站内工艺管道工程 Code for quality acceptance of oil and gas construction engineering station procedure pipeline project	SY 4203-2007	53347-2016	中国石油天然气管道局
175	SY 4204-2016	石油天然气建设工程施工质量验收规范 油气田集输管道工程 Code for quality acceptance of oil and gas construction engineering Oil-gas field gathering and transportation pipeline engineering	SY 4204-2007	53348-2016	四川天然气建设工程有限责任公司

石油天然气工程

序号	标准编号	标准名称	被代替标准编号	备案号	主编单位
176	SY 4205-2016	石油天然气建设工程施工质量验收规范 自动化仪表工程 Code for quality acceptance of oil and gas construction engineering automation instrumentation engineering	SY 4205-2007	53349-2016	中国石油天然气第一建设公司
177	SY 4206-2007	石油天然气建设工程施工质量验收规范 电气工程 Code for quality acceptance of oil and gas construction-electrotechnical engineering		21989-2007	胜利油田胜利石油化工建设有限责任公司
178	SY 4207-2007	石油天然气建设工程施工质量验收规范 管道穿跨越工程 Code for quality acceptance of oil and gas construction-engineering pipeline crossing and aerial crossing engineering		21990-2007	中国石油天然气管道工程有限公司
179	SY 4208-2008	石油天然气建设工程施工质量验收规范 输油输气管道线路工程 Code for quality acceptance of oil and gas construction-engineering oil and gas transmission pipeline project		24251-2008	中国石油天然气管道局
180	SY 4209-2008	石油天然气建设工程施工质量验收规范 天然气净化厂建设工程 Code for quality acceptance of oil and gas construction-engineering natural gas conditioning plant		24252-2008	石油天然长庆工程质量监督站
181	SY 4210-2009	石油天然气建设工程施工质量验收规范 道路工程 Code for quality acceptance of oil and gas construction engineering-road engineering		27527-2010	大庆油田路桥工程有限责任公司
182	SY 4211-2009	石油天然气建设工程施工质量验收规范 桥梁工程 Code for quality acceptance of oil and gas construction engineering-bridge engineering		27528-2010	胜利油田胜利工程建设（集团）有限责任公司
183	SY 4212-2010	石油天然气建设工程施工质量验收规范 高含硫化氢气田集输场站工程 Code for quality acceptance of oil and gas construction engineering-gathering & transmission distributing station projects in highly hydrogen sulfide gas field		29743-2010	四川石油天然气建设工程有限责任公司
184	SY 4213-2010	石油天然气建设工程施工质量验收规范 高含硫化氢气田集输管道工程 Code for quality acceptance of oil and gas construction engineering gathering and transmission pipelining projects in highly hydrogen sulfide gas field		29744-2010	四川石油天然气建设工程有限责任公司
185	SY/T 5262-2009	火筒式加热炉规范 Specification for fire tube type heater	SY/T 5262-2000 SY/T 5261-1991 SY/T 0419-1997	27443-2010	大庆油田工程有限公司
186	SY/T 5270-2000	高压注水管路配件设计技术规定 Technical regulation of design for high pressure water injection fittings	SY 5270-1991	8185-2001	中国石油天然气股份有限公司规划总院

石油天然气工程

序号	标准编号	标准名称	被代替标准编号	备案号	主编单位
187	SY/T 6535-2002	高压气地下储气井 The underground storage wells for high-compression gas		10472-2002	四川省川油天然气科技发展有限公司
188	SY/T 6536-2012	钢质水罐内壁阴极保护技术规范 Specification of cathodic protection of internal submerged surfaces of steel water storage tanks	SY/T 6536-2002	37548-2012	华北石油勘察设计研究院
189	SY/T 6706-2007	油气田及管道岩土工程勘察质量评定要求 Quality evaluation specification for oil and gas field and pipeline investigation of geotechnical engineering		22085-2007	中国石油天然气管道工程有限公司
190	SY/T 6769.1-2010	非金属管道设计、施工及验收规范 第1部分：高压玻璃纤维管线管 Specification for the design, construction and acceptance of non-metallic pipeline Part 1: high pressure fiberglass linepipe		29435-2010	大庆油田工程有限公司
191	SY/T 6769.2-2010	非金属管道设计、施工及验收规范 第2部分：钢骨架聚乙烯塑料复合管 Specification for the design, construction and acceptance of non-metallic pipeline Part 2: steel framed polyethylene composite pipe		29436-2010	大庆油田工程有限公司
192	SY/T 6769.3-2010	非金属管道设计、施工及验收规范 第3部分：塑料合金防腐蚀复合管 Specification for the design, construction and acceptance of non-metallic pipeline Part 3: anticorrosion plastic alloy composite pipe		29437-2010	大庆油田工程有限公司
193	SY/T 6769.4-2012	非金属管道设计、施工及验收规范 第4部分：钢骨架增强塑料复合连续管 Specifications for the design, construction and acceptance of non-metallic pipelines Part 4: steel skeleton reinforced thermoplastic resin coiled composite pipes		29438-2010	大庆油田工程有限公司
194	SY/T 6770.1-2010	非金属管材质量验收规范 第1部分：高压玻璃纤维管线管 Code for acceptance of quality for non-metalic pipe Part 1: high-pressure fiberglass line pipe		29438-2010	大庆油田工程有限公司
195	SY/T 6770.2-2010	非金属管材质量验收规范 第2部分：钢骨架聚乙烯塑料复合管 Code for acceptance of quality for non-metalic pipe Part 2: steel skeleton polyethylene plastic composite pipe		29439-2010	大庆油田工程有限公司
196	SY/T 6770.3-2010	非金属管材质量验收规范 第3部分：塑料合金防腐蚀复合管 Code for acceptance of quality for non-metalic pipe Part 3: anticorrosion plastic alloy composite pipe		29440-2010	大庆油田工程有限公司
197	SY/T 6770.4-2012	非金属管材质量验收规范 第4部分：钢骨架增强塑料复合连续管 Quality acceptance code of non-metalic pipes Part 4: steel skeleton reinforced plastic coiled composite pipes		38029-2013	大庆油田工程有限公司

石油天然气工程

序号	标准编号	标准名称	被代替标准编号	备案号	主编单位
198	SY/T 6771-2009	砂石人工岛总平面设计规范 Code for general layout of gravel artificial islands		27525-2010	中油辽河工程有限公司
199	SY/T 6772-2009	气体防护站设计规范 Code of design for gas protection station		27526-2010	中国寰球工程公司
200	SY/T 6784-2010	钢质储罐腐蚀控制标准 Standard of corrosion control for steel storage tank		29833-2010	中国石油集团工程技术研究院
201	SY/T 6793-2010	油气输送管道线路工程水工保护设计规范 Code for design of oil and gas transportation pipeline hydraulic protection		29805-2010	中国石油天然气管道工程有限公司
202	SY/T 6848-2012	地下储气库设计规范 Specification for design of underground gas storage		35146-2012	中国石油天然气管道工程有限公司天津分公司
203	SY/T 6849-2012	滩海漫水路及井场结构设计规范 Design code of structures for the road of allow over water and the well field in beach-shallow sea		35147-2012	胜利油田胜利工程设计咨询有限责任公司
204	SY/T 6850-2012	油气田及管道工程测量质量评定 Specification for evaluation of quality of oil and gas field and pipeline engineering survey		35148-2012	中国石油天然气管道工程有限公司
205	SY/T 6851-2012	油田含油污泥处理设计规范 Code for design of oil field oily sludge treatment		35149-2012	胜利油田胜利勘察设计研究院有限公司
206	SY/T 6852-2012	油田采出水生物处理工程设计规范 Code for design of oilfield produced water biological treatment		35150-2012	河南石油勘探局勘察设计研究院
207	SY/T 6853-2012	油气输送管道隧道设计规范 Code for design on tunnel of oil and gas transportation pipeline		35151-2012	中国石油天然气管道工程有限公司
208	SY/T 6854-2012	埋地钢质管道液体环氧外防腐层技术标准 Technical standard of liquid epoxy external coating for buried steel pipeline		35152-2012	中油管道防腐工程公司
209	SY 6879-2012	石油天然气建设工程施工质量验收规范 滩海海堤工程 Code for suality acceptance of oil and gas construction engineering seawall engineering in beach-shallow sea		37569-2012	胜利油田胜利工程建设（集团）有限责任公司
210	SY/T 6880-2012	高含硫化氢气田钢质材料光谱检测技术规范 Technical code for spectrometric detection of steel material in high hydrogen sulfide gas field		37567-2012	中国石油天然气股份有限公司西南油气田分公司
211	SY/T 6881-2012	高含硫气田水处理及回注工程设计规范 Design code for water treatment and reinjection engineering in highly hydrogen sulfide gas field		37568-2012	胜利油田胜利勘察设计研究院有限公司

石油天然气工程

序号	标准编号	标准名称	被代替标准编号	备案号	主编单位
212	SY/T 6882-2012	石油天然气建设工程交工技术文件编制规范 Compiling code of technical document for handing over a completed oil and gas construction project		37569-2012	四川石油天然气建设工程有限责任公司
213	SY/T 6883-2012	输气管道工程过滤分离设备规范 Specification for filtration & separation equipment of gas transmission pipeline engineering		37570-2012	中国石油天然气管道工程有限公司
214	SY/T 6884-2012	油气管道穿越工程竖井设计规范 Code for design of shaft in oil and gas pipeline crossing engineering		37571-2012	中国石油集团工程设计有限责任公司西南分公司
215	SY/T 6885-2012	油气田及管道工程雷电防护设计规范 Code for design protection against lightning of oil-gas field and pipeline engineering		37572-2012	中国石油集团工程设计有限责任公司西南分公司
216	SY/T 6886-2012	油田含聚及强腐蚀性采出水处理设计规范 Code for design of oil field treatment polymer-bearing and high corrosive produced water		37573-2012	胜利油田胜利勘察设计研究院有限公司
217	SY/T 6964-2013	石油天然气站场阴极保护技术规范 Specification of cathodic protection for petroleum & gas station		43271-2014	中国石油天然气管道工程有限公司
218	SY/T 6965-2013	石油天然气工程建设遥感技术规范 Specifications of remote sensing for oil and gas engineering construction		43272-2014	西安长庆科技工程有限责任公司
219	SY/T 6966-2013	输油气管道工程安全仪表系统设计规范 Design code for safety instrumented system in oil/gas transportation pipeline engineering		43273-2014	中国石油天然气管道工程有限公司
220	SY/T 6967-2013	油气管道工程数字化系统设计规范 Design code for digitization system in oil and gas pipeline project		43274-2014	中国石油天然气管道工程有限公司
221	SY/T 6968-2013	油气输送管道工程水平定向钻穿越设计规范 Design code for oil and gas transmission pipeline installation by horizontal directional drilling		43275-2014	中国石油天然气管道工程有限公司
222	SY/T 6969-2013	沿海滩涂地区油田10（6）kV架空配电线路设计规范 Code for designing over-head transmission line 10（6）kV of coastal beach areas oilfield		43276-2014	胜利油田胜利勘察设计研究院有限公司
223	SY/T 6970-2013	高含硫化氢气田地面集输系统在线腐蚀监测技术规范 Technical code of on-line corrosion monitoring in gathering system for highly hydrogen sulfide gas field		43277-2014	中国石油集团工程设计有限责任公司西南分公司
224	SY/T 6979-2014	立式圆筒形钢制焊接储罐自动焊技术规范 Technical specification for automatic welding of vertical circular welding steel storage tanks		46018-2014	中国石油集团工程技术研究院

石油天然气工程

序号	标准编号	标准名称	被代替标准编号	备案号	主编单位
225	SY/T 7020-2014	油田采出水注入低渗与特低渗油藏精细处理设计规范 Code for design of oilfield injection produced water advanced treatment for low and special low permeability reservoir		48270-2015	中石化石油工程设计有限公司
226	SY/T 7021-2014	石油天然气地面建设工程供暖通风与空气调节设计规范 Design code for heating ventilation and air conditioning of petroleum and natural gas surface engineering		48271-2015	中国石油集团工程设计有限责任公司西南分公司
227	SY/T 7022-2014	油气输送管道工程水域顶管法隧道穿越设计规范 Code for design of oil and gas transmission pipeline crossing by pipe jacking in water areas		48272-2015	中国石油天然气管道工程有限公司
228	SY/T 7023-2014	油气输送管道工程水域盾构法隧道穿越设计规范 Code for design of oil and gas transmission pipeline crossing by shield tunneling		48273-2015	中国石油天然气管道工程有限公司
229	SY/T 7024-2014	高含硫化氢气田金属材料现场硬度检验技术规范 Technical code for field hardness test of metal materials used in highly hydrogen sulfide gas field		48274-2015	中国石油集团工程设计有限责任公司西南分公司
230	SY/T 7025-2014	酸性油气田用缓蚀剂性能实验室评价方法 Laboratory evaluation methods for performance of corrosion inhibitors used in sour oil and gas fields		48275-2015	中国石油集团工程设计有限责任公司西南分公司
231	SY/T 7036-2016	石油天然气站场管道及设备外防腐层技术规范 Specification of external coating for piping and equipments in petroleum and gas stations		53452-2016	中国石油集团工程设计有限责任公司西南分公司
232	SY/T 7037-2016	油气输送管道监控与数据采集（SCADA）系统安全防护规范 Pipeline SCADA System Security		53453-2016	中国石油天然气管道工程有限公司
233	SY/T 7038-2016	油气田及管道专用道路设计规范 Specifications for special road design of oil-gas fields and pipelines		53454-2016	西安长庆科技工程有限责任公司
234	SY/T 7039-2016	油气厂站钢管架结构设计规范 Code for design of steel pipe rack structure in petroleum or natural gas factory		53455-2016	中国石油集团工程设计有限责任公司西南分公司
235	SY/T 7040-2016	油气输送管道工程地质灾害防治设计规范 Code for design of prevention engineering for oil and gas transportation pipeline geological hazard		53456-2016	中国石油天然气管道工程有限公司
236	SY/T 7041-2016	钢质管道聚丙烯防腐层技术规范 Specification of polypropylene coating for steel pipeline		53457-2016	中国石油集团工程技术研究院

海洋石油工程

序号	标准编号	标准名称	被代替标准编号	备案号	主编单位
1	SY/T 0305-2012	滩海管道系统技术规范 Technical specification of pipeline system for beach-shallow sea	SY/T 0305-1996	35087-2012	胜利油田胜利石油化工建设有限公司
2	SY/T 0306-1996	滩海石油工程热工采暖技术规范 Technical specification for heating engineering and heating of petroleum engineering			胜利石油管理局勘察设计研究院
3	SY/T 0307-1996	滩海石油工程立式圆筒形钢制焊接固定顶储罐技术规范 Technical specification for upright column steel welding top-fixed storage tank of petroleum engineering in beach-shallow sea			胜利石油管理局油建一公司
4	SY/T 0308-2016	滩海石油工程注水设计规范 Design code for water injection on beach-shallow sea petroleum engineering	SY/T 0308-1996	53365-2016	中油辽河工程有限公司
5	SY/T 0309-1996	滩海石油工程采出水处理技术规范 Technical specification of produced water treatment for petroleum engineering in beach-shallow sea			辽河石油勘探局勘察设计研究院
6	SY/T 0310-1996	滩海石油工程仪表与自动控制技术规范 Technical specification for instrument and automatic control of petroleum engineering in beach-shallow sea			辽河石油勘探局勘察设计研究院
7	SY/T 0311-2016	滩海石油工程通信技术规范 Design code for water injection on beach-shallow sea petroleum engineering	SY/T 0311-1996	53366-2016	中油辽河工程有限公司
8	SY/T 0312-1996	滩海石油工程舾装技术规范 Technical specification of outfitting for petroleum engineering in beach-shallow sea			辽河石油勘探局勘察设计研究院
9	SY/T 0313-2010	滩海石油工程码头设计与建造技术规范 Specification for design and construction of wharfs for petroleum engineering in beach-shallow sea	SY/T 0313-1996	29407-2010	中油辽河工程有限公司
10	SY/T 0314-1996	滩海混凝土平台结构设计与建造技术规范 Technical specification of structure design and construction for concrete platform in beach-shallow sea			辽河石油勘探局勘察设计研究院、辽河石油勘探局基建处
11	SY/T 4084-2010	滩海环境条件与载荷技术规范 Technical specification for environmental conditions and loads in shallow water	SY/T 4084-1995	29823-2010	中油辽河工程有限公司
12	SY/T 4085-2012	滩海油田油气集输设计规范 Code for design of oil-gas gathering and transportation for oil field in beach-shallow sea	SY/T 4085-1995	37997-2013	胜利油田胜利勘探设计研究院有限公司
13	SY/T 4086-2012	滩海结构物上管网设计与施工技术规范 Design and construction technical specification of pipe network in structures of shallow sea	SY/T 4086-1995	37998-2013	胜利油田胜利勘探设计研究院有限公司
14	SY/T 4087-1995	滩海石油工程通风空调技术规范 Technical specification of ventilation and air conditioning for petroleum engineering in beach-shallow sea		0045-1996	大港石油管理局勘察设计研究院

海洋石油工程

序号	标准编号	标准名称	被代替标准编号	备案号	主编单位
15	SY/T 4088-1995	滩海石油工程给水排水技术规范 Technical specification of water supply and sewerage for petroleum engineering in beach-shallow sea		0045-1996	大港石油管理局勘察设计研究院
16	SY/T 4089-1995	滩海石油工程电气技术规范 Technical specification of electric installation for petroleum engineering in beach-shallow sea		0045-1996	大港石油管理局勘察设计研究院
17	SY/T 4090-1995	滩海石油工程发电设施技术规范 Technical specification of generating equipment for petroleum engineering in beach-shallow sea		0045-1996	大港石油管理局勘察设计研究院
18	SY/T 4091-1995	滩海石油工程防腐蚀技术规范 Technical specification of anticorrosion for petroleum engineering in beach-shallow sea		0045-1996	辽河石油勘探局勘察设计研究院
19	SY/T 4092-1995	滩海石油工程保温技术规范 Technical specification of insulation for petroleum engineering in beach-shallow sea		0045-1996	大港石油管理局勘察设计研究院
20	SY/T 4093-1995	滩海石油设施上起重机选用与安装技术规范 Technical specification of installation for cranes on petroleum facilities in beach-shallow sea		0045-1996	中国石油天然气总公司基建工程局
21	SY/T 4094-2012	浅海钢质固定平台结构设计与建造技术规范 Technical specification for structure designing and constructing for fixed steel-platforms in shallow-sea	SY/T 4094-1995	37999-2013	胜利油田胜利石油化工建设有限公司
22	SY/T 4095-2012	浅海钢质移动平台结构设计与建造技术规范 Technical specification for structure design and construction of mobile steel platforms in beach-shallow sea	SY/T 4095-1995	35091-2012	中国石化集团胜利石油管理局钻井工艺研究院
23	SY/T 4096-2012	滩海油田井口保护装置技术规范 Technical specification for wellhead protection apparatus in beach-shallow sea	SY/T 4096-1995	38000-2013	胜利油田胜利勘探设计研究院有限公司
24	SY/T 4097-2010	滩海斜坡式砂石人工岛结构设计与施工技术规范 Technical specification of designing and constructing sloping artificial island made of sand and stone for petroleum engineering in beach-shallow sea	SY/T 4097-1995	29414-2010	中油辽河工程有限公司
25	SY/T 4098-1995	滩海环壁式钢模—混凝土人工岛结构设计与施工技术规范 Technical specification for design and construction of ringwall steel module-concrete artificial island in beach-shallow sea		0045-1996	中国石油天然气总公司基建工程局
26	SY/T 4099-2010	滩海海堤设计与施工技术规范 Technical specification of designing and constructing seawall in beach-shallow sea	SY/T 4099-1995	29415-2010	胜利油田胜利工程建设（集团）有限责任公司
27	SY/T 4100-2012	滩海工程测量技术规范 Technical specifications of engineering survey in beach-shallow sea	SY/T 4100-1995	38001-2013	中油辽河工程有限公司

海洋石油工程

序号	标准编号	标准名称	被代替标准编号	备案号	主编单位
28	SY/T 4101-2012	滩海岩土工程勘察技术规范 Technical specification of geotechnical investigation in beach-shallow sea	SY/T 4101-1995	38002-2013	中油辽河工程有限公司
29	SY/T 6773-2010	海上结构热机械控轧（TMCP）钢板规范 Specification for steel plates for offshore structures using thermo-mechanical control processing（TMCP）		29441-2010	中海石油（中国）有限公司研究中心
30	SY/T 6774-2010	海上结构调质钢板规范 Specification for steel plates, quenched-and-tempered, for offshore structure		29442-2010	中海石油（中国）有限公司北京研究中心
31	SY/T 6775-2010	滩海堤坝坝体检测规程（瑞雷波法） Shallow sea petroleum engineering, standard of seabank embankment inspection（Rayleigh wave method）		29827-2010	中国石化股份胜利油田分公司技术检测中心
32	SY/T 6776-2010	海上生产设施设计及危险性分析推荐作法 Recommended practice for design and hazards analysis for offshore production facilities		29828-2010	海洋石油工程股份有限公司
33	SY/T 6874-2012	张力腿平台规划、设计和建造的推荐做法 Bulkling strength of plated structures		35174-2012	中国海洋石油总公司海洋工程股份有限公司
34	SY/T 6875-2012	板式结构屈曲强度 Bulkling strength of plated structures		35175-2012	中国海洋石油总公司海洋工程股份有限公司
35	SY/T 6876-2012	壳的屈曲强度 Bulkling strength of shells		35176-2012	中国海洋石油总公司海洋工程股份有限公司
36	SY/T 6877-2012	杆件、框架和球型壳的屈曲强度分析 Buckling strength analysis of bars and frames, and spherical shells		35177-2012	中国海洋石油总公司海洋工程股份有限公司
37	SY/T 6878-2012	海底管道牺牲阳极阴极保护 Cathodic protection of submarine pipelines by galvanic anodes		35178-2012	中国海洋石油总公司海洋工程股份有限公司
38	SY/T 6930-2012	海上构筑物的保护涂层腐蚀控制 Corrosion control of offshore structures by protective coatings		38025-2013	中海油研究总院
39	SY/T 6980-2014	海上油气生产设施的废弃处置 Guidelines for abandonment of offshore oil and gas production facilities		46019-2014	中海油研究总院
40	SY/T 6981-2014	暖风加热和空气调节系统安装标准 Standard for the installation of warm air heating and air-conditioning systems		46020-2014	中海石油（中国）有限公司北京研究总院
41	SY/T 6982-2014	通风空调系统的安装 Standard for the installation of air-conditioning and ventilating systems		46021-2014	海油石油工程股份有限公司设计公司
42	SY/T 7045-2016	离心泵和回转泵轴封系统 Pumps-shaft sealing systems for centrifugal and rotary pumps		53461-2016	中国海洋石油总公司海洋石油工程股份有限公司
43	SY/T 7046-2016	制冷系统安全标准 Safety standard for refrigeration systems		53462-2016	中海石油（中国）有限公司北京研究总院

海洋石油工程

序号	标准编号	标准名称	被代替标准编号	备案号	主编单位
44	SY/T 7047-2016	锚链规格书 Specification for mooring chain		53463-2016	海洋石油工程股份有限公司
45	SY/T 7048-2016	海洋结构用钢板生产资格预评定推荐作法 Recommended practice for preproduction qualification for steel plates for offshore structures		53464-2016	海洋石油工程股份有限公司
46	SY/T 7049-2016	滩海海底管道检验技术规范 Inspection technical specification of pipelines in beach-shallow sea		53465-2016	中国石化集团胜利石油管理局海上石油工程技术检验中心
47	SY/T 7050-2016	滩海陆岸石油设施检验技术规范 Technical code for survey of offshore and coastal structures on foreshore		53466-2016	中国石化海上石油工程技术检验中心
48	SY/T 7051-2016	人工岛石油设施检验技术规范 Technical code for survey of artificial island		53467-2016	中国石化海上石油工程技术检验中心
49	SY/T 7052-2016	滩海漫水路及井场结构施工技术规范 Construction code of structures for the road of allow over water and the well field in beach-shallow sea		53468-2016	中石化胜利建设工程有限公司
50	SY/T 7053-2016	海底管道总体屈曲——高温/高压下的结构设计 Global buckling of submarine pipelines structural design due to high temperature/high pressure		53469-2016	中海油研究总院
51	SY/T 7054-2016	海底管道维修推荐作法		53470-2016	海洋石油工程股份有限公司
52	SY/T 7055-2016	水下分离器结构设计推荐做法 Subsea separator structural design-recommended practice		53471-2016	海洋石油工程股份有限公司
53	SY/T 7056-2016	海底管道自由悬跨 Free spanning pipelines		53472-2016	中海油研究总院
54	SY/T 7057-2016	动态立管 Dynamic risers		53473-2016	中海油研究总院
55	SY/T 7058-2016	海底管道阀门规范 Specification for subsea pipeline valves		53474-2016	海洋石油工程股份有限公司
56	SY/T 7059-2016	浮式生产系统和张力腿平台的立管设计 Design of risers for floating production systems (FPSs) and tension-leg platforms (TLPs)		53475-2016	中海油研究总院
57	SY/T 7060-2016	海底管道稳定性设计 On-bottom stability design of submarine pipelines		53476-2016	中海油研究总院
58	SY/T 7061-2016	水下高完整性压力保护系统（HIPPS）推荐做法 Recommended practice for subsea high integrity pressure protection system (HIPPS)		53477-2016	中海油研究总院

海洋石油工程

序号	标准编号	标准名称	被代替标准编号	备案号	主编单位
59	SY/T 7062-2016	水下生产系统可靠性及技术风险管理推荐做法 Recommended practice for subsea production system reliability and technical risk management		53478-2016	中海油研究总院
60	SY/T 7063-2016	海底管道风险评估推荐作法 Recommended practice for risk assessment of submarine pipeline		53479-2016	海洋石油工程股份有限公司
61	SY/T 7064.5-2016	石油天然气工业 海洋结构物特殊要求 第5部分：设计和施工过程中的重量控制 Petroleum and natural gas industries-specific requirements for offshore structures-Part 5: Weight control during engineering and construction		53480-2016	中国海洋石油总公司海洋石油工程股份有限公司
62	SY/T 10001-1996	原油生产与储存装置入级规范 Specification for classification of oil production and storage vessels			海洋石油工程总公司海洋石油工程标准化技术委员会
63	SY/T 10002-2000	结构钢管制造规范 Specification for the fabrication of structural steel pipe	SY/T 10002-1996	6949-2000	中国海洋石油生产研究中心
64	SY/T 10003-1996	海上平台起重机规范 Specification for offshore cranes			中国海洋石油总公司海洋石油标准化技术委员会
65	SY/T 10004-2010	海上平台管节点碳锰钢板规范 Specifications for carbon manganese steel plate for offshore platform tubular joints	SY/T 10004-1996	29411-2010	海洋石油工程股份有限公司
66	SY/T 10005-1996	海上结构建造的超声检验推荐作法和超声技师资格的考试指南 Recommended practice for ultrasonic examination of offshore structural fabrication and guidelines for qualification of technicians			中国海洋石油总公司海洋石油标准化技术委员会
67	SY/T 10006-2000	海上井口地面安全阀和水下安全阀规范 Specifications for offshore wellhead surface safety valves and subsea safety valves	SY/T 10006-1996	6950-2000	中国海洋石油生产研究中心
68	SY/T 10007-1996	海底管道的稳定性设计 Stability design of subsea pipelines			中国海洋石油总公司海洋石油标准化技术委员会
69	SY/T 10008-2016	海上钢质固定石油生产构筑物全浸区的腐蚀控制 Corrosion control of submerged areas of permanently installed steel offshore structures associated with petroleum production	SY/T 10008-2010	53506-2016	中海油研究总院
70	SY/T 10009-2002	海上固定平台规划、设计和建造的推荐作法 荷载和抗力系数设计法（增补1） Recommended practice for planning, designing and constructing fixed offshore platforms-load and resistance factor design (supplement 1)		10478-2002	中国海洋石油总公司海洋石油研究中心开发设计院

海洋石油工程

序号	标准编号	标准名称	被代替标准编号	备案号	主编单位
71	SY/T 10010-2012	非分类区域和I级1类及2类区域的固定及浮式海上石油设施的电气系统设计与安装推荐作法 Recommended practice for design and installation of electrical systems for fixed and floating offshore petroleum facilities for unclassified and class I, division 1 and division 2 locations	SY/T 10010-1996	35179-2012	中国海洋石油总公司海洋工程股份有限公司
72	SY/T 10012-1998	海上油气钻井井名命名规范 Recommended practice for naming offshore oil and gas wells		2121-1998	中国海洋石油生产研究中心
73	SY/T 10023.1-2012	海上油（气）田开发项目经济评价方法 第1部分：自营油（气）田 Method of offshore oil/gas field development project economic evaluation Part 1：self-financed offshore oil/gas field	SY/T 10023.1-1999	38030-2013	中国海洋石油研究总院
74	SY/T 10023.2-2012	海上油（气）田开发项目经济评价方法 第2部分：合作油（气）田 Method of offshore oil/gas field development project economic evaluation Part 2：co-operative offshore oil/gas field	SY/T 10023.2-2000	38031-2013	中国海洋石油研究总院
75	SY/T 10024-1998	井下安全阀系统的设计、安装、修理和操作的推荐作法 Recommended practice for design, installation, repair and operation of subsurface safety valve systems	SY/T 4807-1992	2051-1998	海油标委秘书处
76	SY/T 10028-2002	海洋石油工程制图规范 Specification for drawings of offshore oil development engineering		10479-2002	中海石油研究中心开发设计院
77	SY/T 10029-2016	浮式生产系统规划、设计及建造的推荐作法 Recommended practice for planning, designing, and constructing floating production systems	SY/T 10029-2004	53505-2016	中海油研究总院
78	SY/T 10030-2004	海上固定平台规划、设计和建造的推荐作法 工作应力设计法 Recommended practice for planning, designing and constructing fixed offshore platforms-working stress design	SY/T 10030-2002	14095-2004	中海石油研究中心
79	SY/T 10031-2000	寒冷条件下结构和海管规划、设计和建造的推荐作法 Recommended practice for planning, designing and constructing structures and pipelines for arctic conditions	SY/T 4803-1992	6958-2000	中国海洋石油生产研究中心
80	SY/T 10032-2000	单点系泊装置建造与入级规范 Rules for building and classing single point moorings	SY/T 4801-1992	6955-2000	中国海洋石油生产研究中心

海洋石油工程

序号	标准编号	标准名称	被代替标准编号	备案号	主编单位
81	SY/T 10033-2000	海上生产平台基本上部设施安全系统的分析、设计、安装和测试的推荐作法 Recommended practice for analysis, design, installation and testing of basic surface safety systems on offshore production platforms	SY/T 4808-1992	6953-2000	中国海洋石油生产研究中心
82	SY/T 10034-2000	敞开式海上生产平台防火与消防的推荐作法 Recommended practice for fire prevention and control on open-type offshore production platforms	SY/T 4810-1992	6956-2000	中国海洋石油生产研究中心
83	SY/T 10035-2010	钻井平台拖航与就位作业规范 Operational specification for MODU towing and positioning	SY/T 10035-2000	29413-2010	中海油田服务股份有限公司
84	SY/T 10036-2001	海洋石油工程设计文件编制规程 Specification for compiling the design documents of offshore oil development engineering		7581-2000	中海石油海洋工程公司
85	SY/T 10037-2010	海底管道系统 Submarine pipeline system	SY/T 10037-2002	29822-2010	海洋石油工程股份有限公司设计分公司
86	SY/T 10038-2002	海上固定平台直升机场规划、设计和建造的推荐作法 Recommended practice for planning, designing, and constructing heliports for fixed offshore platforms	SY/T 4806-1992	10482-2002	中国海洋石油总公司研究中心开发设计院
87	SY/T 10040-2016	浮式结构物定位系统设计与分析 Design and analysis of stationkeeping systems for floating structures	SY/T 10040-2002	53507-2016	中海油研究总院
88	SY/T 10041-2002	石油设施电气设备安装一级一类和二类区域划分的推荐作法 Recommended practice for classification of locations for electrical installations at petroleum facilities classified as class I, division 1 and division 2	SY/T 4811-1992	10484-2002	中国海洋石油总公司研究中心开发设计院
89	SY/T 10042-2002	海上生产平台管道系统设计和安装的推荐作法 Recommended practice for design and installation of offshore production platform piping systems	SY/T 4809-1992	10485-2002	中国海洋石油总公司研究中心开发设计院
90	SY/T 10043-2002	卸压和减压系统指南 Guide for pressure-relieving and depressuring systems	SY/T 4812-1992	10486-2002	中国海洋石油总公司研究中心开发设计院
91	SY/T 10044-2002	炼油厂压力泄放装置的尺寸确定、选择和安装的推荐作法 Sizing, selection and installation of pressure-relieving devices in refinerie		10487-2002	中国海洋石油总公司中海石油研究中心开发设计院
92	SY/T 10045-2003	工业生产过程中安全仪表系统的应用 Application of safety instrumented systems for the process industries		11646-2003	中海石油研究中心开发设计院

海洋石油工程

序号	标准编号	标准名称	被代替标准编号	备案号	主编单位
93	SY/T 10046-2012	船舶靠泊海上设施作业规范 Operation rules for vessel berthing offshore installation	SY/T 10046-2003	38032-2013	中海油田服务股份有限公司
94	SY/T 10047-2003	海上油（气）田开发工程环境保护设计规范 Specifications for environmental protection design of offshore oil/gas field development		11648-2003	中海石油研究中心
95	SY/T 10048-2003	腐蚀管道评估的推荐作法 Recommended practice for evaluation of corroded pipelines		11649-2003	中国海洋石油总公司海洋石油工程股份有限公司
96	SY/T 10049-2004	海上钢结构疲劳强度分析推荐作法 Recommended practice for fatigue strength analysis of offshore steel structures		14096-2004	中油辽河工程有限公司
97	SY/T 10050-2004	环境条件和环境荷载规范 The specification of environmental conditions and environmental loads		14097-2004	中油辽河工程有限公司
98	SY/T 10051.1-2004	海上结构用改良韧性的碳锰钢板规范 Specification for carbon manganese steel plate with improved toughness for offshore structur		14098-2004	中海石油研究中心开发设计院
99	SY/T 10051.2-2005	改良缺口韧性的轧制型钢规范 Rolled shapes with improved notch toughness		16515-2005	中海石油研究中心

石油化工工程

序号	标准编号	标准名称	被代替标准编号	备案号	主编单位
1	SH/T 3001-2005	石油化工设备抗震鉴定标准 Standard for aseismic appraisal of petrochemical equipment	SH 3001-1992		中国石化工程建设公司
2	SH/T 3002-2000	石油库节能设计导则 Design guideline for energy conservation of petroleum depots	SHJ 2-87		中国石化工程建设公司
3	SH/T 3003-2000	石油化工合理利用能源设计导则 Design guideline for effectively utilizing energy sources in petrochemical industry	SHJ 3-88		中国石化集团兰州设计院
4	SH/T 3004-2011	石油化工采暖通风与空气调节设计规范 Design specification for heating, ventilation and air conditioning in petrochemical engineering	SH 3004-1999		中国石化集团宁波工程有限公司
5	SH/T 3005-2016	石油化工自动化仪表选型设计规范 Design specification for instrument selection in petrochemical industry	SH 3005-1999		中国石化工程建设有限公司
6	SH/T 3006-2012	石油化工控制室设计规范 Design specification for control room in petrochemical industry	SH 3006-1999		中石化宁波工程有限公司
7	SH/T 3007-2014	石油化工储运系统罐区设计规范 Design specification for tank farms of storage and transportation system in petrochemical industry	SH/T 3007-2007		中国石化工程建设有限公司
8	SH 3008-2000	石油化工厂区绿化设计规范 Design specification for green areas in plant areas of petrochemical industry	SHJ 8-89		中国石化工程建设公司
9	SH 3009-2013	石油化工可燃性气体排放系统设计规范 Design specification for combustible gas discharge system in petrochemical industry	SH 3009-2001		中国石化工程建设有限公司
10	SH/T 3010-2013	石油化工设备和管道绝热工程设计规范 Design specification for the thermal insulation of equipment and pipe in petrochemical industry	SH 3010-2000		中石化洛阳工程有限公司
11	SH 3011-2011	石油化工工艺装置布置设计规范 Specification for design of process plant layout in petrochemical engineering	SH 3011-2000		中国石化工程建设公司
12	SH 3012-2011	石油化工金属管道布置设计规范 Specification for design of metallic piping layout in petrochemical engineering	SH 3012-2000		中国石化工程建设公司
13	SH/T 3013-2000	石油化工厂区竖向布置设计规范 Vertical layout design code in petrochemical industry	SHJ 13-89		中国石化工程建设公司
14	SH/T 3014-2012	石油化工储运系统泵区设计规范 Design specification for pump area of storage and transportation system in petrochemical industry	SH/T 3014-2002		中石化洛阳工程有限公司

石油化工工程

序号	标准编号	标准名称	被代替标准编号	备案号	主编单位
15	SH 3015-2003	石油化工给水排水系统设计规范 Design code for water & waster-water in petrochemical industry	SH 3015-1990		中国石化集团兰州设计院
16	SH 3016-1990	石油化工企业循环水场设计规范 Design code of circulating water plants in petrochemical industry	SHJ 16-90		北京石化工程公司
17	SH/T 3017-2013	石油化工生产建筑设计规范 Design specification for production building in petrochemical industry	SH 3017-1999		中石化宁波工程有限公司
18	SH/T 3018-2003	石油化工安全仪表系统设计规范 Design code for safety instrumented system in petrochemical industry	SH 3018-1990		中国石化工程建设公司
19	SH/T 3019-2016	石油化工仪表管道线路设计规范 Design specification for instrument tubing and wiring in petrochemical industry	SH/T 3019-2003		中石化宁波工程有限公司
20	SH/T 3020-2013	石油化工仪表供气设计规范 Design specification for instrument air supply system in petrochemical industry	SH 3020-2001		中石化宁波工程有限公司
21	SH/T 3021-2013	石油化工仪表及管道隔离和吹洗设计规范 Design specification for sealing and purging of instrument and tubing in petrochemical industry	SH 3021-2001		中石化宁波工程有限公司
22	SH/T 3022-2011	石油化工设备和管道涂料防腐蚀设计规范 Design specification for anticorrosion coating of equipment and piping in petrochemical engineering	SH 3022-1999		中国石化集团宁波工程有限公司
23	SH/T 3023-2005	石油化工厂内道路设计规范 Design specification for road in petrochemical engineering	SH 3023-1990		中国石化工程建设公司
24	SH 3024-1995	石油化工企业环境保护设计规范 Design specification for environmental protection enterprises used in petrochemical industry	SHJ 24-90		中国石化集团北京设计院
25	SH 3025-1990	合成纤维厂环境保护设计规范 Design code of environmental protection for synthetic fiber plants			上海金山石化工程公司
26	SH/T 3026-2005	钢制常压立式圆筒形储罐抗震鉴定标准 Standard for aseismic appraisal of atmospheric vertical cylimdrical storage steel tanks	SH 3026-1990		上海高桥石化公司/上海工程有限公司
27	SH/T 3027-2003	石油化工企业照度设计标准 Design specification for illumination level in petrochemical industry	SH 3027-1990		中国石化集团洛阳石油化工工程公司
28	SH/T 3028-2007	石油化工装置电信设计规范 Specification for design of telecommunication for unit in petrochemical industry	SH 3028-1990		中国石化集团宁波工程有限公司
29	SH/T 3029-2014	石油化工排气筒和火炬塔架设计规范 Design specification for derrick of exhaust and flarestack in petrochemical industry	SH 3029-1991		中国石化工程建设有限公司

石油化工工程

序号	标准编号	标准名称	被代替标准编号	备案号	主编单位
30	SH/T 3030-2009	石油化工塔型设备基础设计规范 Specification for design of tower-type equipment foundation in petrochemical industry	SH 3030-1997		中国石化集团洛阳石油化工工程公司
31	SH/T 3031-2013	石油化工逆流式机械通风冷却塔结构设计规范 Design specification for counterflow-type mechanical draught cooling tower structure in petrochemical industry	SH 3031-1997		中石化洛阳工程有限公司
32	SH/T 3032-2002	石油化工企业总体布置设计规范 Design specification for site plan in petrochemical engineering	SH 3032-1992		中国石化集团洛阳石油化工工程公司
33	SH/T 3033-2007	石油化工企业汽车、叉车运输设施设计规范 Specification for design of automobile & forklift conveyance in petrochemical industry	SH/T 3033-1991		中国石化工程建设公司
34	SH 3034-2012	石油化工给水排水管道设计规范 Specification for design of water supply and wastewater piping in petrochemical industry	SH 3034-1999		中石化宁波工程有限公司
35	SH/T 3035-2007	石油化工工艺装置管径选择导则 Guide for pipeline sizing in petrochemical industry	SH 3035-1991		中国石化工程建设公司
36	SH/T 3036-2012	一般炼油装置用火焰加热炉 Fired heaters for general refinery service	SH/T 3036-2003		中国石化工程建设有限公司/中石化洛阳工程有限公司
37	SH/T 3037-2016	炼油厂加热炉炉管壁厚计算方法 Calculation of heater-tube thickness in petroleum refineries	SH/T 3037-2002		中石化洛阳工程有限公司
38	SH/T 3039-2003	石油化工非埋地管道抗震设计通则 General seismic design rule for non-buried pipe in petrochemical industry	SH 3039-1991		中国石化工程建设公司
39	SH/T 3040-2012	石油化工管道伴管和夹套管设计规范 Specification for design of tracing piping and jacket piping in petrochemical industry	SH/T 3040-2002		中石化上海工程有限公司
40	SH/T 3041-2016	石油化工管道柔性设计规范 Specification for design of piping flexibility in petrochemical industry	SH/T 3041-2002		中国石化工程建设有限公司
41	SH/T 3042-2007	合成纤维厂采暖通风与空气调节设计规范 Design specification of heating ventilation and air conditioning for synthetic fiber	SH 3042-1991		中国石化集团上海工程有限公司
42	SH/T 3043-2014	石油化工设备管道钢结构表面色和标志规定 Specification of surface color and mark on equipment, piping and steel structure in petrochemical industry	SH 3043-2003		中石化上海工程有限公司
43	SH/T 3044-2016	石油化工精密仪器抗震鉴定标准 Standard of seismic appraisal for precision instrument in petrochemical industry	SH/T 3044-2004		中国石油化工股份有限公司石油化工科学研究院

石油化工工程

序号	标准编号	标准名称	被代替标准编号	备案号	主编单位
44	SH/T 3045-2003	石油化工管式炉热效率设计计算 Calculation of thermal efficiency for the design of petrochemical tubular heaters	SH 3045-1992		中国石化工程建设公司
45	SH 3046-1992	石油化工立式圆筒形钢制焊接储罐设计规范 Petro-chemical design specification for vertical cylindrical steel welded storage tanks	SYJ 1016-82		中国石化集团北京设计院
46	SH 3047-1993	石油化工企业职业安全卫生设计规范 Design specification for occupational safety and health in petrochemical industry			中国石化集团北京设计院
47	SH 3048-1999	石油化工钢制设备抗震设计规范 Seismic design specification for petrochemical steel facilities	SH 3048-93		中国石化集团北京设计院
48	SH/T 3051-2014	石油化工配管工程术语 Terms of piping engineering in petrochemical industry	SH/T 3051-2004		中石化洛阳工程有限公司
49	SH/T 3052-2014	石油化工配管工程设计图例 Symbols of piping engineering in petrochemical industry	SH/T 3052-2004		中石化洛阳工程有限公司
50	SH/T 3053-2002	石油化工企业厂区总平面布置设计规范 General plot plan design code of petrochemical enterprise	SH 3053-1993		中国石化工程建设公司
51	SH/T 3054-2005	石油化工厂区管线综合设计规范 Coordination of pipelines design code for petrochemical plant	SH 3054-1993		中国石化工程建设公司
52	SH/T 3055-2007	石油化工管架设计规范 Specification for design of pipe supports in petrochemical industry	SH 3055-1993		中国石化工程建设公司
53	SH 3056-1994	石油化工企业排气筒（管）采样口设计规范 Design specification for sampling port of exhaust stack in petrochemical industry			中国石化集团北京设计院
54	SH/T 3057-2007	石油化工落地式离心泵基础设计规范 Specification for design of ground type centrifugal pump foundation in petrochemical industry	SH 3057-1994		中国石化集团洛阳石油化工工程公司
55	SH/T 3058-2016	石油化工冷换设备和容器基础设计规范 Design specification for exchanger and vessel foundations in petrochemical industry	SH/T 3058-2005		中石化洛阳工程有限公司
56	SH/T 3059-2012	石油化工管道设计器材选用规范 Specification for piping material design selection in petrochemical industry	SH 3059-2001		中国石化工程建设有限公司
57	SH/T 3060-2013	石油化工企业供电系统设计规范 Design specification for electric power supply system in petrochemical enterprises	SH 3060-1994		中国石化工程建设有限公司
58	SH/T 3061-2009	石油化工管式炉基础设计规范 Specification for design of pipesill foundation in petrochemical industry	SH 3061-1994		中国石化工程建设公司

石油化工工程

序号	标准编号	标准名称	被代替标准编号	备案号	主编单位
59	SH/T 3062-2007	石油化工球罐基础设计规范 Specification for design of sphere tank foundation in petrochemical industry	SH 3062-1994		中国石化工程建设公司
60	SH/T 3064-2003	石油化工钢制通用阀门选用、检验及验收 Specification for selection, inspection and acceptance of steel valves used in petrochemical industry	SH 3064-1994		中国石化工程建设公司
61	SH/T 3065-2005	石油化工管式炉急弯弯管技术标准 Technical standard for return bends and elbows of tubular heater in petrochemical industry	SH/T 3065-1994		中国石化集团洛阳石油化工工程公司
62	SH/T 3066-2005	石油化工反应器再生器框架设计规范 Design specification for frame of reactor and regenerator in pertrochemical industry	SH 3066-1995		中国石化集团洛阳石油化工工程公司
63	SH/T 3067-2007	石油化工钢筋混凝土冷换框架设计规范 Specification for design of reinforced concrete frame of coolers and exchangers in petrochemical industry	SH 3067-1995		中国石化集团洛阳石油化工工程公司
64	SH/T 3068-2007	石油化工钢储罐地基与基础设计规范 Specification for design of steel storage tanks subgrade and foundation in petrochemical industry	SH 3068-1995		中国石化工程建设公司
65	SH/T 3070-2005	石油化工管式炉钢结构设计规范 Specification for design of steel structure of tubular heater in petrochemical industry	SH 3070-1995		中国石化集团洛阳石油化工工程公司
66	SH 3071-2013	石油化工电气设备抗震鉴定标准 Standard of aseismic appraisal for electrical equipment in petrochemical industry	SH 3071-1995		中国石化工程建设有限公司
67	SH/T 3073-2016	石油化工管道支吊架设计规范 Specification for design of piping support in petrochemical industry	SH/T 3073-2004		中石化洛阳工程有限公司
68	SH/T 3074-2007	石油化工钢制压力容器 Steel pressure vessels in petrochemical industry	SH 3074-1995		中国石化工程建设公司
69	SH/T 3075-2009	石油化工钢制压力容器材料选用规范 Specification of material selections for steel pressure vessels in petrochemical industry	SH 3075-1995		中国石化工程建设公司
70	SH 3076-2013	石油化工建筑物结构设计规范 Design specification for building structure in petrochemical industry	SH 3076-1996		中国石化工程建设有限公司
71	SH/T 3077-2012	石油化工钢结构冷换框架设计规范 Design specification for steel frames supporting coolers and exchangers in petrochemical industry	SH 3077-96		中国石化工程建设有限公司
72	SH 3078-96	立式圆筒形钢制和铝制料仓设计规范 Specification for the design of vertical cylindrical steel and aluminum silos			上海金山石化工程公司

石油化工工程

序号	标准编号	标准名称	被代替标准编号	备案号	主编单位
73	SH/T 3079-2012	石油化工焦炭塔框架设计规范 Design specification for frame structure of coke column in petrochemical industry	SH 3079-96		中国石化工程建设有限公司
74	SH/T 3080-2013	石油化工横流式机械通风冷却塔结构设计规范 Specification for design of cross-flow mechanical ventilation cooling tower structure in petrochemical industry	SH 3080-1997		中国石化工程建设有限公司
75	SH/T 3081-2003	石油化工仪表接地设计规范 Design code for instrument grounding in petrochemical industry	SH 3081-1997		中国石化工程建设公司
76	SH/T 3082-2003	石油化工仪表供电设计规范 Design code for instrument power supply in petrochemical industry	SH 3082-1997		中国石化集团洛阳石油化工工程公司
77	SH/T 3083-1997	石油化工钢储罐地基处理技术规范 Petro-chemical technical specification for steel storage tanks subgrade treatment			上海高桥石化设计院
78	SH 3084-1997	石油化工总图运输设计图例 Legend for design of general layout and transportation for petro-chemical industry	SYJ 1006-81		中国石化集团北京设计院
79	SH/T 3085-2016	石油化工管式炉碳钢和铬钼钢炉管焊接技术条件 Technical specification for welding of carbon steel and chrome-molybdenum alloy steel tube of petrochemical tubular heater	SH 3085-1997		中石化洛阳工程有限公司
80	SH 3086-1998	石油化工管式炉钢结构工程及部件安装技术条件 Technical specification for steel structure engineering and parts installation of petrochemical tubular heater	SHJ 1037-84		中国石化集团洛阳石油化工工程公司
81	SH 3087-1997	石油化工管式炉耐热钢铸件技术标准 Technical standard for heat resistant steel castings of petrochemical tubular heater	SHJ 1041-84		中国石化集团洛阳石油化工工程公司
82	SH/T 3088-2012	石油化工塔盘技术规范 Technical specification of tower trays in petrochemical industry	SH 3088-1998		中国石化工程建设有限公司
83	SH 3089-1998	石油化工给水排水管道设计图例 Design legend of water supply and drainage piping for petrochemical industry	SYJ 1004-81		中国石化集团北京设计院
84	SH 3090-1998	石油化工铁路设计规范 Code for the design of railway for petrochemical industry	SHJ 1056-84		中国石化集团北京设计院
85	SH/T 3091-2012	石油化工压缩机基础设计规范 Design specification for compressor foundation in petrochemical industry	SH 3091-1998		中国石化工程建设有限公司
86	SH/T 3092-2013	石油化工分散控制系统设计规范 Specification of design for distributed control system in petrochemical industry	SH/T 3092-1999		中国石化工程建设有限公司

石油化工工程

序号	标准编号	标准名称	被代替标准编号	备案号	主编单位
87	SH 3093-1999	石油化工企业卫生防护距离 Health protection zone standard for petrochemical industry	SHJ 1070-86		中国石化集团北京设计院
88	SH/T 3094-2013	石油化工厂区雨水明沟设计规范 Design specification for rainwater ditch in petrochemical industry	SH 3094-1999		中国石化工程建设有限公司
89	SH 3095-2000	石油化工企业污水处理设计规范 Design code for wastewater treatment in petrochemical industry			中国石化集团洛阳石油化工工程公司
90	SH/T 3096-2012	高硫原油加工装置设备和管道设计选材导则 Material selection guideline for design of equipment and piping in units processing sulfur crude oils	SH/T 3096-2001, SH/T 3129-2002		中石化洛阳工程有限公司
91	SH 3097-2000	石油化工静电接地设计规范 Code for the design of static electricity grounding for petrochemical industry	—		北京石化工程公司
92	SH/T 3098-2011	石油化工塔器设计规范 Specification for design of column in petrochemical engineering	SH 3098-2000		中国石化集团宁波工程有限公司
93	SH 3099-2000	石油化工给排水水质标准 Water quality standard of water supply and drainage in petrochemical industry	SHJ 1080-91		中国石化集团北京设计院
94	SH/T 3100-2013	石油化工工程测量规范 Specification for the engineering survey in petrochemical industry	SH 3100-2000		北京东方新星石化工程股份有限公司
95	SH/T 3101-2000	炼油厂流程图图例 Process flow diagram legend for refinery	SYJ 1002-81		中国石化工程建设公司
96	SH/T 3102-2000	石油化工采暖通风与空气调节设计图例 Design symbols and alphabet code for heating, ventilation and air conditioning in petrochemical industry	SYJ 1005-81		中国石化工程建设公司
97	SH/T 3103-2009	石油化工中心化验室设计规范 Specification for design of central laboratory in petrochemical industry	SH/T 3103-2000		中国石化集团洛阳石油化工工程公司
98	SH/T 3104-2013	石油化工仪表安装设计规范 Specification for design of instrument installation in petrochemical industry	SH/T 3104-2000		中石化洛阳工程有限公司
99	SH/T 3105-2000	炼油厂自动化仪表管线平面布置图图例及文字代号 Instrument location plan symbols and alphabet code for refinery	SYJ 1012-82		中国石化工程建设公司
100	SH/T 3106-2009	石油化工企业氮氧系统设计规范 Specification for design of nitrogen and oxygen system in petrochemical industry	SYJ 1017-82		中国石化工程建设公司
101	SH/T 3107-2000	石油化工液体物料铁路装卸车设施设计规范 Specification for the design of tank-car loading and unloading facilities liquid stocks in petrochemical industry	SYJ 1020-82		中石化洛阳工程有限公司

石油化工工程

序号	标准编号	标准名称	被代替标准编号	备案号	主编单位
102	SH/T 3108-2000	炼油厂全厂性工艺及热力管道设计规范 Specification for design of plant-wide process and thermal piping of refineries	SYJ 1024-83		中国石化集团北京设计院
103	SH/T 3109-2001	炼油厂添加剂设施设计规范 Specification for design of additive facilities in refineries	SH/T 3109-2000		中国石化工程建设公司
104	SH/T 3113-2016	石油化工管式炉燃烧器工程技术条件 Engineering specification for burners of tubular heater in petrochemical industry	SH/T 3113-2000		中石化洛阳工程有限公司
105	SH/T 3114-2000	石油化工管式炉耐热铸铁件工程技术条件 Engineering specification for heat resistant iron castings of petrochemical tubular fired heater	SHJ 1043-84		中国石化集团洛阳石油化工工程公司
106	SH/T 3115-2000	石油化工管式炉轻质浇注料衬里工程技术条件 Engineering specification for light castable lining in petrochemical tubular fired heater	SHJ 1045-84		中国石化集团洛阳石油化工工程公司
107	SH/T 3116-2000	炼油厂用电负荷计算方法 Calculation method for electricity load design in petroleum refineries	SHJ 1067-85		中国石化集团洛阳石油化工工程公司
108	SH/T 3117-2013	石油化工设计热力工质消耗量计算方法 Calculation method for design of thermal medium consumption in petrochemical industry	SH/T 3117-2000		中国石化工程建设有限公司
109	SH/T 3118-2000	石油化工蒸汽喷射式抽空器设计规范 Specification for design of steam-jet vacuum pump for petrochemical	SHJ 1073-86		中国石化工程建设公司
110	SH/T 3119-2016	石油化工钢制套管换热器技术规范 Technical specification for design of steel hairpin heat exchangers in petrochemical industry	SH/T 3119-2000		中国石化工程建设有限公司
111	SH/T 3120-2000	石油化工喷射式混合器设计规范 Specification for design of jet mixer for petrochemical	SHJ 1075-86		中国石化集团洛阳石油化工工程公司
112	SH/T 3121-2000	炼油装置工艺设计技术规定 Specification of process design in petroleum processing units	SHJ 1076-86		中国石化工程建设公司
113	SH/T 3122-2013	炼油装置工艺管道流程设计规范 Design specification for process pipeline diagram in refining units	SH/T 3122-2000		中石化洛阳工程有限公司
114	SH/T 3123-2001	石油化工钢储罐地基充水预压监测规程 Petrochemical monitoring code for controlled water test to preload steel storage tanks subgrade			中国石化集团勘察设计院
115	SH/T 3124-2001	石油化工给水排水工艺流程设计图例 Legend of water supply and drainage process flow in petrochemical industry			中国石化工程建设公司

石油化工工程

序号	标准编号	标准名称	被代替标准编号	备案号	主编单位
116	SH/T 3126-2013	石油化工仪表及管道伴热和绝热设计规范 Design specification for tracing and insulation of instrument and piping in petrochemical industry	SH 3126-2001		中石化洛阳工程有限公司
117	SH/T 3127-2001	石油化工管式炉铬钼钢焊接回弯头技术规范 Technical standard for chrome-molybdenum alloy steel plug-welding return of petrochemical tubular heater			中国石化工程建设公司
118	SH/T 3128-2002	一般炼油装置火焰加热炉陶瓷纤维衬里 Ceramic fibre linings of fired heaters for general refinery service			大庆石油化工设计院
119	SH/T 3129-2012	高酸原油加工装置设备和管道设计选材导则 Material selection guideline for design of equipment and piping in units processing acid crude oils	SH/T 3096-2001, SH/T 3129-2002		中石化洛阳工程有限公司
120	SH 3130-2013	石油化工建筑抗震鉴定标准 Standard for seismic appraiser of building in petrochemical industry	SH/T 3130-2002		中国石化工程建设有限公司
121	SH/T 3131-2002	石油化工电气设备抗震设计规范 Seismic design specification for electrical equipment in petrochemical industry			中国石化工程建设公司
122	SH/T 3132-2013	石油化工钢筋混凝土水池结构设计规范 Design specification for reinforced concrete tanks in petrochemical industry	SH/T 3132-2002		中石化洛阳工程有限公司
123	SH/T 3133-2002	石油化工企业现状图图式 Specification for cartographic symbols of present maps in petrochemical engineering			中国石化集团勘察设计院
124	SH/T 3134-2002	采用橇装式加油装置的汽车加油站技术规范 Technical specification for portable filling device in automobile filling station			中国石化工程建设公司
125	SH/T 3135-2003	石油化工工程地震破坏鉴定标准 Authentication standard for earthquake destruction in petrochemical industry			中国石化镇海炼化工程公司
126	SH 3136-2003	液化烃球形储罐安全设计规范 Design specification for safety of liquefied spherical industry			中国石化集团上海工程有限公司
127	SH 3137-2013	石油化工钢结构防火保护技术规范 Technical specification of fire protection for steel structures in petrochemical industry	SH 3137-2003		中石化洛阳石化工程有限公司
128	SH/T 3138-2003	球形储罐整体补强凸缘 Integrated forging reinforcement for spherical tanks			大庆石化设计院
129	SH/T 3139-2011	石油化工重载荷离心泵工程技术规范 Technical specification of centrifugal pumps for heavy duty services in petrochemical engineering	SH/T 3139-2004		中国石化集团上海工程有限公司

石油化工工程

序号	标准编号	标准名称	被代替标准编号	备案号	主编单位
130	SH/T 3140-2011	石油化工中、轻载荷离心泵工程技术规范 Technical specification of centrifugal pumps for medium and light duty services in petrochemical engineering	SH/T 3140-2004		中国石化集团上海工程有限公司
131	SH/T 3141-2013	石油化工往复泵工程技术规范 Technical specification for reciprocating pumps in petrochemical industry	SH/T 3141-2004		中石化上海工程有限公司
132	SH/T 3142-2016	石油化工计量泵工程技术规范 Technical specification of controlled volume pumps in petrochemical industry	SH/T 3142-2004		中石化上海工程有限公司
133	SH/T 3143-2012	石油化工往复压缩机工程技术规范 Technical specification for reciprocating compressor in petrochemical industry	SH/T 3143-2004		中石化洛阳工程有限公司
134	SH/T 3144-2012	石油化工离心、轴流压缩机工程技术规范 Technical specification for centrifugal and axial compressor in petrochemical industry	SH/T 3144-2004		中国石化工程建设有限公司
135	SH/T 3145-2012	石油化工特殊用途汽轮机工程技术规范 Technical specification for special purpose steam turbines in petrochemical industry	SH/T 3145-2004		中石化宁波工程有限公司
136	SH/T 3146-2004	石油化工噪声控制设计规范 Design specification for noise control in petrochemical industry			中国石化集团洛阳石油化工工程公司
137	SH 3147-2014	石油化工构筑物抗震设计规范 Specification for seismic design of special structures in petrochemical industry	SH/T 3147-2004		中国石化工程建设有限公司
138	SH/T 3148-2016	石油化工无密封离心泵工程技术规范 Technical specification of sealless centrifugal pumps in petrochemical industry	SH/T 3148-2007		中石化上海工程有限公司
139	SH/T 3149-2016	石油化工一般用途汽轮机工程技术规范 Technical specification of general purpose steam turbines in petrochemical industry	SH/T 3149-2007		中石化宁波工程有限公司
140	SH/T 3150-2016	石油化工搅拌器工程技术规范 Technical specification of agitators in petrochemical industry	SH/T 3150-2007		中国石化集团上海工程有限公司
141	SH/T 3151-2013	石油化工转子泵工程技术规范 Technical specification for rotary pumps in petrochemical industry	SH/T 3151-2007		中石化上海工程有限公司
142	SH/T 3152-2007	石油化工粉粒产品气力输送工程技术规范 Engineering specification for bulk solid pneumatic conveying in petrochemical industry			中国石化集团南京设计院
143	SH/T 3153-2007	石油化工企业电信设计规范 Specification for design of telecommunication for petrochemical factory			中国石化工程建设公司
144	SH/T 3154-2009	石油化工非金属衬里管道技术规范 Specification of lined steel pipe in petrochemical industry			中国石化集团宁波工程有限公司

石油化工工程

序号	标准编号	标准名称	被代替标准编号	备案号	主编单位
145	SH/T 3156-2009	石油化工离心泵和转子泵用轴封系统工程技术规范 Engineering specification of pumps-SH aft sealing systems of centrifugal and rotary pumps for petrochemical industry			中国石化集团上海工程有限公司
146	SH/T 3157-2009	石油化工回转式压缩机工程技术规范 Engineering specification of rotary-type compressor in petrochemical industry			中国石化集团洛阳石油化工程公司
147	SH/T 3158-2009	石油化工管壳式余热锅炉 Tubular waste heat boiler in petrochemical industry			山东三维石化工程股份有限公司
148	SH/T 3159-2009	石油化工岩土工程勘察规范 Specification of geotechnical engineering investigation in petrochemical industry			北京东方新星石化工程股份有限公司
149	SH/T 3160-2009	石油化工控制室抗爆设计规范 Specification for design of blast resistant control building in petrochemical industry			中国石化集团洛阳石油化工程公司
150	SH/T 3161-2011	石油化工非金属管道技术规范 Technical specification of nonmetallic piping in petrochemical engineering			中国石化集团南京工程有限公司
151	SH/T 3162-2011	石油化工液环真空泵和压缩机工程技术规范 Technical specification of liquid ring vacuum pumps and compressors in petrochemical engineering			中国石化集团宁波工程有限公司
152	SH/T 3163-2011	石油化工静设备分类标准 Specification for classification of static equipment in petrochemical engineering			大庆石化工程有限公司
153	SH/T 3164-2012	石油化工仪表系统防雷设计规范 Specification for design of instrument system lightning surge protection in petrochemical industry			中国石化工程建设有限公司
154	SH/T 3165-2011	石油化工粉体工程设计规范 Design specification of material handling in petrochemical engineering			中国石化集团上海工程有限公司
155	SH/T 3166-2011	石油化工管式炉烟风道结构设计规范 Design specification for flue gas & air ducts of tubular heater in petrochemical engineering			南京金凌石化工程设计有限公司
156	SH/T 3167-2012	钢制焊接低压储罐 Steel welded low-pressure storage tanks			中国石化工程建设有限公司
157	SH/T 3168-2011	石油化工装置（单元）竖向设计规范 Design specification of vertical for unit in petrochemical engineering			大庆石化工程有限公司
158	SH/T 3169-2012	长输油气管道站场布置规范 Specification for general layout of oil and gas pipeline station			中国石油天然气管道工程有限公司

石油化工工程

序号	标准编号	标准名称	被代替标准编号	备案号	主编单位
159	SH/T 3170-2011	石油化工离心风机工程技术规范 Technical specification of centrifugal fan in petrochemical engineering			中国石化集团宁波工程有限公司
160	SH/T 3171-2011	石油化工挠性联轴器工程技术规范 Technical specification of flexible coupling in petrochemical engineering			中国石化集团洛阳石油化工工程公司
161	SH/T 3172-2012	石油化工总图运输术语 Terminology standard for general planning and industrial transportation in petrochemical industry			大庆石化工程有限公司
162	SH 3173-2013	石油化工污水再生利用设计规范 Design specification for wastewater reclamation and reutilization in petrochemical industry			中石化洛阳工程有限公司
163	SH/T 3174-2013	石油化工在线分析仪系统设计规范 Design specification for on-line analyzer systems in petrochemical industry			中国石化工程建设有限公司
164	SH/T 3175-2013	固体工业硫磺储存输送设计规范 Design specification for storage and conveying of solid industrial sulfur			中国石化工程建设有限公司/中石化南京工程有限公司
165	SH/T 3176-2014	石油化工工厂系统工程设计文件编制标准 Compilation specification for the design documents of the plant system in petrochemical industry			中石化洛阳工程有限公司
166	SH/T 3177-2015	加油站用埋地玻璃纤维增强塑料双层油罐工程技术规范 Technical specification for buried glass fiber reinforced plastic double skin oil tanks in fuel filling station			镇海石化工程股份有限公司
167	SH/T 3178-2015	加油站用埋地钢-玻璃纤维增强塑料双层油罐工程技术规范 Technical specification for buried steel-glass fiber reinforced plastics double skin oil tanks in fuel filling station			镇海石化工程股份有限公司
168	SH/T 3179-2016	石油化工管式炉炉衬设计规范 Design specification for linings of tubular heater in petrochemical industry			中国石化工程建设有限公司
169	SH/T 3180-2016	石油化工机泵配套用油系统及其辅助设备工程技术规范 Technical specification for lubrication and oil-control systems and auxiliaries of machinery in petrochemical industry			中石化洛阳工程有限公司
170	SH/T 3181-2016	石油化工仪表远程监控及数据采集系统设计规范 Design specification for instrument remote supervisory control and data acquisition system in petrochemical industry			中石化洛阳工程有限公司

石油化工工程

序号	标准编号	标准名称	被代替标准编号	备案号	主编单位
171	SH/T 3182-2016	石油化工低温余热回收系统设计规范 Design specification for low-temperature waste heat recovery system in petrochemical industry			中石化洛阳工程有限公司
172	SH/T 3401-2013	石油化工钢制管法兰用非金属平垫片 Nonmetallic flat gaskets for steel pipe flanges in petrochemical industry	SH 3401-1996		中国石化工程建设有限公司
173	SH/T 3402-2013	石油化工钢制管法兰用聚四氟乙烯包覆垫片 Polytetrafluoroethylene envelope gaskets for steel pipe flanges in petrochemical industry	SH 3402-1996		中国石化工程建设有限公司
174	SH/T 3403-2013	石油化工钢制管法兰用金属环垫 Metallic ring joint gaskets for steel pipe flanges in petrochemical industry	SH 3403-1996		中国石化工程建设有限公司
175	SH/T 3404-2013	石油化工钢制管法兰用紧固件 Fasteners for steel pipe flanges in petrochemical industry	SH 3404-1996		中国石化工程建设有限公司
176	SH/T 3405-2012	石油化工钢管尺寸系列 Series of steel pipe size in petrochemical industry	SH 3405-1996		中石化洛阳工程有限公司
177	SH/T 3406-2013	石油化工钢制管法兰 Steel pipe flanges in petrochemical industry	SH 3406-1996		中国石化工程建设有限公司
178	SH/T 3407-2013	石油化工钢制管法兰用缠绕式垫片 Spiral wound gaskets for steel pipe flanges in petrochemical industry	SH 3407-1996		中国石化工程建设有限公司
179	SH/T 3408-2012	石油化工钢制对焊管件 Steel butt-welding pipe fittings in petrochemical industry	SH 3408-1996 SH 3409-1996		中石化洛阳工程有限公司
180	SH 3409-1996	钢板制对焊管件 Steel plate butt-welded fittings	SHJ 409-90		中国石化集团洛阳石油化工工程公司
181	SH/T 3410-2012	石油化工锻钢制承插焊和螺纹管件 Forged steel socket-welded and threaded fittings in petrochemical industry	SH 3410-1996		中石化洛阳工程有限公司
182	SH/T 3411-1999	石油化工泵用过滤器选用、检验及验收 Specification for selection, inspection and acceptance of strainer for pump in petrochemical industry			天津石化设计院
183	SH/T 3413-1999	石油化工石油气管道阻火器选用、检验及验收 Specification for selection, inspection and acceptance of pipeline flame arresters for petroleum gas in petrochemical industry			中国石化集团北京设计院
184	SH/T 3415-2005	高频电阻焊螺旋翅片管 High frequency resistance welded helically finned tubes			茂名石化机械厂

石油化工工程

序号	标准编号	标准名称	被代替标准编号	备案号	主编单位
185	SH/T 3416-2005	石油化工用套管结晶器 Double pipe crystallizers in petrochemical industry			茂名石化机械厂
186	SH/T 3417-2007	石油化工管式炉高合金炉管焊接工程技术条件 Technical specification of weld engineering for high alloy steel tube of petrochemical tubular heater			中国石化集团宁波工程有限公司
187	SH/T 3418-2007	石油化工换热器钢制鞍式支座技术条件 Technical specification of saddle for heat exchanger in petrochemical industry			中国石化集团上海工程有限公司
188	SH/T 3419-2007	钢制异径短节 Steel swage（d）nipples			中国石化工程建设公司
189	SH/T 3420-2007	石油化工管式炉用空气预热器通用技术条件 General specification for air preheater of tubular heater in petrochemical industry			中国石化集团洛阳石油化工工程公司
190	SH/T 3421-2009	金属波纹管膨胀节设置和选用通则 General rules for arrangement and selection of metal bellows expansion joint			中国石化集团宁波工程有限公司
191	SH/T 3422-2011	石油化工管式炉钉头管技术标准 Technical specification for stud tubes of tubular heater in petrochemical engineering			中国石化集团宁波工程有限公司
192	SH/T 3423-2011	石油化工管式炉用铸造高合金炉管及管件技术条件 Technical specification for high alloy cast tube and fitting of tubular heater in petrochemical engineering			中国石化工程建设公司
193	SH/T 3424-2011	石油化工锻钢制承插焊和螺纹活接头 Forged steel, socket-welding and threaded unions in petrochemical engineering			中国石化工程建设公司
194	SH/T 3425-2011	石油化工钢制管道用盲板 Steel line blanks in petrochemical engineering			中国石化工程建设公司
195	SH/T 3426-2014	石油化工钢制夹套管法兰 Steel jacketed pipe flanges in petrochemical industry			中石化上海工程有限公司
196	SH/T 3500-2016	石油化工工程质量监督规范 Specification for quality supervision of petrochemical project			石油化工工程质量监督总站
197	SH 3501-2011	石油化工有毒、可燃介质钢制管道工程施工及验收规范 Construction and acceptance specification for toxic and combustible medium steel piping engineering in petrochemical industry	SH 3501-2002		中国石化集团第十建设公司/中国石化集团第五建设公司
198	SH/T 3502-2009	钛和锆管道施工及验收规范 Specification for construction and acceptance of titanium and zirconium piping	SH 3502-2000		北京燕华建筑安装工程有限责任公司

石油化工工程

序号	标准编号	标准名称	被代替标准编号	备案号	主编单位
199	SH/T 3503-2007	石油化工建设工程项目交工技术文件规定（中英文表格） Regulation for technical document of construction completion for petrochemical project	SH 3503-2001		北京燕山石化工程质量监督站/中国石化集团第四建设公司
200	SH/T 3504-2014	石油化工隔热耐磨衬里设备和管道施工质量验收规范 Acceptance specification for construction quality of equipment and piping with thermal insulating and abrasion resistant refractory lining in petrochemial industry	SH/T 3504-2009		中石化第四建设有限公司
201	SH/T 3506-2007	管式炉安装工程施工及验收规范 Specification for construction and acceptance for installation work of tubular fired heater	SH 3506-2000		中国石化集团第十建设公司
202	SH/T 3507-2011	石油化工钢结构工程施工质量验收规范 Acceptance specification for construction quality of steel structure in petrochemical industry	SH/T 3507-2005		中国石化集团宁波工程有限公司
203	SH/T 3508-2011	石油化工安装工程施工质量验收统一标准 Unification criterion of construction quality check and acceptance for petrochemical engineering project installation			石油化工工程质量监督总站燕山石化分站/中国石化集团第四建设公司
204	SH/T 3510-2011	石油化工设备混凝土基础工程施工质量验收规范 Acceptance specification for consruction quality of equipment concrete foundations in petrochemical industry	SH 3510-2000		中国石化集团第十建设公司
205	SH/T 3511-2007	石油化工乙烯裂解炉和制氢转化炉施工技术规程 Technical specification for construction of ethylene cracking heater and hydrogen product reformer in petrochemical industry	SH/T 3511-2000		中国石化集团第十建设公司
206	SH/T 3512-2011	石油化工球形储罐施工技术规程 Technical specification for construction of spherical tanks in petrochemical industry	SH/T 3512-2002		中国石化集团第十建设公司
207	SH/T 3513-2009	石油化工铝制料仓施工质量验收规范 Construction quality inspection & acceptance specification for petrochemical aluminum silos	SH 3513-2000		中国石化集团第四建设公司
208	SH/T 3515-2003	大型设备吊装工程施工工艺标准 Construction standard for large-size equipment hoisting engineering	SH/T 3515-1990		中国石化集团第十建设公司
209	SH/T 3516-2012	催化裂化装置轴流压缩机-烟气轮机能量回收机组施工及验收规范 Specification for construction and acceptance of axial compressor-flue gas expander energy recovery unit in flow catalytic cracking unit	SH/T 3516-2001		中石化第十建设有限公司
210	SH/T 3517-2013	石油化工钢制管道工程施工技术规范 Technical specification for construction of steel piping in petrochemical industry	SH/T 3517-2001		中石化第五建设有限公司

石油化工工程

序号	标准编号	标准名称	被代替标准编号	备案号	主编单位
211	SH 3518-2013	石油化工阀门检验与管理规范 Specification for valves inspection and management in petrochemical industry	SH 3518-2000		中石化第十建设有限公司
212	SH/T 3519-2013	乙烯装置离心压缩机组施工及验收规范 Specification for construction and acceptance of centrifugal compressor unit for ethylene plant	SH/T 3519-2002		中石化第十建设有限公司
213	SH/T 3520-2015	石油化工铬钼钢焊接规范 Welding specification of chrome molybdenum steel in petrochemical industry	SH/T 3520-2004		北京燕华工程建设有限公司
214	SH/T 3521-2013	石油化工仪表工程施工技术规程 Technical specification for construction of instrumentation engineering in petrochemical industry	SH/T 3521-2007		中石化第十建设有限公司
215	SH/T 3522-2003	石油化工隔热工程施工工艺标准 Construction standard for heat insulation work in petrochemical industry	SH/T 3522-1991		中国石化集团第四建设公司
216	SH/T 3523-2009	石油化工铬镍不锈钢、铁镍合金和镍合金焊接规程 Specification for welding of Cr-Ni stainless steel, Fe-Ni and Ni alloy in petrochemical industry	SH/T 3523-1999		北京燕华建筑安装工程有限责任公司
217	SH/T 3524-2009	石油化工静设备现场组焊技术规程 Technical specification of site assembling and welding for petrochemical static equipment	SH 3524-1999		中国石化集团第二建设公司
218	SH/T 3525-2015	石油化工低温钢焊接规范 Welding specification of low temperature steel in petrochemical industry	SH/T 3525-2004		中石化第十建设有限公司 惠生工程（中国）有限公司
219	SH/T 3526-2015	石油化工异种钢焊接规范 Welding specification of dissimilar steels in petrochemical industry	SH/T 3526-2004		中石化第四建设有限公司
220	SH/T 3527-2009	石油化工不锈钢复合钢焊接规程 Specification for welding of stainless clad steel in petrochemical industry	SH 3527-1999		中国石化集团第四建设公司
221	SH/T 3528-2014	石油化工钢制储罐地基与基础施工及验收规范 Specification for construction and acceptance of steel storage tank subgrade & foundation in petrochemical industry	SH/T 3528-2005		中石化第四建设有限公司
222	SH/T 3529-2005	石油化工厂区竖向工程施工及验收规范 Specification of construction and acceptance for vertical engineering in petrochemical plant	SH 3529-1993		中国石化集团第二建设公司
223	SH/T 3530-2011	石油化工立式圆筒形钢制储罐施工技术规程 Technical specification for vertical cylindrical steel storage tank in petrochemical industry	SH/T 3530-2001		中国石化集团第四建设公司

石油化工工程

序号	标准编号	标准名称	被代替标准编号	备案号	主编单位
224	SH/T 3533-2013	石油化工给水排水管道工程施工及验收规范 Specification for construction and acceptance of water and sewerage pipeline in petrochemical industry	SH 3533-2003		中石化南京工程有限公司
225	SH/T 3534-2012	石油化工筑炉工程施工质量验收规范 Specification for quality inspection and acceptance of furnace brick lining engineering in petrochemical industry	SH 3534-2001		天津金耐达筑炉衬里有限公司
226	SH/T 3535-2012	石油化工混凝土水池工程施工及验收规范 Specification for construction and acceptance of concrete water pit engineering in petrochemical industry	SH/T 3535-2002		天津众业石化建筑安装工程有限公司
227	SH/T 3536-2011	石油化工工程起重施工规范 Specification for lifting in petrochemical industry	SH/T 3536-2002		中国石化集团第四建设公司
228	SH/T 3537-2009	立式圆筒形低温储罐施工技术规程 Technical specification for construction of vertical cylindrical low temperature storage tank	SH 3537-2002		中国石化集团第二建设公司
229	SH/T 3539-2007	石油化工离心式压缩机组施工及验收规范 Specification for construction and acceptance of petrochemical centifugal compressor unit			中国石化集团第二建设公司
230	SH/T 3540-2007	钢制换热设备管束复合涂层施工及验收规范 Specification for construction and acceptance of complex coating for steel heat exchanger bundle			中国石化工程建设公司
231	SH/T 3541-2007	石油化工泵组施工及验收规范 Specification for construction and acceptance of pump unit installation in petrochemical industry			中国石化集团公司第十建设公司
232	SH/T 3542-2007	石油化工静设备安装工程施工技术规程 Technical procedure for construction of static equipment installation in petrochemical industry			中国石化集团第四建设公司
233	SH/T 3543-2007	石油化工建设工程项目施工过程技术文件规定 Specification for process technical document of petrochemical project construction			北京燕山石化工程质量监督站/中国石化集团第四建设公司
234	SH/T 3544-2009	石油化工对置式往复压缩机组施工及验收规范 Specification for construction and acceptance of opposed reciprocating compressor unit in petrochemical industry			中国石化集团宁波工程有限公司
235	SH/T 3545-2011	石油化工管道无损检测标准 Specification for NDT on pipelines in petrochemical industry			南京金陵检测工程有限公司

石油化工工程

序号	标准编号	标准名称	被代替标准编号	备案号	主编单位
236	SH/T 3546-2011	石油化工夹套管施工及验收规范 Construction and acceptance specifications for jacket piping in petrochemical industry			中国石化集团第十建设公司
237	SH/T 3547-2011	石油化工设备和管道化学清洗施工及验收规范 Construction and acceptance specifications for chemistry cleaning of equipment and piping in petrochemical industry			中国石化集团第十建设公司
238	SH/T 3548-2011	石油化工涂料防腐蚀工程施工质量验收规范 Acceptance specificationfor construction quality of anticorrosive coating in petrochemical industry			中国石化集团宁波工程有限公司
239	SH/T 3549-2012	酸性环境可燃流体输送管道施工及验收规范 Specification for construction and acceptance of combustible fluid pipage in acid service condition			中石化第十建设有限公司
240	SH/T 3550-2012	石油化工建设工程项目施工技术文件编制规范 Specification for preparation of technical documentation in petrochemical construction projects			中石化第四建设有限公司
241	SH/T 3551-2013	石油化工仪表工程施工质量验收规范 Specification for quality inspection and acceptance of instrumentation engineering in petrochemical industry			中石化第十建设有限公司
242	SH/T 3552-2013	石油化工电气工程施工质量验收规范 Specification for quality inspection and acceptance of electric engineering in petrochemical industry			中石化宁波工程有限公司
243	SH/T 3553-2013	石油化工汽轮机施工及验收规范 Specification for construction and acceptance of steam turbine installation in petrochemical industry			中石化南京工程有限公司
244	SH/T 3554-2013	石油化工钢制管道焊接热处理规范 Specification for heat treatment of steel piping in petrochemical industry			中石化宁波工程有限公司
245	SH/T 3555-2014	石油化工工程钢脚手架搭设安全技术规范 Technical specification for construction safety of steel scaffold building in petrochemical industry			中石化第四建设有限公司 天津星源石化工程有限公司
246	SH/T 3556-2015	石油化工工程临时用电配电箱安全技术规范 Technical specification for electricity safety of distribution box in petrochemical industry			中石化炼化工程（集团）股份有限公司
247	SH/T 3557-2015	石油化工大型设备运输施工规范 Construction specification for large equipment transporting in petrochemical industry			中石化宁波工程有限公司 中创物流股份有限公司

石油化工工程

序号	标准编号	标准名称	被代替标准编号	备案号	主编单位
248	SH/T 3558-2016	石油化工工程焊接通用规范 General specification of welding for engineering in petrochemical industry			石油化工工程质量监督总站燕山石化分站
249	SH/T 3601-2009	催化裂化装置反应再生系统设备施工技术规程 Regulation for construction technology of reactor-regenerator system equipment of catalytic cracking unit	SH 3601-2000		中国石化集团第四建设公司
250	SH/T 3602-2009	石油化工管式炉用燃烧器试验检测规程 Testing and measurements code of burner tubular heater in petrochemical industry			中国石化工程建设公司
251	SH/T 3603-2009	石油化工钢结构防腐蚀涂料应用技术规程 Technical specification for the coating anticorrosion of steel structures in petrochemical industry			中国石化集团上海工程有限公司
252	SH/T 3604-2009	石油化工水泥基无收缩灌浆材料应用技术规程 Technical specification for the application of cement-based nonSH rinkage grouting materials in petrochemical industry			中国石化集团上海工程有限公司
253	SH/T 3605-2009	石油化工铝制料仓施工技术规程 Construction technical specification for petrochemical aluminum silos			中国石化集团第四建设公司
254	SH/T 3606-2011	石油化工涂料防腐蚀工程施工技术规程 Technical specification for construction of anticorrosive coating in petrochemical industry			中国石化集团宁波工程有限公司
255	SH/T 3607-2011	石油化工钢结构工程施工技术规程 Technical specification for construction of steel structure in petrochemical industry			中国石化集团宁波工程有限公司
256	SH/T 3608-2011	石油化工设备混凝土基础工程施工技术规程 Technical specification for concrete foundation of equipment in petrochemical industry			中国石化集团第十建设公司
257	SH/T 3609-2011	石油化工隔热耐磨衬里施工技术规程 Technical specification for heat-insulated and wear-resistant linings in petrochemical industry			中国石化集团第四建设公司/天津金耐达筑炉衬里有限公司
258	SH/T 3610-2012	石油化工筑炉工程施工技术规程 Technical specification for construction of furnace brick lining engineering in petrochemical industry			天津金耐达筑炉衬里有限公司
259	SH/T 3611-2012	酸性环境可燃流体输送管道焊接规程 Welding specification of combustible medium piping in acid service condition			中石化第十建设有限公司
260	SH 3612-2013	石油化工电气工程施工技术规程 Technical specification for construction of electric engineering in petrochemical industry			中石化第四建设有限公司

石油化工工程

序号	标准编号	标准名称	被代替标准编号	备案号	主编单位
261	SH/T 3613-2013	石油化工非金属管道工程施工技术规程 Technical specification for construction of nonmetallic piping in petrochemical industry			中石化第五建设有限公司
262	SH/T 3901-1994	工程设计计算机软件开发及文档编制规范 Procedure for computer software development and documentation			中国石化集团北京设计院
263	SH/T 3902-2014	石油化工配管工程常用缩略语 Abbreviations of piping engineering in petrochemical industry	SH/T 3902-2004		中石化洛阳工程有限公司
264	SH/T 3903-2004	石油化工建设工程项目监理规范 Code petrochemical construction project management			南京扬子石化工程监理有限责任公司
265	SH/T 3904-2014	石油化工建设工程项目竣工验收规定 Regulation for acceptance of completed construction projects in petrochemical industry	SH/T 3904-2005		石油化工工程质量监督总站燕山石化分站
266	SH/T 3905-2007	石油化工企业地下管网管理工作导则 Management guideline for underground pipe works of petrochemical factory			中国石油化工股份公司洛阳分公司

化工工程

序号	标准编号	标准名称	被代替标准编号	备案号	主编单位
1	HGJ 205-92	化工机器安装工程施工及验收规范（离心式压缩机）			中国化学工程集团公司 中国化学工程第十四建设有限公司 全国化工施工标准化管理中心站
2	HGJ 206-92	化工机器安装工程施工及验收规范（中小型活塞式压缩机）			中国化学工程集团公司 中国化学工程第十四建设有限公司 全国化工施工标准化管理中心站
3	HGJ 212-83	金属焊接结构湿式气柜施工及验收规范			中国化学工程集团公司 中化二建集团有限公司 全国化工施工标准化管理中心站
4	HGJ 222-92	铝及铝合金焊接技术规程			中国化学工程集团公司 中国化学工程第十三建设有限公司 全国化工施工标准化管理中心站
5	HGJ 223-92	铜及铜合金焊接及钎焊技术规程			中国化学工程集团公司 中国化学工程第十三建设有限公司 全国化工施工标准化管理中心站
6	HGJ 229-91	工业设备、管道防腐蚀工程施工及验收规范			中国化学工程集团公司 中化二建集团有限公司 全国化工施工标准化管理中心站
7	HGJ 232-92	化学工业大、中型装置生产准备工作规范			中国化学工程总公司
8	HG 20201-2000	工程建设安装工程起重施工规范			中国化学工程总公司 中国化学工程第七建设有限公司 全国化工施工标准化管理中心站 陕西化建工程有限责任公司
9	HG/T 20201-2007	带压密封技术规范		J666-2007	全国化工施工标准化管理中心站 北京巨业华能科技开发有限责任公司
10	HG 20202-2014	脱脂工程施工及验收规范		J1973-2015	中国化学工程集团公司 中国化学工程第三建设有限公司 全国化工施工标准化管理中心站
11	HG 20203-2000	化工机器安装工程施工及验收通用规范			中国化学工程总公司 中国化学工程第十三建设有限公司 全国化工施工标准化管理中心站

化工工程

序号	标准编号	标准名称	被代替标准编号	备案号	主编单位
12	HG 20226-2015	管式炉安装工程施工及验收规范		J2034-2015	中国化学工程集团公司 中国化学工程第三建设公司 全国化工施工标准化管理中心站
13	HG 20231-2014	化学工业建设项目试车规范		J1974-2015	中国化学工程集团公司 五环科技股份有限公司 中国天辰工程有限公司 全国化工施工标准化管理中心站
14	HG 20234-93	化工建设项目进口设备、材料检验大纲			中国化学工程第十三建设有限公司
15	HG 20235-2014	化工建设项目施工组织设计标准		J1975-2015	中国化学工程集团公司 中国化学工程第三建设有限公司 全国化工施工标准化管理中心站
16	HG 20236-93	化工设备安装工程质量检验评定标准			中国化学工程第十四建设有限公司 全国化工施工标准化管理中心站
17	HG/T 20237-2014	化学工业工程建设交工技术文件规定		J1976-2015	中国化学工程集团公司 中国化学工程第十一建设有限公司 全国化工施工标准化管理中心站
18	HG/T 20256-2016	化工高压管道通用技术规范		J2139-2016	中国化学工程第四建设公司 岳阳筑盛阀门管道有限责任公司 中国化学工程股份有限公司
19	HG/T 20272-2011	镁钢制品绝热工程施工技术规范			全国化工施工标准化管理中心站
20	HG/T 20273-2011	喷涂型聚脲防护材料涂装工程技术规范			上海瑞鹏化工材料科技有限公司 华东理工大学华昌聚合物有限公司 全国化工施工标准化管理中心站
21	HG/T 20501-2013	化工建设项目环境保护监测站设计规定		J1657-2013	全国环境保护设计技术中心站
22	HG/T 20504-2013	化工危险废物填埋场设计规定		J1660-2013	全国环境保护设计技术中心站
23	HG/T 20505-2014	过程检测和控制仪表的功能标识及图形符号 Functional identification and symbols for process measuring and controlling instrumentation		J1824-2014	中国成达工程有限公司
24	HG/T 20507-2014	自动化仪表选型设计规范 Design code for instrument selection		J1809-2014	中国寰球工程公司 中国五环工程公司
25	HG/T 20508-2014	控制室设计规范 Design code for control room		J1808-2014	中石化宁波工程有限公司
26	HG/T 20509-2014	仪表供电设计规范 Design code for instrument power supply system		J1823-2014	中国成达工程有限公司

化工工程

序号	标准编号	标准名称	被代替标准编号	备案号	主编单位
27	HG/T 20510-2014	仪表供气设计规范 Design code for instrument air engineering		J1822-2014	赛鼎工程有限公司
28	HG/T 20511-2014	信号报警及安全及联锁系统设计规范 Design code for signal alarm and interlock system engineering		J1821-2014	东华工程科技股份有限公司
29	HG/T 20512-2014	仪表配管配线设计规范 Design code for instrument piping and wiring		J1820-2014	中石化南京工程有限公司
30	HG/T 20513-2014	仪表系统接地设计规范 Design code of instrument grounding		J1819-2014	华陆工程科技有限责任公司
31	HG/T 20514-2014	仪表及管线伴热和绝热保温设计规范 Design code for tracing and insulation of instrument and impulse line			中国石油集团东北炼化工程有限公司吉林设计院
32	HG/T 20515-2014	仪表隔离和吹洗设计规范 Design code for instrument seal and purge		J1817-2014	中国五环工程有限公司
33	HG/T 20516-2014	自动分析器室设计规范 Design code for analyzer room		J1816-2014	中石化宁波工程有限公司
34	HG 20517-1992	钢制低压湿式气柜			化工部第三设计院
35	HG/T 20518-2008	化工粉体工程设计通用规范 Code for design of material handling of chemical industry			五环科技股份有限公司 中国石化集团南京设计院
36	HG/T 20519-2009	化工工艺设计施工图内容和深度统一规定 Specification on content and procedure of detail design for chemical plant			全国化工工艺配管设计技术中心站 东华工程科技股份有限公司
37	HG 20520-1992	玻璃钢/聚氯乙烯（FRP/PVC）复合管道设计规定			化工部第三设计院
38	HG/T 20524-2006	化工企业循环冷却水处理加药装置设计统一规定 Specification on design of chemical feed system of recirculating cooling water treatment for chemical enterprises			华陆工程科技有限责任公司
39	HG/T 20525-2006	化学工业管式炉传热计算设计规定 Design specification on heat transfer calculation of tubular furnace in chemical industry			化学工业第二设计院
40	HG 20528-92	衬里钢管用承插环松套钢管法兰			化工部设备设计技术中心站
41	HG 20530-92	钢制管法兰用焊唇密封环			化工部设备设计技术中心站
42	HG 20531-1993	铸钢、铸铁容器			化工部第四设计院
43	HG 20532-93	化工粉体工程设计安全卫生规定 Safety and hygiene specifications for the powder engineering of chemical industry			化工部第八设计院
44	HG 20533-93	化工机械化运输设计常用名词术语统一规定 The specification of the terms for chemical bulk material handling			中国寰球化学工程公司
45	HG 20534-93	化工固体原、燃料制备设计规定 Design regulations concerning preparation of chemical solid feed stocks and fuels			化工部第二设计院

化工工程

序号	标准编号	标准名称	被代替标准编号	备案号	主编单位
46	HG 20535-93	化工固体物料装卸系统设计规定 Design regulations for loading and unloading system of chemical solid materials			化工部第六设计院
47	HG/T 20536-1993	聚四氟乙烯衬里设备			化工部设备设计技术中心站
48	HG/T 20537.1-92	奥氏体不锈钢焊接钢管选用规定			化工部设备设计技术中心站
49	HG/T 20537.2-92	管壳式换热器用奥氏体不锈钢焊接钢管技术要求			化工部设备设计技术中心站
50	HG/T 20537.3-92	化工装置用奥氏体不锈钢焊接钢管技术要求			化工部设备设计技术中心站
51	HG/T 20537.4-92	化工装置用奥氏体不锈钢大口径焊接钢管技术要求			化工部设备设计技术中心站
52	HG/T 20538-2016	衬塑钢管和管件选用系列 Selection series of plastic-lined pipe and fittings		J2141-2016	中国寰球工程公司
53	HG 20539-1992	增强聚丙烯（FRPP）管和管件			中国寰球化学工程公司
54	HG 20540-1992	化肥厂电力设计技术规定			中国五环化学工程公司
55	HG/T 20541-2006	化学工业炉结构设计规定 Design specification of structure of chemical industrial fuenace			全国化工工业炉设计技术中心站
56	HG/T 20542-2006	电石炉砌筑技术条件 Technical specification of linling layling for calcium carbide furnace			全国化工工业炉设计技术中心站
57	HG/T 20543-2006	化学工业炉砌筑技术条件 Technical specification of lining for chemical industrial furnace			上海化工设计院有限公司 全国化工工业炉设计技术中心站
58	HG/T 20544-2006	化学工业炉结构安装技术条件 Technical specifiction on erection of structure parts for chemical industrial furnace			华陆工程科技有限责任公司
59	HG/T 20545-92	化学工业炉受压元件制造技术规范			化工部第八设计院
60	HG/T 20546-2009	化工装置设备布置设计规定 Design specification of equipment layout for chemical plant		J980-2010	全国化工工艺配管设计技术中心站 中国天辰工程有限公司 华陆工程科技有限责任公司
61	HG/T 20547-2000	化工粉粒产品包装计量准确度规定 Specification on Accuracy of Package and Measure for Chemical Powder-Granule Products			中国五环化学工程公司
62	HG/T 20549-1998	化工装置管道布置设计规定 Design standard piping layout for chemical plant			全国化工工艺配管设计技术中心站 中国寰球化学工程公司
63	HG 20551-93	化工厂电力设计常用计算规定			中国五环化学工程公司
64	HG/T 20552-2016	化工企业化学水处理设计计算标准 Calculation standard for design of chemical water treatment in chemical plant		J2138-2016	中石化南京工程有限公司 中石化宁波工程有限公司
65	HG/T 20553-2011	化工配管用无缝钢管及焊接钢管尺寸选用系列 Selection series of steel pipe size for chemical piping		J1207-2011	全国化工工艺配管设计技术中心站 湖南化工医药设计院

化工工程

序号	标准编号	标准名称	被代替标准编号	备案号	主编单位
66	HG/T 20554-93	活塞式压缩机基础设计规定			中国五环化学工程公司
67	HG/T 20555-2006	离心式压缩机基础设计规定			吉化公司设计院
68	HG 20556-93	化工厂控制室建筑设计规定			中石化兰州设计院
69	HG 20557-93	工艺系统设计管理规定			化工部工艺系统设计技术中心站 中国寰球化学工程公司
70	HG 20558-93	工艺系统设计文件内容的规定			化工部工艺系统设计技术中心站 中国寰球化学工程公司
71	HG 20559-93	管道仪表流程图设计规定			化工部工艺系统设计技术中心站 中国寰球化学工程公司
72	HG/T 20560-97	化工机械化运输工艺设计施工图内容和深度规定 The Process design specification of the detail design content and procedure for chemical bulk material handling			化工部粉体工程设计技术中心站
73	HG/T 20561-1994	化工工厂总图运输施工图设计文件编制深度规定			中国华陆工程公司
74	HG/T 20566-2011	化工回转窑设计规定 Design specification for chemical rotary kiln			东华工程科技股份有限公司
75	HG/T 20567-2013	热油炉技术条件 Techical specification of Heat Carrier Furnacegxb		J1666-2013	全国化工工业炉设计技术中心站
76	HG/T 20568-2014	化工粉体物料堆场及仓库设计规定 The design specification of the ground and storehouse for chemical solid material		J1886-2014	中石化南京工程有限公司
77	HG/T 20569-2013	机械搅拌设备 Mechanical Mixing Equipment		J1663-2013	全国化工设备设计技术中心站
78	HG/T 20570-95	工艺系统工程设计技术规定			中国寰球化学工程公司
79	HG 20571-2014	化工企业安全卫生设计规范 Code for safety and hygiene design of chemical enterprise		J1815-2014	中国天辰工程有限公司
80	HG/T 20572-2007	化工企业给排水专业施工图设计深度统一规定 Rules of content depth for design of construction drawing of water supply and drainage in chemical industry		J664-2007	化学工业部第二设计院 全国化工给排水设计中心站
81	HG/T 20573-2012	分散型控制系统工程设计规范 Code for distributed control system engineering design		J1433-2012	中国成达工程有限公司
82	HG/T 20575-95	化学工业炉阻力计算规定 Specification of resistance calculation for chemical industrial furnace			中石化北京石化工程公司
83	HG/T 20576-98	粉粒体静壁面摩擦系数的测定 Measurement of Friction Coefficients of Powder/Particle and Static Wall			化工部化学工程设计技术中心站
84	HG/T 20577-2013	塔填料流体力学及传质性能测试规范		J1658-2013	化学工程中心站

化工工程

序号	标准编号	标准名称	被代替标准编号	备案号	主编单位
85	HG/T 20578-2013	真空预压法加固软土地基施工技术规程 Technical Specification of Vacuum Preloading Construction		J1661-2013	中国化学工程第一岩土工程有限公司
86	HG/T 20580-2011	钢制化工容器设计基础规定 Specification for design base of steel chemical vessels			中国石化集团宁波工程有限公司
87	HG/T 20581-2011	钢制化工容器材料选用规定 Specification for materials selected of steel chemical vessels			北京冠天能工程科技有限公司
88	HG/T 20582-2011	钢制化工容器强度计算规定 Specification for stress calculation of steel chemical vessels			上海工程化学设计院有限公司
89	HG/T 20583-2011	钢制化工容器结构设计规定 Specification for structural design of steel chemical vessels			赛鼎工程有限公司
90	HG/T 20584-2011	钢制化工容器制造技术要求 Technical requirements for fabrication of steel chemical vessels			锦西化工机械（集团）有限责任公司
91	HG/T 20585-2011	钢制低温压力容器技术规定 Technical specification for steel low temperature pressure vessels			中石油东北炼化工程有限公司吉林设计院
92	HG/T 20586-1996	化工企业照明设计技术规定			中国天辰化学工程公司
93	HG/T 20587-96	化工建筑涂装设计规定			中国成达工程公司
94	HG/T 20588-2012	化工建筑、结构施工图内容、深度统一规定		J1436-2012	中国天辰化学工程公司
95	HG/T 20589-2011	化学工业炉受压元件强度计算规定 Specification of strength calculation on pressure parts for chemical industrial furnace			中国石化集团洛阳石油化工工程公司
96	HG/T 20590-1997	硫酸、磷肥生产污水处理设计技术规定 Design Technical Regulation for the treatment of wastewater from sulphuric acid & phosphate fertilizer industries			南化集团设计院
97	HG/T 20592-2009	钢制管法兰（PN系列） Steel pipe flanges (PN designated)		J883-2009	全国化工设备设计技术中心站
98	HG/T 20593-2014	钢制化工设备焊接与检验工程技术规范		J1978-2015	惠生工程（中国）有限公司
99	HG/T 20606-2009	钢制管法兰用非金属平垫片（PN系列） Non-metallic flat gaskets for use with steel pipe flanges (PN designated)			全国化工设备设计技术中心站
100	HG/T 20607-2009	钢制管法兰用聚四氟乙烯包覆垫片（PN系列） PTFE envelope gaskets for use with steel pipe flanges (PN designated)			全国化工设备设计技术中心站
101	HG/T 20609-2009	钢制管法兰用金属包覆垫片（PN系列） Metal jacketed gaskets for use with steel pipe flanges (PN designated)			全国化工设备设计技术中心站
102	HG/T 20610-2009	钢制管法兰用缠绕式垫片（PN系列） Spiral wound gaskets for use with steel pipe flanges (PN designated)			全国化工设备设计技术中心站

化工工程

序号	标准编号	标准名称	被代替标准编号	备案号	主编单位
103	HG/T 20611-2009	钢制管法兰用具有覆盖层的齿形组合垫（PN 系列） Covered serrated metal gaskets for use with steel pipe flanges（PN designated）			全国化工设备设计技术中心站
104	HG/T 20612-2009	钢制管法兰用金属环形垫（PN 系列） Metallic ring joint gaskets for use with steel pipe flanges（PN designated）			全国化工设备设计技术中心站
105	HG/T 20613-2009	钢制管法兰用紧固件（PN 系列） Bolting for use with steel pipe flanges（PN designated）			全国化工设备设计技术中心站
106	HG/T 20614-2009	钢制管法兰、垫片、紧固件选用配合规定（PN 系列） Specification for selection of Steel pipe flanges, gaskets, bolting（PN designated）			全国化工设备设计技术中心站
107	HG/T 20615-2009	钢制管法兰（Class 系列） Steel pipe flanges（Class designated）			全国化工设备设计技术中心站
108	HG/T 20623-2009	大直径钢制管法兰（Class 系列） Large diameter steel pipe flanges（Class designated）			全国化工设备设计技术中心站
109	HG/T 20627-2009	钢制管法兰用非金属平垫片（Class 系列） Non-metallic flat gaskets for use with steel pipe flanges（Class designated）			全国化工设备设计技术中心站
110	HG/T 20628-2009	钢制管法兰用聚四氟乙烯包覆垫片（Class 系列） PTFE envelope gaskets for use with steel pipe flanges（Class designated）			全国化工设备设计技术中心站
111	HG/T 20630-2009	钢制管法兰用金属包覆垫片（Class 系列） Metal jacketed gaskets for use with steel pipe flanges（Class designated）			全国化工设备设计技术中心站
112	HG/T 20631-2009	钢制管法兰用缠绕式垫片（Class 系列） Spiral wound gaskets for use with steel pipe flanges（Class designated）			全国化工设备设计技术中心站
113	HG/T 20632-2009	钢制管法兰用具有覆盖层的齿形组合垫（Class 系列） Covered serrated metal gaskets for use with steel pipe flanges（Class designated）			全国化工设备设计技术中心站
114	HG/T 20633-2009	钢制管法兰用金属环垫（Class 系列） Metallic ring joint gaskets for use with steel pipe flanges（Class designated）			全国化工设备设计技术中心站
115	HG/T 20634-2009	钢制管法兰用紧固件（Class 系列） Bolting for use with steel pipe flanges（Class designated）			全国化工设备设计技术中心站
116	HG/T 20635-2009	钢制管法兰、垫片、紧固件选配规定（美洲体系）			全国化工设备设计技术中心站
117	HG/T 20636-1998	化工装置自控工程设计管理规范 Code for engineering design management of instrumentation in chemical industry			中国寰球工程公司

化工工程

序号	标准编号	标准名称	被代替标准编号	备案号	主编单位
118	HG/T 20637-1998	化工装置自控工程设计文件编制规范 Code for preparation of engineering design documents of instrumentation in chemical industry			中国成达工程有限公司
119	HG/T 20638-1998	化工装置自控工程设计文件深度规范 Code for details of engineering design of instrumentation in chemical industry			中国五环工程有限公司 中石化宁波工程有限公司
120	HG/T 20639-1998	化工装置自控工程设计用典型图表 Typical tables for engineering design of instrumentation in chemical plant			中国成达工程有限公司 赛鼎工程有限公司 东华工程科技股份有限公司
121	HG 20640-1997	塑料设备 Plastic equipment			化工部第三设计院
122	HG/T 20641-1998	石灰窑砌筑技术条件 Technical specification of linling laying for lime-kiln			全国化工工业炉设计技术中心站
123	HG/T 20642-2011	化学工业炉耐火陶瓷纤维炉衬设计技术规定 Specification of design in refractory fibre linings for chemical industrial furnace			中国石化集团上海工程有限公司
124	HG/T 20643-2012	化工设备基础设计规定		J 1437-2012	中国寰球工程公司 湖南化工医药设计院
125	HG/T 20644-1998	变力弹簧支吊架 Variable spring supports			上海化工设计院
126	HG/T 20645-1998	化工装置管道机械设计规定			中国成达工程公司
127	HG/T 20646-1999	化工装置管道材料设计规定 Design standard piping material for for chemical plant			中国五环化学工程公司
128	HG 20652-1998	塔器设计技术规定 Specification of Coiumns Design			中国石化集团兰州设计院
129	HG/T 20653-2011	化工企业化学水处理设计技术规定 Technical specification for design of chemical water treatment in chemical plant			中国五环工程有限公司
130	HG/T 20654-98	化工企业化学水处理系统监测与控制设计条件技术规定 Technical code of design requirements for monitoring and control of chemical water treatment system in chemical plant			中海油山东化学工程有限责任公司
131	HG/T 20655-2016	化工企业锅炉装置及汽轮机组热工监测与控制设计条件技术规范 Technical regulation of design requirements for monitoring and control of heat supply unit and steam turbine unit in chemical enterprise		J2140-2016	中国寰球工程公司
132	HG/T 20656-1998	化工采暖通风与空气调节详细设计内容和深度的规定 Specification of contents and procedure on heating, ventilating and air conditioning detailed design for chemical plant			化工暖通设计技术委员会
133	HG/T 20657-2013	化工采暖通风与空气调节术语		J1659-2013	全国采暖通风设计技术中心站

化工工程

序号	标准编号	标准名称	被代替标准编号	备案号	主编单位
134	HG/T 20658-2014	熔盐炉技术规范		J1814-2014	全国化工工业炉设计技术中心站
135	HG/T 20659-2014	化学工业管式炉对流段模块技术规范 Technical specification of convection section module for chemical industial tubular furnace		J1977-2015	中国寰球工程公司
136	HG 20660-2000	压力容器中化学介质毒性危害和爆炸危险程度分类			赛鼎工程有限公司
137	HG/T 20661-2011	硫酸沸腾炉砌筑技术条件 Technical specification of linling lsying for fluidized bed furnace of sulphuric acid			中国石化集团南京设计院
138	HG/T 20662-1999	化工粉体物料机械输送设计技术规定 Specification on Technology Design of Mechanical Conveying for Powder-Granule Products			中石化上海金山工程公司
139	HG/T 20663-1999	化工粉粒产品计量、包装及码垛系统设计规定 Specification on Design of Metro-Measuring and Palletizing System for Chemical Powder-Granule Products			中国五环化学工程公司
140	HG/T 20664-1999	化工企业供电设计技术规定			中国成达工程公司
141	HG/T 20665-1999	化工建、构筑物抗震设防分类标准			中国寰球工程公司
142	HG/T 20666-1999	化工企业腐蚀环境电力设计规程			南化公司设计院
143	HG/T 20667-2005	化工建设项目环境保护设计规定			吉化公司设计院 全国化工环境保护设计技术中心站
144	HG/T 20668-2000	化工设备设计文件编制规定 Specification for organization of chemical equipment design documents			中国寰球化学工程公司
145	HG/T 20670-2000	化工、石油化工管架、管墩设计规定			中国石化工程建设公司
146	HG/T 20671-1989	铅衬里化工设备			南京化学工业公司设计院
147	HG/T 20672-2005	尿素造粒塔设计规定			中国五环化学工程公司
148	HG/T 20673-2005	压缩机厂房建筑设计规定			中国成达工程公司
149	HG/T 20674-2005	化工、石化建（构）筑物荷载设计规定			中石化北京石油化工工程公司
150	HG/T 20675-1990	化工企业静电接地设计规程			吉化公司设计院
151	HG/T 20676-1990	砖板衬里化工设备			吉林化学工业公司设计院
152	HG/T 20677-2013	橡胶衬里化工设备设计规范 Design code for chemical equipment with rubber lining		J1665-2013	山东齐鲁石化工程有限公司
153	HG/T 20678-2000	衬里钢壳设计技术规定 Specification on design of steel shell with liner			化学工业第二设计院
154	HG/T 20679-2014	化工设备、管道外防腐设计规定 Design code for external corrosion protection of chemical equipment and piping		J1854-2014	东华工程科技股份有限公司
155	HG/T 20680-2011	锅炉房设计工艺计算规定 Process calculation code for design of boiler plant			中国石化集团宁波工程公司

化工工程

序号	标准编号	标准名称	被代替标准编号	备案号	主编单位
156	HG/T 20681-2005	锅炉房、汽机房土建荷载设计条件技术规范 Technical stipulation of civil-load requirements for the design of boiler and turbine house			中国天辰工程有限公司
157	HG/T 20682-2005	化学工业炉燃料燃烧设计计算规定 Design calculation specification for fuel combustion of chemical industrial furnace			上海工程化学设计院
158	HG/T 20683-2005	化学工业炉耐火、隔热材料设计选用规定 Selected specification of design in refractory & insulating material for chemical industrial furnace			全国化工工业炉设计技术中心站
159	HG/T 20684-2012	化学工业炉金属材料设计选用规定 Specification for design selected metal material in chemical industrial furnace		J1438-2012	中国成达工程有限公司
160	HG/T 20685-2005	化学工业炉名词术语统一规定 Unified specification of terminology for chemical industrial furnace			全国化工工业炉设计技术中心站
161	HG/T 20686-90	化工企业电力设计图形和文字符号统一规定			化工部电气设计技术中心站 华泰工程公司
162	HG/T 20688-2000	化工工厂初步设计文件内容深度规定 Specification on content and procedure of preliminary design file for chemical plant			中国化工勘察设计协会 山东省化工规划设计院
163	HG/T 20689-2007	化工装置基础工程设计深度规定 Specification on content and procedure of basic engineering design for chemical plant		J727-2007	中国天辰化学工程公司
164	HG/T 20691-2006	高压喷射注浆施工技术规程 Standard of chemistrys industry trade of the peoples republic of china regulations for operative technique of high pressures spurtal pour thick liquid construction			湖南化工地质工程勘察院
165	HG/T 20692-2014	化工企业热工设计施工图内容和深度统一规定 Specification on content and procedure of detail design of thermal engineering for chemical plant		J1813-2014	中国五环工程公司
166	HG/T 20693-2006	岩土体现场直剪试验规程 Specification for field direct shear test of rock and soil mass			中国化学工程南京岩土工程公司
167	HG/T 20694-2006	振动沉管灌注桩低强度混凝土桩施工技术规程 Constrction techmical regulation for low strength concrete pile cast in tube by vibrating			山西华晋岩土工程勘察有限公司
168	HG/T 20696-1999	纤维增强塑料化工设备技术规范 Specification of FRP equipment design for chemical industry			中国五环化学工程公司
169	HG/T 20697-2007	化工暖通空调设备采购规定 Purchase specification on heating, ventilating and air conditioning equipment for chemical plant		J665-2007	化工暖通设计技术委员会

化工工程

序号	标准编号	标准名称	被代替标准编号	备案号	主编单位
170	HG/T 20698-2009	化工采暖通风与空气调节设计规范 Code for design of heating ventilation and air conditioning for chemical plant			化工暖通设计技术委员会
171	HG/T 20699-2014	自控设计常用名词术语 Common terms and definition of measurement and control system		J1812-2014	中国寰球工程公司
172	HG/T 20700-2014	可编程控制器系统工程设计规范		J1811-2014	中国天辰工程有限公司
173	HG/T 20701-2000	容器、换热器专业工程设计管理规定			中国五环化学工程公司
174	HG/T 20702-2000	特殊设备专业工程设计管理规定			中国五环化学工程公司
175	HG/T 20703-2000	材料专业工程设计管理规定			上海工程化学设计院
176	HG/T 20704-2000	机泵专业工程设计管理规定			中国成达化学工程公司
177	HG/T 20705-2009	石油和化学工业工程建设项目管理规范 Code for petroleum and chemical engineering project management		J979-2010	中国石油和化学工业协会项目管理委员会 中国石油和化工勘察设计协会
178	HG 20706-2013	化工建设项目废物焚烧处置工程设计规范		J1656-2013	全国环境保护设计技术中心站
179	HG/T 20707-2014	化工行业岩土工程勘察成果质量检查与评定标准 Inspection and Evaluation Standard for Achievements Quality		J1810-2014	化工部福州地质工程勘察院
180	HG/T 20708-2011	化工建构筑物地基加筋垫层技术规程 Technical regulation of reinforced cushion for chemical constructions foundation			山西华晋岩土工程勘察有限公司 中化二建集团有限公司
181	HG 21501-1993	衬胶钢管和管件			中国寰球化学工程公司
182	HG/T 21502.1-92	钢制立式圆筒形固定顶储罐系列			化工部第六设计院
183	HG/T 21502.2-92	钢制立式圆筒形内浮顶储罐系列			化工部第六设计院
184	HG 21503-92	钢制固定式薄管板列管换热器			国家医药管理局上海医药设计院
185	HG/T 21504.1-92	玻璃钢储槽标准系列			化工部第六设计院
186	HG/T 21504.2-92	拼装式玻璃钢储罐标准系列			化工部第六设计院
187	HG 21505-92	组合式视镜			化工部设备设计技术中心站
188	HG 21506-92	补强圈			化工部第三设计院
189	HG/T 21507-1992	化工企业电力设计施工图内容深度统一规定			中国天辰化学工程公司
190	HG/T 21509.1-95	化工企业电缆隧道敷设通用图（电气部分）			中国华陆工程公司
191	HG/T 21510-2005	橡胶工厂初步设计文件内容深度规定 Specification on content and procedure of preliminary design file for rubber plant		J474-2005	北京橡胶工业研究设计院 中国化学工业桂林工程公司
192	HG/T 21511-2011	橡胶工厂施工图设计文件内容和深度规定 Code for file content and procedure of detail design of rubber factory		J1210-2011	全国橡胶塑料设计技术中心 昊华工程有限公司 中国化学工业桂林工程公司
193	HG/T 21512-95	梁型气体喷射式填料支撑板			全国化工化学工程设计技术中心站
194	HG/T 21514-2014	钢制人孔和手孔的类型与技术条件 Technical specification of steel manholes and handholes		J1885-2014	中国天辰工程有限公司
195	HG/T 21515-2014	常压人孔 Manholes for atmospheric use		J1884-2014	中国天辰工程有限公司

化工工程

序号	标准编号	标准名称	被代替标准编号	备案号	主编单位
196	HG/T 21516-2014	回转盖板式平焊法兰人孔 Plate flange manholes with hinged cover		J1883-2014	中国天辰工程有限公司
197	HG/T 21517-2014	回转盖带颈平焊法兰人孔 Slip on flange manholes with hinged cover		J1882-2014	中国天辰工程有限公司
198	HG/T 21518-2014	回转盖带颈对焊法兰人孔 Welding neck flange manholes with hinged cove		J1881-2014	中国天辰工程有限公司
199	HG/T 21519-2014	垂直吊盖板式平焊法兰人孔 Plate flange manholes with davit to vertical cover		J1880-2014	中国天辰工程有限公司
200	HG/T 21520-2014	垂直吊盖带颈平焊法兰人孔 Slip on flange manholes with davit to vertical cover		J1879-2014	中国天辰工程有限公司
201	HG/T 21521-2014	垂直吊盖带颈对焊法兰人孔 Welding neck flange manholes with davit to vertical cover		J1878-2014	中国天辰工程有限公司
202	HG/T 21522-2014	水平吊盖板式平焊法兰人孔 Plate flange manholes with davit to horizontal cover		J1877-2014	中国天辰工程有限公司
203	HG/T 21523-2014	水平吊盖带颈平焊法兰人孔 Slip on flange manholes with davit to horizontal cover		J1876-2014	中国天辰工程有限公司
204	HG/T 21524-2014	水平吊盖带颈对焊法兰人孔 Welding neck flange manholes with davit to horizontal cover		J1875-2014	中国天辰工程有限公司
205	HG/T 21525-2014	常压旋柄快开人孔 Quick open manholes for atmospheric use		J1874-2014	中国天辰工程有限公司
206	HG/T 21526-2014	椭圆形回转盖快开人孔 Elliptical quick open manholes with hinged cover		J1873-2014	中国天辰工程有限公司
207	HG/T 21527-2014	回转拱盖快开人孔 Quick open manholes with hinged dome cover		J1872-2014	中国天辰工程有限公司
208	HG/T 21528-2014	常压手孔 Handholes for atmospheric use		J1871-2014	中国天辰工程有限公司
209	HG/T 21529-2014	板式平焊法兰手孔 Plate flange handholes		J1870-2014	中国天辰工程有限公司
210	HG/T 21530-2014	带颈平焊法兰手孔 Slip on flange handholes		J1869-2014	中国天辰工程有限公司
211	HG/T 21531-2014	带颈对焊法兰手孔 Welding neck flange handholes		J1868-2014	中国天辰工程有限公司
212	HG/T 21532-2014	回转盖带颈对焊法兰手孔 Welding neck flange handholes with hinged cover		J1867-2014	中国天辰工程有限公司
213	HG/T 21533-2014	常压快开手孔 Quick open handholes for atmospheric use		J1866-2014	中国天辰工程有限公司
214	HG/T 21534-2014	旋柄快开手孔 Quick open handholes with rotating handle		J1865-2014	中国天辰工程有限公司

化工工程

序号	标准编号	标准名称	被代替标准编号	备案号	主编单位
215	HG/T 21535-2014	回转盖快开手孔 Quick open handholes with hinged cover		J1864-2014	中国天辰化学工程公司
216	HG/T 21536-2011	化工工厂工业炉设计施工图内容深度统一规定 Unified specification on contents and procedure of working drawing for design of industrial furnace for chemical plants		J1196-2011	全国化工工业炉设计技术中心站
217	HG 21537-92	填料箱			化工部第一设计院
218	HG 21538-92	化工机械化运输工艺设计流程图图例 The process legend for chemical bulk material handling			化工部第一设计院
219	HG 21539-1992	钢筋混凝土独立式管架通用图			吉化公司设计院
220	HG 21540-1992	钢筋混凝土纵梁式管架通用图			吉化公司设计院
221	HG/T 21541-92	焊接H型钢标准节点通用图（一）高强螺栓（二）焊接连接			中国华陆工程公司
222	HG/T 21542-92	单轨、悬挂吊车梁通用图			中国寰球化学工程公司
223	HG/T 21543-2009	圆形塔平台通用图			中国五环工程有限公司
224	HG/T 21544-2006	预埋件通用图			中石化宁波工程有限公司
225	HG/T 21545-2006	地脚螺栓（锚栓）通用图			中石化宁波工程有限公司
226	HG/T 21546.1-93	回转圆筒用托轮、挡轮类型与技术条件			南京化学工业（集团）公司设计院
227	HG/T 21546.2-93	回转圆筒用托轮			南京化学工业（集团）公司设计院
228	HG/T 21546.3-93	回转圆筒用挡轮			南京化学工业（集团）公司设计院
229	HG/T 21546.4-93	回转圆筒用进出料箱装置类型与技术条件			南京化学工业（集团）公司设计院
230	HG/T 21546.5-93	回转圆筒用进出料箱密封装置（MG型）与技术条件			南京化学工业（集团）公司设计院
231	HG/T 21546.6-93	回转圆筒用进出料箱密封装置（JC型）与技术条件			南京化学工业（集团）公司设计院
232	HG/T 21547-1993	管道用钢制插板、垫环、8字盲板			化工部第六设计院
233	HG 21548-93	辐流式二次沉淀池吸泥机标准系列			吉化公司设计院
234	HG/T 21549-95	钢制低压湿式气柜系列			化工部第三设计院
235	HG/T 21550-93	防霜液面计			中国石化兰州设计院
236	HG/T 21551-95	柱塞式放料阀			中国石化兰州设计院
237	HG/T 21552-93	钢筋混凝土桁架式管架通用图			中国石化上海工程有限公司
238	HG/T 21553-93	钢铺板通用图			中国寰球工程公司
239	HG/T 21554-95	矩鞍环填料			全国化工化学工程设计技术中心站
240	HG/T 21556-95	鲍尔环填料			全国化工化学工程设计技术中心站
241	HG/T 21557-95	阶梯环填料			全国化工化学工程设计技术中心站
242	HG/T 21557.3-2006	塑料阶梯环填料			全国化工化学工程设计技术中心站

化工工程

序号	标准编号	标准名称	被代替标准编号	备案号	主编单位
243	HG/T 21558-2011	橡胶工厂工艺设计技术规定 Code for design on process technics of rubber factory		J1211-2011	全国橡胶塑料设计技术中心 昊华工程有限公司
244	HG/T 21559.1-2013	不锈钢网孔板填料工程技术规范		J1664-2013	化学工程中心站
245	HG/T 21559.2-2005	不锈钢孔板波纹填料			全国化工化学工程设计技术中心站
246	HG/T 21559.3-2005	不锈钢丝网波纹填料			全国化工化学工程设计技术中心站
247	HG/T 21561-1994	丙烯腈—丁二烯—苯乙烯（ABS）管和管件 The ABS pipe and fittings series			中国五环化学工程公司
248	HG/T 21563-1995	搅拌传动装置系统组合、选用及技术条件			化工部设备设计技术中心站
249	HG/T 21564-95	搅拌传动装置-凸缘法兰			化工部设备设计技术中心站
250	HG/T 21565-95	搅拌传动装置-安装底盖			化工部设备设计技术中心站
251	HG/T 21566-95	搅拌传动装置-单支点机架			南化公司设计院
252	HG/T 21567-95	搅拌传动装置-双支点机架			化工部设备设计技术中心站
253	HG/T 21568-95	搅拌传动装置-传动轴			化工部设备设计技术中心站
254	HG/T 21569.1-95	搅拌传动装置-带短节联轴器			上海医药设计院
255	HG/T 21569.2-95	搅拌传动装置-块式弹性联轴器			上海医药设计院
256	HG/T 21570-1995	搅拌传动装置—联轴器			上海医药设计院
257	HG/T 21571-95	搅拌传动装置-机械密封			化工部设备设计技术中心站
258	HG/T 21572-95	搅拌传动装置-机械密封循环保护系统			合肥通用机械研究所
259	HG/T 21573-95	高压螺栓和螺栓液压上紧装置			中国五环化学工程公司
260	HG/T 21574-2008	化工设备吊耳设计选用规范 Chemical equipment lifting lugs and engineering technical specification		J795-2008	中国成达工程公司
261	HG/T 21575-94	带灯视镜			上海医药设计院
262	HG/T 21579-1995	聚丙烯/玻璃钢（PP/FRP）复合管及管件			化工部化工工艺配管设计技术中心站
263	HG/T 21581-2012	自控安装图册（共12个分册） Instrument hook-up		J1434-2012	中国成达工程有限公司 全国化工自控设计技术中心站 中国寰球工程公司 东华工程科技股份有限公司 中石化宁波工程有限公司 中石化上海工程有限公司
264	HG/T 21583-95	快开不锈钢活动盖			化工部设备设计技术中心站
265	HG/T 21584-96	磁性液位计			化工部设备设计技术中心站
266	HG/T 21585.1-1998	可拆型槽盘气液分布器			全国化工化学工程设计技术中心站
267	HG/T 21586-1998	抽屉式丝网除沫器			全国化工化学工程设计技术中心站
268	HG 21588-95	玻璃板液面计标准系列及技术要求			中国寰球化学工程公司
269	HG 21589.1-95	透光式玻璃板液面计（PN2.5）			中国寰球化学工程公司
270	HG 21589.2-95	透光式玻璃板液面计（PN6.3）			中国寰球化学工程公司
271	HG 21590-95	反射式玻璃板液面计			中国寰球化学工程公司
272	HG 21591.1-95	视镜式玻璃板液面计（常压）			中国寰球化学工程公司
273	HG 21591.2-95	视镜式玻璃板液面计（PN0.6）			中国寰球化学工程公司
274	HG 21592-95	玻璃管液面计标准系列及技术要求			中国寰球化学工程公司

化工工程

序号	标准编号	标准名称	被代替标准编号	备案号	主编单位
275	HG/T 21594-2014	衬不锈钢人、手孔分类与技术条件 Classification and technical specification for lining stainless steel manholes and handholes		J1863-2014	中国石油集团东北炼化工程有限公司吉林设计院
276	HG/T 21596-2014	回转盖衬不锈钢人孔 Lining stainless steel rotary cover manholes		J1862-2014	中国石油集团东北炼化工程有限公司吉林设计院
277	HG/T 21597-2014	回转盖快开衬不锈钢人孔 Lining stainless steel rotary arch cover of quick opening manhole		J1861-2014	中国石油集团东北炼化工程有限公司吉林设计院
278	HG/T 21598-2014	水平吊盖衬不锈钢人孔 Lining stainless steel horizontal hanging cover manhole		J1860-2014	中国石油集团东北炼化工程有限公司吉林设计院
279	HG/T 21599-2014	垂直吊盖衬不锈钢人孔 Lining stainless steel vertical hanging cover manhole		J1859-2014	中国石油集团东北炼化工程有限公司吉林设计院
280	HG/T 21600-2014	椭圆快开衬不锈钢人孔 Lining stainless steel elliptic quick open manhole		J1858-2014	中国石油集团东北炼化工程有限公司吉林设计院
281	HG/T 21602-2014	平盖衬不锈钢手孔 Lining stainless steel flat cover handhole		J1857-2014	中国石油集团东北炼化工程有限公司吉林设计院
282	HG/T 21603-2014	回转盖快开衬不锈钢手孔 Lining stainless steel rotary cover quick opening handhole		J1856-2014	中国石油集团东北炼化工程有限公司吉林设计院
283	HG/T 21604-2014	旋柄快开衬不锈钢手孔 Lining stainless steel swing-type quick opening handhole		J1855-2014	中国石油集团东北炼化工程有限公司吉林设计院
284	HG 21605-1995	钢与玻璃烧结视镜			化工部设备设计技术中心站
285	HG 21606-1995	钢与玻璃烧结液位计			化工部设备设计技术中心站
286	HG 21607-1996	异型筒体和封头			化工部设备设计技术中心站
287	HG/T 21608-2012	液体装卸臂工程技术要求 Engineering technical requirements for liquid loading arm		J1435-2012	华陆工程科技有限责任公司
288	HG/T 21610-96	热轧普通型钢标准节点通用图（焊接连接）			中国华陆工程公司
289	HG/T 21611.1-96	钢筋混凝土带式输送机栈桥通用图			中国五环工程公司
290	HG/T 21612-96	压焊钢格板图集			全国化工建筑设计技术中心站
291	HG/T 21613-96	钢梯及钢栏杆通用图			全国化工建筑设计技术中心站
292	HG/T 21615-97	改性聚丙烯厢式和板框式压滤机系列			上海化工设计院
293	HG/T 21616-97	化工厂常用设备消声器标准系列			吉化公司设计院 化工部环境保护设计技术中心站
294	HG/T 21617-1998	槽型锚定轨预埋件通用图			全国化工建筑设计技术中心站 中国天辰化学工程公司
295	HG/T 21618-1998	丝网除沫器			化工部设备设计技术中心站
296	HG/T 21619-1986	压力容器视镜			齐鲁石油化工公司设计院
297	HG/T 21620-1986	带颈视镜			齐鲁石油化工公司设计院
298	HG/T 21621-1991	化工企业电缆直埋和电缆沟敷设通用图			中国华陆工程公司
299	HG/T 21622-1990	衬里视镜			齐鲁石油化工公司设计院
300	HG/T 21623-1990	硬聚氯乙烯视镜			齐鲁石油化工公司设计院

化工工程

序号	标准编号	标准名称	被代替标准编号	备案号	主编单位
301	HG/T 21629-1999	管架标准图			中国成达化学工程公司
302	HG/T 21630-1990	补强管			化工部设备设计技术中心站
303	HG/T 21633-1991	玻璃钢管和管件			化工部工艺配管设计技术中心站
304	HG/T 21636-1987	玻璃钢/聚氯乙烯（FRP/PVC）复合管和管件			中国寰球化学工程公司
305	HG/T 21637-1991	化工管道过滤器			化工部第六设计院
306	HG/T 21638-2005	树脂整体地面通用图			中国寰球工程公司
307	HG/T 21639-2005	塔顶吊柱			中国石化集团上海工程有限公司
308	HG/T 21640-2000	H 型钢钢结构管架通用图			中国成达工程公司
309	HG/T 21641-2013	管道工厂化预制技术规范 Technical code for metal piping factory prefabrication		J1662-2013	全国化工工艺配管设计技术中心站 江阴东联高压管件有限公司
310	HG/T 22801-2013	化工矿山企业初步设计内容和深度的规定 Regulations on Content and Depth		J1667-2013	中蓝连海设计研究院 中国寰球工程公司华北规划设计院
311	HG/T 22802-2014	化工矿山矿区总体规划内容和深度的规定 Regulations on Content and depth of general plan for chemical mine		J1979-2015	中国寰球工程公司华北规划设计院
312	HG/T 22803-2016	化工矿山工程设计三（二）级矿量原则规定 The principle and rule for ore volume of Grade 3（2）for chemical mine project		J2135-2016	化学矿山规划设计院
313	HG/T 22804-93	化工矿山土地复垦规划设计内容和深度的规定 Regulation on content and depth of planning and design of land reclamation for chemical mines			化工矿山设计研究院
314	HG/T 22805.1-2016	化工矿山企业施工图设计内容和深度的规定——地质·采矿专业 Regulations for content and depth of detail engineering for chemical mine enterprise-geological and mining		J2136-2016	化工矿山设计研究院
315	HG/T 22805.2-2016	化工矿山企业施工图设计内容和深度的规定——选矿专业 Regulations for content and depth of detail engineering for chemical mine enterprise-mineral processing		J2137-2016	化工矿山设计研究院
316	HG 22805.3-93	化工矿山企业施工图设计内容和深度的规定——矿山机械专业 Regulations for content and depth of detail engineering for chemical mine enterprise-mining machine			化工矿山设计研究院
317	HG 22805.4-93	化工矿山企业施工图设计内容和深度的规定——尾矿专业 Regulation on content and depth of planning and design of land reclamation for chemical-tailings disposal			化学矿山规划设计院

化工工程

序号	标准编号	标准名称	被代替标准编号	备案号	主编单位
318	HG 22805.5-93	化工矿山企业施工图设计内容和深度的规定——总图运输专业 Regulation on content and depth of detal engineering of chemical mine enterprise—general layout and transportation			化工矿山设计研究院
319	HG 22805.6-93	化工矿山企业施工图设计内容和深度的规定——建筑专业 Regulation on content and depth of detal engineering design of chemical mine enterprise——architecture			化工矿山设计研究院
320	HG 22805.7-93	化工矿山企业施工图设计内容和深度的规定——结构专业 Regulations for content and depth of detail engineering for chemical mine enterprise-structure			化工矿山设计研究院
321	HG 22805.8-93	化工矿山企业施工图设计内容和深度的规定——热工专业 Regulation on content and depth of detal engineering design of chemical mine enterprise——thermotechnical			化学矿山规划设计院
322	HG 22805.9-93	化工矿山企业施工图设计内容和深度的规定——采暖通风专业 Regulation on content and depth of detal engineering design of chemical mine enterprise-heating, ventilation			化工矿山设计研究院
323	HG 22805.10-93	化工矿山企业施工图设计内容和深度的规定——给水排水专业 Regulation on content and depth of detal engineering design of chemical mine enterprise-water supply and drainage			长沙化学矿山设计研究院
324	HG 22805.11-93	化工矿山企业施工图设计内容和深度的规定——电气专业 Regulations for content and depth of detail engineering for chemical mine enterprise-Electrical			化工矿山设计研究院
325	HG 22805.12-93	化工矿山企业施工图设计内容和深度的规定——电信专业 Regulations for content and depth of detail engineering for chemical mine enterprise-Telecommunication			化工矿山设计研究院
326	HG 22805.13-93	化工矿山企业施工图设计内容和深度的规定——自控专业 Regulations for content and depth of detail engineering for chemical mine enterprise-Autocontrol			化工矿山设计研究院
327	HG 22805.14-93	化工矿山企业施工图设计内容和深度的规定——机修专业 Regulations for content and depth of detail engineering for chemical mine enterprise-machine maintenance			化工矿山设计研究院
328	HG 22806-94	化工矿山建设项目环境保护设计规定 Regulations for design of environmental protection for chemical mine project			化工矿山设计研究院

化工工程

序号	标准编号	标准名称	被代替标准编号	备案号	主编单位
329	HG/T 22808-2016	化工矿山选矿厂工艺设计规范 Codes for process design of beneficiation plant for chemical mine		J2134-2016	化工部连云港设计研究院
330	HG/T 22809-97	化工矿山地下采矿设计规范 Codes for underground mining design of chemical mines			化工部连云港设计研究院
331	HG/T 22810-97	化工矿山露天采矿设计规范 Codes for opencast mining design of chemical mines			化工矿山设计技术中心站
332	HG/T 22811-1998	化工矿山机械设计规范 Codes for mechanical design for chemical mine			化工部连云港设计研究院
333	HG/T 22813-1998	化工矿山机汽修工艺设计规范 Codes for process design of mechanical maintenance for chemical mines			化工部连云港设计研究院
334	HG/T 22814-1999	化工矿山井巷工程设计规范 Codes for design of shaft sinking and drifting for chemical mines			化工部连云港设计研究院

煤炭工业工程

序号	标准编号	标准名称	被代替标准编号	备案号	主编单位
1	NB/T 51014-2014	煤炭建设工程监理与项目管理规范 Coal plant engineering consulting service specification		J1803-2014	中煤陕西中安项目管理有限责任公司 煤炭工业济南设计研究院有限公司
2	NB/T 51015-2014	煤炭设备工程监理规范 Coal plant engineering consulting service specification		J1804-2014	中煤设备工程咨询公司 山西辰诚建设工程有限公司
3	NB/T 51028-2015	煤炭工业矿井施工组织设计规范 Code for coal mine construction organization plan of coal industry		J2035-2015	陕西煤业化工集团有限公司
4	NB/T 51029-2015	煤矿井巷工程质量评价标准 Evaluating standard for excellent quality of shaft sinking and drifting of coal mine		J2036-2015	煤炭工业河北建投建设工程质量监督站
5	NB/T 51030-2015	煤矿井巷工作面注浆工程施工与验收规范 Grouting engineering construction and acceptance specification of working face in coal mine		J2037-2015	河南能源化工建设集团有限公司
6	NB/T 51042-2015	选煤厂建筑工程施工与验收规范 The specification for construction and acceptance of architectural engineering in coal preparation plant		J2111-2016	中煤建筑安装工程集团有限公司 淮北矿业（集团）工程建设有限责任公司
7	NB/T 51043-2015	生产矿井立井垮塌修复治理规范 Technical specification for repairing collapse shaft of production mine		J2112-2016	湖南楚湘建设工程有限公司 西山煤电建筑工程集团有限公司 河南能源化工建设集团有限公司
8	NB/T 51051-2016	煤炭建设工程资料管理标准 Standard for data management of construction project in coal mine		J2206-2016	重庆巨能建设（集团）有限公司 国电建投内蒙古能源有限公司 北京筑业志远软件开发有限公司
9	MT 5010-95	煤矿安装工程质量检验评定标准			
10	MT/T 5017-2009	煤矿井筒装备防腐蚀技术规范			原中煤第三建设公司（现中煤矿山建设集团有限责任公司）
11	AQ 1049-2008	煤矿建设项目安全核准基本要求			
12	AQ 1055-2008	煤矿建设项目安全设施设计审查和竣工验收规范			
13	AQ 1083-2011	煤矿建设安全规范			
14	AQ 1095-2014	煤矿建设项目安全预评价实施细则			
15	AQ 1096-2014	煤矿建设项目安全验收评价实施细则			

水利工程

序号	标准编号	标准名称	被代替标准编号	备案号	主编单位
1	SL 16-2010	小水电建设项目经济评价规程 Economic evaluation code for small hydropower projects		J1286-2011	水利部农村电气化研究所
2	SL 19-2014	水利基本建设项目竣工财务决算编制规程 Compilation rules of final account for completion of water capital construction project		J837-2009	水利部淮河水利委员会
3	SL 41-2011	水利水电工程启闭机设计规范 Design code for gate hoist in water resources and hydropower projects		J1295-2011	湖南省水利水电勘测设计研究总院 黄河勘测规划设计有限公司
4	SL 45-2006	江河流域规划环境影响评价规范 Regulation for environmental impact assessment of river basin planning	SL 45-92	J732-2007	中水淮河工程有限责任公司和水利部水利水电规划设计总院
5	SL 140-2006	水泵模型及装置模型验收试验规程 Code of practice for model pump and its installation acceptance tests	SL 140-97	J733-2007	中水北方勘测设计有限责任公司 水利部水利水电规划设计总院
6	SL 142-2008	水轮机模型浑水验收试验规程 Code for model acceptance tests of hydraulic turbine with sediment water		J838-2009	中国水利水电科学研究院
7	SL 156-2010	水流空化模型试验规程 Specification for model test of flow cavitation		J1291-2011	中国水利水电科学研究院
8	SL 158-2010	水工建筑物水流压力脉动和流激振动模型试验规程		J1300-2011	中国水利水电科学研究院
9	SL 162-2010	水电站有压输水系统模型试验规程		J1301-2011	中国水利水电科学研究院
10	SL 163-2010	水利水电工程施工导流和截流模型试验规程		J1294-2011	长江水利委员会长江科学院
11	SL 176-2007	水利水电工程施工质量评定规程 Inspection and assessment specification for construction quality of hydraulic and hydroelectric engineering	SL 176—1996	J750-2007	四川省水利科学研究院 水利部建管司
12	SL 179-2011	小型水电站初步设计报告编制规程 Specification on compiling preliminary design report of small hydropower station		J1285-2011	水利部山西水利水电勘测设计研究院
13	SL 190-2007	土壤侵蚀分类分级标准 Standards for classification and gradation of soil erosion		J839-2009	水利部水保司
14	SL 191-2008	水工混凝土结构设计规范 Design code for hydraulic concrete structures		J840-2009	水利部长江水利委员会长江勘测规划设计研究院
15	SL 223-2008	水利水电建设工程验收规程 Acceptance code of practice on water resources and hydroelectric engineering		J841-2009	中水淮河工程有限责任公司（水利部淮委规划设计研究院）
16	SL 345-2007	水利水电工程注水试验规程 Code of water injection test for water resources and hydropower engineering		J755-2007	水利部水利水电规划设计总院
17	SL 348-2006	水域纳污能力计算规程 Code of practice for computation on permissible pollution bearing capacity of water bodies		J728-2007	长江流域水资源保护局
18	SL/Z 349-2015	水资源实时监控系统建设技术导则 Guidelines for constructing real-time monitoring system of water resources		J729-2007	中国水利水电科学研究院，水利部海委 水利部信息中心

水利工程

序号	标准编号	标准名称	被代替标准编号	备案号	主编单位
19	SL 350-2006	沙棘生态建设工程技术规程 Technical code on seabuckthorn for eco-engineering construction		J730-2007	水利部沙棘开发管理中心
20	SL 352-2006	水工混凝土试验规程 Test code for hydraulic concrete	SD 105-82 SL 48-94	J731-2007	中国水利水电科学研究院 南京水利水电科学研究院
21	SL 356-2006	小型水电站建设项目建议书编制规程 Code of compiling proposal of construction project for small hydropower station		J734-2007	水利部水利水电规划设计总院
22	SL 357-2006	农村水电站可行性研究报告编制规程 Code of feasibility study report of rural hydropower station		J735-2007	水利部水利水电规划设计总院
23	SL 358-2006	农村水电站施工环境保护导则 Guidelines of enviroment profection for the construction of rural hydropower station		J736-2007	水利部水利水电规划设计总院
24	SL 359-2006	水利水电工程环境保护概估算编制规程 Cost estimate compilation code environmental protection of water resources and hydropower projects		J737-2007	水利部水利水电规划设计总院
25	SL 360-2006	地下水监测站建设技术规范 Technical specification for the construction of groundwater monitoring station		J738-2007	南京水利科学研究院 水利部水文局
26	SL 364-2015	土壤墒情监测规范 Technical standard for soil moisture monitoring		J739-2007	河海大学 水利部水文局
27	SL 365-2015	水资源水量监测技术导则 Technique guideline for water quantity measurement		J748-2007	水利部长江水利委员会水文局 水利部水文局
28	SL 367-2006	城市综合用水量标准 Standards of integreted quota urban water use		J740-2007	水利部水利水电规划设计总院
29	SL 368-2006	再生水水质标准 Standards of reclaimed water quality		J741-2007	水利部水利水电规划设计总院
30	SL 371-2006	农田水利示范园区建设标准 Construction standard of demonstration farmland in irrigation and drainage		J742-2007	扬州大学 中国灌溉排水发展中心
31	SL 372-2006	节水灌溉设备现场验收规程 Acceptance code of practice for water-saving irrigation equipment on the site		J743-2007	中国灌溉排水发展中心
32	SL 373-2007	水利水电工程水文地质勘察规范 Water conservation water and electricity project hydrology geologicalprospecting standard		J744-2007	黄河勘测规划设计有限公司
33	SL 377-2007	水利水电工程锚喷支护技术规范 Technical specification of shotcrete and rock boltfor water resources and hydropower project	SDJS7-85	J746-2007	水利部松辽水利委员会
34	SL 378-2007	水工建筑物地下开挖工程施工规范 Construction specifications on underground excavating engineering of hydraulic structures	SDJ 212-83	J747-2007	中水东北勘测设计研究有限责任公司

水利工程

序号	标准编号	标准名称	被代替标准编号	备案号	主编单位
35	SL 379-2007	水工挡土墙设计规范 Design specification for hydraulic retainling wall		J745-2007	水利部水利水电规划设计总院
36	SL 383-2007	河道演变勘测调查规范 Code for river channel change survey		J751-2007	水利部长江水利委员会水文局
37	SL 384-2007	水位观测平台技术标准 Technical standard for stage observation platform		J753-2007	水利部长江水利委员会水文局
38	SL 386-2007	水利水电工程边坡设计规范 Water conservation water and electricity project side SL ope design standard		J752-2007	黄河勘测设计有限公司 水规总院
39	SL 387-2007	开发建设项目水土保持设施验收技术规程 Acceptance specification for soil and water conservation engineering of development and construction projects		J749-2007	水利部水土保持监测中心
40	SL/Z 388-2007	实时水情交换协议 Protocol for exchange of real-time hydrological information		J756-2007	水利部水利信息中心
41	SL 389-2008	滩涂治理工程技术规范 Technical code for tidal zone regulation project		J842-2009	中国水利学会滩涂湿地保护与利用专业委员会
42	SL 395-2007	地表水资源质量评价技术规程 Technological regulations for surface water resources quality assessment		J754-2007	水利部水环境监测评价研究中心
43	SL 396-2011	水利水电工程水质分析规程 Code of water quality analysis for water conservancy and hydropower development		J1293-2011	中国水利水电科学研究院
44	SL 398-2007	水利水电工程施工通用安全技术规程 General technical specification for safety of hydraulic and hydroelectric engineering construction	SD 267-1988	J757-2007	三峡大学 中国水利水电建设集团公司
45	SL 399-2007	水利水电工程土建施工安全技术规程 Technical specification of safety for civil construction of hydraulic and hydroelectric engineering	SD 267-1988	J758-2007	三峡大学 中国水利水电建设集团公司
46	SL 400-2016	水利水电工程金属结构与机电设备安装安全技术规程 Technical specification for safety of installation of metal structure and mechanical & electrical equipment of hydraulic and hydroelectric engineering	SD 267-1988	J759-2007	三峡大学 中国水利水电建设集团公司
47	SL 401-2007	水利水电工程施工作业人员安全技术操作规程 Technical operation specification of safety for workmen of hydraulic and hydroelectric engineering	SD 267-1988	J760-2007	三峡大学 中国水利水电建设集团公司
48	SL 418-2008	大型灌区技术改造规程 Code for amelioration of large-sized irrigation and drainage scheme		J843-2009	中国灌溉排水发展中心

水利工程

序号	标准编号	标准名称	被代替标准编号	备案号	主编单位
49	SL 419-2007	水土保持试验规范 Test specification of soil and water conservation		J844-2009	水利部水保中心
50	SL 423-2008	河道采砂规划编制规程 Code of practice for compilation of river sand-mining planning		J845-2009	水利部长江水利委员会
51	SL 428-2008	凌汛计算规范 Specification for ice flood computation		J846-2009	黄河勘测设计有限公司
52	SL 429-2008	水资源供需预测分析技术规范 Technical specification for the analysis of supply and demand balance of water resources		J847-2009	水利部南水北调规划设计管理局
53	SL 430-2008	调水工程设计导则 Design guideline for water diversion project		J848-2009	水利部南水北调规划设计管理局
54	SL 431-2008	城市水系规划导则 Guidelines for urban river and lake systems planning		J849-2009	河海大学
55	SL 434-2008	水利信息网建设指南 Guide of network establishment on the national information network for water resources		J850-2009	水利部海河水利委员会
56	SL 435-2008	海堤工程设计规范 Code for design of sea dike project		J851-2009	水利部水利水电规划设计总院
57	SL 436-2008	堤防隐患探测规程 Specification of exploration for dike hidden trouble		J852-2009	黄河水利委员会黄河水利科学研究院
58	SL 438-2008	水利水电工程二次接线设计规范 Specifications for design secondary circuit of hydroengineering		J853-2009	中水北方勘测设计研究有限责任公司
59	SL 454-2010	地下水资源勘察规范 Code for investigation of ground water resources		J1283-2011	中水北方勘测设计研究有限责任公司
60	SL 456-2010	水利水电工程电气测量设计规范		J1296-2011	长江水利委员会长江勘测规划设计研究院
61	SL 482-2011	灌溉与排水渠系建筑物设计规范 Design Standard of Irrigation and Drainage Systems Building		J1284-2011	陕西省水利电力勘测设计研究院
62	SL 484-2010	水利水电工程施工机械设备选择设计导则		J1282-2011	中水东北勘测设计研究有限责任公司
63	SL 485-2010	水利水电工程厂（站）用电系统设计规范 Water resources and hydropower engineering design code for auxiliary power system		J1297-2011	中水东北勘测设计研究有限责任公司
64	SL 489-2010	水利建设项目后评价报告编制规程		J1299-2011	水利部水利建设与管理总站
65	SL 490-2010	水利水电工程采暖通风与空气调节设计规范 Design code for heating ventilation and air conditioning of water resources and		J1298-2011	长江勘测规划设计研究院

水利工程

序号	标准编号	标准名称	被代替标准编号	备案号	主编单位
66	SL 501-2010	土石坝沥青混凝土面板和心墙设计规范 Design code of asphalt concrete facings and cores for embankment dams		J1287-2011	水利部水利水电规划设计总院 陕西省水利电力勘测设计研究院
67	SL 504-2011	水文设施工程项目建议书编制规程 Code of practice on proposal of hydrological infrastructure project		J1288-2011	黄河水利委员会水文局
68	SL 505-2011	水文设施工程可行性研究报告编制规程 Code of practice on feasibility study report of hydrological infrastructure project		J1289-2011	黄河水利委员会水文局
69	SL 506-2011	水文设施工程初步设计报告编制规程		J1290-2011	黄河水利委员会水文局
70	SL 535-2011	水利水电工程施工压缩空气及供水供电系统设计规范		J1292-2011	长江水利委员会长江勘测规划设计研究院

铁路工程

序号	标准编号	标准名称	被代替标准编号	备案号	主编单位
1	TB 10001-2016	铁路路基设计规范 Code for design on subgrade of railway	TB 10001-2005	J447-2017	中铁第一勘察设计院集团有限公司
2	TB 10002.1-2005	铁路桥涵设计基本规范 Fundamental code for design on railway bridge and culvert	TB 10002.1-99	J460-2005	铁道第三勘察设计院集团有限公司
3	TB 10002.2-2005	铁路桥梁钢结构设计规范 Code for design on steel structure of railway bridge	TB 10002.2-99	J461-2005	中铁大桥勘测设计院集团有限公司
4	TB 10002.3-2005	铁路桥涵钢筋混凝土和预应力混凝土结构设计规范 Code for design on reinforced and prestressed concrete structure of railway bridge and culvert	TB 10002.3-99	J462-2005	中铁工程设计咨询集团有限公司
5	TB 10002.4-2005	铁路桥涵混凝土和砌体结构设计规范 Code for design on concrete and block masonry structure of railway bridge and culvert	TB 10002.4-99	J463-2005	铁道第三勘察设计院集团有限公司
6	TB 10002.5-2005	铁路桥涵地基和基础设计规范 Code for design on subsoil and foundation of railway bridge and culvert	TB 10002.5-99	J464-2005	铁道第三勘察设计院集团有限公司
7	TB 10003-2016	铁路隧道设计规范 Code for design on tunnel of railway	TB 10003-2005	J449-2016	中铁二院工程集团有限责任公司
8	TB 10004-2008	铁路机务设备设计规范 Code for design of railway locomotive facilities	TB 10004-1998	J833-2009	中铁第四勘察设计院集团有限公司
9	TB 10005-2010	铁路混凝土结构耐久性设计规范 Code for durability design of concrete structures of railway	铁建设〔2005〕157号	J1167-2011	中国铁道科学研究院
10	TB 10006-2016	铁路运输通信设计规范 Code for design of railway transportation communication	TB 10006-2005	J451-2016	中铁二院工程集团有限责任公司
11	TB 10007-2006	铁路信号设计规范 Code for design of railway signaling	TB 10007-99	J529-2006	北京全路通信信号研究设计院有限公司
12	TB 10008-2015	铁路电力设计规范 Code for design of railway electric power	TB 10008-99, 2006	J660-2016	铁道第三勘察设计院集团有限公司
13	TB 10009-2016	铁路电力牵引供电设计规范 Code for design of railway traction power supply	TB 10009-99, 2005	J452-2016	中铁电气化勘测设计研究院有限公司 中铁电气化局集团公司
14	TB 10010-2016	铁路给水排水设计规范 Code for design of water supply and sewerage of railway	TB 10010-2008	J832-2016	中铁第四勘察设计院集团有限公司
15	TB 10012-2007	铁路工程地质勘察规范 Code for geology investigation of railway engineering	TB 10012-2001	J124-2007	中铁第一勘察设计院集团有限公司
16	TB 10013-2010	铁路工程物理勘探规范 Code for geophysical prospecting of railway engineering	TB 10013-2004	J1089-2010	中铁第四勘察设计院集团有限公司

铁路工程

序号	标准编号	标准名称	被代替标准编号	备案号	主编单位
17	TB 10014-2012	铁路工程地质钻探规程 Code fora geological drilling of railway engineering	TB 10014-98	J1413-2012	中铁二院工程集团有限责任公司
18	TB 10015-2012	铁路无缝线路设计规范 Code for design of railway continuously welded rail	铁建设函〔2003〕205号	J1586-2013	中铁第四勘察设计院集团有限公司
19	TB 10016-2016	铁路工程节能设计规范 Code for design of energy conservation of railway engineering	TB 10016-2002,2006	J2199-2016	铁道第三勘察设计院集团有限公司
20	TB 10017-99	铁路工程水文勘测规范 Code for hydrogeological investigation of railway engineering	TBJ 17-86		铁道第三勘察设计院集团有限公司
21	TB 10018-2003	铁路工程地质原位测试规程 Codo for in-sita measunement of railway engineering geology	(多本合一)	J261-2003	中铁第四勘察设计院集团有限公司
22	TB 10020-2012	铁路隧道防灾救援疏散工程设计规范 Code for design on evacuation engineering for disaster prevention and rescue of railway tunnel		J1455-2012	铁道第三勘察设计院集团有限公司
23	TBJ 24-89	铁路结合梁设计规定			铁道部专业设计院
24	TB 10025-2006	铁路路基支挡结构设计规范 Code for design on retaining structure of railway subgrade	TB 10025-2001	J127-2006	铁道第二勘察设计院
25	TB 10027-2012	铁路工程不良地质勘察规程 Code for unfavorable geological condition investigation of railway engineering	TB 10027-2001	J1407-2012	中铁二院工程集团有限责任公司
26	TB 10028-2016	铁路动车组设备设计规范 Code for design of electric multiple unit facility	铁建设〔2007〕89号	J2223-2016	中铁第四勘察设计院集团有限公司
27	TB 10029-2009	铁路客车车辆设备设计规范 Code for design of railway passenger car rolling stock facilities	TB 10029-2002	J165-2009	中铁二院工程集团有限责任公司
28	TB 10031-2009	铁路货车车辆设备设计规范 Code for design of railway freight car rolling stock facilities	TB 10031-2000	J73-2009	中铁第四勘察设计院集团有限公司
29	TB 10035-2006	铁路特殊路基设计规范 Code for design on special subgrade of railway	TB 10035-2002	J158-2006	中铁第四勘察设计院集团有限公司
30	TB 10038-2012	铁路工程特殊岩土勘察规程 Code for special soil and rock investigation of railway engineering	TB 10038-2001	J1408-2012	中铁第一勘察设计院集团有限公司
31	TB 10041-2003	铁路工程地质遥感技术规程 Technical specification for geological remote sensing of railway engineering	TB 10041-95	J262-2003	中铁工程设计咨询集团有限公司
32	TB 10049-2014	铁路工程水文地质勘察规范 Code for hydrogeological investigation of railway engineering	TB 10049-2004	J339-2015	中铁第一勘察设计院集团有限公司

铁路工程

序号	标准编号	标准名称	被代替标准编号	备案号	主编单位
33	TB 10050-2010	铁路工程摄影测量规范 Photogrammetry code for railway engineering	TB 10050-97	J1087-2010	中铁工程设计咨询集团有限公司
34	TB 10054-2010	铁路工程卫星定位测量规范 Satellite positioning system survey specifications for railway engineering	TB 10054-97	J1088-2010	中铁第一勘察设计院集团有限公司
35	TB 10056-98	铁路房屋暖通空调设计标准 Standard for design of heating, ventilation and air conditioning in railway building	TBJ 11-85		铁道第三勘察设计院集团有限公司
36	TB 10057-2010	铁路车辆运行安全监控系统设计规范 Design specification for running safety monitoring system of rolling stock	TB 10057-1998	J989-2010	中铁二院工程集团有限责任公司
37	TB/T 10058-2015	铁路工程制图标准 Drawing standards of railway engineering	TB/T 10058-98	J2092-2015	中铁第一勘察设计院集团有限公司
38	TB/T 10059-2015	铁路工程制图图形符号标准 Standard for graphical symbol of railway engineering	TB/T 10059-98	J2093-2015	中铁第一勘察设计院集团有限公司
39	TB 10061-98	铁路工程劳动安全卫生设计规范 Code for design of working security and sanitation for railway engineering			铁道第三勘察设计院集团有限公司
40	TB 10062-99	铁路驼峰及调车场设计规范 Code for design on hump and marshalling yard of railway			铁道第三勘察设计院集团有限公司
41	TB 10063-2016	铁路工程设计防火规范 Code for design on fire prevention of railway engineering	TB 10063-2007	J2180-2016	铁道第三勘察设计院集团有限公司
42	TB 10066-2000	铁路站场道路和排水设计规范 Code for design of road and drainage for railway station and yard		J69-2001	中铁二院工程集团有限责任公司
43	TB 10067-2000	铁路站场客货运设备设计规范 Code for design of passenger & freight equipments for railway station and yard		J70-2001	中铁第一勘察设计院集团有限公司
44	TB 10068-2010	铁路隧道运营通风设计规范 Code for design on operating ventilation of railway tunnel	TB 10068-2000	J1123-2010	中铁二院工程集团有限责任公司
45	TB 10069-2000	铁路驼峰信号设计规范 Code for design of railway hump signaling		J74-2001	北京全路通信信号研究设计院有限公司
46	TB 10074-2016	铁路客运服务信息系统设计规范 Code for design of railway passenger service information system	TB10074-2000、 TB 10074-2007	J81-2016	铁道第三勘察设计院集团有限公司
47	TB 10077-2001	铁路工程岩土分类标准 Code for rcok and soil classification of railway engineering		J123-2001	中铁第一勘察设计院集团有限公司
48	TB 10082-2005	铁路轨道设计规范 Code for design of railway track		J448-2005	中国铁路经济规划研究院

铁路工程

序号	标准编号	标准名称	被代替标准编号	备案号	主编单位
49	TB 10083-2005	铁路旅客车站无障碍设计规范 Code for design on accessibility of railway passenger station buildings		J458-2005	铁道第三勘察设计院集团有限公司
50	TB 10084-2007	铁路天然建筑材料工程地质勘察规程 Code for geology investigation of natural building of railway engineering		J722-2007	中铁第一勘察设计院集团有限公司
51	TB 10088-2015	铁路数字移动通信系统（GSM-R）设计规范 Code for design of railway digital mobile communication system（GSM-R）	铁建设〔2007〕92号	J2108-2015	北京全路通信信号研究设计院有限公司
52	TB 10089-2015	铁路照明设计规范 Code for design of railway lighting		J2142-2016	中国铁路经济规划研究院
53	TB 10101-2009	铁路工程测量规范 Code for railway engineering survey	TB 10101-99	J961-2009	中铁二院工程集团有限责任公司
54	TB 10102-2010	铁路工程土工试验规程 Code for soil tests of railway engineering	TB 10102-2004	J1135-2010	中铁第一勘察设计院集团有限公司
55	TB 10103-2008	铁路工程岩土化学分析规程 Regulations for rock and soil chemical analysis of railway engineering	TBJ 103-87	J862-2009	中铁二院工程集团有限责任公司
56	TB 10104-2003	铁路工程水质分析规程 Code for water analysis of railway engineering	TBJ 104-87	J263-2003	中铁二院工程集团有限责任公司
57	TB 10105-2009	改建铁路工程测量规范 Code for engineering survey of railway reconstruction project	TBJ 105-88	J963-2009	中铁第四勘察设计院集团有限公司
58	TB 10106-2010	铁路工程地基处理技术规程 Technical code for ground treatment of railway engineering	TB 10113-96	J1078-2010	中铁二院工程集团有限责任公司
59	TB 10115-2014	铁路工程岩石试验规程 Code for rock test of railway engineering	TB 10115-98	J1943-2015	中铁第一勘察设计院集团有限公司
60	TB 10120-2002	铁路瓦斯隧道技术规范 Technical code for railway tunnel with gas		J160-2002	中铁五局集团有限公司
61	TB 10180-2016	铁路防雷及接地工程技术规范 Technical code for lightning protection and earthing of railway	铁建设〔2007〕39号	J2179-2016	中铁二院工程集团有限责任公司
62	TB 10218-2008	铁路工程基桩检测技术规程 Technical specification for railway piles	TB 10218-99	J808-2008	中国铁道科学研究院
63	TB 10223-2004	铁路隧道衬砌质量无损检测规程 Code for undestructive detecting of railway tunnel lining		J341-2004	中国铁路工程总公司
64	TB 10301-2009	铁路工程基本作业施工安全技术规程 Safety constructional regulations on basic work for railway engineering		J944-2009	中铁九局集团有限公司
65	TB 10302-2009	铁路路基工程施工安全技术规程 Safety constructional regulations for railway subgrade engineering		J945-2009	中铁二十一局集团有限公司局
66	TB 10303-2009	铁路桥涵工程施工安全技术规程 Safety construction regulations for railway bridge and culvert engineering		J946-2009	中铁十局集团有限公司

铁路工程

序号	标准编号	标准名称	被代替标准编号	备案号	主编单位
67	TB 10304-2009	铁路隧道工程施工安全技术规程 Safety constructional regulations for railway tunnel engineering		J947-2009	中铁二局股份有限公司
68	TB 10305-2009	铁路轨道工程施工安全技术规程 Safety constructional regulations for railway track engineering		J948-2009	中铁一局集团有限公司
69	TB 10306-2009	铁路通信、信号、电力、电力牵引供电工程施工安全技术规程 Safety constructional regulations for railway communication, signal, electric power and electric traction feeding engineering		J949-2009	中国中铁电气化局集团公司 中铁六局集团有限公司集团有限公司
70	TB 10402-2007	铁路建设工程监理规范 Code for construction project management of railway	TB 10402-2003	J269-2007	西南交通大学
71	TB/T 10403-2004	铁路工程地质勘察监理规程 Code for supervision on geology investigation of railway engineering			中铁第一勘察设计院集团有限公司
72	TB 10413-2003	铁路轨道工程施工质量验收标准 Standard for constructional quality acceptance of railway track engineering	TB 10413-98	J284-2004	中铁一局集团有限公司
73	TB 10414-2003	铁路路基工程施工质量验收标准 Standard for constructional quality acceptance of railway subgrade engineering	TB 10414-98	J285-2004	中铁二局股份有限公司
74	TB 10415-2003	铁路桥涵工程施工质量验收标准 Standard for constructional quality acceptance of railway bridge and culvert engineering	TB 10415-98	J286-2004	中铁三局集团有限公司
75	TB 10417-2003	铁路隧道工程施工质量验收标准 Standard for constructional quality acceptance of railway tunnel engineering	TB 10417-98	J287-2004	中国中铁二局股份有限公司
76	TB 10418-2003	铁路运输通信工程施工质量验收标准 Standard for construction quality acceptance of railway transportation communication engineering	TB 10418-2000	J288-2004	中国铁路通信信号上海工程局集团有限公司
77	TB 10419-2003	铁路信号工程施工质量验收标准 Standard for construction quality acceptance of railway signaling engineering	TB 10419-2000	J289-2004	中铁二局集团有限公司
78	TB 10420-2003	铁路电力工程施工质量验收标准 Standard for constructional quality acceptance of railway electric power engineering	TB 10420-2000	J290-2004	中铁十一局集团有限公司
79	TB 10421-2003	铁路电力牵引供电工程施工质量验收标准 Standard for constructional quality acceptance of railway electric traction feeding engineering	TB 10421-2000	J291-2004	中铁电气化局集团公司
80	TB 10422-2011	铁路给水排水工程施工质量验收标准 Standard for constructional quality acceptance of railway water supply and sewerage engineering	TB 10422-2003	J1192-2011	中铁四局集团有限公司 中铁二十二局集团有限公司
81	TB 10423-2014	铁路站场工程施工质量验收标准 Standard for construction quality acceptance of railway station and yard engineering	TB 10423-2003	J1827-2014	中铁五局集团有限公司

铁路工程

序号	标准编号	标准名称	被代替标准编号	备案号	主编单位
82	TB 10424-2010	铁路混凝土工程施工质量验收标准 Standard for construction quality acceptance of railway concrete engineering	TB 10424-2003	J1155-2011	中铁三局集团有限公司
83	TB 10425-94	铁路混凝土强度检验评定标准 Standard for check and accept concrete strength of railway			中国铁道科学研究院
84	TB 10426-2004	铁路工程结构混凝土强度检测规程 Inspection specification for structure concrete strength of railway engineering		J342-2004	中铁三局集团有限公司
85	TB 10427-2011	铁路旅客车站客运服务信息系统工程施工质量验收标准 Standard for construction quality acceptance of passenger transport service information system of railway passenger station engineering		J1226-2011	中国铁道科学研究院
86	TB 10428-2012	铁路声屏障工程施工质量验收标准 Standard for constructional quality acceptance of railway sound barrier		J1499-2013	中铁二院工程集团有限责任公司
87	TB/T 10429-2014	绿色铁路客站评价标准 Evaluation standard for green railway stations		J1828-2014	中国铁路经济规划研究院 清华大学
88	TB 10430-2014	铁路数字移动通信系统（GSM-R）工程检测规程 Specification for engineering test of railway digital mobile communication system (GSM-R)		J1944-2015	中国铁路通信信号上海工程局集团有限公司
89	TB 10443-2010	铁路建设项目资料管理规程 Code for records management of railway construction project		J978-2010	中国铁路经济规划研究院
90	TB 10501-2016	铁路工程环境保护设计规范 Code for environmental protection design of railway engineering	TBJ 501-87, TB 10501-98	J2216-2016	中铁第四勘察设计院集团有限公司
91	TB 10504-2007	铁路建设项目预可行性研究、可行性研究和设计文件编制办法 Guidelines for preparing pre-feasibility study, feasibility study and design documents for railway engineering projects	铁建设〔1999〕99号		中铁第一勘察设计院集团有限公司
92	TB 10601-2009	高速铁路工程测量规范 Code for engineering survey of high speed railway	铁建设〔2006〕189号 铁建设〔2007〕76号	J962-2009	中铁二院工程集团有限责任公司
93	TB 10621-2014	高速铁路设计规范 Code for design of high speed railway	TB 10621-2009	J1942-2014	铁道第三勘察设计院集团有限公司 中铁第四勘察设计院集团有限公司
94	TB 10623-2014	城际铁路设计规范 Code for design of intercity railway		J1980-2015	铁道第三勘察设计院集团有限公司 中铁第四勘察设计院集团有限公司

铁路工程

序号	标准编号	标准名称	被代替标准编号	备案号	主编单位
95	TB 10751-2010	高速铁路路基工程施工质量验收标准 Standard for constructional quality acceptance of hight-speed railway earth structure engineering	铁建设〔2005〕160号	J1147-2011	中铁十二局集团有限公司
96	TB 10752-2010	高速铁路桥涵工程施工质量验收标准 Standard for constructional quality acceptance of hight-speed railway bridge and culvert engineering	铁建设〔2005〕160号	J1148-2011	中铁三局集团有限公司、中铁六局集团有限公司
97	TB 10753-2010	高速铁路隧道工程施工质量验收标准 Standard for constructional quality acceptance of high-speed railway tunnel engineering	铁建设〔2005〕160号	J1149-2011	中铁一局集团有限公司
98	TB 10754-2010	高速铁路轨道工程施工质量验收标准 Standard for constructional quality acceptance of hight-speed railway track engineering	（多本合一）	J1150-2011	中铁八局集团有限公司
99	TB 10755-2010	高速铁路通信工程施工质量验收标准 Standard for constructional quality acceptance of hight-speed railway communication engineering	铁建设〔2007〕251号	J1151-2011	中国铁路通信信号上海工程集团有限公司
100	TB 10756-2010	高速铁路信号工程施工质量验收标准 Standard for constructional quality acceptance of hight-speed railway signaling engineering	铁建设〔2007〕213号	J1152-2011	中国铁路通信信号集团公司
101	TB 10757-2010	高速铁路电力工程施工质量验收标准 Standard for constructional quality acceptance of hight-speed railway electric power supply engineering		J1153-2011	中铁建电化局 中铁十一局集团有限公司 中铁电气化局集团公司
102	TB 10758-2010	高速铁路电力牵引供电工程施工质量验收标准 Standard for constructional quality acceptance of hight-speed railway traction power supply engineering	铁建设〔2006〕167号	J1154-2011	中国铁建电气化局集团有限公司 中铁电气化局集团公司
103	TB 10760-2013	高速铁路工程静态验收技术规范 Technical regulations for static acceptance for high-speed railways construction	铁建设〔2009〕183号	J1534-2013	中铁电气化局集团有限公司
104	TB 10761-2013	高速铁路工程动态验收技术规范 Technical regulations for dynamic acceptance of high-speed railways construction	铁建设〔2010〕214号	J1535-2013	中国铁道科学研究院

公路工程

序号	标准编号	标准名称	被代替标准编号	备案号	主编单位
1	JTG A01-2002	公路工程标准体系			中国工程建设标准化协会公路工程委员会
2	JTG A02-2013	公路工程行业标准制修订管理导则 Management regulation for compilation and revision of highway engineering standards			交通运输部公路局 中国工程建设标准化协会公路分会
3	JTG A04-2013	公路工程标准编写导则 Compilation regulations for highway engineering standards			交通运输部公路局 中国工程建设标准化协会公路分会
4	JTG B01-2014	公路工程技术标准 Technical standard of highway engineering	JTG B01-2003		交通运输部公路局 中交第一公路勘察设计研究院有限公司
5	JTJ 002-87	公路工程名词术语 Standard of technical terms for highway engineering			交通部公路规划设计院
6	JTJ 003-86	公路自然区划标准 Standard of climatic zoning for highway			交通部公路规划设计院
7	JTG B02-2013	公路工程抗震规范 Specification of seismic design for highway engineering	JTJ 004-89		中交路桥技术有限公司
8	JTG/T B02-01-2008	公路桥梁抗震设计细则 Guidelines for seismic design of highway bridges	JTJ 004-89		重庆交通科研设计院
9	JTG B03-2006	公路建设项目环境影响评价规范 Specifications for environmental impact assessment of highways	JTJ 005-96		交通部公路科学研究院
10	JTG B04-2010	公路环境保护设计规范 Design specifications of highway environmental protection	JTJ/T 006-98		中交第一公路勘察设计研究院有限公司
11	JTG B05-2015	公路项目安全性评价规范 Specifications for highway safety audit	JTG/T B05-2004		华杰工程咨询有限公司
12	JTG B05-01-2013	公路护栏安全性能评价标准 Standard for safety performance evaluation of highway barriers	JTG/T F83-01-2004		北京深华达交通工程检测有限公司
13	JTG B06-2007	公路工程基本建设项目概算预算编制办法			交通公路工程定额站
14	JTG/T B06-01-2007	公路工程概算定额			交通公路工程定额站
15	JTG/T B06-02-2007	公路工程预算定额			交通公路工程定额站
16	JTG/T B06-03-2007	公路工程机械台班费用定额			交通公路工程定额站
17	JTG/T B07-01-2006	公路工程混凝土结构防腐蚀技术规范 Specifications for seterioration prevention of highway concrete structures			长沙理工大学、清华大学
18	JTG B10-01-2014	公路电子不停车收费联网运营和服务规范 Operation and service specification for highway unified electronic toll collection			交通运输部公路科学研究院 交通运输部路网监测与应急处置中心

公路工程

序号	标准编号	标准名称	被代替标准编号	备案号	主编单位
19	JTG C10-2007	公路勘测规范 Specifications for highway reconnaissance	JTJ 061-85 JTJ 061-99 JTJ 062-91 JTJ 063-85 JTJ 065-97 JTJ/T 066-98		中交第一公路勘察设计研究院
20	JTG/T C10-2007	公路勘测细则 Guidelines for highway survey			中交第一公路勘察设计研究院
21	JTG C20-2011	公路工程地质勘察规范 Code for highway engineering geological investigation	JTJ 064-98		中交第一公路勘察设计研究院有限公司
22	JTG/T C21-01-2005	公路工程地质遥感勘察规范 Specifications for highway engineering geological remote sensing			中交第二公路勘察设计研究院
23	JTG/T C21-02-2014	公路工程卫星图像测绘技术规程 Specifications of satellite imagery mapping in highway engineering			中交第二公路勘察设计研究院有限公司
24	JTG/T C22-2009	公路工程物探规程 Guidelines for highway engineering geophysical exploration			中交第一公路勘察设计研究院有限公司
25	JTG C30-2015	公路工程水文勘测设计规范 Hydrological specifications for survey and design of highway engineering	JTG C30-2002		河北省交通规划设计院
26	JTG D20-2006	公路路线设计规范 Design specification for highway alignment	JTJ 011-1994		中交第一公路勘察设计研究院
27	JTG/T D21-2014	公路立体交叉设计细则 Guidelines for design of highway grade-separated intersections			中国公路工程咨询集团有限公司
28	JTG D30-2015	公路路基设计规范 Specifications for design of highway subgrades	JTG D30-2004		中交第二公路勘察设计研究院有限公司
29	JTG/T D31-2008	沙漠地区公路设计与施工指南 Guidelines for highway design and construction in desert area			新疆交通科学研究院
30	JTG/T D31-02-2013	公路软土地基路堤设计与施工技术细则 Technical cuidelines for design and construction of highway embankment on soft ground	JTJ 017-96		中交第一公路勘察设计研究院有限公司
31	JTG/T D31-03-2011	采空区公路设计与施工技术细则 Guidelines for design and construction of highway engineering in the mined-out area			山西省交通规划勘察设计院
32	JTG/T D31-04-2012	多年冻土地区公路设计与施工技术细则 Guidelines for design and consturction of highway in permafrost area			中交第一公路勘察设计研究院有限公司
33	JTG/T D32-2012	公路土工合成材料应用技术规范 Technical Specifications for application of geosynthetics in highway	JTJ/T 019-98		招商局重庆交通科研设计院有限公司
34	JTG/T D33-2012	公路排水设计规范 Specifications for drainage design of highway	JTJ/T 018-97		中交路桥技术有限公司

公路工程

序号	标准编号	标准名称	被代替标准编号	备案号	主编单位
35	JTG D40-2011	公路水泥混凝土路面设计规范 Specifications for design of highway cement concrete pavement	JTG D40-2002		中交公路规划设计院有限公司
36	JTG D50-2006	公路沥青路面设计规范 Specifications for design of highway asphalt pavement	JTJ 014-97		中交公路规划设计院
37	JTG D60-2015	公路桥涵设计通用规范 General specifications for design of highway bridges and culverts	JTG D60-2004		中交公路规划设计院有限公司
38	JTG/T D60-01-2004	公路桥梁抗风设计规范 Wind-resistent design specification for highway bridges			中交公路规划设计院
39	JTG D61-2005	公路圬工桥涵设计规范 Code for Design of Highway Masonry Bridges and Culverts	JTJ 022-85		中交公路规划设计院
40	JTG D62-2004	公路钢筋混凝土及预应力混凝土桥涵设计规范 Code for design of highway reinforced concrete and prestressed concrete bridges and culverts	JTJ 023-85		中交公路规划设计院
41	JTG D63-2007	公路桥涵地基与基础设计规范 Code for design of ground base and foundation of highway bridges and culverts	JTJ 024-85		中交公路规划设计院有限公司
42	JTG D64-2015	公路钢结构桥梁设计规范 Specifications for design of steel bridge	JTJ 025-86		中交公路规划设计院有限公司
43	JTG/T D64-01-2015	公路钢混组合桥梁设计与施工规范 Specifications for design and construction of highway composite steel and concrete bridges			中交公路规划设计院有限公司
44	JTG/T D65-01-2007	公路斜拉桥设计细则 Guidelines for design of highway cable-stayed bridge	JTJ 027-96		重庆交通科研设计院
45	JTG/T D65-04-2007	公路涵洞设计细则 Guidelines for design of highway culvert			河北省交通规划设计院
46	JTG/T D65-05-2015	公路悬索桥设计规范 Specifications for design of highway suspension Bridge			中交公路规划设计院有限公司
47	JTG/T D65-06-2015	公路钢管混凝土拱桥设计规范 Specifications for design of highway concrete-filled steel tubular arch bridges			四川省交通运输厅公路规划勘察设计研究院
48	JTG D70-2004	公路隧道设计规范 Code for design of road tunnel	JTJ 026-90		重庆交通科研设计院
49	JTG/T D70-2010	公路隧道设计细则 Guidelines for design of highway tunnel			中交第二公路勘察设计研究院有限公司
50	JTG D70/2-2014	公路隧道设计规范 第二分册（交通工程与附属设施） Specifications for design of highway tunnels section 2 traffic engineering and affiliated facilities			招商局重庆交通科研设计院有限公司

公路工程

序号	标准编号	标准名称	被代替标准编号	备案号	主编单位
51	JTG/T D70/2-01-2014	公路隧道照明设计细则 Guidelines for design of lighting of highway tunnels	JTJ 026.1-1999		招商局重庆交通科研设计院有限公司
52	JTG/T D70/2-02-2014	公路隧道通风设计细则 Guidelines for design of ventilation of highway tunnels	JTJ 026.1-1999		招商局重庆交通科研设计院有限公司
53	JTG D80-2006	高速公路交通工程及沿线设施设计通用规范 General specification of freeway traffic engineering and roadside facilities			中交第一公路勘察设计研究院
54	JTG D81-2006	公路交通安全设施设计规范 Specifications for design of highway safety facilities	JTJ 074-94		交通部公路科学研究院
55	JTG/T D81-2006	公路交通安全设施设计细则 Guidelines for design of highway safety facilities			交通部公路科学研究院
56	JTG D82-2009	公路交通标志和标线设置规范 Specifications for layout of highway traffic signs and markings			交通部公路科学研究院
57	JTG E20-2011	公路工程沥青及沥青混合料试验规程 Standard test methods of btumen and bituminous mixtures for highway engineering	JTJ 052-2000		交通运输部公路科学研究院
58	JTG E30-2005	公路工程水泥及水泥混凝土试验规程 Test methods of cement and concrete for highway engineering	JTJ 053-94		交通部公路科学研究所
59	JTG E40-2007	公路土工试验规程 Test methods of soils for highway engineering	JTJ 051-93		交通部公路科学研究院
60	JTG E41-2005	公路工程岩石试验规程 Test methods of rock for highway engineering	JTJ 054-94		中交第二公路勘察设计研究院
61	JTG E42-2005	公路工程集料试验规程 Test methods of aggregate for highway engineering	JTJ 058-2000		交通部公路科学研究所
62	JTG E50-2006	公路工程土工合成材料试验规程 Test methods of geosynthetics for highway engineering	JTJ/T 060-98		交通部公路科学研究院
63	JTG E51-2009	公路工程无机结合料稳定材料试验规程 Test methods of materials stabilized with inorganic binders for highway engineering	JTJ 057-94		交通部公路科学研究院
64	JTG E60-2008	公路路基路面现场测试规程 Field test methods of subgrade and pavement for highway engineering	JTJ 059-95		交通部公路科学研究院
65	JTG/T E61-2014	公路路面技术状况自动化检测规程 Specifications of automated pavement condition survey			交通运输部公路科学研究院

公路工程

序号	标准编号	标准名称	被代替标准编号	备案号	主编单位
66	JTG F10-2006	公路路基施工技术规范 Technical specifications for construction of highway subgrades	JTJ 033-95		中交第一公路工程局有限公司
67	JTG/T F20-2015	公路路面基层施工技术细则 Technical guidelines for construction of highway roadbases	JTJ 034-2000		交通运输部公路科学研究院
68	JTG/T F30-2014	公路水泥混凝土路面施工技术细则 Technical guidelines for construction of highway cement concrete pavements	JTG F30-2003		交通运输部公路科学研究院
69	JTG/T F31-2014	公路水泥混凝土路面再生利用技术细则 Technical Guidelines for recycling of highway cement concrete pavement			交通运输部公路科学研究院
70	JTG F40-2004	公路沥青路面施工技术规范 Technical specifications for construction of highway asphalt pavements	JTJ 032-94		交通部公路科学研究所
71	JTG F41-2008	公路沥青路面再生技术规范 Technical specifications for highway asphalt pavement recycling			交通部公路科学研究院
72	JTG/T F50-2011	公路桥涵施工技术规范 Technical specification for construction of highway bridge and culvert	JTJ 041-2000		中交第一公路工程局有限公司
73	JTG F60-2009	公路隧道施工技术规范 Technical specifications for construction of highway tunnel	JTJ 042-94		中交第一公路工程局有限公司
74	JTG/T F60-2009	公路隧道施工技术细则 Technical guidelines for construction of highway tunnel			中交第一公路工程局有限公司
75	JTG F71-2006	公路交通安全设施施工技术规范 Technical specification for construction of highway safety facilities	JTJ 074-94		交通部公路科学研究院
76	JTG/T F72-2011	公路隧道交通工程与附属设施施工技术规范 Technical specifications for construction of traffic engineering and affiliated facilities of highway tunnel			重庆市交通委员会
77	JTG F80/1-2004	公路工程质量检验评定标准 第一册（土建工程） Quality inspection and evaluation standards for highway engineering section 1 civil engineering	JTJ 071-98		交通部公路科学研究所
78	JTG F80/2-2004	公路工程质量检验评定标准 第二册（机电工程） Quality inspection and evaluation standards for highway engineering section 2 electrical and mechanical engineering	JTJ 071-98		交通部公路科学研究所
79	JTG/T F81-01-2004	公路工程基桩动测技术规程 Technical specification of dynamic pile tests for highway engineering			浙江省交通厅工程质量监督站

公路工程

序号	标准编号	标准名称	被代替标准编号	备案号	主编单位
80	JTG F90-2015	公路工程施工安全技术规范 Safety technical specifications for highway engineering construction	JTJ 076-95		中国交通建设股份有限公司 中交第四公路工程局有限公司
81	JTG G10-2016	公路工程施工监理规范 Specifications for highway construction supervision	JTG G10-2006		北京市道路工程质量监督站
82	JTG H10-2009	公路养护技术规范 Technical specifications of maintenance for highway	JTJ 073-96		浙江省公路管理局
83	JTG H11-2004	公路桥涵养护规范 Code for maintenance of highway bridges and culverts	JTJ 073-96		陕西省公路局
84	JTG H12-2015	公路隧道养护技术规范 Technical specifications of maintenance for highway tunnel	JTG H12-2003		重庆市交通委员会
85	JTJ 073.1-2001	公路水泥混凝土路面养护技术规范 Technical specifications of cement concrete pavement maintenance for highway			江苏省交通厅公路局 水泥混凝土路面技术委员会
86	JTJ 073.2-2001	公路沥青路面养护技术规范 Technical specifications for maintenance of highway asphalt pavement			上海市公路管理处
87	JTG H20-2007	公路技术状况评定标准 Highway performance assessment standards	JTJ 075-94		交通部公路科学研究院 上海市公路管理处
88	JTG/T H21-2011	公路桥梁技术状况评定标准 Standards for technical condition evaluation of highway bridges			交通运输部公路科学研究院
89	JTG H30-2015	公路养护安全作业规程 Safety work rules for highway maintenance	JTG H30-2004		交通运输部公路科学研究院
90	JTG H40-2002	公路工程养护预算编制导则 Guidelines for budgeting of highway maintenance			交通公路工程定额站
91	JTG/T J21-2011	公路桥梁承载能力检测评定规程 Specification for inspection and evaluation of load-bearing capacity of highway bridges			交通运输部公路科学研究院
92	JTG/T J21-01-2015	公路桥梁荷载试验规程 Load test methods of highway bridge			长安大学
93	JTG/T J22-2008	公路桥梁加固设计规范 Specifications for strengthening design of highway bridges			中交第一公路勘察设计研究院有限公司
94	JTG/T J23-2008	公路桥梁加固施工技术规范 Technical specifications for strengthening construction of highway bridges			中交第一公路勘察设计研究院有限公司
95	JTG/T L11-2014	高速公路改扩建设计细则 Guidelines for design of expressway reconstruction and extension			浙江省交通运输厅
96	JTG/T L80-2014	高速公路改扩建交通工程及沿线设施设计细则 Guidlines for design of expressway traffic engineering and facilities of reconstruction and extension			中交第二公路勘察设计研究院有限公司

公路工程

序号	标准编号	标准名称	被代替标准编号	备案号	主编单位
97	JTG M20-2011	公路工程基本建设项目投资估算编制办法 Standard method of cost estimation for highway infrastructure projects			中交公路规划设计院有限公司
98	JTG/T M21-2011	公路工程估算指标 Initial cost estimation quotas for highway projects			中交公路规划设计院有限公司

水路工程

序号	标准编号	标准名称	被代替标准编号	备案号	主编单位
1	JTS 101-2014	水运工程标准编写规定 Stipulations on compiling technical standards of port and waterway engineering			交通运输部水运局 中国工程建设标准化协会水运专业委员会
2	JTJ/T 204-96	航道工程基本术语标准 The standard of basic technical terms for waterway engineering			长江航道局
3	JTS 105-1-2011	港口建设项目环境影响评价规范 Specifications for environmental impact assessment of port engineering			中交第二航务工程勘察设计院有限公司
4	JTJ 227-2001	内河航运建设项目环境影响评价规范 Environmental impact assessment specifications for inland waterway project			中交第二航务工程勘察设计院
5	JTS 110-4-2008	港口工程初步设计文件编制规定 The stipulation for the preparation of preliminary design documents of port engineering			中交第一航务工程勘察设计院有限公司
6	JTS 110-5-2008	航道工程初步设计文件编制规定 The Stipulation for the preparation of preliminary design documents of waterway engineering			四川省交通厅交通勘察设计研究院
7	JTS 110-6-2013	水运支持系统工程初步设计文件编制规定 The Stipulation for the preparation of preliminary deign documents for waterway support system engineering			北京金交信息通信导航设计院
8	JTS 110-8-2008	水运工程标准施工招标文件 Standard for bid document of construction of port and waterway engineering			中交第一航务工程局有限公司
9	JTS 110-9-2012	桥梁通航安全影响论证研究报告编制规定 The stipulation for report compilating of bridge's influence demonstration on navigation safety			交通运输部长江航务管理局
10	JTS 110-10-2012	水运工程标准施工监理招标文件 Standard bidding documents for construction supervision works in water transport engineering			中交水规院京华工程监理有限公司 广州南华工程管理有限公司 广西八桂工程监理咨询有限公司
11	JTS 110-11-2013	水运工程标准勘察设计招标文件 Standard bidding document for investment and design works in water transport engineering			中交第一航务工程勘察设计院有限公司 长江航道规划设计研究院
12	JTS/T 170-2-2012	港口建设项目安全预评价规范 Code for safety assessment prior to port construction project			交通运输部水运科学研究院 交通运输部天津水运工程科学研究院
13	JTS/T 170-3-2012	港口建设项目安全验收评价规范 Code for safety assessment upon completion of port construction project			交通运输部水运科学研究院 交通运输部天津水运工程科学研究院
14	JTS 195-3-2012	水运支持保障系统工程设计总体技术要求 Technical requirements for water transportation supporting system			交通运输部水运科学研究院

水路工程

序号	标准编号	标准名称	被代替标准编号	备案号	主编单位
15	JTS/T 105-4-2013	绿色港口等级评价标准 Standard for green port grade evaluation			交通运输部水运科学研究院
16	JTS 110-7-2013	水运工程施工图文件编制规定 The stipulation for preparation of detail design documents of port and waterway engineering			中交第一航务工程勘察设计院有限公司 四川省交通运输厅交通勘察设计研究院
17	JTS 111-2013	水运工程定额编写规定 Stipulations on compiling quora of port and waterway engineering			交通部疏浚工程定额站
18	JTS 115-2014	水运工程建设项目投资估算编制规定 Regulations on preparation of cost estimate for maritime structure construction project			交通部水运工程定额站
19	JTS 116-1-2014	内河航运建设工程概算预算编制规定			交通部水运工程定额站
20	JTS 116-4-2014	水运工程测量概算预算编制规定 Stipulations on compiling estimate and budget for port and waterway engineering survey			交通部疏浚工程定额站
21	JTS/T 106-2016	水运工程建设项目节能评估规范 Specifications on energy conservation assessment for port and waterway engineering			中交水运规划设计院有限公司 交通运输部天津水运工程科学研究所
22	JTS/T 105-3-2016	水运工程竣工验收环境保护调查技术规程 Specification on environmental investigation technical of port and waterway engineering			交通运输部天津水运工程科学研究所
23	JTS 131-2012	水运工程测量规范 Specifications for port and waterway engineering survey			中交天津航道局有限公司 中交天津港航勘察设计研究院有限公司
24	JTS 133-2013	水运工程岩土勘察规范 Code for geotechnical investigation on port and waterway engineering			中交第二航务工程勘察设计院有限公司 长江航道规划设计院研究院
25	JTS 132-2015	水运工程水文观测规范 Specifications of hydrometry for port and waterway engineering			交通运输部天津水运工程科学研究所 天津水运工程勘察设计院 中交天津港航勘察设计院有限公司
26	JTS 141-2011	水运工程设计通则 General rules for design of port and waterway works			中国交通建设股份有限公司
27	JTJ 206-96	港口工程制图标准 Harbour engineering graphics standard			河海大学 中交水运规划设计院
28	JTS 144-1-2010	港口工程荷载规范 Load code for harbour engineering			中交第一航务工程勘察设计院有限公司 中交第二航务工程勘察设计院有限公司
29	JTS 145-2015	港口与航道水文规范 Code of hydrology for harbour and waterway			中交第一航务工程勘察设计院有限公司

水路工程

序号	标准编号	标准名称	被代替标准编号	备案号	主编单位
30	JTS 146-2012	水运工程抗震设计规范 Code of earthquake resistant design for water transport engineering			中交水运规划设计院有限公司
31	JTS 147-1-2010	港口工程地基规范 Code for soil foundations of port engineering			中交天津港湾工程研究院有限公司
32	JTJ/T 260-97	港口工程粉煤灰填筑技术规程 Code for pulverized-fuel ash backfill technique in harbour engineering			交通部第三航务工程局科学研究所
33	JTJ/T 259-2004	水下深层水泥搅拌法加固软土地基技术规程 Technical specification for offshore cement deep mixing technique to consolidate soft soils			中港第一航务工程局 中交第一航务工程勘察设计院
34	JTJ 246-2004	港口工程碎石桩复合地基设计与施工规程 Specification for design and construction of composite foundation of stone column in harbour engineering			武汉港湾工程设计研究院
35	JTJ 239-2005	水运工程土工合成材料应用技术规范 Technical code for application of geosynthetics for port and waterway engineering			天津港湾工程研究所
36	JTS 149-1-2007	港口工程环境保护设计规范 Design code of environment protection for port engineering			中交第二航务工程勘察设计院有限公司
37	JTS 150-2007	水运工程节能设计规范 Code of energy-saving edesign for port and waterway engineering			中交水运规划设计院有限公司
38	JTS 151-2011	水运工程混凝土结构设计规范 Design code for concrete structures of port and waterway engineering			中交水运规划设计院有限公司
39	JTS 152-2012	水运工程钢结构设计规范 Code for design of steel structures in port and waterway engineering			中交水运规划设计院有限公司
40	JTJ 275-2000	海港工程混凝土结构防腐蚀技术规范 Corrosion prevention technical specifications for concrete structures of marine harbour engineering			广州四航工程技术研究院
41	JTS 153-2-2012	海港工程钢筋混凝土结构电化学防腐蚀技术规范 Technical specification for electrochemical anticorrosion of reirnforcement concrete structures in harbour and marine engineering			南京水利科学研究院
42	JTS 153-3-2007	海港工程钢结构防腐蚀技术规范 Technical specification for corrosion protection of steel structures for sea port construction			中交天津港湾工程研究院有限公司
43	JTS 153-2015	水运工程结构耐久性设计标准 Standard for durability design of port and waterway enginnering structure			中交四航工程研究院有限公司 中交水运规划设计院有限公司

水路工程

序号	标准编号	标准名称	被代替标准编号	备案号	主编单位
44	JTS 154-1-2011	防波堤设计与施工规范 Code of design and construction of breakwaters			中交第一航务工程勘察设计院有限公司
45	JTJ 300-2000	港口及航道护岸工程设计与施工规范 Code for design and construction of port and waterway revetment engineering			中交水运规划设计院
46	JTS 155-2012	码头船舶岸电设施建设技术规范 Technical code of shore-to-ship power supply system			交通运输部水运科学研究院
47	JTS 165-2013	海港总体设计规范 Design code of general layout for sea port			中交水运规划设计有限公司 中交第一航务工程勘察设计院有限公司
48	JTS 165-5-2016	液化天然气码头设计规范 Code for design of liquefied natural gas port and jetty			中交第四航务工程勘察设计院有限公司
49	JTS 165-7-2014	游艇码头设计规范 Code for design of marinas			中交第四航务工程勘察设计院有限公司 重庆市交通规划勘察设计院
50	JTJ 237-99	装卸油品码头防火设计规范 Code for fire-prevention design of oil loading/unloading terminals			交通部公安局 中交第一航务工程勘察设计院
51	JTJ 212-2006	河港工程总体设计规范 Code for master design of river port engineering			中交第二航务工程勘察设计院有限公司
52	JTS 167-1-2010	高桩码头设计与施工规范 Design and construction code for open type wharf on piles			中交第三航务工程勘察设计院有限公司
53	JTS 167-2-2009	重力式码头设计与施工规范 Design and construction code for gravity quay			中交第四航务工程局有限公司 中交四航局港湾工程设计院有限公司
54	JTS 167-3-2009	板桩码头设计与施工规范 Code for design and construction for quay wall of sheet pile			中交第一航务工程勘察设计院有限公司
55	JTS 167-4-2012	港口工程桩基规范 Code for pile foundation of harbor engineering			中交第三航务工程勘察设计院有限公司
56	JTS 167-6-2011	港口工程后张法预应力混凝土大管桩设计与施工规程 Port engineering technical code of design and construction for large dimeter post tensioned prestress concrete cylinder Pile			中交上海三航科学研究院有限公司
57	JTJ 294-98	斜坡码头及浮码头设计与施工规范 Code for design and construction of sloping wharf and floating wharf			交通部第二航务工程勘察设计院
58	JTJ 279-2005	港口工程桩式柔性靠船设施设计与施工技术规程 Design and construction technical code for resilient pile facilities for port engineering			中交水运规划设计院

水路工程

序号	标准编号	标准名称	被代替标准编号	备案号	主编单位
59	JTJ 303-2003	港口工程地下连续墙结构设计与施工规程 Design and construction technical code for diaphragm wall structure of port engineering			中交第一航务工程勘察设计院
60	JTJ 293-98	格型钢板桩码头设计与施工规程 Technical standard for cellular steel sheetpile wharf			中港第二航务工程局
61	JTJ 296-96	港口道路、堆场铺面设计与施工规范 Technical code of road and stockyard pavement for the port area			交通部第四航务工程勘察设计院
62	JTJ 297-2001	码头附属设施技术规范 The technical code of subsidiary facilities for wharf			中交水运规划设计院
63	JTS 180-2-2011	运河通航标准 Navigation standard of canal			交通运输部天津水运工程科学研究所
64	JTS 180-4-2015	长江干线通航标准 Navigation standard of the trunk stream of Changjiang River			长江航道局 中交水运规划设计院有限公司
65	JTJ 311-97	通航海轮桥梁通航标准 Bridge navigation standard for seagoing vessel			中交水运规划设计院
66	JTS 181-5-2012	疏浚与吹填工程设计规范 Design code for dredging and reclamation works			中交上海航道勘察设计研究院有限公司 中交天津港航勘察设计研究院有限公司
67	JTS 182-1-2009	渠化工程枢纽总体设计规范 Design code for hydrojunction general layout of canalization works			中交水运规划设计院有限公司
68	JTJ 305-2001	船闸总体设计规范 Code for master design of shiplocks			中交水运规划设计院
69		《船闸总体设计规范》(JTJ 305-2001)局部修订（船闸附属设施设计部分） Code for master design of shiplocks partially revised (ancillary facilities design)			江苏省交通规划设计院股份有限公司
70	JTJ 306-2001	船闸输水系统设计规范 Design code for filling and emptying system of shiplocks			南京水利科学研究院 天津水运工程科学研究所
71	JTJ 307-2001	船闸水工建筑物设计规范 Code for design of hydraulic structures of shiplocks			中交水运规划设计院
72	JTJ 308-2003	船闸闸阀门设计规范 Code for design of lock gates and valves of shiplocks			四川省交通厅内河勘察规划设计院
73	JTJ 309-2005	船闸启闭机设计规范 Code for design of headstock gears of shiplocks			中交水运规划设计院
74	JTJ 310-2004	船闸电气设计规范 Code for electrical dsign of shiplocks			四川省交通厅交通勘察设计研究院
75	JTJ 251-87	干船坞设计规范（工艺设计）			交通部水运规划设计院

水路工程

序号	标准编号	标准名称	被代替标准编号	备案号	主编单位
76	JTJ 252-87	干船坞设计规范（水工结构）			交通部水运规划设计院
77	JTJ 253-87	干船坞设计规范（坞门及灌水排水系统）			交通部水运规划设计院
78	JTJ/T 351-96	船舶交通管理系统工程技术规范 Code for engineering technology of vessel traffic services			中交水运规划设计院
79	JTJ/T 343-96	港口地区有线电话通信系统工程设计规范 Code of engineering for local wire telephone communication system in harbor			中交水运规划设计院
80	JTJ/T 345-99	甚高频海岸电台工程设计规范 Code for design of engineering for VHF coast station			交通部规划研究院
81	JTJ/T 341-96	海岸电台总体及工艺设计规范 Code for design of general planning and technology for coast stations			中交水运规划设计院
82	JTJ/T 282-2006	集装箱码头计算机管理控制系统设计规范 Design code for computer management and control system of container terminal			中交水运规划设计院有限公司
83	JTS 196-1-2009	海港集装箱码头建设标准 Construction standard of container terminal for sea port			中交水运规划设计院有限公司
84	JTS 196-5-2009	三峡船闸设施安全检测技术规程 Technical code of safety inspection for Three Gorges shiplock			长江三峡通航管理局
85	JTS 196-6-2012	三峡船闸通航调度技术规程 Technical code of navigation scheduling for the Three Gorges shiplock			长江三峡通航管理局
86	JTS 196-7-2007	长江三峡库区港口客运缆车安全设施技术规范 Technical code for the safety facilities of passenger cable cars in the ports of Three Gorges reservoir area			中交水运规划设计院有限公司
87	JTS 197-2011	港口货运缆车安全设施技术规范 Technical code for the safety facilities of freight cable cars of port			中交水运规划设计院有限公司
88	JTS 167-8-2013	水运工程先张法预应力高强混凝土管桩设计与施工规程 Code for design and construction for pretensioned spun high-strengh concrete pile of port and waterway engineering			中交第三航务工程局有限公司
89	JTS 196-9-2014	集装箱码头堆场装卸设备供电设施建设技术规范 Technical code of power supply facilities for handling equipment in container yard			中交水运规划设计院有限公司
90	JTS 196-10-2015	长江干线桥区和航道整治建筑物助航标志 Aids to navigation for bridge area and training structures on the trunk waterway of Changjiang river			长江航道局
91	JTS 170-2015	邮轮码头设计规范 Design code for cruise terminal			中交第三航务工程勘察设计院有限公司

水路工程

序号	标准编号	标准名称	被代替标准编号	备案号	主编单位
92	JTS 171-2016	海上固定转载平台设计规范 Design code of fixed offshore platform for transshipment			中交第三航务工程勘察设计院有限公司
93	JTS/T 172-2016	码头结构加固改造技术指南 Guidance for strengthening and renovation of wharf structure			中交第三航务工程勘察设计院有限公司 中国工程建设标准化协会水运专业委员会 中交上海三航科学研究院有限公司
94	JTS 181-2016	航道工程设计规范 Design code for waterway engineering			长江航道规划设计研究院 中交天津港航勘察设计院有限公司
95	JTS 201-2011	水运工程施工通则 General rules for construction of port and waterway works			中国交通建设股份有限公司 中国港湾工程有限责任公司
96	JTS 202-2011	水运工程混凝土施工规范 Specifications for concrete construction of port and waterway engineering			中交天津港湾工程研究院有限公司
97	JTS 202-1-2010	水运工程大体积混凝土温度裂缝控制技术规程 Technical specification for thermal craking control of mass concrete of port and waterway engineering			中交武汉港湾工程设计研究院有限公司
98	JTS 311-2011	港口水工建筑物修补加固技术规范 Techinical code for repair and strengthening of harbor and marine structures			中交四航工程研究院有限公司
99	JTS 204-2008	水运工程爆破技术规范 Technical code of blasting for port and waterway engineering			长江重庆航道工程局
100	JTS 205-1-2008	水运工程施工安全防护技术规范 Safety protection standard for port and waterway engineering construction			中国水运建设行业协会 中国交通建设股份有限公司 中交第一航务工程局有限公司
101	JTS 206-1-2009	水运工程塑料排水板应用技术规程 Technical specification for application of plastic drainboard for port and waterway engineering			中交天津港湾工程研究院有限公司
102	JTS 207-2012	疏浚与吹填工程施工规范 Construction code for dredging and reclamation works			中交天津航道局有限公司 中国水运建设行业协会 中交上海航道局有限公司 中交广州航道局有限公司
103	JTJ 280-2002	港口设备安装工程技术规范 Technical code for port equipment installation construction			中港第三航务工程局
104	JTS 147-2-2009	真空预压加固软土地基技术规程 Technical specification for vacuum preloading technique to improve soft soils			中交天津港湾工程研究院有限公司
105	JTS 218-2014	船闸工程施工规范 Code for construction of shiplock engineering			中交第二航务工程局有限公司

水路工程

序号	标准编号	标准名称	被代替标准编号	备案号	主编单位
106	JTS 224-2016	航道整治工程施工规范 Construction code of channel regulation works			长江航道局
107	JTS/T 231-2-2010	海岸与河口潮流泥沙模拟技术规程 Technical regulation of modelling for tidal current and sediment on coast and estuary			交通运输部天津水运工程科学研究所
108	JTJ/T 234-2001	波浪模型试验规程 Wave model test regulation			南京水利科学研究院
109	JTJ/T 232-98	内河航道与港口水流泥沙模拟技术规程 Technical regulation of modelling for flow and sediment in inland waterway and harbour			交通部天津水运工程科学研究所
110	JTJ/T 235-2003	通航建筑物水力学模拟技术规程 Technical regulation of modeling for hydraulics of navigation structures			交通部天津水运工程科学研究所
111	JTJ 270-98	水运工程混凝土试验规程 Testing code of concrete for port and waterway engineering			天津港湾工程研究所 南京水利科学研究院
112	JTJ/T 272-99	港口工程混凝土非破损检测技术规程 Technical specifications for non-destructive inspection of concrete structures in port engineering			天津港湾工程研究所
113	JTJ 255-2002	港口工程基桩静载荷试验规程 Specification for testing of pile under static load in harbour engineering			武汉港湾工程设计研究院
114	JTJ 249-2001	港口工程桩基动力检测规程 Specification for dynamic testing of piles in port engineering			武汉港湾工程设计研究院
115	JTS/T 231-7-2013	港口工程离心模型试验技术规程 Code for centrifugal model test for port engineering			南京水利科学研究院
116	JTS 239-2015	水运工程混凝土结构实体检测技术规程 Technical specification for solid inspection of concrete structure in port and waterway engineering			中交天津港湾工程研究院有限公司
117	JTS 238-2016	水运工程试验检测仪器设备技术标准 Technical standard of test detection instrument and equipment for water transport engineering			交通运输部天津水运工程科学研究所
118	JTS 235-2016	水运工程水工建筑物原型观测技术规范 Technical code for hydraulic structures prototype observation of port and waterway engineering			中交上海三航科学研究院有限公司
119	JTS 252-2015	水运工程施工监理规范 Construction supervision code for port and waterway engineering			北京水规院京华工程管理有限公司 天津中北港湾工程建设监理有限公司

水路工程

序号	标准编号	标准名称	被代替标准编号	备案号	主编单位
120	JTS 252-1-2013	水运工程机电专项监理规范 Supervision specifications for construction of water transport engineering mechanical and eletrical special			厦门港湾咨询监理有限公司
121	JTS 202-2-2011	水运工程混凝土质量控制标准 Quality control standard of concrete for port and waterway engineering			中交四航工程研究院有限公司
122	JTS 257-2-2012	海港工程高性能混凝土质量控制标准 Quality control standard of high performance concrete for sea port engineering			中交四航工程研究院有限公司 中交第三航务工程局有限公司
123	JTS 258-2008	水运工程测量质量检验标准 Standard for quality inspection of port and waterway engineering survey			长江航道局
124	JTS 257-2008	水运工程质量检验标准 Standard for quality inspection of port and waterway engineering construction			中交第一航务工程局有限公司 福建省交通基本建设工程质量监督检测站
125		《水运工程质量检验标准》（JTS 257-2008）局部修订（航道整治工程质量检验部分） Standard for quality inspection of port and waterway engineering construction partially revised（part of quality inspection for waterway requlation engineering）			江苏省交通规划设计院股份有限公司
126	JTS 271-2008	水运工程工程量清单计价规范 Code of valuation with bill quantity of port and waterway engineering			交通部水运工程定额站
127	JTS/T 274-1-2011	水运工程数学模型试验研究参考定额 Reference norm on numerical model test of port and waterway engineering			南京水利科学研究院
128	JTS 275-1-2014	内河航运水工建筑工程定额 Quota for water-side structure works for inland river transportation			交通部水运工程定额站
129	JTS 275-2-2014	内河航运工程船舶机械艘（台）班费用定额 Quota for per-shift cost of construction plant and equipment used in construction project of inland river transportation			交通部水运工程定额站
130	JTS 275-3-2014	内河航运设备安装工程定额 Quota for costing of equipment installation works of inland river transportation			交通部水运工程定额站
131	JTS/T 275-4-2014	内河航运工程参考定额 Referece quota for inland river transportation works			交通部水运工程定额站
132	JTS 277-2014	水运工程混凝土和砂浆材料用量定额 Quota for consumption of concrete and mortar in construction works for water transportation terminals			交通部水运工程定额站
133	JTS 273-2014	水运工程测量定额 Quota for port and waterway engineering survey			交通部疏浚工程定额站

水路工程

序号	标准编号	标准名称	被代替标准编号	备案号	主编单位
134		远海区域疏浚工程计价暂行办法			交通部疏浚工程定额站
135		远海区域疏浚工程船舶艘班费用参考定额			交通部疏浚工程定额站
136		远海区域疏浚工程参考定额			交通部疏浚工程定额站
137		远海区域水运工程计价暂行办法			交通部水运工程定额站
138		远海区域水运工程船舶机械艘（台）班费用参考定额			交通部水运工程定额站
139		远海区域水运工程参考定额			交通部水运工程定额站
140	JTS/T 272-1-2014	沿海港口建设工程投资估算指标 Indexes for estimation of cost of seaport construction project			交通部水运工程定额站
141	JTS 117-1-2016	港口设施维护工程预算编制规定 Regulations on preparation of budget for maintenance works for port facilities			交通部水运工程定额站
142	JTJ 302-2006	港口水工建筑物检测与评估技术规范 Technical specification for detection and assessment of harbour and marine structures			中交四航工程研究院有限公司
143	JTS 310-2013	港口设施维护技术规范 Technical code for maintenance of port facilities			中国水运建设行业协会 中交天津港湾工程研究院有限公司 天津港（集团）有限公司
144	JTJ 287-2005	内河航道维护技术规范 Technical code of inland waterway maintenance			长江航道局 南京水利科学研究院
145	JTS 320-3-2013	船闸检修技术规程 Technical code of overhaulfor shiplocks			长江三峡通航管理局

航空工业工程

序号	标准编号	标准名称	被代替标准编号	备案号	主编单位
1	HBJ 1-1980	航空工业镁合金铸造厂房建筑结设计规定			中华人民共和国第三机械工业部第四规划设计研究院（现更名为：中国航空规划设计研究总院有限公司）
2	HBJ/T 5-2014	航空工业工厂照度标准	HBJ 5-1986		中国航空规划建设发展有限公司（现更名为：中国航空规划设计研究总院有限公司）
3	HBJ/T 14-2014	航空工业建设项目投资估算和设计概算编制办法	HBJ 14-1997		中国航空规划建设发展有限公司（现更名为：中国航空规划设计研究总院有限公司）
4	HBJ 15-2005	航空工业精密铸造车间设计规定			中国航空工业规划设计研究院（现更名为：中国航空规划设计研究总院有限公司）
5	HBJ 16-2006	航空工业复合材料车间和金属胶结车间设计规程			中国航空工业规划设计研究院（现更名为：中国航空规划设计研究总院有限公司）
6	HBJ 17-2006	航空工业特种焊接车间设计规程			中国航空工业规划设计研究院（现更名为：中国航空规划设计研究总院有限公司）

冶金工业工程

序号	标准编号	标准名称	被代替标准编号	备案号	主编单位
1	YB/T 4184-2009	钢渣混合料路面基层施工技术规程			中冶建筑研究总院有限公司
2	YB/T 4185-2009	尾矿砂浆技术规程			中冶建筑研究总院有限公司
3	YB/T 4227-2010 节能司	不锈钢钢渣中金属含量测定方法			中冶建筑研究总院有限公司
4	YB/T 4252-2011	耐热混凝土应用技术规程			中冶建筑研究总院有限公司
5	YB/T 4253-2011	冶金设备工程安装质量评定标准			中国一冶集团有限公司
6	YB/T 4279-2012	水泥基耐磨材料应用技术规程			中冶建筑研究总院有限公司
7	YB/T 4280-2013	焊管设备安装工程施工验收规范			天工冶天工集团有限公司
8	YB/T 4280-2013	焊管设备安装工程施工验收规范		J1807-2014	中国华冶科工集团有限公司 中冶天工集团有限公司
9	YB 4353-2013	栓钉焊机技术规程			中冶建筑研究总院有限公司
10	YB 4354-2013	冶金工业自动化仪表验收规范			中国二十冶集团有限公司
11	YB 4355-2013	炼钢连铸机械设备检修技术标准			中冶天工集团有限公司
12	YB/T 4356-2013	钢铁企业电气火灾监控系统设计规范			中冶华天工程技术有限公司
13	YB 4357-2013	线型光纤感温火灾探测报警系统设计及施工规范			中冶南方工程技术有限公司
14	YB 4358-2013	钢铁企业胶带机钢结构通廊设计规范			中冶赛迪工程技术股份有限公司
15	YB 4359-2013	钢铁企业通风除尘设计规范			中冶赛迪工程技术股份有限公司
16	YB/T 4385-2013	冶金矿山井巷工程测量规范		J1704-2014	中国华冶科工集团有限公司 新七建设有限公司
17	YB/T 4386-2013	冶金电气工程通讯、网络施工及验收规范			中国二十冶集团有限公司
18	YB/T 4387-2013	冶金工程基坑降水技术规程			中国二十冶集团有限公司
19	YB/T 4388-2013	钢管挤压设备安装验收规范			天工冶天工集团有限公司
20	YB/T 4389-2013	冶金低压变频传动成套设备规范			中冶南方工程技术有限公司
21	YB/T 4390-2013	工业建（构）筑物钢结构防腐蚀涂装质量检测、评定标准			中冶建筑研究总院有限公司
22	YB/T 4390-2013	冶金矿山井巷工程施工质量验收规范		J1705-2014	中国华冶科工集团有限公司
23	YB 4407-2014	冶金矿山井巷安装工程质量验收规范		J1825-2014	中国华冶科工集团有限公司
24	YB 4408-2014	高炉TRT系统电气设备安装工程施工验收规范			中冶天工集团有限公司
25	YB 4409-2014	环形加热炉砌筑工程质量验收规范			中冶天工集团有限公司
26	YB 4410-2014	煤气柜工程施工及验收规范			中冶天工集团有限公司
27	YB 4441-2014	钢铁企业除尘工程施工及验收规范			中冶南方工程技术有限公司
28	YB/T 4476-2014	不锈钢复合板球形储罐施工验收规范			天工冶天工集团有限公司
29	YB/T 4483-2015	余热利用设备设计技术规定			中冶华天工程技术有限公司
30	YB/T 4504-2016	钢铁企业煤气—蒸汽联合循环电厂设计规范			中冶华天工程技术有限公司
31	YB/T 4505-2016	冶金行业设备基础后置锚栓技术规范			中国十九冶集团有限公司
32	YB 9057-1993	高炉炼铁工艺设计技术规定			中冶赛迪工程技术股份有限公司
33	YB 9063-2000	钢铁企业电信设计规范			中冶京诚工程技术有限公司（原北京钢铁设计研究总院）
34	YB 9070-2014	冶金行业压力容器设计管理规定			中冶华天工程技术有限公司
35	YB/T 9071-2015	余热利用设备设计管理规定			中冶华天工程技术有限公司
36	YB 9073-2014	钢制压力容器设计技术规定			中冶华天工程技术有限公司

冶金工业工程

序号	标准编号	标准名称	被代替标准编号	备案号	主编单位
37	YB 9078-1999	钢铁企业铁路信号设计规范			中冶京诚工程技术有限公司（原北京钢铁设计研究总院）
38	YB 9081-1997	冶金建筑抗震设计规范			中冶京诚工程技术有限公司（原北京钢铁设计研究总院）
39	YB/T 9231-2009	钢筋阻锈剂应用技术规程			中冶建筑研究总院有限公司
40	YB/T 9251-1994	组合钢模板质量检验评定标准			中冶建筑研究总院有限公司
41	YB 9258-1997	建筑基坑工程技术规范			中冶建筑研究总院有限公司
42	YB/T 9259-1998	冶金工程建设焊工考试规程			中冶建筑研究总院有限公司
43	YB/T 9260-1998	冶金工业设备抗震鉴定标准			中冶建筑研究总院有限公司
44	YB/T 9261-98	水泥基灌浆材料施工技术规程			中冶建筑研究总院有限公司

有色金属工业工程

序号	标准编号	标准名称	被代替标准编号	备案号	主编单位
1	YSJ 003-1988	有色金属工业劳动设计规程			北京有色冶金设计研究总院（中国恩菲工程技术有限公司）
2	YSJ 008-1990	有色金属矿山生活福利设施建筑设计标准			长沙有色冶金设计研究院（长沙有色冶金设计研究院有限公司）
3	YSJ 009-1990	机器动荷载作用下建筑物承重结构的振动计算和隔振设计规程			北京有色冶金设计研究总院（中国恩菲工程技术有限公司）
4	YSJ 016-1992	有色金属矿山机修与汽修设施工艺设计标准			昆明有色冶金设计研究院（昆明有色冶金设计研究院股份公司）
5	YSJ 020-1993	重有色金属冶炼术语标准			北京有色冶金设计研究总院（中国恩菲工程技术有限公司）
6	YSJ 217-1989	工程测量资料整理规程			中国有色金属工业总公司西安勘察院（中国有色金属工业西安勘察设计研究院）
7	YSJ 401-1989	土方与爆破工程施工操作规程			兰州有色金属建筑研究所（甘肃土木工程科学研究院）
8	YSJ 402-1989	地基与基础工程施工操作规程			兰州有色金属建筑研究所（甘肃土木工程科学研究院）
9	YSJ 403-1989	钢筋混凝土工程施工操作规程			兰州有色金属建筑研究所（甘肃土木工程科学研究院）
10	YSJ 404-1989	结构安装工程施工操作规程			兰州有色金属建筑研究所（甘肃土木工程科学研究院）
11	YSJ 405-1989	特种结构工程施工操作规程			兰州有色金属建筑研究所（甘肃土木工程科学研究院）
12	YSJ 406-1989	砌筑工程施工操作规程			兰州有色金属建筑研究所（甘肃土木工程科学研究院）
13	YSJ 409-1989	装饰工程施工操作规程			兰州有色金属建筑研究所（甘肃土木工程科学研究院）
14	YSJ 411-1989	防腐蚀工程施工操作规程			兰州有色金属建筑研究所（甘肃土木工程科学研究院）
15	YSJ 415-1993	有色金属矿山井巷工程测量规程			中国有色金属工业第十四冶金建设公司
16	YSJ 416-1993	有色金属地下开采矿山基建地质规程			中国有色金属工业第十四冶金建设公司
17	YS 5018-1996	有色金属工业技术经济设计规范			北京有色冶金设计研究总院（中国恩菲工程技术有限公司）
18	YS/T 5022-94	冶金矿山采矿术语标准			北京有色冶金设计研究总院（中国恩菲工程技术有限公司）
19	YS/T 5023-94	有色金属选矿厂工艺设计制图标准			北京有色冶金设计研究总院（中国恩菲工程技术有限公司）
20	YS/T 5027-95	有色金属加工术语标准			洛阳有色金属加工设计研究院（中色科技股份有限公司）
21	YS/T 5028-96	有色金属选矿术语标准			北京有色冶金设计研究总院（中国恩菲工程技术有限公司）
22	YS 5029-1996	铝矿山土地复垦工程设计规程			沈阳铝镁设计研究院（沈阳铝镁设计研究院有限公司）

有色金属工业工程

序号	标准编号	标准名称	被代替标准编号	备案号	主编单位
23	YS 5030-1996	有色金属矿山电力设计规范			北京有色冶金设计研究总院（中国恩菲工程技术有限公司）
24	YS 5031-1997	有色金属加工厂电力设计规范			洛阳有色金属加工设计研究院（中色科技股份有限公司）
25	YS/T 5035-2016	铝电解槽带电焊接技术规范			贵阳铝镁设计研究院有限公司
26	YS/T 5225-2016	土工试验规程 Code for soil test		J2195-2016	中国有色金属长沙勘察设计研究院有限公司
27	YS/T 5226-2016	水质分析规程 Code for water analysis		J2196-2016	中国有色金属长沙勘察设计研究院有限公司
28	YS/T 5419-2013	有色金属工业安装工程质量检验评定统一标准			有色金属工业建设工程质量监督总站
29	YS/T 5420-2014	有色金属工业通用设备安装工程质量检验评定标准 Inspection and evaluation standards of project quality of general equipment installation in non-ferrous industry		J1831-2014	有色金属工业建设工程质量监督总站 中国第四冶金建设有限责任公司
30	YS/T 5421-2014	有色金属矿山设备安装工程质量的检验评定 Inspection and evaluation standards of project quality of mining equipment installation in non-ferrous industry		J1832-2014	有色金属工业建设工程质量监督总站 中十冶集团有限公司
31	YS/T 5422-2014	轻金属冶炼设备安装工程质量检验评定标准 Inspection and evaluation standards of quality of smelting equipment installation for light metal		J1833-2014	有色金属工业建设工程质量监督总站 九冶建设有限公司
32	YS/T 5423-2014	重有色金属冶炼设备安装工程质量检验评定标准 Inspection and evaluation standards of project quality of installation of heavy non-ferrous metal smelting equipment		J1834-2014	有色金属工业建设工程质量监督总站 八冶建设集团有限公司
33	YS/T 5424-2014	有色金属工业炉窑砌筑工程质量检验评定标准 Inspection and evaluation standards of project quality of furnace masonry in non-ferrous industry		J1835-2014	有色金属工业建设工程质量监督总站 中国第四冶金建设有限责任公司
34	YS/T 5425-2014	有色金属加工设备安装工程质量检验评定标准 Inspection and evaluation standards of project quality of processing equipment installation in non-ferrous industry		J1836-2014	有色金属工业建设工程质量监督总站 十一冶建设集团有限公司
35	YS 5426-2014	铜母线焊接施工及验收规范			中国十五冶金建设集团有限公司、金光道环境建设集团有限公司
36	YS/T 5427-2015	铝电解多功能机组安装技术规程 Technical Specification of Pot Tending Machine Installation		J2032-2015	中十冶集团有限公司 中国黄金集团建设有限公司
37	YS/T 5428-2015	混凝土电解槽施工技术规程 Technical specification for construction of concrete electrolytic tank		J2033-2015	金川集团工程建设有限公司

有色金属工业工程

序号	标准编号	标准名称	被代替标准编号	备案号	主编单位
38	YS/T 5429-2016	酸性烟气输送管道及设备内衬施工技术规程 Technical specification for construction of lining of acidic gas pipeline and equipment		J2197-2016	金川集团工程建设有限公司
39	YS/T 5430-2016	有色金属工业建筑工程质量检验评定统一标准 Unified standards for constructional quality inspection and assessment of non-ferrous metal industrial building engineering		J2198-2016	有色金属工业建设工程质量监督总站
40	YS/T 5431-2016	钢制焊接立式锥形容器施工及验收规范			中色十二冶金建设有限公司
41	2015-1020T-YS	氧化铝厂通风除尘与烟气净化设计规范			贵阳铝镁设计研究院有限公司
42	2015-1021T-YS	有色金属工业建筑工程绿色施工评价标准			兰州有色金属建筑研究所（甘肃土木工程科学研究院）
43	2015-1027T-YS	边坡工程勘察规范			中国有色金属工业昆明勘察设计研究院
44	2015-1028T-YS	标准贯入试验规程			中国有色金属工业西安勘察设计研究院
45	2015-1029T-YS	抽水试验规程			中国有色金属长沙勘察设计研究院有限公司
46	2015-1030T-YS	地面与楼面工程施工操作规程			兰州有色金属建筑研究所（甘肃土木工程科学研究院）
47	2015-1031T-YS	电测十字板剪切试验规程			中国有色金属工业昆明勘察设计研究院
48	2015-1032T-YS	动力机械基础地基动力特性测试规程			中国有色金属工业西安勘察设计研究院
49	2015-1033T-YS	工程测量作业规程			中国有色金属工业西安勘察设计研究院
50	2015-1034T-YS	工程地质测绘规程			中国有色金属长沙勘察设计研究院有限公司
51	2015-1035T-YS	灌注桩基础技术规程			中国有色金属工业昆明勘察设计研究院
52	2015-1036T-YS	静力触探试验规程			中国有色金属工业昆明勘察设计研究院
53	2015-1037T-YS	门窗安装工程施工操作规程			兰州有色金属建筑研究所（甘肃土木工程科学研究院）
54	2015-1038T-YS	旁压试验规程			中国有色金属长沙勘察设计研究院有限公司
55	2015-1039T-YS	强夯地基技术规程			中国有色金属工业西安勘察设计研究院
56	2015-1040T-YS	湿陷性土起始压力测试规程			中国有色金属工业西安勘察设计研究院
57	2015-1041T-YS	天然建筑材料勘探规程			中国有色金属工业昆明勘察设计研究院
58	2015-1042T-YS	屋面工程施工操作规程			兰州有色金属建筑研究所（甘肃土木工程科学研究院）
59	2015-1043T-YS	现场直剪试验规程			中国有色金属工业昆明勘察设计研究院
60	2015-1044T-YS	压水试验规程			中国有色金属长沙勘察设计研究院有限公司

有色金属工业工程

序号	标准编号	标准名称	被代替标准编号	备案号	主编单位
61	2015-1045T-YS	岩土工程监测规范			中国有色金属工业昆明勘察设计研究院
62	2015-1046T-YS	岩土工程勘察报告书编制规程			中国有色金属工业西安勘察设计研究院
63	2015-1047T-YS	岩土工程勘察图式图例规程			中国有色金属工业西安勘察设计研究院
64	2015-1048T-YS	岩土工程现场描述规程			中国有色金属工业西安勘察设计研究院
65	2015-1049T-YS	岩土静力载荷试验规程			中国有色金属工业西安勘察设计研究院
66	2015-1050T-YS	圆锥动力触探试验规程			中国有色金属工业昆明勘察设计研究院
67	2015-1051T-YS	注浆技术规程			中国有色金属工业西安勘察设计研究院
68	2015-1052T-YS	注水试验规程			中国有色金属长沙勘察设计研究院有限公司
69	2015-1053T-YS	钻探、井探、槽探操作规程			中国有色金属工业西安勘察设计研究院

民航工程

序号	标准编号	标准名称	被代替标准编号	备案号	主编单位
1	MH 5001-2013	民用机场飞行区技术标准 Aerodrome technical standards	MH 5001-2006		中国民航机场建设集团公司
2	MH 5002-1999	民用机场总体规划规范 Code for master planning of civil airport			中国民航机场建设总公司
3	MH/T 5003-2004	民用机场航站楼离港系统工程设计规范			中航机场设备有限公司
4	MH/T 5004-2010	民用机场水泥混凝土道面设计规范 Specifications for airport cement concrete pavement design			中国民航机场建设集团公司
5	MH 5005-2002	民用机场飞行区排水工程施工技术规范 Technical specifications for construction of drainage engineering for airfield area of civil airports			中国民用航空总局机场司
6	MH 5006-2015	民用机场水泥混凝土面层施工技术规范 Specifications for construction of aerodrome cement concrete pavement	MH 5006-2002		中国民航机场建设集团公司
7	MH 5007-2000	民用机场飞行区工程竣工验收质量检验评定标准 Quality inspection and evaluation standards for acceptance of airfield area engineering of civil airports			中国民用航空总局机场司
8	MH 5008-2005	民用机场供油工程建设技术规范 Technical code for the construction of civil airport fuel system			中国航空油料总公司
9	MH/T 5009-2004	民用机场航站楼楼宇自控系统工程设计规范 Design code for building automatic control system engineering of civil airport terminal building			中航机场设备有限公司
10	MH 5010-1999	民用机场沥青混凝土道面设计规范 Specifications for asphalt concrete pavement design of civil airports			中国民航中南机场设计研究院
11	MH 5011-1999	民用机场沥青混凝土道面施工技术规范 Specifications for asphalt concrete pavement construction of civil airports			中国民航中南机场设计研究院
12	MH 5012-2010	民用机场目视助航设施工质量验收规范 Code of acceptance of construction quality of civil airport visual aids	MH 5012-1999		上海民航新时代机场设计研究院有限公司 北京中航空港建设工程有限公司
13	MH 5013-2014	民用直升机场飞行场地技术标准 Technical standards of civil heliports	MH 5013-2008		上海民航新时代机场设计研究院有限公司
14	MH 5014-2002	民用机场飞行区土（石）方与道面基础施工技术规范 Technical specifications for construction of earthwork（rockwork）and pavement foundation for airfield area of civil airports			中国民用航空总局机场司
15	MH/T 5015-2004	民用机场航站楼航班信息显示系统工程设计规范 Design code for flight information display system engineering of civil airport terminal building			中航机场设备有限公司

民航工程

序号	标准编号	标准名称	被代替标准编号	备案号	主编单位
16	MH 5016-2001	民用机场工程初步设计文件编制内容及深度要求 The preliminary design documents on compiling content and depth demand of civil airport project			中国民航机场建设总公司
17	MH/T 5017-2004	民用机场航站楼闭路电视监控系统工程设计规范 Design code for CCTV system engineering of civil airport terminal building			中航机场设备有限公司
18	MH/T 5018-2004	民用机场航站楼计算机信息管理系统工程设计规范 Design code for information management system engineering of civil airport terminal building			中航机场设备有限公司
19	MH/T 5019-2004	民用机场航站楼时钟系统工程设计规范 Design code for clock system engineering of civil airport terminal building			中航机场设备有限公司
20	MH/T 5020-2004	民用机场航站楼广播系统工程设计规范 Design code for broadcast system engineering of civil airport terminal building			中航机场设备有限公司
21	MH/T 5021-2004	民用机场航站楼综合布线系统工程设计规范			中航机场设备有限公司
22	MH 5022-2005	民用机场工程施工图设计文件编制内容及深度要求 The compiling content and depth demand for detail design documents of civil airport project			中国民航机场建设总公司
23	MH 5023-2006	民用航空支线机场建设标准 Construction standards of civil feeder airports			中国民用航空总局机场司
24	MH/T 5024-2009	民用机场道面评价管理技术规范 Technical specifications of aerodrome pavement evaluation and management			同济大学
25	MH/T 5025-2011	民用机场勘测规范 Specifications for geotechnical investigation and surveying of airpor			中国民航机场建设集团公司
26	MH/T 5026-2012	通用机场建设规范 General aviation airport construction specification			中国民用航空局机场司 中国民航工程咨询公司
27	MH/T 5027-2013	民用机场岩土工程设计规范 Code for geotechnical engineering design of airport			中国民航机场建设集团公司
28	MH 5028-2014	民航专业工程工程量清单计价规范 Code of bills of quantities and valuation for civil aviation specialized engineering			中国民航工程咨询公司
29	MH 5029-2014	小型民用运输机场供油工程设计规范 Design for small civil airport fuelling system construction			中国航空油料有限责任公司

民航工程

序号	标准编号	标准名称	被代替标准编号	备案号	主编单位
30	MH/T 5030-2014	通用航空供油工程建设规范 Code for general aviation fuelling system construction			中国航空油料有限责任公司
31	MH 5031-2015	民航专业工程施工监理规范 Specifications for construction supervision of civil aviation specialty projects			上海华东民航机场建设监理有限公司
32	MH/T 5032-2015	民用运输机场航班信息显示系统检测规范 Detecting specification of flight information display system for civil airport			民航专业工程质量监督总站

建材工业工程

序号	标准编号	标准名称	被代替标准编号	备案号	主编单位
1	JCJ 01-89	钢管混凝土结构设计施工及验收规程			苏州混凝土水泥制品研究院、中国船舶总公司第九设计研究院
2	JCJ 02-90	钢管混凝土构件N-M相关设计计算图表			南京水泥工业设计研究院、苏州混凝土水泥制品研究院
3	JCJ 03-90	水泥工厂机械设备安装工程施工及验收规范			中国建筑材料工业建设总公司
4	JCJ 04-90	平板玻璃工厂工艺设计规范			蚌埠玻璃工业设计研究院
5	JCJ 05-92	水泥工厂建设项目初步设计内容及深度规定			天津水泥工业设计研究院
6	JCJ 06-92	平板玻璃厂建设项目初步设计内容深度规定			国家建材局蚌埠玻璃工业设计研究院
7	JCJ 07-94	建材矿山工程施工及验收规范			中国建筑材料工业建设总公司、苏州非金属矿设计研究院
8	JCJ 08-95	平板玻璃工厂环境保护设计规定			秦皇岛玻璃工业设计研究院
9	JCJ 09-95	平板玻璃工厂职业安全卫生设计规定			蚌埠玻璃工业设计研究院
10	JCJ 10-97	水泥工业劳动安全卫生设计规定			国家建材局天津水泥工业设计研究院
11	JCJ 11-97	水泥工业环境保护设计规定			国家建材局天津水泥工业设计研究院
12	JCJ 12-97	GREC屋面防水施工技术规程			中国建筑材料科学研究院、中研益工程技术开发中心
13	JCJ 13-98	浮法玻璃熔窑砌筑工程施工及验收规程			国家建材局蚌埠玻璃工业设计研究院
14	JCJ 14-1999	聚氨酯硬泡体防水保温工程技术规范			国家建筑材料工业局标准定额中心站
15	JCJ 15-2000	铸石制品应用工程技术规范			中国建筑材料科学研究院
16	JCJ/T 16-2012	建材工业建设项目投资估算编审规程		J1485-2012	国家建筑材料工业标准定额总站、蚌埠玻璃工业设计研究院
17	JCJ/T 17-2012	建材工业建设项目设计概算编审规程		J1484-2012	国家建筑材料工业标准定额总站、蚌埠玻璃工业设计研究院

机械工业工程

序号	标准编号	标准名称	被代替标准编号	备案号	主编单位
1	JBJ 6-96	机械工厂电力设计规范 Code for electrical design of machine factory	JBJ 6-80		中机中电设计研究院有限公司
2	JBJ 12-97	单层工厂厂房抗震设计规范 Code for design of aseismic single story machine factory building	JBJ 12-93		中国中元国际工程公司
3	JBJ/T 33-1999	机械工厂中央实验室设计规范 Design code for central laboratories of machinery plants			中机国际工程设计研究院有限责任公司
4	JBJ/T 34-99	机械工厂热处理车间设计规范 Code for design of machinery factory heat treatment shop			中国联合工程公司

信息产业（邮政、电信、电子）工程

序号	标准编号	标准名称	被代替标准编号	备案号	主编单位
1	SJ/T 11447-2013	电子工业建设项目可行性研究报告编制规定		J1844-2014	信息产业电子第十一设计研究院科技工程股份有限公司
2	SJ/T 11448-2013	软件园区规划设计规范		J1845-2014	中国电子工程设计院
3	SJ/T 11449-2013	集中空调电子计费信息系统工程技术规范		J1846-2014	郑州春泉暖通节能设备有限公司

广播电影电视工程

序号	标准编号	标准名称	被代替标准编号	备案号	主编单位
1	GY/T 5002-2015	广播电影电视工程建设项目前期文件编审规程 Specification for preparation and reviewing rules of the pre-project documents for radio film and television engineering construction projects	GY 5002-1995 GY 5003-1995	J2110-2016	中广电广播电影电视设计研究院
2	GY/T 5006-2010	广播电影电视工程建设项目竣工验收工作规程 Specification for acceptance work of radio film and television construction project completion	GY 5006-1990	J984-2010	国家广电总局无线电台管理局
3	GY 5013-2014	广播电视工程测量规范 Code for radio television engineering survey	GY 5013-2005	J1915-2014	中广电广播电影电视设计研究院
4	GY 5022-2007	广播电视播音（演播）室混响时间测量规范 Code for measurement of reverberation time in radio & television studios	GYJ 22-1985	J775-2008	中广电广播电影电视设计研究院
5	GY/T 5031-2013	广播电视微波站（台）工程设计规范 Code for engineering design of radio and television microwave station	GYJ 31-87	J1674-2013	四川省广播电视发射传输中心
6	GY/T 5032-2012	广播电视 SDH 数字微波工程安装及验收规范 Code for installation and acceptance of radio and TV SDH digital microwave engineering	GYJ 32-87	J1464-2012	江苏省广播电视总台发射传输台
7	GY/T 5034-2015	中、短波广播发射台设计规范 Code for design of MW and SW broadcast transmitting station	GYJ 34-88	J2109-2015	中广电广播电影电视设计研究院
8	GY/T 5039-2011	广播电视卫星地球站场地要求 Requirements for site of satellite earth station transmitting video and audio signals	GYJ 39-89	J1317-2011	国家广播电影电视总局广播科学研究院
9	GY 5040-2009	卫星广播电视地球站系统设备安装调试验收规范 Code for equipment installation, adjustment and acceptance of Satellite Radio & TV Earth Station	GYJ 40-89	J972-2010	国家广播电影电视总局广播科学研究院
10	GY/T 5041-2012	广播电视卫星地球站工程设计规范 Code for engineering design of satellite earth station transmitting video and audio signals	GYJ 41-89	J1498-2013	中广电广播电影电视设计研究院
11	GY/T 5043-2013	广播电视中心技术用房室内环境要求 Standard for indoor environmental requirement of radio & TV	GYJ 43-90	J1641-2013	中广电广播电影电视设计研究院
12	GY 5055-2008	扩声、会议系统安装工程施工及验收规范 Code for installation and acceptance of PA、Conference system	GY 5055-1995	J809-2008	北京中广广播电视工程有限公司
13	GY 5057-2006	中短波广播天线馈线系统安装工程施工及验收规范 Code for construction and acceptance of MW & SW broadcasting antenna and feeder system	GY 5057-1995	J656-2007	北京中广广播电视工程安装公司

广播电影电视工程

序号	标准编号	标准名称	被代替标准编号	备案号	主编单位
14	GY 5060-2008	广播电影电视建筑工程抗震设防分类标准 Standard for classification of seismic protection of radio, film and TV building constructions	GY 5060-1997	J829-2009	中广电广播电影电视设计研究院
15	GY/T 5061-2015	广播电影电视工程技术用房照明设计规范 Code for Lighting Design of Radio, Film & TV Engineering Technology Rooms	GY/T 5061-2007	J776-2016	中广电广播电影电视设计研究院
16	GY 5070-2013	电视演播室灯光系统施工及验收规范 Code for construction and acceptance of TV studio lighting system		J297-2013	中广电广播电影电视设计研究院
17	GY 5077-2007	广播电视微波通信铁塔及桅杆质量验收规范 Requirements for site of satellite earth station transmitting video and audio signals		J777-2008	中广电广播电影电视设计研究院
18	GY 5078-2008	有线电视分配网络工程安全技术规范 Technical code for safety of CATV distributed networks engineering		J796-2008	中广电广播电影电视设计研究院
19	GY 5079-2008	广播电视传输电缆、光缆损坏损失计算标准 Loss calculation standard for broken electric and optical transmission cable of radio & TV		J830-2009	中广电广播电影电视设计研究院
20	GY 5080-2008	广播电视工程监理规范 Code for supervision of broadcast and TV project		J834-2009	国家广播电影电视总局规划财务司 中广电广播电影电视设计研究院
21	GY/T 5081-2009	广播电影电视工程设计文件档案管理规范 Code for design document filing and arrangementof radio, film and TV project		J943-2009	中广电广播电影电视设计研究院
22	GY 5082-2010	有线广播电视网络管理中心设计规范 Code for design of the network management center of cable television		J985-2010	中广电广播电影电视设计研究院
23	GY/T5083-2010	省级广播电视安全播出指挥调度中心工程技术规范 Technical code for engineering of provincial level radio and television safety broadcasting command and control center		J1090-2010	国家广播电视安全播出调度中心
24	GY/T 5084-2011	广播电视工程工艺接地技术规范 Code for technics earthing of radio and television engineering		J1270-2011	中广电广播电影电视设计研究院
25	GY/T 5085-2012	广播电视监测台场地技术要求 Field technical requirements for monitoring station of radio and television		J1406-2012	国家广播电影电视总局监管中心
26	GY/T 5086-2012	广播电视录(播)音室、演播室声学设计规范 Standard for acoustical design of radio & television studio		J1439-2012	中广电广播电影电视设计研究院
27	GY/T 5087-2012	广播电视中心声学装修工程施工及验收规范 Code for construction and acceptance of the acoustical decoration engineering		J1440-2012	中广电广播电影电视设计研究院

广播电影电视工程

序号	标准编号	标准名称	被代替标准编号	备案号	主编单位
28	GY/T 5088-2013	电视和调频广播发射天馈线系统技术指标及测量方法 Technical specification and measurement methods for television and FM transmitting antenna & feeder		J1708-2014	中广电广播电影电视设计研究院
29	GY/T 5089-2014	广播通信钢塔桅可靠性检测鉴定规范 Code for inspection and appraisal of reliability of broadcast communication steel tower and mast		J1924-2014	中广电广播电影电视设计研究院
30	GY/T 5090-2015	广播电视中心制作播出专用局域网工程技术规范 Technical code for program production and broadcasting professional local area network of radio and TV station		J2031-2015	中广电广播电影电视设计研究院
31	GY/T 5091-2015	广播电影电视工程建设项目管理规范 Code for construction project management of radio film and television engineering		J2072-2015	中广电广播电影电视设计研究院
32	GY 5212-2008	广播电视传输网络系统安装工程预算定额	GY 5212-1997		四川省广播电影电视局

国内贸易工程

序号	标准编号	标准名称	被代替标准编号	备案号	主编单位
1	SBJ/T 08-2007	牛羊屠宰与分割车间设计规范 Code for design of cattle and sheep slaughtering and cutting rooms	SBJ 08-94	J773-2008	国内贸易工程设计研究院
2	SBJ 12-2011	氨制冷系统安装工程施工及验收规范 Code for Installation and acceptance specification of ammonia	SBJ 12-2000	J38-2011	国内贸易工程设计研究院国内贸易工程设计研究院
3	SBJ 14-2007	氢氯氟烃、氢氟烃类制冷系统安装工程施工及验收规范 Code for installation and acceptance of refrigerating system with HCFC, HFC refrigerant		J726-2007	国内贸易工程设计研究院
4	SBJ 15-2008	禽类屠宰与分割车间设计规范 Code for design of poultry slaughtering and cutting rooms		J819-2008	

林业工程

序号	标准编号	标准名称	被代替标准编号	备案号	主编单位
1	LY/T 5001-2014	林业工程名词术语及计量标准	LYJ 001-87	J2217-2016	国家林业局昆明勘察设计院
2	LY/T 5002-2014	林业工程制图标准	LYJ 002-87	J2218-2016	国家林业局昆明勘察设计院
3	LY/T 5003-2014	林业工程建设分类标准	LYJ 003-87	J2219-2016	国家林业局昆明勘察设计院
4	LY/T 5004-2014	竹材胶合板工程设计规范		J2220-2016	国家林业局林产工业规划设计院
5	LY/T 5005-2014	林区公路设计规范		J2222-2016	国家林业局昆明勘察设计院
6	LY/T 5008-2014	细木工工程设计规范		J2221-2016	国家林业局林产工业规划设计院

纺织业工程

序号	标准编号	标准名称	被代替标准编号	备案号	主编单位
1	FZJ 116-93	多层织造厂房结构动力设计规范			纺织工业部设计院
2	FZ 211-2013	夹套管施工及验收规范			中国纺织工业设计院

轻工业工程

序号	标准编号	标准名称	被代替标准编号	备案号	主编单位
1	QB 6001-91	制浆造纸设计规范碱回收工艺部分			中轻国际工程有限公司 中国海诚工程科技股份有限公司
2	QB 6002-91	制糖专用设备施工及验收规范			中国轻工建设工程总公司
3	QB 6003-91	真空制盐专用设备施工安装验收规范			中国轻工业长沙工程有限责任公司 湖南省轻工安装工程公司
4	QB 6004-92	啤酒厂设计规范			中国轻工业广州设计工程有限公司 中国中轻国际工程有限公司
5	QB 6005-92	麦芽厂设计规范			中国轻工业广州设计工程有限公司 中国中轻国际工程有限公司
6	QB 6007-93	合成洗涤剂工厂设计规范			中国中轻国际工程有限公司
7	QG 6008-95	真空制盐厂设计规范 Design code for vatuum Salt factory			中国轻工业长沙工程有限公司
8	QB 6009-95	味精厂设计规范 Design code for mono-Sodium factory			中国海诚工程科技股份有限公司 中国中轻国际工程有限公司
9	QB 6010-95	感光材料厂设计规范 Design code for fabricating plant of photosensitive materials			中国海诚工程科技股份有限公司
10	QB/T 6011-95	制革、毛皮厂设计规范 Design code for fabricating plant of leather and fur			中国轻工业成都设计工程有限公司
11	QB/T 6012-96	日用陶瓷窑炉施工验收规范			江西省景德镇陶瓷工业公司
12	QB 6013-96	轻工企业建筑抗震设防分类标准 Standard for classification of seismic protection of buildings for light industry enterprise			中国轻工总会综合计划司组织编制
13	QB/T 6014-96	酒精厂设计规范			中国轻工业广州设计工程有限公司
14	QB/T 6015-96	罐头厂设计规范 Design code for cammery			中国海诚工程科技股份有限公司
15	QB/T 6016-97	家庭装饰工程质量规范			中国室内装饰协会
16	QB/T 6017-97	日用陶瓷厂设计规范 Design code for domestic ceramics			中国轻工业长沙工程有限公司
17	QB/T 6018-98	塑料制品厂设计规范 Design plant of plastic products			中国海诚工程科技股份有限公司
18	QBJ-102G-87	甘蔗糖厂设计规范			中国轻工业广州设计工程有限公司
19	QBJ-103T-88	甜菜糖厂设计规范			中国轻工业西安设计工程有限责任公司
20	QBJ-203-87	井矿盐钻井技术规范			自贡井矿盐钻井大队
21	QBJ-204-88	海盐专用设备安装施工验收规范			天津汉沽盐场
22	QB/6009-2004	制浆造纸安装工程施工验收规范			中国轻工业安装工程公司
23	QBJ S5-2005 (2008)	轻工业建设项目可行性研究报告编制内容深度规定			中国轻工业勘察设计协会

轻工业工程

序号	标准编号	标准名称	被代替标准编号	备案号	主编单位
24	QBJ S6-2005 (2008)	轻工业建设项目初步设计编制内容深度规定			中国轻工业勘察设计协会
25	QBJ S34-2005 (2008)	轻工业建设项目施工图设计编制内容深度规定			中国轻工业勘察设计协会
26	QBJ S10-2005 (2008)	轻工业工程设计概算编制办法			中国轻工业勘察设计协会
27		井矿盐矿山钻井专用设备施工技术规范			自贡井矿盐钻井大队
28		盐化工专业设备施工安装预算定额			湖南湘澧盐矿
29		造纸工业专用设备安装工程预算定额			中国轻工业安装工程公司

船舶工业工程

序号	标准编号	标准名称	被代替标准编号	备案号	主编单位
1	CB 4288-2013	船厂起重设备安全技术要求 Safty requirements for crane equipments			中船第九设计研究院工程有限公司
2	CB/T 8502-2005	纵向倾斜船台及滑道设计规范 Code of design for longitudinal inclined building berth and slipway	CB/T 8502-1992		中国船舶工业第九设计研究院
3	CB/T 8504-2011	船厂门座起重机技术规定 Technical specification of portal cranes for shipyard	CB/T 8504-1995		中船第九设计研究院工程有限公司
4	CB/T 8505-1998	建设工程可行性研究报告编制内容和深度规定			第九设计研究院
5	CB/T 8506-1998	建设工程初步设计编制内容和深度规定			第九设计研究院
6	CB/T 8520-2005	船厂卷扬式垂直升船机设计规范 Code of design for vertical hoisting shiplift of shipyard			中国船舶工业第九设计研究院
7	CB/T 8521-2008	造船门式起重机设计要求 Design requirements of shipbuilding gantry crane			中国船舶工业第九设计研究院
8	CB/T 8522-2011	舾装码头设计规范 Code for design of fitting-out quay			中船第九设计研究院工程有限公司
9	CB/T 8523-2011	机械化滑道设计规范 Code for design of mechanical slipway			中船第九设计研究院工程有限公司
10	CB/T 8524-2011	干船坞设计规范 Code for design of dry dock			中船第九设计研究院工程有限公司

四、中国工程建设标准化协会标准

序号	标准编号	标准名称	被代替标准编号	主编单位
1	CECS 01：2004	呋喃树脂防腐蚀工程技术规程 Technical specification for anticorrosion engineering with furan resin	CECS 01：88	中国寰球工程公司 中冶集团建筑研究总院
2	CECS 02：2005	超声回弹综合法检测混凝土强度技术规程 Technical specification for detecting strength of concrete by ultrasonic-rebound combined method	CECS 02：88	中国建筑科学研究院
3	CECS 03：2007	钻芯法检测混凝土强度技术规程 Technical specification for testing concrete strenth with drilled core	CECS 03：88	中国建筑科学研究院
4	CECS 04：88	静力触探技术标准 Technical specification for electrical cone penetration test		建设部综合勘察研究院 同济大学
5	CECS 11：89	贮藏构筑物常用术语标准 Standard for terms used in the storage structures		北京市市政设计研究院 中国市政工程华北设计院
6	CECS 13：2009	纤维混凝土试验方法标准 Standard test methods for fiber reinforced concrete	CECS 13：89	大连理工大学
7	CECS 17：2000	埋地硬聚氯乙烯给水管道工程技术规程 Technical specification for buried unplasticized polyvinyl chloride (PVC-U) pipeline of water supply engineering		北京市市政工程设计研究院 哈尔滨建筑大学
8	CECS 18：2000	聚合物水泥砂浆防腐蚀工程技术规程 Technical specification for anticorrosion works of polymer cement mortar		中国寰球化学工程公司
9	CECS 21：2000	超声法检测混凝土缺陷技术规程 Technical specification for inspection of concrete defects by ultrasonic method	CECS 21：90	陕西省建筑科学研究设计院 上海同济大学
10	CECS 22：2005	岩土锚杆（索）技术规程 Technical specification for ground anchors	CECS 22：90	中冶集团建筑研究总院
11	CECS 23：90	钢货架结构设计规范 The specification for design of steel storage racks		全国薄壁型钢结构标准技术委员会
12	CECS 24：90	钢结构防火涂料应用技术规程 Technical code for application of fire resistive coating for steel structure		公安部四川消防科学研究所
13	CECS 27：2016	工业炉水泥耐火浇注料冬期施工技术规程 Technical specification for cement castable refractories construction in winter of industrial furnaces	CECS 27：90	中油吉林化建工程有限公司
14	CECS 28：2012	钢管混凝土结构技术规程 Technical specification for concrete-filled steel tubular structures	CECS 28：90	哈尔滨工业大学 中国建筑科学研究院
15	CECS 31：2006	钢制电缆桥架工程设计规范 Code for design of steel cable tray engineering	CECS 31：91	中国工程建设标准化协会电气专业委员会
16	CECS 38：2004	纤维混凝土结构技术规程 Technical specification for fiber reinforced concrete structures		大连理工大学
17	CECS 39：92	钢筋混凝土深梁设计规程 Specification for design of reinforced concrete deep beams		华南理工大学

中国工程建设标准化协会标准

序号	标准编号	标准名称	被代替标准编号	主编单位
18	CECS 41：2004	建筑给水硬聚氯乙烯管管道工程技术规程 Technical specification for PVC-U pipeline engineering for water supply in buildings	CECS 41：92	广西建筑综合设计研究院 广东联塑科技实业有限公司
19	CECS 42：92	深井曝气设计规范 Code for design of deep well aeration		北京市市政设计研究院
20	CECS 43：92	钢筋混凝土装配整体式框架节点与连接设计规程 Specification for design of joints and connections of precast monolithic reinforced concrete frames		北京市建筑设计研究院
21	CECS 45：92	地下建筑照明设计标准 Standard for design of underground building lighting		中国建筑科学研究院
22	CECS 48：93	砂、石碱活性快速试验方法 A rapid test method for determining the alkali reactivity of sands and rocks		南京化工学院无机非金属材料研究所
23	CECS 51：93	钢筋混凝土连续梁和框架考虑内力重分布设计规程 Specification for design of reinforced concrete continuous beams and frames considering redistribution of internal forces		重庆建筑大学
24	CECS 52：2010	整体预应力装配式板柱结构技术规程 Technical specification for integral prefabricated pre-stressed concrete slab-column structure	CECS 52：93	北京中建建筑科学研究院有限公司 四川省建筑科学研究院
25	CECS 55：93	孔隙水压力测试规程 Specification for measurement of porewater pressure		上海岩土工程勘察设计研究院
26	CECS 56：94	室内灯具光分布分类和照明设计参数标准 Standard for classification of light distribution and for design parameters of lighting for interior lamps		采光照明委员会
27	CECS 59：94	水泵隔振技术规程 Technical specification for vibration isolation of water pump		上海建筑设计研究院
28	CECS 68：94	氢氧化钠溶液（碱液）加固湿陷性黄土地基技术规程 Technical specification for strengthening collapsible loess foundation by sodium bydroxide solution grouting		西安建筑科技大学
29	CECS 69：2011	拔出法检测混凝土强度技术规程 Technical specification for test concrete strength by pullout method	CECS 69：94	中国建筑科学研究院 哈尔滨工业大学等
30	CECS 77：96	钢结构加固技术规范 Technical code for strengthening steel structures		清华大学土木工程系
31	CECS 79：2011	特殊单立管排水系统技术规程 Technical specification for special single stack drainage system	CECS 79：96	湖南大学 中建（北京）国际设计顾问有限公司

中国工程建设标准化协会标准

序号	标准编号	标准名称	被代替标准编号	主编单位
32	CECS 80：2006	塔桅钢结构工程施工质量验收规程 Specification for constructional quality acceptance of steel tower and mast structures	CECS 80：96	同济大学
33	CECS 83：96	管道工程结构常用术语 Standard for terms used in pipeline engineering structures		北京市市政工程设计研究总院
34	CECS 86：2015	混凝土水池软弱地基处理设计规范 Design code of soft ground treatment for concrete water tank	CECS 86：96	上海市政工程设计研究总院（集团）有限公司
35	CECS 88：97	钢筋混凝土承台设计规程 Specification for design of reinforced concrete pile caps		同济大学
36	CECS 91：97	合流制系统污水截流井设计规程 Specification for design of combined sewage intercepting well		北京建筑工程学院
37	CECS 92：97	重金属污水化学法处理设计规范 Code for design of treatment of wastewater containing heavy metals with chemical method		中国有色金属工业总公司长沙有色冶金设计研究院
38	CECS 94：2002	建筑排水用硬聚氯乙烯内螺旋管管道工程技术规程 Technical specification of PVC-U inner spiral rib pipe for building drainage	CECS 94：97	北京市市政工程设计研究总院
39	CECS 95：97	玻璃纤维氯氧镁水泥通风管道技术规程 Technical standard for glass fiber magnesium oxychloride cement and composite material ventilation duct		中国人民解放军总参工程兵第四设计研究院
40	CECS 96：97	基坑土钉支护技术规程 Specification for soil nailing in foundation excavations		清华大学土木工程系 总参工程兵科研三所
41	CECS 97：97	鼓风曝气系统设计规程 Specification for design of aeration blowing system		北京建筑工程学院
42	CECS 98：98	浆体长距离管道输送工程设计规程 Specification for design of long distance pipeline engineering for slurry transportation		冶金部长沙冶金设计研究院
43	CECS 99：98	岩土工程勘察报告编制标准 Standard for geotechnical investigation report		建设部综合勘察研究设计院
44	CECS 104：99	高强混凝土结构技术规程 Technical specification for high-strength concrete structures		中国土木工程学会高强与高性能混凝土委员会
45	CECS 105：2000	建筑给水铝塑复合管道工程技术规程 Technical specification for polythylene-aluminum composite pipeline engineering of building water supply		广东省建筑设计研究院
46	CECS106：2000	铝合金电缆桥架技术规程 Technical specification for aluminum-alloy cable tray		中国工程建设标准化协会电气工程委员会电缆分委员会 江苏华威电气（集团）公司 上海立新电讯器材股份有限公司

中国工程建设标准化协会标准

序号	标准编号	标准名称	被代替标准编号	主编单位
47	CECS 107：2000	终端电器选用及验收规程 Specification for the control reception and selection of terminal electrical equipment		中国工程建设标准化协会电气工程委员会终端用电分委员会
48	CECS 109：2013	建筑给水减压阀应用技术规程 Technical specification for application of pressure reducing valve in building water supply system	CECS 109：2000	中建（北京）国际设计顾问有限公司 上海上龙供水设备有限公司
49	CECS 110：2000	低温低浊水给水处理设计规程 Specification for design of water supply treatment of low temperature and turbidity water		中国市政工程东北设计研究院
50	CECS 111：2000	寒冷地区污水活性污泥法处理设计规程 Specification for design of actived sludge treatment of waste water in cold regions		中国市政工程东北设计研究院
51	CECS 114：2000	氧气曝气设计规程 Specification for design of oxygen aeration		北京市市政工程设计研究总院
52	CECS 116：2000	钾水玻璃防腐蚀工程技术规程 Technical specification for anticorrosion engineering of potassium silicate		华东工程公司（原化工部第三设计院） 中国寰球化学工程公司
53	CECS 117：2000	给水排水工程混凝土构筑物变形缝设计规程 Specification for the deformation joint design of concrete structures in water work engineering		北京市市政工程设计研究总院
54	CECS 118：2000	冷却塔验收测试规程 Specification for acceptance test of water-cooling tower		西安建筑科技大学
55	CECS 119：2000	城市住宅建筑综合布线系统工程设计规程 Code for engineering design of generic cabling system for civic residential buildings		中国工程建设标准化协会通信工程委员会
56	CECS 120：2007	套接紧定式钢导管电线管路施工及验收规程 Specification for construction and acceptance of wire pipelines with fastening connection steel conduit	CECS 120：2000	中国工程建设标准化协会电气专业委员会
57	CECS 122：2001	埋地硬聚氯乙烯排水管道工程技术规程 Technical specification of buried PVC-U pipeline for sewer engineering		天津市市政工程研究院
58	CECS 125：2001	建筑给水钢塑复合管管道工程技术规程 Technical specification for steel-plastic complex pipeline engineering of water supply in building		上海建筑设计科技发展中心
59	CECS 126：2001	叠层橡胶支座隔震技术规程 Technical specification for seismic-isolation with laminated rubber bearing isolators		广州大学 中国建筑科学研究院
60	CECS 127：2001	点支式玻璃幕墙工程技术规程 Technical specification for point supported glass curtain wall		同济大学 汕头经济特区金刚玻璃幕墙有限公司
61	CECS 129：2001	埋地给水排水玻璃纤维增强热固性树脂夹砂管管道工程施工及验收规程 Specification for construction and acceptance of water supply and sewerage engineering with underground glass fiber reinforced thermosetting resin mortar pipes		中国市政工程东北设计研究院

中国工程建设标准化协会标准

序号	标准编号	标准名称	被代替标准编号	主编单位
62	CECS 130：2001	混凝沉淀烧杯试验方法 Standard for coagulation-flocculation and sedimentation beaker test method		山东建筑工程学院
63	CECS 131：2002	埋地钢骨架聚乙烯复合管燃气管道工程技术规程 Technical specification for buried steel skeleton polyethylene fuel gas pipeline engineering		中国建筑设计研究院
64	CECS 132：2002	给水排水多功能水泵控制阀应用技术规程 Technical specification for application of multi-function control valve for pumping systems of water and wastewater engineering		湖南大学 株洲市南方阀门制造有限公司
65	CECS 134：2002	燃油、燃气热水机组生活热水供应设计规程 Specification for design of hot water supply with burning oil and gas hot water heater		中国建筑设计研究院
66	CECS 135：2002	建筑给水超薄壁不锈钢塑料复合管管道工程技术规程 Technical specification for extra-thin-wall stainless steel and plastic composite pipeline engineering of building water supply		广西建筑综合设计研究院
67	CECS 136：2002	建筑给水氯化聚氯乙烯（PVC-C）管管道工程技术规程 Technical specification for chlorinated poly (vinyl chloride) pipeline engineering of building water supply		上海建筑设计科技发展中心
68	CECS 137：2015	给水排水工程钢筋混凝土沉井结构设计规程 Specification for structural design of reinforced concrete sinking well of water supply and sewerage engineering	CECS 137：2002	上海市政工程设计研究总院（集团）有限公司
69	CECS 138：2002	给水排水工程钢筋混凝土水池结构设计规程 Specification for structural design of reinforced concrete water tank of water supply and sewerage engineering		北京市市政工程设计研究总院
70	CECS 139：2002	给水排水工程水塔结构设计规程 Specification for structural design of water tower of water supply and sewerage engineering		铁道专业设计院
71	CECS 140：2011	给水排水工程埋地预应力混凝土管和预应力钢筒混凝土管管道结构设计规程 Specification for structural design of buried pre-stressed concrete pipeline and pre-stressed concrete cylinder pipeline of water supply and sewerage engineering	CECS 140：2002、 CECS 16：90	北京市市政工程设计研究总院
72	CECS 141：2002	给水排水工程埋地钢管管道结构设计规程 Specification for structural design of buried steel pipeline of water supply and sewerage engineering		北京市市政工程设计研究总院
73	CECS 142：2002	给水排水工程埋地铸铁管管道结构设计规程 Specification for structural design of buried cast-iron pipeline of water supply and sewerage engineering		北京市市政工程设计研究总院

中国工程建设标准化协会标准

序号	标准编号	标准名称	被代替标准编号	主编单位
74	CECS 143：2002	给水排水工程埋地预制混凝土圆形管管道结构设计规程 Specification for structural design of buried premade concrete round pipeline of water supply and sewerage engineering		北京市市政工程设计研究总院
75	CECS 144：2002	水力控制阀应用设计规程 Specification for applied design of hydraulic control valves		上海沪标工程建设咨询有限公司 上海精嘉阀门制造有限公司
76	CECS 145：2002	给水排水工程埋地矩形管管道结构设计规程 Specification for structural design of buried rectangular pipeline of water supply and sewerage engineering		北京市市政工程设计研究总院
77	CECS 146：2003	碳纤维片材加固混凝土结构技术规程（2007年版） Technical specification for strengthening concrete structures with carbon fiber reinforced polymer laminate		国家工业建筑诊断与改造工程技术研究中心
78	CECS 147：2016	加筋水泥土桩锚支护技术规程 Technical specification for retaining and protection with reinforced cement soil piles and anchors	CECS 147：2004	北京交通大学隧道与岩土工程研究所
79	CECS 148：2003	户外广告设施钢结构技术规程 Technical specification for steel structures of outdoor advertisement facility		中国工程建设标准化协会高耸构筑物委员会 上海市市容环境卫生管理局
80	CECS 150：2003	高效燃煤锅炉房设计规程 Specification for design of efficient coal-fired boiler houses		中元国际工程设计研究院
81	CECS 151：2003	沟槽式连接管道工程技术规程 Technical specification for grooved coupling of pipeline engineering		上海沪标工程建设咨询有限公司
82	CECS 152：2003	一体式膜生物反应器污水处理应用技术规程 Technical specification for application of integrative submerged membrane bioreactor for wastewater treatment		北京碧水源科技发展有限公司 北京中关村国际环保产业促进中心
83	CECS 153：2003	建筑给水薄壁不锈钢管管道工程技术规程 Technical specification for light gauge stainless steel pipeline engineering of building water supply		中国建筑设计研究院 江苏金羊集团有限公司
84	CECS 154：2003	建筑防火封堵应用技术规程 Technical specification for application of fire stopping in buildings		公安部天津消防研究所
85	CECS 155：2003	防静电瓷质地板地面工程技术规程 Technical specification for engineering of antistatic tiled floor		北京惠华防静电技术研究所
86	CECS 157：2004	合成树脂幕墙装饰工程施工及验收规 Specification for construction and acceptance of vinylite decorating imitated screen wall		深圳嘉达化工有限公司
87	CECS 158：2004	膜结构技术规程 Technical specification for membrane structures		中国钢结构协会空间结构分会 北京工业大学

中国工程建设标准化协会标准

序号	标准编号	标准名称	被代替标准编号	主编单位
88	CECS 159:2004	矩形钢管混凝土结构技术规程 Technical specification for structures with concrete-filled rectangular steel tube members		同济大学 浙江杭萧钢构股份有限公司
89	CECS 160:2004	建筑工程抗震性态设计通则（试用） General rule for performance-based seismic design of buildings		中国地震局工程力学研究所 中国建筑科学研究院工程抗震研究所 哈尔滨工业大学
90	CECS 161:2004	喷射混凝土加固技术规程 Technical specification for structural strengthening with sprayed concrete		国家工业建筑诊断与改造工程技术研究中心
91	CECS 162:2004	给水排水仪表自动化控制工程施工及验收规程 Specification for construction and acceptance of instrumentation automation control engineering in water supply and wastewater		哈尔滨工业大学 上海市政工程设计研究院
92	CECS 163:2004	建筑用省电装置应用技术规程 Technical specification for application of power saving unit in buildings		北京建标科技发展有限公司 高和机电设备有限公司
93	CECS 164:2004	埋地聚乙烯排水管管道工程技术规程 Technical specification for buried PE pipeline of sewer engineering		上海市城市建设设计研究院
94	CECS 165:2004	城市地下通信塑料管道工程设计规范 Code for engineering design of buried telecommunication plastic conduit in city area		中京邮电通信设计院 中国工程建设标准化协会信息通信专业委员会
95	CECS 166:2004	陶瓷工业窑炉施工及验收规程 Specification for construction and acceptance of ceramic industrial kilns		国家日用及建筑陶瓷工程技术研究中心
96	CECS 167:2004	拱形波纹钢屋盖结构技术规程（试用） Technical specification for arched corrugated steel roof		天津大学
97	CECS 168:2004	建筑排水柔性接口铸铁管管道工程技术规程 Technical specification for flexible joint cast iron pipe system of building drainage		上海沪标工程建设咨询有限公司
98	CECS 169:2015	烟雾灭火系统技术规程 Technical specification for smoke fire extinguishing systems	CECS 169:2004	公安部天津消防研究所
99	CECS 170:2004	低压母线槽选用、安装及验收规程 Specification for selection, installation and acceptance of low-voltage busways		中国工程建设标准化协会电气专业委员会 杰帝母线（上海）有限公司 上海海外精成电气有限公司
100	CECS 171:2004	建筑给水铜管管道工程技术规程 Technical specification for copper pipeline engineering of building water supply		上海沪标工程建设咨询有限公司 国际铜业协会（中国）
101	CECS 172:2004	排水系统水封保护设计规程 Specification for design of water-seal protection in drainage systems		湖南大学土木工程学院 上海沪标工程建设咨询有限公司
102	CECS 173:2004	水泥聚苯模壳格构式混凝土墙体住宅技术规程 Technical specification for dwelling houses with EPSC form latticed concrete wall		中国建筑科学研究院

中国工程建设标准化协会标准

序号	标准编号	标准名称	被代替标准编号	主编单位
103	CECS 175：2004	现浇混凝土空心楼盖结构技术规程 Technical specification for cast-in-situ concrete hollow floor structure		中国建筑科学研究院
104	CECS 176：2005	聚丙烯复合塑料隔离护栏应用技术规程 Technical specification for application of PP-N plastic fence		上海市政工程设计研究院
105	CECS 177：2005	城市地下通信塑料管道工程施工及验收规范 Code for construction and acceptance of buried telecommunication plastic conduit engineering in city area		中国工程建设标准化协会信息通信专业委员会
106	CECS 178：2009	气水冲洗滤池整体浇筑滤板及可调式滤头技术规程 Technical specification for integral pouring concrete filter floor and adjustable nozzle of air-water washing filter	CECS 178：2005	上海市政工程设计研究总院
107	CECS 179：2009	健康住宅建设技术规程 Technical specification for construction of healthy housing	CECS 179：2005	国家住宅与居住环境工程技术研究中心
108	CECS 180：2005	建筑工程预应力施工规程 Specification for pre-stressed construction of building engineering		东南大学华东预应力技术联合开发中心 中国建筑科学研究院结构所
109	CECS 181：2005	给水钢丝网骨架塑料（聚乙烯）复合管管道工程技术规程 Technical specification for steel wire mesh and plastic（PE）composite pipe of water supply pipeline		上海沪标工程建设咨询有限公司
110	CECS 182：2005	智能建筑工程检测规程 Specification for checking and measuring of intelligent building		中国建筑业协会工程质量监督分会
111	CECS 183：2015	虹吸式屋面雨水排水系统技术规程 Technical specification for siphonic drainage system of roof	CECS 183：2005	同济大学建筑设计研究院（集团）有限公司 上海吉博力房屋卫生设备工程技术有限公司
112	CECS 184：2005	给水系统防回流污染技术规程 Technical specification for prevention of backflow pollution in water supply system		上海沪标工程建设咨询有限公司
113	CECS 185：2005	建筑排水中空壁消音硬聚氯乙烯管管道工程技术规程 Technical specification for unplasticized polyvinyl chloride double-wall noise-reducing pipeline engineering of water sewerage in buildings		上海沪标工程建设咨询有限公司
114	CECS 186：2005	多边形稀油密封储气柜工程施工质量验收规程 Specification for construction quality acceptance of polygon oil seal gasholder		中国市政工程华北设计研究院
115	CECS 187：2005	油浸变压器排油注氮装置技术规程 Technical specification for oil evacuation and nitrogen injection equipment of transformer		公安部天津消防研究所

中国工程建设标准化协会标准

序号	标准编号	标准名称	被代替标准编号	主编单位
116	CECS 188：2005	钢管混凝土叠合柱结构技术规程 Technical specification for steel tube-reinforced concrete column structure		清华大学 辽宁省建筑设计研究院
117	CECS 189：2005	注氮控氧防火系统技术规程 Technical specification for nitrogen injection/oxygen control fire prevention system		公安部四川消防研究所 天津易可大科技有限公司
118	CECS 190：2005	给水排水工程埋地玻璃纤维增强塑料夹砂管管道结构设计规程 Specification for structural design of buried glass fiber reinforced plastic mortar pipeline of water supply and sewerage engineering		北京市市政工程设计研究总院
119	CECS 191：2005	木质地板铺装工程技术规程 Technical specification for installation works of woodfloor		建设部住宅产业化促进中心 中国林产工业协会地板专业委员会 上海名企住宅装饰咨询服务有限公司
120	CECS 192：2005	挤扩支盘灌注桩技术规程 Technical specification for cast-in-situ pile with expanded branches and plates		北京交通大学
121	CECS 193：2005	城镇供水长距离输水管（渠）道工程技术规程 Technical specification for long distance water transmission pipeline engineering of urban water supply		中国市政工程东北设计研究院 长安大学
122	CECS 195：2006	聚合物水泥、渗透结晶型防水材料应用技术规程 Technical specification for polymer modified cementitious waterproofing materials and penetrating and crystalline waterproofing materials		中国建筑科学研究院
123	CECS 196：2006	建筑室内防水工程技术规程 Technical specification for interior waterproof engineering of building		中国建筑标准设计研究院
124	CECS 197：2006	孔内深层强夯法技术规程 Technical specification for down-hole dynamic compaction		北京交通大学
125	CECS 198：2006	空调用无规共聚聚丙烯（PP-R）塑铝稳态复合管管道工程技术规程 Technical specification of random polypropylene/Aluminum stable composite pipeline for air-conditioning		建设部科技发展促进中心 武汉金牛经济发展有限公司
126	CECS 199：2006	聚乙烯丙纶卷材复合防水工程技术规程 Technical specification for composite waterproof engineering with polyethylene polyproplylene sheet		中国建筑标准设计研究院
127	CECS 200：2006	建筑钢结构防火技术规范 Technical code for fire safety of steel structure in buildings		同济大学 中国钢结构协会防火与防腐分会
128	CECS 202：2006	轻骨料混凝土桥梁技术规程 Technical specification for lightweight aggregate concrete bridges		中国建筑学会建筑材料分会 轻骨料及轻骨料混凝土专业委员会 铁道部专业设计院

中国工程建设标准化协会标准

序号	标准编号	标准名称	被代替标准编号	主编单位
129	CECS 203：2006	自密实混凝土应用技术规程 Technical specifications for self compacting concrete application		中国建筑标准设计研究院 清华大学
130	CECS 204：2006	红外热像法检测建筑外墙饰面层粘结缺陷技术规程 Technical specification for inspecting the defects of exterior walls cementing coats of building using infrared thermograph method		上海市房地产科学研究院
131	CECS 205：2006	内衬（覆）不锈钢复合钢管管道工程技术规程 Technical specification for stainless stell lined or claded composite steel pipeline engineering	CECS 205：2006	悉地国际设计顾问（深圳）有限公司 江苏众信绿色管业科技有限公司
132	CECS 206：2006	钢外护管真空复合保温预制直埋管道技术规程 Technical specification for steel jacket pre-insulated pipeline with vacuum layer		北京豪特耐管道设备有限公司
133	CECS 207：2006	高性能混凝土应用技术规程 Technical specification for application of high performance concrete		清华大学老科技工作者协会 北京交通大学土建学院
134	CECS 208：2006	泳池用聚氯乙烯膜片应用技术规程 Technical specification for application of PVC membrane on swimming pools		中国建筑标准设计研究院 中国游泳运动协会
135	CECS 210：2006	埋地聚乙烯钢肋复合缠绕排水管管道工程技术规范 Technical specification for helical winding polyethylene profile wall with external steel rib pipeline o buried sewer engineering		南京市市政设计研究院有限责任公司 福建亚通新材料科技股份有限公司
136	CECS 211：2006	自动门应用技术规程 Technical specification for application of automatic door for pedestrian use		北京凯必盛自动门技术有限公司
137	CECS 212：2006	预应力钢结构技术规程 Technical specification for pre-stressed steel structures		北京工业大学 中国钢结构协会专家委员会
138	CECS 213：2012	旋转型喷头自动喷水灭火系统技术规程 Technical specification for automatic fire suppression rotary sprinkler systems	CECS 213：2006	公安部四川消防研究所 广州龙雨消防设备有限公司
139	CECS 214：2006	自承式给水钢管跨越结构设计规程 Design specification for self-sustaining longitudinally over-crossing structure of steel pipe of water supply		上海市政工程设计研究院
140	CECS 215：2006	燃气采暖热水炉应用技术规程 Technical specification for application of gas-fired heating and hot water combi-boilor		中国市政工程华北设计研究院
141	CECS216：2006	给水排水工程预应力混凝土圆形水池结构技术规程 Technical specification for structure of circular pre-stressed concrete water tanks of water supply and sewerage engineering		中国市政工程华北设计研究院

中国工程建设标准化协会标准

序号	标准编号	标准名称	被代替标准编号	主编单位
142	CECS 217：2006	聚硫、聚氨酯密封胶给水排水工程应用技术规程 Technical specification for application of polysulfide and polyurethane sealant in water supply and sewerage engineering		苏州非金属矿工业设计研究院防水材料设计研究所 北京市市政工程设计研究总院
143	CECS 219：2007	简易自动喷水灭火系统应用技术规程 Technical specification for application of simple sprinkler systems		公安部四川消防研究所
144	CECS 220：2007	混凝土结构耐久性评定标准 Standard for durability assessment of concrete structures		西安建筑科技大学
145	CECS 221：2012	叠压供水技术规程 Technical specification for pressure-superposed water supply	CECS 221：2007	中国建筑金属结构协会给水排水设备分会 上海熊猫机械（集团）有限公司
146	CECS 222：2007	小区集中生活热水供应设计规程 Specification for design of central hot water supply system in sub-district		中国建筑设计研究院
147	CECS 223：2007	埋地排水用钢带增强聚乙烯螺旋波纹管管道工程技术规程 Technical specification for buried spirally wound corrugated polyethylene with ribbed steel reinforced pipes of sewerage engineering		中国市政工程西南设计研究院 四川森普管材股份有限公司
148	CECS 224：2007	节能型双向集热卫浴间应用技术规程 Technical specification for application of energy-saving bathroom with two-way heat collecting technology		中国房地产及住宅研究会住宅设施委员会
149	CECS 225：2007	建筑物移位纠倾增层改造技术规范 Technical code for moving, incline-rectifying, story-increasing and reconstruction of buildings		北京交通大学
150	CECS 226：2007	栓钉焊接技术规程 Technical specification for welding of stud		中冶集团建筑研究总院
151	CECS 227：2007	建筑小区塑料排水检查井应用技术规程 Technical specification for application of plastic inspection chambers for sewerage in building area		上海现代建筑设计（集团）有限公司
152	CECS 228：2007	建筑铜管管道工程连接技术规程 Technical specification for connection of copper pipeline engineering		中国建筑科学研究院
153	CECS 229：2008	自动水灭火系统薄壁不锈钢管管道工程技术规程 Technical specification for light gauge stainless steel pipeline engineering of automation water-extinguishing systems		公安部四川消防研究所
154	CECS 230：2008	高层建筑钢-混凝土混合结构设计规程 Specification for design of steel-concrete mixed structure of tall buildings		中国建筑标准设计研究院

中国工程建设标准化协会标准

序号	标准编号	标准名称	被代替标准编号	主编单位
155	CECS 231：2007	铝塑复合板幕墙工程施工及验收规程 Specification for construction and acceptance of aluminum-plastic composite panel curtain walls engineering		国家建筑材料测试中心 中国建筑材料工业协会铝塑复合材料分会
156	CECS 232：2007	AD型特殊单立管排水系统技术规程（2011年版） Technical specification for AD single stack drainage system（2011）		积水（上海）国际贸易有限公司
157	CECS 233：2007	厨房设备灭火装置技术规程 Technical specification for fire extinguishing equipment of cooking appliance		公安部天津消防研究所
158	CECS 234：2008	自动喷水灭火系统CPVC管管道工程技术规程 Technical specification for CPVC pipeline engineering on sprinkler systems		公安部天津消防研究所
159	CECS 235：2008	铸钢节点应用技术规程 Technical specification for application of connections of structural steel casting		同济大学 清华大学 中国钢结构协会专家委员会
160	CECS 236：2008	钢结构单管通信塔技术规程 Technical specification for steel communication monopole		同济大学 上海市金属结构行业协会
161	CECS 237：2008	给水钢塑复合压力管管道工程技术规程 Technical specification for plastic-steel-plastic composite pressure pipeline of water supply engineering		中国市政工程华北设计研究院 新兴铸管股份有限公司
162	CECS 238：2008	工程地质测绘标准 Standard for engineering geological survey and mapping		建设综合勘察研究设计院
163	CECS 239：2008	岩石与岩体签定和描述标准 Standard for identification and description of rock & rock mass		建设综合勘察研究设计院
164	CECS 240：2008	工程地质钻探标准 Standard for engineering geological drilling		建设综合勘察研究设计院
165	CECS 241：2008	工程建设水文地质勘察标准 Standard hydrogeological investigation for civil engineering		建设综合勘察研究设计院
166	CECS 242：2008	水泥复合砂浆钢筋网加固混凝土结构技术规程 Technical specification for strengthening concrete structures with grid rebar and mortar		湖南大学
167	CECS243：2008	园林绿地灌溉工程技术规程 Technical specification for landscape irrigation engineering		中国农业大学
168	CECS 244：2008	砂基透水砖工程施工及验收规程 Specification for construction and acceptance of sand-based water permeable brick		北京仁创科技集团有限公司
169	CECS 245：2008	自动消防炮灭火系统技术规程 Technical specification for automatic fire monitor extinguishing systems		公安部四川消防研究所 中国科技大学火灾科学国家重点实验室

中国工程建设标准化协会标准

序号	标准编号	标准名称	被代替标准编号	主编单位
170	CECS 246:2008	给水排水工程顶管技术规程 Technical specification for pipe jacking of water supply and sewerage engineering		上海市政工程设计研究总院
171	CECS 248:2008	聚乙烯塑钢缠绕排水管管道工程技术规程 Technical specification for steel-reinforced spirally wound polyethylene of drainage pipeline engineering		哈尔滨工业大学
172	CECS 249:2008	现浇泡沫轻质土技术规程 Technical specification for cast-in-situ foamed lightweight soil		广州大学 华鑫博越国际土木建筑工程技术（北京）有限公司
173	CECS 250:2008	城镇污水污泥流化床干化焚烧技术规程 Technical specification for fluidized-bed drying and incineration for sewage sludge		上海城环水务运营有限公司 上海市政工程设计研究总院
174	CECS 251:2009	钢水罐砌筑工程施工及验收规程 Specification for construction and acceptance of ladle building		武汉钢铁集团精鼎工业炉有限责任公司
175	CECS 252:2009	火灾后建筑结构鉴定标准 Standard for building structural assessment after fire		中冶建筑研究总院有限公司 上海市建筑科学研究院
176	CECS 253:2009	基桩孔内摄像检测技术规程 Technical specification for testing method with video monitor through the hole of foundation pile		福建省建筑科学研究院
177	CECS 254:2012	实心与空心钢管混凝土结构技术规程 Technical specification for solid and hollow concrete-filled steel tubular structure	CECS 254:2009	哈尔滨工业大学
178	CECS 255:2009	建筑室内吊顶工程技术规程 Technical specification for installation of ceiling systems in buildings		中国建筑标准设计研究院
179	CECS 256:2009	蒸压粉煤灰砖建筑技术规范 Technical code for autoclaved fly ash brick buildings		中国建筑东北设计研究院有限公司 长沙理工大学
180	CECS 257:2009	混凝土砖建筑技术规范 Technical code for concrete brick masonry structures building		中国建筑东北设计研究院有限公司 长沙理工大学
181	CECS 258:2009	轻质复合板应用技术规程 Technical specification for application of lightweight and multiple plates		北京绿华园科技有限公司 建设部科技发展促进中心
182	CECS 259:2009	低阻力倒流防止器应用技术规程 Technical specification for application of low-resistance backflow preventer		中国中元国际工程公司 上海上龙供水设备有限公司
183	CECS 260:2009	端板式半刚性连接钢结构技术规程 Technical specification for steel structures with end-plate semi-rigid connection		同济大学
184	CECS 261:2009	钢结构住宅设计规范 Code for design of steel structure residential buildings		中国建筑金属结构协会建筑钢结构委员会 住房和城乡建设部科技发展促进中心

中国工程建设标准化协会标准

序号	标准编号	标准名称	被代替标准编号	主编单位
185	CECS 262:2009	发泡水泥绝热层与水泥砂浆填充层地面辐射供暖工程技术规程 Technical specification for foamed cement insulating course and cement mortar filler course of floor radiant heating		中国建筑金属结构协会 威海嘉中进出口有限公司
186	CECS 263:2009	大空间智能型主动喷水灭火系统技术规程 Technical specification for large-space intelligent active control sprinkler systems		广州市设计院 佛山市南海天雨智能灭火装置有限公司
187	CECS 264:2009	建筑燃气铝塑复合管管道工程技术规程 Technical specification for polyethylene-aluminum composite pipeline engineering for building gas supply		佛山市日丰企业有限公司
188	CECS 265:2009	曝气生物滤池工程技术规程 Technical specification for biological aerated filter engineering		中国市政工程华北设计研究总院 马鞍山市华骐环保科技发展有限公司
189	CECS 266:2009	建设工程施工现场安全资料管理规程 Management specification for construction engineering safety document on construction site		北京市建设工程安全质量监督总站 福建省九龙建设集团有限公司
190	CECS 267:2009	橡胶膜密封储气柜工程施工质量验收规程 Specification for construction quality acceptance of rubber seal gasholder		中国市政工程华北设计研究总院
191	CECS 268:2010	水工混凝土外保温聚苯板施工技术规范 Technical code for polystyrene board construction in hydraulic concrete external insulation		宜昌瑞派尔特种工程技术有限责任公司 北京万澎科技有限公司
192	CECS 269:2010	灾损建（构）筑物处理技术规范 Technical code for treatment of disaster damaged buildings and structures		北京交通大学 广东金辉华集团有限公司
193	CECS 270:2010	给水排水丙烯腈-丁二烯-苯乙烯（ABS）管管道工程技术规程 Technical specification for ABS plastical pipeline of water supply and drainage engineering		中建（北京）国际设计顾问有限公司 住房和城乡建设部给水排水产品标准化技术委员会
194	CECS 271:2013	旋流加强（CHT）型特殊单立管排水系统技术规程 Technical specification for CHT special single stack drainage system	CECS 271:2010	悉地国际设计顾问（深圳）有限公司 青岛嘉泓建材有限公司
195	CECS 272:2010	预制塑筋水泥聚苯保温墙板应用技术规程 Technical specification for application of prefabricated cement polystyrene insulation panel with PET reinforcement		江苏顺通建设工程有限公司 湖北力特土工材料有限公司
196	CECS 273:2010	组合楼板设计与施工规范 Code for composite slabs design and construction		中冶建筑研究总院有限公司
197	CECS 274:2010	真空破坏器应用技术规程 Technical specification for application of vacuum breakers		上海现代建筑设计（集团）有限公司技术中心
198	CECS 275:2010	苏维托单立管排水系统技术规程 Technical specification for Sovent single stack drainage system		中建（北京）国际设计顾问有限公司 上海吉博力房屋卫生设备工程技术有限公司

中国工程建设标准化协会标准

序号	标准编号	标准名称	被代替标准编号	主编单位
199	CECS 276：2010	彗星式纤维滤池工程技术规程 Technical specification of comet type fibrous filter engineering		中国市政工程华北设计研究总院 浙江省德安集团
200	CECS 277：2010	建筑给水排水薄壁不锈钢管连接技术规程 Technical specification for connecting of light gauge stainless steel pipes of building water supply and drainage		中建（北京）国际设计顾问有限公司 深圳雅昌管业有限公司
201	CECS 278：2010	剪压法检测混凝土抗压强度技术规程 Technical specification for testing of concrete compressive strength by shear-pressure method		中国建筑科学研究院
202	CECS 279：2010	强夯地基处理技术规程 Technical specification for ground treatment by heavy tamping		山西省机械施工公司 山西建筑工程（集团）总公司
203	CECS 280：2010	钢管结构技术规程 Technical specification for structures with steel hollow sections		中冶建筑研究总院有限公司 同济大学
204	CECS 281：2010	自承重砌体墙技术规程 Technical specification for non-bearing masonry walls		长沙理工大学 中国轻工业长沙工程有限公司
205	CECS 282：2010	建筑排水高密度聚乙烯（HDPE）管道工程技术规程 Technical specification for HDPE pipeline engineering of building drainage		上海建筑设计研究院有限公司 上海吉博力房屋卫生设备工程技术有限公司
206	CECS 283：2010	轻钢构架固模剪力墙结构技术规程 Technical specification for shearwall structures with lightweight-steel framework and fixed form		清华大学建筑设计研究院
207	CECS 284：2010	中小套型住宅厨房和卫生间工程技术规程 Technical specification for kitchen and bathroom of medium-small apartments		住房和城乡建设部政策研究中心厨房卫生间研究所 北京都龙科技发展有限公司
208	CECS 285：2011	村庄景观环境工程技术规程 Technical specification for village landscape environmental engineering		中国建筑设计研究院 国家住宅与居住环境工程技术研究中心
209	CECS 286：2015	建筑用无机集料阻燃木塑复合墙板应用技术规程 Technical specification for application of inorganic aggregate fire retardant wood-plastic composite wall-panel in buildings		中国建筑标准设计研究院有限公司 北京恒通创新赛木科技股份有限公司
210	CECS 287：2011	漩流降噪特殊单立管排水系统技术规程 Technical specification for special single stack drainage system with cyclone noise-reducing joints		中建（北京）国际设计顾问有限公司 浙江光华塑业有限公司
211	CECS 288：2011	建筑装饰工程木制品制作与安装技术规程 Technical specification for fabrication and installation of wooden product in building decoration engineering		上海市建筑装饰工程有限公司 上海建工（集团）总公司
212	CECS 289：2011	蒸压加气混凝土砌块砌体结构技术规范 Technical code for masonry structure of autoclaved aerated concrete block		中国建筑东北设计研究院有限公司 沈阳建筑大学

中国工程建设标准化协会标准

序号	标准编号	标准名称	被代替标准编号	主编单位
213	CECS 290：2011	波浪腹板钢结构应用技术规程 Technical specification for application of sinusoidal web steel structures		清华大学 巴特勒（上海）有限公司
214	CECS 291：2011	波纹腹板钢结构技术规程 Technical specification for steel structures with corrugated webs		同济大学
215	CECS 292：2011	气体消防设施选型配置设计规程 Technical specification for design of lectotype and distribution on gas fire-suppression facilities		中船第九设计研究院工程有限公司
216	CECS 293：2011	房屋裂缝检测与处理技术规程 Technical specification for inspection and treatment of cracks in buildings		湖南大学 福建省建筑科学研究院
217	CECS 294：2011	雨、污水分层生物滴滤处理（MBTF）技术规程 Technical specification for treatment of rainwater and wastewater with Multilayer Bio-Tricking Filter (MBTF)		上海交通大学 上海明谛科技实业有限公司
218	CECS 295：2011	建（构）筑物托换技术规程 Technical specification for underpinning for buildings and structures		广东金辉华集团有限公司 北京交通大学
219	CECS 296：2011	无源无线智能控制系统技术规程 Technical specification for self-powered wireless intelligent control system		中国建筑标准设计研究院 成都英泰力科技有限公司
220	CECS 297：2011	乡村建筑外墙无机保温砂浆应用技术规程 Technical specification for application of inorganic insulation mortar for external wall of rural residence		中国建筑科学研究院
221	CECS 298：2011	乡村建筑混凝土瓦应用技术规程 Technical specification for application of concrete tiles of rural residence		中国建筑科学研究院
222	CECS 299：2011	乡村建筑屋面泡沫混凝土应用技术规程 Technical specification for application of foamed concrete of rural residential roof		中国建筑科学研究院
223	CECS 300：2011	钢结构钢材选用与检验技术规程 Technical specification for selection and inspection of steel products for steel structure		中国钢结构协会 中冶建筑研究总院有限公司
224	CECS 301：2011	乡村建筑内隔墙板应用技术规程 Technical specification for application of inner wall panel of rural residence		中国建筑科学研究院
225	CECS 302：201	乡村建筑外墙板应用技术规程 Technical specification for application of external partition wall-panel of rural residence		中国建筑科学研究院
226	CECS 303：2011	住宅远传抄表系统应用技术规程 Technical specification for application of remote transmission meter reading system for residence		住房和城乡建设部政策研究中心厨房卫生间研究所 北京化工大学
227	CECS 304：2011	建筑用金属面绝热夹芯板安装及验收规程 Specification for installation and accceptance of double skin metal faced insulating sandwich panels for building		国家住宅与居住环境工程技术研究中心 中国绝热节能材料协会

中国工程建设标准化协会标准

序号	标准编号	标准名称	被代替标准编号	主编单位
228	CECS 305:2011	环压连接管道工程技术规程 Technical specification for ring-compression-connection pipeline engineering		中国建筑金属结构协会给水排水设备分会
229	CECS 306:2012	超高分子量聚乙烯钢骨架复合管管道施工及验收规程 Specification for construction and acceptance of UHMWPE steel skeleton composite pipeline engineering		常州大学设计研究院
230	CECS 307:2012	加强型旋流器特殊单立管排水系统技术规程 Technical specification for strengthening cyclone special single stack drainage system		中建(北京)国际设计顾问有限公司
231	CECS 308:2012	农村住宅用能测试标准 Standard for measuring domestic energy consumption of rural dwellings		中国建筑设计研究院
232	CECS 309:2012	农村住宅用能核算标准 Standard for reporting and verifying domestic energy consumption of rural dwellings		中国建筑设计研究院
233	CECS 310:2012	炼钢连铸中间包砌筑施工及验收规程 Specification for construction and acceptance of continuous casting tundish		武汉钢铁集团精鼎工业炉有限责任公司
234	CECS 311:2012	非烧结块材砌体专用砂浆技术规程 Technical specification for special mortar in non-fired bulk materials masonry wall		中国建筑东北设计研究院有限公司 北京建筑材料科学研究总院有限公司 沈阳建筑大学
235	CECS 312:2012	惰性气体灭火系统技术规程 Technical specification for inert gas extinguishing systems		公安部天津消防研究所
236	CECS 313:2012	农村小型地源热泵供暖供冷工程技术规程 Technical specification for ground source heat pump in village and small township		天津大学 山东海利丰地源热泵有限公司
237	CECS 314:2012	高差仪应用技术规程 Technical specification for application of cathetometer		浙江浙大之光照明技术有限公司 通化恒星高差仪研制有限公司
238	CECS 315:2012	钢骨架聚乙烯塑料复合管管道工程技术规程 Technical specification for steel skeleton polyethylene (PE) composite pipeline engineering		华创天元实业发展有限责任公司
239	CECS 316:2012	室外真空排水系统工程技术规程 Technical specification for outdoor vacuum sewage system engineering		上海建筑设计研究院有限公司
240	CECS 317:2012	村镇住宅建筑材料选择与性能测试标准 Standard for selection and test methods of building material in villages and towns		中国建筑科学研究院
241	CECS 318:2012	轻质芯模混凝土叠合密肋楼板技术规程 Technical specification for lightweight embedded formwork concrete composite rib floor		陕西省建筑科学研究院 陕西建研结构工程股份有限公司
242	CECS 319:2012	双曲拱桥加固改造技术规程 Technical specification for strengthening and reconstruction of two-way curved arch bridge		江西中煤建设集团有限公司 中际联发交通建设有限公司

中国工程建设标准化协会标准

序号	标准编号	标准名称	被代替标准编号	主编单位
243	CECS 320：2012	模块化同层排水节水系统应用技术规程 Technical specification for application of modular same-floor drainage & water-saving system		中国石化集团中原石油勘探局勘察设计研究院 山东聊建集团有限公司
244	CECS 321：2012	翻板滤池设计规程 Design specification for shutter filter		上海市政工程设计研究总院（集团）有限公司
245	CECS 322：2012	干粉灭火装置技术规程 Technical specification for dry powder fire extinguishing equipment		公安部天津消防研究所
246	CECS 323：2012	交错桁架钢框架结构技术规程编辑 Technical specification for staggered truss steel framing system		兰州大学
247	CECS 324：2012	既有村镇住宅功能评价标准 Evaluation standard for function of existing rural houses		哈尔滨工业大学
248	CECS 325：2012	既有村镇住宅建筑抗震鉴定和加固技术规程 Technical specification for seismic appraisal and strengthening of existing residence in village and town		建研科技股份有限公司
249	CECS 326：2012	既有村镇住宅建筑安全性评定标准 Standard for safety assessment of existing residence in village and town		上海市建筑科学研究院（集团）有限公司
250	CECS 327：2012	集合管型特殊单立管排水系统技术规程 Technical specification for shugokan special single stack drainage system		中国建筑设计研究院
251	CECS 328：2012	整体地坪工程技术规程 Technical specification for seamless flooring engineering		中国建筑材料联合会地坪材料分会 中国建材检验认证集团股份有限公司
252	CECS 329：2012	钢制承插口预应力混凝土管管道工程技术规程 Technical specification for steel ring pre-stressed concrete pipeline engineering		中国市政工程华北设计研究总院
253	CECS 330：2013	钢结构焊接热处理技术规程 Technical specification for welding heat treatment of steel structure		中国电力科学研究院 中冶建筑研究总院有限公司
254	CECS 331：2013	钢结构焊接从业人员资格认证标准 Standard for qualification and certification of welding personnel of steel structural		中冶建筑研究总院有限公司 河南鼎力钢结构检测有限公司
255	CECS 332：2012	农村单体居住建筑节能设计标准 Design standard for energy-efficient of detached rural housing		中国建筑标准设计研究院 清华大学
256	CECS 333：2012	结构健康监测系统设计标准 Design standard for structural health monitoring system		大连理工大学 大连金广建设集团有限公司
257	CECS 334：2013	集装箱模块化组合房屋技术规程 Technical specification for modular freight container building		中国钢结构协会 中国国际海运集装箱（集团）股份有限公司
258	CECS 335：2013	酚醛泡沫板薄抹灰外墙外保温工程技术规程 Technical specification for phenolic foam panel external thermal insulation with thin rendering		中国建筑科学研究院

中国工程建设标准化协会标准

序号	标准编号	标准名称	被代替标准编号	主编单位
259	CECS 336：2013	住宅生活排水系统立管排水能力测试标准 Standard for test on stack drainage capacity of domestic drainage system		悉地（北京）国际建筑设计顾问有限公司 华东建筑设计研究院有限公司
260	CECS 337：2013	建筑给水纤维增强无规共聚聚丙烯复合管道工程技术规程 Technical specification for fiber reinforced polypropylene random copolymer composite pipeline of building water supply		华东建筑设计研究院有限公司 浙江伟星新型建材股份有限公司
261	CECS 338：2013	多层保温砌模混凝土网格墙建筑技术规程 Technical specification for multi-story buildings of reinforced concrete grillage shear wall formed with lightweight insulation hollow blocks		清华大学 北京大兴宏光新型保温建筑材料厂
262	CECS 339：2013	地源热泵式沼气发酵池加热技术规程 Technical specification for heating biogas reactor by ground source heat pump		同济大学
263	CECS 340：2013	地道风建筑降温技术规程 Technical specification for space cooling by earth-cooled air		同济大学
264	CECS 341：2013	电力通信系统防雷技术规程 Technical specification for lightning protection of electric power communication systems		深圳供电局有限公司 四川中光防雷科技股份有限公司
265	CECS 342：2013	丙烯酸盐喷膜防水应用技术规程 Technical specification for application of spray-applied acrylate waterproofing		西南交通大学
266	CECS 343：2013	钢结构防腐蚀涂装技术规程 Technical specification for painting and anticorrosion of steel structures		中国钢结构协会 中冶建筑研究总院有限公司
267	CECS 344：2013	地源热泵系统地埋管换热器施工技术规程 Technical specification or construction of ground heat exchanger in ground-source heat pump system		同济大学
268	CECS 345：2013	探火管灭火装置技术规程 Technical specification for extinguishing equipment with fire detection tube		公安部天津消防研究所
269	CECS 346：2013	厚层腻子墙体隔热保温系统应用技术规程 Technical specification for wall thermal insulation system with thermal insulating thick putty		住房和城乡建设部科技发展促进中心 江苏塞尚低碳科技有限公司
270	CECS 347：2013	约束混凝土柱组合梁框架结构技术规程 Technical specification for frame structure of confined RC columns and composite beams		河北合创建筑节能科技有限公司
271	CECS 348：2013	平板太阳能热水系统与建筑一体化技术规程 Technical specification for integration of building with flat plate solar water heating system		住房和城乡建设部住宅产业化促进中心 北京昌日新能源科技有限公司
272	CECS 349：2013	一体化给水处理装置应用技术规程 Technical specification for application of integrated water treatment device		中国市政工程华北设计研究总院 西门子（天津）水技术工程有限公司
273	CECS 350：2013	火墙通用技术规程 General technical specification for fire wall		哈尔滨工业大学

中国工程建设标准化协会标准

序号	标准编号	标准名称	被代替标准编号	主编单位
274	CECS 351：2013	聚氨酯硬泡复合保温板应用技术规程 Technical specification for application of polyurethane rigid foam composite insulation panel		中国建筑科学研究院
275	CECS 352：2013	聚氨酯硬泡外墙外保温技术规程 Technical specification for external thermal insulation of polyurethane rigid foam on walls		中国建筑科学研究院
276	CECS 353：2013	生态格网结构技术规程 Technical specification for application of eco-mesh structure		无锡金利达生态科技有限公司 北京万澎科技有限公司
277	CECS 354：2013	乡村公共服务设施规划标准 Standard for planning of rural area public facilities		天津市城市规划设计研究院
278	CECS 355：2013	模塑聚苯（EPS）模块外保温工程技术规程 Technical specification for external thermal insulation system with EPS module		哈尔滨鸿盛房屋节能体系研发中心
279	CECS 356：2013	高强箍筋混凝土结构技术规程 Technical specification for high-strength stirrup concrete structure		中国京冶工程技术有限公司 西安建筑科技大学
280	CECS 357：2013	模块化水处理系统技术规程 Technical specification for modular water treatment system		中国市政工程华北设计研究总院 山东华通环境科技股份有限公司
281	CECS 358：2013	电壁炉应用技术规程 Technical specification for application of electric fireplace		中国建筑标准设计研究院 福建亚伦电子电器科技有限公司
282	CECS 359：2013	三氟甲烷灭火系统技术规程 Technical specification or trifluoromethane extinguishing systems		公安部天津消防研究所 天津盛达安全科技有限责任公司
283	CECS 360：2013	村镇传统住宅设计规范 Design code for traditional residence in towns and villages		中国建筑标准设计研究院 中国建筑设计研究院 清华大学
284	CECS 361：2013	生态混凝土应用技术规程 Technical specification for application of ecological concrete		上海嘉洁生态科技有限公司 北京万澎科技有限公司
285	CECS 362：2014	热源塔热泵系统应用技术规程 Technical specification for heat-source-tower heat pump system		中国建筑科学研究院 江苏辛普森新能源有限公司
286	CECS 363：2014	建筑同层检修（WAB）排水系统技术规程 Technical specification for WAB building same-floor maintenance drainage system		悉地国际设计顾问（深圳）有限公司 昆明群之英科技有限公司
287	CECS 364：2014	建筑燃气安全应用技术导则 Technical guideline for building gas safety application		中国市政工程华北设计研究总院
288	CECS 365：2014	夹模喷涂混凝土夹芯剪力墙建筑技术规程 Technical specification for clamp mould sprayed concrete sandwich shear wall building		清华大学建筑设计研究院有限公司 北京华美科博科技发展有限公司
289	CECS 366：2014	乡规划标准 Standard for planning of township		中国建筑设计研究院

中国工程建设标准化协会标准

序号	标准编号	标准名称	被代替标准编号	主编单位
290	CECS 367：2014	合建式氧化沟技术规程 Technical specification for integrated oxidation ditch		凌志环保股份有限公司 上海市政工程设计研究总院（集团）有限公司
291	CECS 368：2014	商务写字楼等级评价标准 Grade evaluation standard for business office building		中国房地产业协会商业地产专业委员会
292	CECS 369：2014	滑动测微测试规程 Specification for test of sliding micrometer		机械工业勘察设计研究院
293	CECS 370：2014	隧道工程防水技术规范 Technical code for tunnel engineering waterproofing		上海市隧道工程轨道交通设计研究院 清华大学
294	CECS 371：2014	地下建筑空间声环境控制标准 Acoustic environment control standard for underground architectural spaces		哈尔滨工业大学
295	CECS 372：2014	石灰粉料投加系统技术规程 Technical specification for lime powder dosing system		北京工业大学 北京中电加美环保科技股份有限公司
296	CECS 373：2014	附着式升降脚手架升降及同步控制系统应用技术规程 Technical specification for application of inserted false-work and synchronous control system		四川华山建筑有限公司
297	CECS 374：2014	建筑碳排放计量标准 Standard for measuring, accounting and reporting of carbon emission from buildings		中国建筑设计研究院
298	CECS 375：2014	一体化生物转盘污水处理装置技术规程 Technical specification for integrated rotating biological contactor sewage treatment device		中国市政工程华北设计研究总院有限公司 浙江德安科技股份有限公司
299	CECS 376：2014	改性无机粉复合建筑饰面片材应用技术规程 Technical specification for application of modified inorganic powder composite building decorative material		广东省建筑设计研究院 广东福美软瓷有限公司
300	CECS 377：2014	绿色住区标准 Standard of green residential areas		中国房地产研究会人居环境委员会
301	CECS 378：2014	聚苯模板混凝土楼盖技术规程 Technical specification for polystyrene form concrete floor		中国建筑技术集团有限公司
302	CECS 379：2014	硫铝酸盐水泥基发泡保温板外墙外保温工程技术规程 Technical specification for sulphoaluminate cement-based foamed thermal insulation board in exterior wall thermal insulation engineering		中国建筑材料工业规划研究院 唐山北极熊建材有限公司
303	CECS 380：2014	膨胀珍珠岩保温板薄抹灰外墙外保温工程技术规程 Technical specification for external thermal insulation on walls with thin-plastered expanded perlite insulation board		住房和城乡建设部科技发展促进中心 大连铭源全建材有限公司

中国工程建设标准化协会标准

序号	标准编号	标准名称	被代替标准编号	主编单位
304	CECS 381：2014	硅砂雨水利用工程技术规程 Technical specification for silicon sand storm water utilization engineering		北京仁创科技集团有限公司
305	CECS 382：2014	水平定向钻法管道穿越工程技术规程 Technical specification for pipeline crossing by horizontal directional drilling		中国地质大学（武汉）
306	CECS 383：2015	插合自锁卡簧式管道连接技术规程 Technical specification for push-fit self-lock connection		悉地国际设计顾问（深圳）有限公司 广东捷荣管道科技发展有限公司
307	CECS 384：2014	陶瓷工业窑炉工程质量验收规范 Code for construction quality acceptance of ceramic industrial kilns		国家日用及建筑陶瓷工程技术研究中心 广东中窑窑业股份有限公司
308	CECS 385：2014	再生骨料混凝土耐久性控制技术规程 Technical specification for durability control in recycled aggregate concrete		中国建筑科学研究院
309	CECS 386：2014	外储压七氟丙烷灭火系统技术规程 Technical specification for external stored pressure heptafluoropropane extinguishing systems		公安部天津消防研究所 天津盛达安全科技有限责任公司
310	CECS 387：2014	既有砌体结构隔震支座托换技术规程 Technical specification for underpinning of seismic isolators to existing masonry structure		北京筑福建筑事务有限责任公司 四川省建筑科学研究院
311	女 388：2014	水平管沉淀池工程技术规程 Technical specification for horizontal-tube sedimentation tank engineering		中国市政工程华北设计研究总院
312	CECS 389：2014	拔出法检测水泥砂浆和纤维水泥砂浆强度技术规程 Technical specification for inspection of strength of cement mortar and fiber reinforced cement mortar by pullout method		湖南大学
313	CECS 390：2014	住宅排气道系统应用技术规程 Technical specification for application of exhaust duct system in residential buildings		中国建筑标准设计研究院有限公司 北京丹轩厨卫技术中心
314	CECS 391：2014	风力发电机组消防系统技术规程 Technical specification for fire protection system of wind turbine generator system		公安部天津消防研究所
315	CECS 392：2014	建筑结构抗倒塌设计规范 Code for anti-collapse design of building structures		清华大学 中国建筑科学研究院
316	CECS 393：2014	数字集成全变频控制恒压供水设备应用技术规程 Technical specification for application of digital integrated full frequency control constant pressure water supply equipment		悉地国际设计顾问（深圳）有限公司 上海中韩杜科泵业制造有限公司
317	CECS 394：2015	七氟丙烷泡沫灭火系统技术规程 Technical specification for heptafluoropropane foam extinguishing system		公安部天津消防研究所

中国工程建设标准化协会标准

序号	标准编号	标准名称	被代替标准编号	主编单位
318	CECS 395：2015	胶圈电熔双密封聚乙烯复合供水管道工程技术规程 Technical specification for polyethylene composite water supply pipeline with double seals joint of rubber ring and electric melting		悉地国际设计顾问（深圳）有限公司 江苏狼博管道制造有限公司
319	CECS 396：2015	装配式玻纤增强无机材料复合 保温墙板应用技术规程 Technical specification for application of assembled glass fiber reinforced inorganic materials composite insulation wallboard		中国建筑标准设计研究院有限公司 卓达新材料科技集团有限公司
320	CECS 397：2015	水泥基再生材料的环境安全性检测标准 Standard for environmental safety testing of cement-based recycled materials		中国建筑科学研究院 中国路桥工程有限责任公司
321	CECS 398：2015	硅藻泥装饰壁材应用技术规程 Technical specification for application of diatomaceous decorative interior wall materials		辽宁省建筑节能环保协会 湖南蓝天豚硅藻泥新材料有限公司
322	CECS 399：2015	铜包铝电力电缆工程技术规范 Technical code for copper-clad aluminum power cables engineering		中国建筑标准设计研究院有限公司 上海胜华电气股份有限公司
323	CECS 400：2015	自攻型锚栓应用技术规程 Technical specification for screw anchor		同济大学 浙江上锚建筑科技有限公司
324	CECS 401：2015	城市地下空间开发建设管理标准 Standard for development construction management of urban underground space		广州大学 广东省基础工程公司
325	CECS 402：2015	城市地下空间运营管理标准 Standard for operation management of urban underground space		广州大学 广东省基础工程公司
326	CECS 403：2015	建筑排水不锈钢管道工程技术规程 Technical specification for stainless pipeline of building drainage		中国建筑金属结构协会给水排水设备分会 深圳市民乐管业有限公司
327	CECS 404：2015	建筑排水聚丙烯静音管道工程技术规程 Technical specification for polypropylene sound insulating pipeline of building drainage system		福建省建筑设计研究院 中国建筑装饰装修材料协会建筑塑料分会
328	CECS 405：2015	建设工程质量检测机构检测技术管理规范 Testing technology management code for engineering quality test organization		湖北省建设工程质量安全监督总站 中国建筑业协会工程建设质量监督与检测分会
329	CECS 406：2015	现浇泡沫混凝土轻钢龙骨复合墙体应用技术规程 Technical specification for application of composite wall of cast-in-place foamed concrete with light steel keel		中国建筑标准设计研究院有限公司 北京优耐德新型建材有限公司
330	CECS 407：2015	一体化预制泵站应用技术规程 Technical specification for application of integrated prefabricated pumping station		上海市政工程设计研究总院（集团）有限公司 格兰富水泵（上海）有限公司
331	CECS 408：2015	特殊钢管混凝土构件设计规程 Design specification for specified concrete filled steel tubular members		哈尔滨工业大学深圳研究生院 中国建筑第五工程局有限公司

中国工程建设标准化协会标准

序号	标准编号	标准名称	被代替标准编号	主编单位
332	CECS 409：2015	模塑聚苯模块混凝土剪力墙建筑技术规程 Technical specification for concrete shear wall building with molding EPS module		哈尔滨鸿盛房屋节能体系研发中心 中国建筑标准设计研究院有限公司
333	CECS 410：2015	不锈钢结构技术规程 Technical specification for stainless steel structures		东南大学
334	CECS 411：2015	金属面绝热夹芯板技术规程 Technical specification for double metal faced insulating sandwich panel		哈尔滨工业大学深圳研究生院 浙江东南网架股份有限公司
335	CECS 412：2015	整体爬模安全技术规程 Technical specification for safety of integral climbing formwork in construction		深圳市特辰科技股份有限公司、 甘肃第七建设集团股份有限公司
336	CECS 413：2015	升降式物料平台安全技术规程 Technical specification for safety of lifting material platform		深圳市特辰科技股份有限公司、 甘肃第六建设集团股份有限公司
337	CECS 414：2015	复合β晶型无规共聚聚丙烯（NFPP-RCT）管道工程技术规程 Technical specification for composite beta crystalline polypropylene random copolymer (NFPP-RCT) pipeline engineering		中国建筑标准设计研究院有限公司 上海瑞河企业集团有限公司
338	CECS 415：2015	预制双层不锈钢烟道及烟囱技术规程 Technical specification for factory-made double stainless steel flues and chimneys		苏州云白环境设备制造有限公司
339	CECS 416：2015	城镇径流污染控制调蓄池技术规程 Technical specification for detention tank controlling urban stormwater pollution		上海市政工程设计研究总院（集团）有限公司
340	CECS 417：2015	开合屋盖结构技术规程 Technical specification for retractable roof structures		中国建筑设计院有限公司
341	CECS 418：2015	太阳能光伏发电系统与建筑一体化技术规程 Technical specification for integration of building and solar photovoltaic system		住房和城乡建设部住宅产业化促进中心 浙江合大太阳能科技有限公司
342	CECS 419：2015	中小型给水泵站设计规程 Design specification for small-and-medium-sized water supply pumping station		悉地国际设计顾问（深圳）有限公司 株洲南方阀门股份有限公司
343	CECS 420：2015	抗震支吊架安装及验收规程 Specification for seismic bracing installation and acceptance		深圳优力可科技有限公司
344	CECS 421：2015	建筑电气细导线连接器应用技术规程 Technical specification for application of fine wire connector in building electrical installation		中国工程建设标准化协会电气专业委员会 万可电子（天津）有限公司
345	CECS 422：2015	建筑装饰室内石材工程技术规程 Technical specification for stone upholstery engineering		苏州金螳螂建筑装饰股份有限公司
346	CECS 423：2016	卡粘式连接薄壁不锈钢管道工程技术规程 Technical specification for compression and adhesive connected light gauge stainless steel pipes		悉地国际设计顾问（深圳）有限公司 中国建筑西北设计研究院有限公司

中国工程建设标准化协会标准

序号	标准编号	标准名称	被代替标准编号	主编单位
347	CECS 424：2016	预制沟槽保温模块地面辐射供暖系统应用技术规程 Technical specification for application of pre-grooved insulation board floor radiant heating		中国建筑标准设计研究院有限公司 东营瑞源特种建筑材料有限公司
348	CECS 425：2016	水锤吸纳器应用技术规程 Technical specification for application of water hammer arrestor		中国建筑金属结构协会给水排水设备分会 广东永泉阀门科技有限公司
349	CECS 426：2016	减压型倒流防止器应用技术规程 Technical specification for application of reduced-pressure backflow preventer		中国建筑金属结构协会给水排水设备分会 广东永泉阀门科技有限公司
350	CECS 427：2016	接地装置放热焊接技术规程 Technical specification for grounding exothermic welding		浙江华甸防雷科技有限公司 浙江人通电力科技有限公司
351	CECS 428：2016	电铸铜接地棒（线）技术规程 Technical specification for copper electroformed ground rod（wire）		浙江华甸防雷科技有限公司 浙江雷泰电气有限公司
352	CECS 429：2016	城市轨道用槽型钢轨闪光焊接质量检验标准 Standard for quality inspection of groove rail flash-butt welding for urban rail transit		中国铁道科学研究院金属及化学研究所
353	CECS 430：2016	城市轨道用槽型钢轨铝热焊接质量检验标准 Standard for quality inspection of groove rail thermit welding for urban rail transit		中国铁道科学研究院金属及化学研究所
354	CECS 431：2016	低热硅酸盐水泥应用技术规程 Technical specification for application of low heat Portland cement		中国建筑科学研究院 嘉华特种水泥股份有限公司
355	CECS 432：2016	蒸压硅酸盐企口小型砌块应用技术规程 Technical specification for application of autoclaved silicate tongue-and-groove small block		沈阳建筑大学 中国建筑东北设计研究院有限公司
356	CECS 433：2016	毛细管网辐射供暖供冷施工技术规程 Technical specification for construction of capillary mat and heating radiant cooling system		同济大学
357	CECS 434：2016	圆竹结构建筑技术规程 Technical specification for round bamboo-structure building		住房和城乡建设部住宅产业化促进中心 成都市无比节能科技有限公司
358	CECS 435：2016	排烟系统组合风阀应用技术规程 Technical specification for application of multi-damper on smoke extraction system		公安部天津消防研究所
359	CECS 436：2016	连锁混凝土预制桩墙支护技术规范 Technical code for chained wall of concrete precast piles		中国建筑科学研究院 浙江华展工程研究设计有限公司
360	CECS 437：2016	工业化住宅建筑外窗系统技术规程 Technical specification for exterior window system in industrialized residential building		中国建筑科学研究院 四川省建筑科学研究院
361	CECS 438：2016	住宅卫生间建筑装修一体化技术规程 Technical specification for integration of construction and decoration in residential bathroom		中国建筑装饰协会厨卫工程委员会 北京工业大学

中国工程建设标准化协会标准

序号	标准编号	标准名称	被代替标准编号	主编单位
362	CECS 439：2016	民用建筑新风系统工程技术规程 Technical specification for fresh air system of civil buildings		辽宁省建筑节能环保协会 北京建筑节能与环境工程协会
363	CECS 440：2016	建筑排水用机械式连接高密度聚乙烯（HDPE）管道工程技术规程 Technical specification for high density polypropylene (HDPE) pipeline of mechanical connection for building drainage system		中国建筑标准设计研究院有限公司 上海佐逸管业有限公司
364	CECS 441：2016	城市地下空间内部环境设计标准 Design standard for indoor environment of urban underground space		中国人民解放军理工大学 中国建筑标准设计研究院有限公司
365	CECS 442：2016	防气蚀大压差可调减压阀应用技术规程 Technical specification for application of anti-cavitation pressure regulation valve		中国建筑设计院有限公司 广东永泉阀门科技有限公司
366	CECS 443：2016	泡沫玻璃板薄抹灰外墙外保温工程技术规程 Technical specification for external thermal insulation composite system based on cellular glass board		中国建筑标准设计研究院有限公司 浙江振申绝热科技有限公司
367	CECS 444：2016	钢筋机械连接装配式混凝土结构技术规程 Technical specification for precast concrete structures with rebars mechanical splicing		清华大学 内蒙古蒙西工程设计有限公司
368	CECS 445：2016	非金属面结构保温夹芯板设计规程 Specification for design of non-metal face structural insulating sandwich panel		哈尔滨工业大学深圳研究生院 陆宇皇金建材（河源）有限公司
369	CECS 446：2016	双止回阀倒流防止器应用技术规程 Technical specification for application of double check valve backflow prevention assembly		中国建筑设计院有限公司 广东永泉阀门科技有限公司
370	CECS 447：2016	高压冷雾工程技术规程 Technical specification for high pressure cold fog engineering		中国建筑金属结构协会 杭州天腾环境艺术有限公司
371	CECS 448：2016	可视图像早期火灾报警系统技术规程 Technical specification for early fire alarm system via visual image		公安部天津消防研究所

五、工程建设地方标准

北京市

序号	标准编号	标准名称	被代替标准编号	备案号	主编单位
1	DB11/T 316-2015	地下管线探测技术规程 Technical regulations for detecting and surveying underground pipelines and cables		J13233-2015	北京市测绘设计研究院
2	DB11/T 344-2006	陶瓷墙地砖胶粘剂应用技术规程 Technical specification for ceramic wall and floor tile adhesives		J10881-2006	北京市建筑材料质量监督检验站 北京城建科技促进会
3	DB11/T 346-2006	混凝土界面处理剂应用技术规程 Technical specifications of concrete interface treating agent		J10853-2006	北京市建筑材料质量监督检验站 北京城建科技促进会
4	DB11/T 363-2016	建筑工程施工组织设计管理规程 Management specification for construction organization plan of building engineering		10877-2016	北京市建设监理协会
5	DB11/T 364-2006	建筑排水柔性接口铸铁管技术规程 Technical specification for flexible joint cast iron pipe of building drainage		J10879-2006	北京市市政工程研究院
6	DB11/T 365-2016	钢筋保护层厚度和钢筋直径检测技术规程 Technical specification for test of the depth of coverage and diameter of reinforcing bars		J10880-2017	北京市建设工程质量监督总站 北京市建设工程质量检测中心
7	DB11/ 366-2006	种植屋面防水施工技术规程 Construction technical specification of green roofs waterproof		J10888-2006	北京城建科技促进会
8	DB11/367-2006	地下室防水施工技术规程 Construction technical specification of basement waterproof		J10887-2006	北京城建科技促进会
9	DB11/T 380-2006	桥面防水工程技术规程 Bridge waterproof project treating agent		J10889-2006	北京市建筑材料质量监督检验站 北京城建科技促进会 中国土木工程学会市政工程分会
10	DB11/T 381-2016	既有居住建筑节能改造技术规程 Technical specification for energy efficiency renovation of existing residential buildings		J13148-2015	北京城建科技促进会
11	DB11/382-2006	建设工程安全监理规程 The regulation of safety construction supervision		J10859-2006	北京市建设监理协会 北京建工京精大房工程建设监理公司
12	DB11/383-2006	建设工程施工现场安全资料管理规程 Specification for the management of the safety document on engineering construction site		J10864-2006	北京市建设委员会
13	DB11/385-2011	预拌混凝土质量管理规程 Management specification for quality of ready-mixed concrete	DB11/385-2006	J10893-2011	北京市混凝土协会
14	DB11/T 386-2006	建设工程检测试验管理规程 Management specification of inspection and test for construction engineering		J10931-2007	北京市建设工程质量监督总站 北京市建设工程质量检测中心

北京市

序号	标准编号	标准名称	被代替标准编号	备案号	主编单位
15	DB11/T 446-2015	建筑施工测量技术规程 Technical specification for construction survey		J10972-2016	北京测绘学会
16	DB11/T 461-2010	民用建筑太阳能热水系统应用技术规程 Technical specification for solar water heating systems of civil buildings	DB11/T 461-2007	J11006-2010	北京市新能源与可再生能源协会 北京首建标工程技术开发中心
17	DB11/T 463-2012	胶粉聚苯颗粒复合型外墙外保温工程技术规程 Technical specification for external thermal insulation on outer-walls with composited systems based on mineral binder and expanded polystyrene granule plaster	DB11/T 463-2012 DBJ/T 01-102-2005	J12270-2013	北京振利节能环保科技股份有限公司
18	DB11/T 464-2007	建筑工程清水混凝土施工技术规程 Technical specification for architectural concrete construction		J11005-2007	中国建筑一局（集团）有限公司
19	DB11/489-2016	建筑基坑支护技术规程 Technical specification of retaining and protecting for building foundation excavation		J13247-2016	北京城建科技促进会 中国建筑科学研究院 北京市勘察设计研究院
20	DB11/490-2007	地铁工程监控量测技术规程 Technical code for monitoring measurement of subway engineering		J10909-2006	北京市轨道交通建设管理有限公司
21	DB11/T 491-2007	轻骨料混凝土隔墙板施工技术规程 Construction technique specification for lightweight aggregate concrete panel		J11057-2007	北京市建筑材料质量监督检验站 北京城建科技促进会
22	DBJ 11-501-2008	北京地区建筑地基基础勘察设计规范 Code for geotechnical investigation and design of building foundations in Beijing area	DBJ 01-501-92	J11254-2008	北京市勘察设计研究院
23	DB11/509-2007	房屋修缮工程施工质量验收规程 Specification for acceptance of construction quality of building repairing		J11031-2007	北京市房屋修缮工程监督站
24	DB11/510-2016	公共建筑节能施工质量验收规程 Specification for acceptance of public building construction quality of energy efficiency		J11070-2016	北京城建科技促进会
25	DB11/T 511-2007	自流平地面施工技术规程 Technical specification of self-leveling floor		J11123-2008	北京市建筑材料质量监督检验站 北京城建科技促进会
26	DB11/T 512-2007	建筑装饰工程石材应用技术规程 Techinical specification for application of stone in decoration		J11124-2008	北京市建筑装饰协会 北京市建设工程物资协会
27	DB11/513-2015	绿色施工管理规程 Management specification of green construction		J12875-2015	中国建筑一局（集团）有限公司
28	DB11/T 514-2008	市政基础设施长城杯工程质量评审标准 The evaluation and investigation standards for the municipal infrastructure projecting of changcheng cup		J11211-2008	北京市政工程行业协会
29	DB11/T 536-2007	农村民居建筑抗震设计施工规程 Specification for seismio design and construction of country house		J11232-2008	北京市建筑设计研究院

北京市

序号	标准编号	标准名称	被代替标准编号	备案号	主编单位
30	DB11/T 537-2007	墙体内保温施工技术规程（胶粉聚苯颗粒保温浆料玻纤网格布抗裂砂浆做法和增强粉刷石膏聚苯板做法） Construction technical specification of interior insulation for walls (the method for polystyrene foaming granule paste, a glass fiber net and anti-crack mortar system and the method for typsum reinforced with a glass fiber, plaster polystyrene foam board system)		J11212-2008	北京市建筑节能专业委员会
31	DB11/T 555-2015	民用建筑节能现场检验标准 standard for on-site tesing of energy efficient civil buildings engineerng	DB11/T 555-2008	J12993-2015	北京市中建建筑科学研究院有限公司
32	DB11/581-2008	轨道交通地下工程防水技术规程 Technical specification for water proofing of under-ground works construction of rail transportation		J11175-2008	北京城建科技促进会
33	DB11/T 582-2008	长螺旋钻孔压灌混凝土后插钢筋笼灌注桩施工技术规程 Technology specification for excavating piles bored with long auger inserted reinforcing steel cage after jacked concrete		J11267-2008	北京城建科技促进会
34	DB11/T 583-2015	钢管脚手架、模板支架安全选用技术规程 Technical specification for selecting and using of steel tubular scaffold or formwork undercarriage		J11268-2015	中建二局第三建筑工程有限公司
35	DB11/T 584-2013	保温板薄抹灰外墙外保温施工技术规程 Technical specification for external thermal insulation composite systems based on insulation board residential buildings	DB11/T 584-2008	J12370-2013	北京住总集团有限责任公司
36	DBJ/T 11-609-2008	住宅区及住宅楼房邮政信报箱 Private mail box for multi-storied residential building	DBJ 01-609-2002	J10152-2008	北京市邮政管理局
37	DB11/611-2008	施工现场塔式起重机检验规则 Tower crane inspection regulation on construction site		J11228-2008	北京市建设工程安全质量监督总站 北京市特种设备行业标准化技术委员会
38	DBJ 01-622-2005	吸气式烟雾探测火灾报警系统设计、施工及验收规范 Code for design, installation and acceptance of aspirating smoke detection fire alarm system		J10675-2013	北京市公安局消防局 中国建筑设计研究院
39	DBJ 11-624-2006	防火玻璃框架系统设计、施工及验收规范 Code for design, installation and approval of fire-resistant glazed frame system		J10869-2006	北京市公安局消防局 中国建筑科学研究院
40	DBJ/T 11-625-2007	疏散用门安全控制与报警逃生门锁系统设计、施工及验收规程 Code for design, installation and acceptance of security control and alarm system for emergency exit and evacuating doors		J10969-2007	北京市消防局

北京市

序号	标准编号	标准名称	被代替标准编号	备案号	主编单位
41	DBJ/T 11-626-2007	建筑物供配电系统谐波抑制设计规程 Design regulations for harmonics suppression of building power supply and distribution systems		J11018-2007	北京华建标建筑标准技术开发中心
42	DB11/635-2009	村镇住宅太阳能采暖应用技术规程 Technical specification for application of solar space heating for residence in villages and small towns		J11216-2008	北京市新能源与可再生能源协会
43	DB11/T 636-2009	施工现场齿轮齿条式施工升降机检验规程 Inspection regulation for rack and pinion hoist on construction site		J11393-2009	北京市建设工程安全质量监督总站
44	DB11/637-2015	房屋结构综合安全性鉴定标准 Standard for structure comprehensive safety appraisal of buildings	DB11/T 637-2009	J12935-2015	北京三茂建筑工程检测鉴定有限公司
45	DB11/T 638-2016	房屋修缮工程工程量清单计价规范 Specification of valuation with bill quantity in house renovation engineering		J11394-2017	北京市建设工程造价管理处
46	DB11/641-2009	商品住宅工程质量保修规程 Home protection specification		J11329-2009	北京市房地产科学技术研究所
47	DB11/T 642-2014	预拌混凝土绿色生产管理规程 Management specification of green production for ready-mixed concrete		J12712-2014	北京市混凝土协会
48	DB11/T 643-2009	屋面保温隔热工程施工技术规程 Construction technical specification of thermal preservation and heat insulation for roof		J11465-2009	北京市建设工程物资协会建筑节能专业委员会
49	DB11/T 644-2009	外墙外保温技术规程（现浇混凝土模板内置保温板做法） Technical specification for external thermal insulation on outer-wall (insulation board in cast-in-place concrete form)		J11436-2009	北京市建筑设计研究院研究所
50	DB11/T 645-2009	钢绞线网片-聚合物砂浆加固混凝土结构施工及验收规程 Specification for construction and acceptance of RC structure strengthened with steel stranded wire mesh and polymer mortar		J11437-2009	中国建筑科学研究院工程抗震研究所
51	DB11/ 685-2013	雨水控制与利用工程设计规范 Code for design of stormwater management and harvest engineering		J12366-2013	北京市建筑设计研究院有限公司 北京市市政工程设计研究总院 北京市水科学技术研究院
52	DB11/687-2015	公共建筑节能设计标准 Design standard for energy efficiency of public buildings		J10579-2014	北京市建筑设计研究院有限公司
53	DB11/T 688-2009	城市雕塑工程建设质量技术规范 The quality and technical specification for city sculpture project construction		J11563-2010	北京城市雕塑建设管理办公室
54	DB11/T 689-2009	建筑抗震鉴定与加固技术规程 Technical specification for seismic appraisement and strengthening of buildings in Beijing		J11561-2010	北京市建筑设计研究院

北京市

序号	标准编号	标准名称	被代替标准编号	备案号	主编单位
55	DB11/T 690-2009	城市轨道交通无障碍设施设计规程 The barrier-free facility design instruction for urban track transportation of Beijing		J11562-2010	北京市城建设计研究院
56	DB11/T 691-2009	市政工程通用混凝土模块砌体构筑物结构设计规程 Structural design specification for concrete small-sized hollow model block structures of municipal masonry engineering		J11559-2010	北京市市政工程设计研究总院
57	DB11/T 692-2009	历史文化街区工程管线综合规划规范 Code for engineering pipeline comprehensive planning of historic conservation area		J11560-2010	北京市市政工程设计研究总院
58	DB11/T 693-2009	温拌沥青路面施工及验收规范 Specification for construction and acceptance of warm mix asphalt pavements		J12269-2013	北京市政路桥集团有限公司
59	DB11/694-2009	模板早拆施工技术规程 Technical specification for early striking construction		J11498-2009	北京建工集团有限责任公司
60	DB11/T 695-2009	建筑工程资料管理规程 Management specification of construction engineering documentation	DBJ 01-51-2003	J11581-2010	北京市建设工程安全质量监督总站
61	DB11/T 696-2016	预拌砂浆应用技术规程 Technical specifications for application of pre-mixed mortar	DB11/T 696	J11582-2016	北京住总集团有限责任公司
62	DB11/T 697-2009	外墙外保温施工技术规程（外墙保温装饰板做法） Technical specification for exterior insulation (decorative insulated exterior wall panel method)	DBJ 01-92-2004	J11583-2010	北京市建筑材料质量监督检验站
63	DB11/T 698-2009	清水混凝土预制构件生产与质量验收标准 Standard for production and quality acceptance of precast fair-faced concrete elements		J11584-2010	北京城建建材工业有限公司
64	DB11/729-2010	外墙外保温工程施工防火安全技术规程 Technical specification to fire prevention of exterior insulation construction for outer-wall		J11545-2010	北京市建设工程物资协会建筑节能专业委员会
65	DB11/T 742-2010	框架填充墙（轻集料砌块）设计及施工技术规程 Technical specification for design and construction 0f filler wall with lightweight aggregate concrete hollow blocks		J11707-2010	北京市建筑设计标准化办公室
66	DB11/T 743-2010	膜结构施工质量验收规范 Specification for acceptance of constructional quality of membrane structures		J11708-2010	北京城建集团有限责任公司
67	DB11/T 751-2010	住宅物业服务标准 Residential property service standard		J11724-2010	北京市物业服务指导中心
68	DB11/T 774-2010	新建物业项目交接查验标准 The acceptance standard for the delivery of the new completed property		J11798-2011	北京市物业服务指导中心

北京市

序号	标准编号	标准名称	被代替标准编号	备案号	主编单位
69	DB11/T 775-2010	透水混凝土路面技术规程 Technical specification of pervious concrete pavement		J11797-2011	中国建筑股份有限公司
70	DB11/T 803-2011	再生混凝土结构设计规程 Code for design of recycled concrete structures		J11866-2011	北京工业大学
71	DB11/804-2015	民用建筑通信及有线广播电视基础设施设计规范 Specification on civil building design for communication and CATV infrastructure	DB11/804-2011	J13140-2015	北京电信规划设计院有限公司 北京歌华有线电视网络股份有限公司
72	DB11/T 805-2011	人行天桥与人行地下通道无障碍设施设计规程 Specification for design of unobstructed pedestrian overpass and underpass		J11868-2011	北京市市政工程设计研究总院
73	DB11/806-2011	地面辐射供暖技术规范 Technical for radiant floor heating specification	DBJ/T 01-49-2000	J11742-2010	北京市建筑设计研究院
74	DB11/807-2011	施工现场钢丝绳式施工升降机检验规程 Rope-style construction hoist inspection procedures at construction site		J11777-2011	北京市建设机械与材料质量监督检验站
75	DB11/T 808-2011	市政基础设施工程资料管理规程 Management specification of municipal infrastructure engineering documentation	DBJ 01-71-2003	J10254-2011	北京市政工程行业协会
76	DB11/T 825-2016	绿色建筑评价标准 Evaluation standard for green building		J11906-2016	北京市住房和城乡建设科技促进中心
77	DB11/T 848-2011	压型金属板屋面工程施工质量验收标准 Code for acceptance of constructional quality of profiled metal sheet roofing		J12020-2012	中国新兴建设开发总公司 北京城建集团有限责任公司
78	DB11/T 849-2011	房屋鉴定与结构检测操作规程 Operating specification for appraisal and structure inspection of building		J12019-2012	北京市房屋安全管理事务中心
79	DB11/T 850-2011	建筑墙体用腻子应用技术规程 Technical specifications of building wall putty		J12018-2012	北京市建筑材料质量监督检验站
80	DB11/T 851-2011	聚脲弹性体防水涂料施工技术规程 Technical specification for polyurea elastomer waterproof engineering		J12017-2012	北京城建科技促进会
81	DB11/T 881-2012	建筑太阳能光伏系统设计规范 Design code for solar photovoltaic system of civil buildings		J12081-2012	北京首建标工程技术开发中心 北京市新能源与可再生能源协会
82	DB11/T 882-2012	房屋建筑安全评估技术规程 Technical specification for safety assessment of buildings		J12108-2012	北京市房地产科学技术研究所
83	DB11/883-2012	建筑弱电工程施工及验收规范 Code for construction and acceptance		J11923-2011	北京市建筑业联合会智能建筑专业委员会
84	DB11/891-2012	居住建筑节能设计标准 Design standard for energy efficiency of residential buildings		J12070-2012	北京市建筑设计研究院
85	DB11/T 913-2012	外墙夹心保温设计规程 Specification for design of external sandwich wall insulation		J12236-2013	北京市建筑设计研究院有限公司 北京首建标工程技术开发中心

北京市

序号	标准编号	标准名称	被代替标准编号	备案号	主编单位
86	DB11/T 938-2012	绿色建筑设计标准 Design standard of green buildings		J12161-2012	北京勘察设计与测绘管理办公室 中国建筑科学研究院 清华大学
87	DB11/T 939-2012	建设工程临建房屋技术标准 Technical specification of temporary building for construction engineering	DBJ 01-98-2005	J10611-2009	中国建筑一局(集团)有限公司
88	DB11/ 940-2012	基坑工程内支撑技术规程 Technical specification of internal bracing for excavation engineering		J12162-2012	北京城建科技促进会
89	DB11/T 941-2012	无机纤维喷涂工程技术规程 Construction technical specification of spray inorganic fiber		J12268-2013	北京城建科技促进会
90	DB11/T 942-2012	居住建筑供热计量施工质量验收规程 Technical specification of acceptance for construction quality of heat metering device of residential buildigs		J12267-2013	北京城建科技促进会
91	DB11/T 943-2012	外墙外保温施工技术规程(复合酚醛保温板聚合物水泥砂浆做法) Technical specification for external thermal insulation on outer-walls (composite phenolic foam board)		J12266-2013	北京建筑材料科学研究总院有限公司
92	DB11/T 944-2012	防滑地面工程施工及验收规程 Specification of construction and acceptance for slip resistance floor engineering		J12265-2013	北京城建科技促进会
93	DB11/ 945-2012	建设工程施工现场安全防护、场容卫生及消防保卫标准 The facilities standards for safety & protection, environmental sanitation and safeguard & fore fighting for construction engineering		J12175-2012	北京市住房和城乡建设委员会
94	DB11/T 967-2013	塑料排水检查井应用技术规程 Technical specification for application of thermoplastics manholes and inspection chambers for underground drainage and seweragepiping systems		J12371-2013	北京市建设工程物资协会
95	DB11/T 968-2013	预制混凝土构件质量检验标准 Quality acceptance inspection for precast concrete components	DBJ 01-1-92	J12369-2013	北京市住房和城乡建设科技促进中心
96	DB11/T 969-2013	城市雨水系统规划设计暴雨径流计算标准 Standard of storm water runoff calculation for urban storm drainage system planning and design		J12340-2013	北京市城市规划设计研究院
97	DB11/T 970-2013	装配式剪力墙住宅建筑设计规程 Regulation for residential building with precast share wall		J12341-2013	北京市建筑设计研究院有限公司 北京市住房和城乡建设委员会科技促进中心
98	DB11/994-2013	平战结合人民防空工程设计规范 Code for design of civil air defence works of dual-utilization of peacetime and wartime		J12288-2013	中国建筑标准设计研究院
99	DB11/995-2013	城市轨道交通工程设计规范 Code for design of urban rail transit		J12329-2013	北京城建设计研究总院有限责任公司

北京市

序号	标准编号	标准名称	被代替标准编号	备案号	主编单位
100	DB11/996-2013	城乡规划用地分类标准 Code for classification of land use for urban & rural planning		J12322-2013	北京市城市规划设计研究院
101	DB11/T 997-2013	城乡规划计算机辅助制图标准 Standard for computer aided drawing in urban & rural planning		J12397-2013	北京市城市规划设计研究院
102	DB11/T 998-2013	基础测绘成果检查验收技术规程 Technical regulations for inspection and acceptance of fundamental survey results		J12398-2013	北京市测绘设计研究院
103	DB11/T 999-2013	城镇道路建筑垃圾再生路面基层施工与质量验收规范 Code for construction and acceptance of recycled construction waste for urban road base		J12431-2013	北京市市政工程研究院
104	DB11/1003-2013	装配式剪力墙结构设计规程 Design specification for precast concrete shear wall structure		J12293-2013	北京市建筑设计研究院有限公司
105	DB11/T 1004-2013	房屋建筑使用安全检查技术规程 Technique specification for service safety check of buildings		J12432-2013	北京三茂建筑工程检测鉴定有限公司
106	DB11/T 1005-2013	公共建筑空调采暖室内温度节能监测标准 Energy saving monitoring standard of indoor temperature for public building with air-conditioning system		J12429-2013	北京市房地产科学技术研究所
107	DB11/T 1006-2013	民用建筑能效测评标识标准 Standard for civil building energy performance evaluation and certification		J12427-2013	北京建筑技术发展有限责任公司
108	DB11/T 1007-2013	公共建筑能源审计技术通则 Technology general clauses of energy audits to public buildings		J12428-2013	中国建筑科学研究院
109	DB11/T 1008-2013	建筑太阳能光伏系统安装及验收规程 Installation and acceptance specification for solar PV system on civil buildings		J12430-2013	北京城建科技促进会
110	DB11/1022-2013	简易自动喷水灭火系统设计规程 Code of design for simplified sprinkler systems		J10478-2013	北京市公安局消防局 公安部天津消防研究所
111	DB11/1023-2013	疏散用门安全控制与报警逃生门锁系统设计、施工及验收规程 Code for design, installation and acceptance of security control and alarm system for emergency exit		J10969-2013	北京市公安局消防局 中国建筑科学研究院建筑防火研究所
112	DB11/1024-2013	消防安全疏散标志设置标准 Standard on fire safety evacuation signs installation		J10195-2013	北京市公安局消防局 中国建筑科学研究院建筑防火研究所
113	DB11/1025-2013	自然排烟系统设计、施工及验收规范 Code for design, installation and acceptance of security control and alarm system for emergency exit		J10820-2013	北京市公安局消防局 中国建筑科学研究院建筑防火研究所

北京市

序号	标准编号	标准名称	被代替标准编号	备案号	主编单位
114	DB11/1027-2013	防火玻璃框架系统设计、施工及验收规范 Code for design, installation and approval of fire-resistant glazed frame system		J10869-2013	北京市公安局消防局 中国建筑科学研究院建筑防火研究所
115	DB11/1028-2013	居住建筑门窗工程技术规范 Technical specification for doors and windows of residential buildings	DBJ 01-79-2004	J12358-2013	国家建筑材料工业建筑五金水暖产品质量监督检验测试中心
116	DB11/T 1029-2013	混凝土矿物掺合料应用技术规程 Technical specification for application of mineral admixtures in concrete		J12534-2014	北京市混凝土协会 北京市建设工程安全质量监督总站 北京市城市规划设计研究院
117	DB11/T 1030-2013	装配式混凝土结构工程施工与质量验收规程 Specification for construction and quality acceptance of precast concrete structures		J12535-2014	北京市住房和城乡建设科技促进中心 北京市建筑工程研究院有限责任公司
118	DB11/T 1031-2013	低层蒸压加气混凝土承重建筑技术规程 Technical specification on building for lower-layer of autoclaved aerated concrete load-carrying		J12536-2014	北京市建筑设计研究院有限公司
119	DB11/1066-2014	供热计量设计技术规程 Technical specification for heat metering of district heating system		J12489-2013	北京市建筑设计研究院有限公司
120	DB11/T 1068-2014	下凹桥区雨水调蓄排放设计规范 Design code for stormwater storing and discharging for underpass road		J12594-2014	北京市市政工程设计研究总院有限公司
121	DB11/T 1069-2014	民用建筑信息模型设计标准 Building information modeling design standard for civil building		J12593-2014	北京市勘察设计和测绘地理信息管理办公室 北京工程勘察设计行业协会
122	DB11/T 1070-2014	市政基础设施工程质量检验与验收标准 Check and accept standard for construction of municipal engineering in Beijing		J12502-2013	北京市政建设集团有限责任公司
123	DB11/1071-2014	排水管（渠）工程施工质量检验标准 Check standard for construction of sewer conduit engineering in Beijing		J12503-2013	北京市政建设集团有限责任公司
124	DB11/1072-2014	城市桥梁工程施工质量检验标准 Check standard for construction of urban bridge engineering in Beijing		J12504-2013	北京市政建设集团有限责任公司 北京市道路工程质量监督站
125	DB11/T 1073-2014	城市道路工程施工质量检验标准 Check standard for construction of urban road engineering in Beijing	DBJ 01-11-2004	J12660-2014	北京市政建设集团有限责任公司
126	DB11/T 1074-2014	建筑结构长城杯工程质量评审标准 The specification of building structure quality evaluation for the great wall cup		J12726-2014	北京市工程建设质量管理协会
127	DB11/T 1075-2014	建筑长城杯工程质量评审标准 The specification of building quality evaluation for the great wall cup	DBJ/T 01-70-2003	J12725-2014	北京市工程建设质量管理协会

北京市

序号	标准编号	标准名称	被代替标准编号	备案号	主编单位
128	DB11/T 1076-2014	居住建筑装修装饰工程质量验收规范 Specification for acceptance of construction quality of residential building decoration project		J12659-2014	北京市建筑装饰协会
129	DB11/T 1079-2014	泡沫水泥保温板外墙外保温工程施工技术规程 Technical specification for external thermal insulation systems of composite foam cement panel		J12691-2014	北京市建设工程物资协会
130	DB11/T 1080-2014	硬泡聚氨酯复合板现抹轻质砂浆外墙外保温工程施工技术规程 Technical specification for external thermal insulation construction based on rigid polyurethane foam composite board with plastering lightweight mortar		J12690-2014	北京建筑技术发展有限责任公司
131	DB11/T 1081-2014	岩棉外墙外保温工程施工技术规程 Technical specification for external thermal insulation composite systems based on rock wool		J12688-2014	北京住总集团有限责任公司
132	DB11/T 1087-2014	公共建筑装饰工程质量验收标准 Standard for construction quality acceptance of public building decoration		J12689-2014	北京市建筑装饰协会
133	DB11/T 1102-2014	城市轨道交通工程规划核验测量规程 Regulation for planning verification survey of urban rail transit engineering		J12752-2014	北京市规划监察执法大队 北京市测绘设计研究院
134	DB11/T 1103-2014	泡沫玻璃板建筑保温工程施工技术规程 Technical specification on construction of thermal insulation of cellular glass panel for building		J12765-2014	北京工业大学
135	DB11/T 1104-2014	地面辐射供暖工程防水施工和验收规程 Specification of construction and acceptance for waterproofing of floor radiant heating building		J12764-2014	北京市建设工程物资协会
136	DB11/T 1105-2014	建筑外遮阳工程施工及验收规程 Specification for construction and acceptance of external solar shading engineering of buildings		J12763-2014	中国建材检验认证集团股份有限公司
137	DB11/T 1106-2014	建筑墙体砌块结构自保温施工和验收规程 Code for consfruction and acceptance for building engineering of walls self-insulation block structure		J12762-2014	北京城建科技促进会
138	DB11/1115-2014	城市建设工程地下水控制技术规范 Technical code for groundwater control in urban construction projects		J12737-2014	北京市勘察设计研究院有限公司 北京城建科技促进会
139	DB11/1116-2014	城市道路空间规划设计规范 Code for planning & design of urban road spac		J12549-2014	北京市城市规划设计研究院
140	DB11/T 1117-2014	玻璃棉板外墙外保温施工技术规程 Technical specification for external thermal insulation systems based on glass wool board		J12808-2014	北京建筑材料科学研究总院有限公司

北京市

序号	标准编号	标准名称	被代替标准编号	备案号	主编单位
141	DB11/T 1130-2014	公共建筑空调制冷系统节能运行管理技术规程 Technical regulation for energy-saving operation management of public building air conditioning and refrigeration system		J12863-2014	北京城建科技促进会
142	DB11/T 1131-2014	公共建筑电气设备节能运行管理技术规程 Technical regulation for energy-saving operation management of public building automation system		J12864-2014	北京城建科技促进会
143	DB11/T 1132-2014	建设工程施工现场生活区设置和管理规范 Living facilities and management standard for construction engineering		J12862-2014	中国建筑一局（集团）有限公司
144	DB11/T 1133-2014	人工砂应用技术规程 Technical specification for application of manufactured sand		J12861-2014	北京建筑大学
145	DB11/T 1196-2015	公共租赁住房内装设计模数协调标准 Standard for interior design modular coordination of public rental housing		J13051-2015	中国建筑设计研究院
146	DB11/T 1197-2015	住宅全装修设计标准 Standard for fully interior fitting out of residential buildings		J13052-2015	中国建筑设计研究院
147	DB11/T 1198-2015	公共建筑节能评价标准 Assessment standard for energy efficiency of public buildings		J13106-2015	中国建筑科学研究院
148	DB11/T 1199-2015	农村既有单层住宅建筑综合改造技术规程 Technical specification for comprehensive retrofiting of existing rural single-story residence building		J13107-2015	北京市房地产科学技术研究所
149	DB11/T 1200-2015	超长大体积混凝土结构跳仓法技术规程 Technical specification for mass and super-length concrete structure with alternative bay construction method		J13108-2015	北京市建筑工程研究院有限责任公司
150	DB11/1222-2015	居住区无障碍设计规程 Code for accessibility design of residential areas		J12978-2015	北京市建筑设计研究院有限公司
151	DB11/T 1223-2015	大型公共建筑用电分项监测技术规程 Technical code of subitem electric consumption monitoring for large public buildings		J13238-2015	中国建筑科学研究院
152	DB11/1245-2015	建筑防火涂料（板）工程设计、施工与验收规程 Code of design, installation and acceptance for fire-retardant coating（sheets）of buildings		J10411-2015	北京市公安局消防局 中国建筑科学研究院建筑防火研究所
153	DB11/T 1246-2015	城市地下联系隧道防火设计规范 Code for fire protection design of city road underground linked tunnel		J13234-2015	北京市公安局消防局 中国建筑科学研究院建筑防火研究所

北京市

序号	标准编号	标准名称	被代替标准编号	备案号	主编单位
154	DB11/T 1247-2015	公共建筑电气设备节能运行管理技术规程 Technical regulation for energy-saving operation management of public building electrical equipment		J13274-2015	北京城建科技促进会
155	DB11/T 1248-2015	公共建筑给水排水系统节能运行管理技术规程 Technical specification for energy efficiency of operation and management in water supply and drainage system of public buildings		J13275-2015	北京城建科技促进会
156	DB11/T 1249-2015	居住建筑节能评价技术规范 Technical specification for evaluation of energy saving in residential buildings		J13418-2016	北京中建建筑科学研究院有限公司
157	DB11/1309-2015	社区养老服务设施设计标准 Design standard for community-based elderly care facilities		J13210-2015	北京市建筑设计研究院有限公司 北京维拓时代建筑设计有限公司
158	DB11/1310-2015	装配式框架及框架-剪力墙结构设计规程 Design specification for precast concrete frame and frame-shear wall structure		J13257-2015	北京市建筑设计研究院有限公司
159	DB11/T 1311-2015	污染场地勘察规范 Code for investigation of contaminated sites		J13344-2016	北京市勘察设计研究院有限公司
160	DB11/T 1312-2015	预制混凝土构件质量控制标准 Standard for quality control of precast reinforced concrete elements		J13446-2016	北京市建设工程安全质量监督总站
161	DB11/T 1313-2015	薄抹灰外墙外保温用聚合物水泥砂浆应用技术规程 Technical specification on polymer modified cement mortar for external thermal insulation composite systems		J10187-2016	北京建筑材料检验研究院有限公司
162	DB11/T 1314-2015	混凝土外加剂应用技术规程 Technical specifications for application of concrete admixtures		J10165-2016	北京建筑材料检验研究院有限公司
163	DB11/T 1315-2015	绿色建筑工程验收规范 Code for acceptance of green building construction		J13382-2016	北京市住房和城乡建设科技促进中心
164	DB11/T 1316-2016	城市轨道交通工程建设安全风险技术管理规范 Code for safety risk management of urban rail transit construction		J13153-2016	北京市轨道交通建设管理有限公司
165	DB11/1339-2016	住宅区及住宅管线综合设计标准 Design standard for pipeline comprehension of residential district and residential building		J13252-2016	北京市建筑设计研究院有限公司 北京市住宅建筑设计研究院有限公司
166	DB11/1340-2016	居住建筑节能工程施工质量验收规程 Specification for insulation constructional quality acceptance of residential building		J13149-2015	北京住总集团有限责任公司

天津市

序号	标准编号	标准名称	被代替标准编号	备案号	主编单位
1	DB 29-1-2013	天津市居住建筑节能设计标准 Tianjin energy efficiency design standard for residential buildings	DB 29-1-2010	J10409-2013	天津市建筑设计院
2	DB/T 29-6-2010	天津市建设项目配建停车场（库）标准 Parking standard for construction projects in Tianjin city	DB 29-6-2004	J10484-2010	天津市规划局
3	DB/T 29-7-2014	天津市居住区公共服务设施配置标准 Tianjin residential public service facilities allocation standard	DB 29-7-2008	J11250-2014	天津市城市规划设计研究院
4	DB 29-20-2000	天津市岩土工程技术规范 Technical code for geotechnic engineering	TBJ 1-88 TBJ 3-91 JJG 3-89		天津大学 天津建筑设计院
5	DB 29-22-2013	天津市住宅设计标准 Design standard for Tianjin urban residential buildings	DB 29-22-2007	J10968-2013	天津市建设工程技术研究所 天津市建筑设计院
6	DB/T 29-23-2016	天津市住宅建设智能化技术规程 Technical code for the construction of residential intelligent system in Tianjin	DB 29-23-2000	J10038-2016	天津市中环系统工程责任有限公司
7	DB 29-26-2008	天津市集中供热住宅计量供热设计规程 Specification for design of heat measurement-based supply system in residential building with central heating of Tianjin		J10067-2008	天津市供热办公室
8	DB 29-35-2010	天津市住宅装饰装修工程技术标准 Technical standard of decaration of housings in Tianjin	DB 29-35-2005	J10181-2010	天津市环境装饰协会
9	DB/T 29-36-2010	天津市园林植物保护技术规程 Technical regulations for plant protection of Tianjin landscape gardening	DB 29-36-2002	J10149-2010	天津市园林绿化研究所
10	DB/T 29-37-2009	天津市草坪建植与养护管理技术规程 Technical standard of Tianjin for turf establishment and maintenance		J11531-2010	天津市园林绿化研究所
11	DB/T 29-38-2015	建筑基桩检测技术规程 Technical specification for testing of building foundation piles	DB 29-38-2002	J10198-2015	天津市勘察院
12	DB 29-42-2009	天津市集中供热住宅计量供热施工质量验收规程 Code for acceptance of construction quality of heat measurement-based supply system in residential buildings with central heating of Tianjin		J10243-2009	天津市供热办公室
13	DB/T 29-47-2012	天津市城市道路养护技术规程 Technical specification for maintenance of urban road of Tianjin	DB 29-47-2003	J10248-2012	天津市道路桥梁管理处
14	DB 29-48-2003	埋地玻璃纤维增强塑料夹砂排水管道技术规程 Technical specification for underguound glass fiber reinforced plastic mortar pipe for sewerage		J10257-2003	天津市创业环保股份有限公司

天津市

序号	标准编号	标准名称	被代替标准编号	备案号	主编单位
15	DB 29-50-2003	城市道路工程质量检验标准 Quality inspection standards of municipal roag engineering		J10273-2003	天津市市政公路工程质量监督站
16	DB 29-51-2003	城市桥梁工程质量检验标准 Quality inspection standards for urban bridge engineering		J10274-2003	天津市市政公路工程质量监督站
17	DB 29-52-2003	城市排水工程质量检验标准 Quality inspection standards for urban drainage engineering		J10275-2003	天津市市政公路工程质量监督站
18	DB 29-53-2003	城市污水处理厂工程质量检验标准 Quality inspection standards for urban sewage treatment plant engineering		J10276-2003	天津市市政公路工程质量监督站
19	DB 29-54-2003	城市地铁工程质量检验标准 Quality inspection and enaluation standards for municipal metro engineering		J10277-2003	天津市市政公路工程质量监督站
20	DB/T 29-57-2016	天津市钢结构住宅设计规程 Tianjin design specification for steel structure of residential buildings	DB 29-57-2003	J10297-2016	天津市建筑设计院 天津大学
21	DB/T 29-58-2010	雷电电磁脉冲建筑防护标准 Standard for protection in buildings against lightning electromagnetic impulse	DB 29-58-2003	J10308-2010	天津市中力防雷技术有限公司 天津市建筑设计院
22	DB 29-59-2003	电气装置消防安全检测技术规程 Technical code of fire-security check for electric devices		J10309-2003	天津市公安消防局
23	DB 29-61-2012	电热辐射供暖技术规程 Technical specification for electrical radiant heating	DB 29-61-2004	J10356-2012	天津市供热办公室
24	DB 29-62-2004	天津市城市桥梁养护技术规程 Technical specification for maintenance of urban bridge		J10357-2004	天津市道路桥梁管理处
25	DB/T 29-64-2007	天津市建设事业IC卡应用标准 Application standard for construction cause IC card of Tianjin	DB/T 29-64-2004	J10950-2007	天津市建设信息技术服务中心
26	DB 29-65-2011	挤扩灌注桩技术规程 Technical specification for cast-in-situ piles with expanded branches and plates	DB 29-65-2004	J10360-2011	天津市勘察院
27	DB/T 29-67-2015	天津市园林绿化养护管理技术规程 The technical regulations of maintaining for open space and parks in Tianjin urban	DB 29-67-2004	J10367-2015	天津市风景园林学会
28	DB 29-68-2004	天津市城市绿化工程施工技术规程 The technical specification of landscapes engineering for Tianjin urbans	TBJ 10-95	J10368-2004	天津市园林局
29	DB 29-69-2016	天津市二次供水工程技术规程 Technical specification for secondary water supply engineering in Tianjin	DB 29-69-2008	J10369-2016	天津市供水管理处

天津市

序号	标准编号	标准名称	被代替标准编号	备案号	主编单位
30	DB 29-71-2004	天津市城市景观照明工程技术规范 Technical standard for urban landscape lighting		J10402-2004	天津市夜景灯光建设管理中心
31	DB/T 29-72-2010	人工砂应用技术规程 Technical specification for application of manufactured sand	DB 29-72-2004	J10403-2010	天津市建筑科学研究院
32	DB 29-74-2004	天津市市政工程施工技术规范（道路工程部分） Technical specification for construction of municipal engineering of Tianjin (road engineering)		J10405-2004	天津市市政工程局
33	DB 29-75-2004	天津市市政工程施工技术规范（桥梁工程部分） Technical specification for construction of municipal engineering of Tianjin (bridge engineering)		J10406-2004	天津市市政工程局
34	DB 29-76-2004	天津市市政工程施工技术规范（排水工程部分） Technical specification for construction of municipal engineering of Tianjin (drainge engineering)		J10407-2004	天津市市政工程局
35	DB 29-77-2004	天津市市政工程施工技术规范（污水处理厂工程部分） Technical specification for construction of municipal engineering of Tianjin (sewage treatment plant engineering)		J10408-2004	天津市市政工程局
36	DB/T 29-78-2013	天津市政公路工程施工组织设计编制标准 Standard of working out for work organization plan of Tianjin municipal and road engineering	DB 29-78-2004	J10414-2013	天津城建集团有限公司
37	DB 29-79-2004	桥用轻骨料混凝土应用技术规程 Technical specifications for application of lightweight aggregate concrete used for bridge		J10415-2004	天津市市政工程研究院
38	DB/T 29-80-2010	天津城市道路绿化建设标准 Standards of planting trees and grasses in roads for Tianjin urban	DB 29-80-2010	J10416-2010	天津市河西区市容和园林管理委员会
39	DB/T 29-81-2010	天津市园林绿化工程施工质量验收标准 The standards of accepting for completed landscaping project in Tianjin	DB 29-81-2004	J10417-2010	天津市园林学会
40	DB 29-83-2004	天津市城市道路工程管网检查井综合设置技术规程 Technical specification of integrated equipment for Tianjin urban road pineline network manhole		J10418-2004	天津市城市规划设计研究院
41	DB/T 29-86-2011	天津市建设工程文件归档整理规程 Tianjin urban code for construction project document filing and arrangement	DB/T 29-86-2004	J10424-2011	天津市城建档案馆

天津市

序号	标准编号	标准名称	被代替标准编号	备案号	主编单位
42	DB/T 29-87-2015	天津市城市排水泵站建设标准 The construction standard of city sewerage pump stations in Tianjin	DB 29-87-2004	J10425-2015	天津市市政工程设计研究院
43	DB/T 29-88-2014	天津市民用建筑围护结构节能检测技术规程 Tianjin technical specification for energy efficiency testing of civil building envelope	DB/T 29-88-2010	J12653-2014	天津建科建筑节能环境检测有限公司 天津市建设工程质量检测试验行业协会
44	DB 29-89-2010	天津市市政工程施工现场安全管理标准 Safety management standard for construction site of municipal engineering in Tianjin	DB 29-89-2004	J10436-2010	天津城建集团有限公司
45	DB 29-91-2004	天津城市大树移植技术规程 The technical specifications of transplanting big trees for Tianjin		J10438-2004	天津市园林局
46	DB 29-92-2004	天津市古树名木保护和复壮技术规程 The technical specifications of protecting and reviving for ancient and historic in Tianjin		J10439-2004	天津市园林局
47	DB/T 29-93-2004	土压平衡和泥水平衡顶管工程技术规程 Technical code for construction of soil pressure blanced pipe-pushing and muddy soil pressure blanced pipe-pushing engineering		J10452-2004	天津市管道工程集团有限公司 城建集团二公司总工办
48	DB 29-94-2004	天津市房屋修缮工程安全技术管理规程 Housing repairing project safety technology management specification in Tianjin		J10453-2004	市房管局科技处
49	DB 29-95-2004	混凝土排水管道工程闭气检验标准 Standard for the air test method of concrete sewer pipe work		J10454-2004	天津市市政工程研究院
50	DB/T 29-98-2010	天津市城市供水服务标准 Standard for city water supply service in Tianjin	DB/T 29-98-2004	J10465-2010	天津市供水管理处
51	DB/T 29-99-2009	天津市城镇燃气供气服务管理标准 Administrative standard for gas supply service in Tianjin	DB/T 29-99-2004	J10466-2009	天津市燃气管理处
52	DB/T 29-100-2014	天津市住宅厨房卫生间防火型变压式排气道应用技术规程 Technicalspecificationof fire-proofing variable pressure ventilation duct for building kitchen and toilet for Tianjin	DB/T 29-100-2004	J10467-2014	天津市建筑设计院 住房和城乡建设部政策研究中心厨房卫生间研究所
53	DB 29-102-2004	劲性搅拌桩技术规程 Technical code for reinforced minxing pile		J10469-2004	天津大学建筑设计研究院
54	DB 29-103-2010	钢筋混凝土地下连续墙施工技术规程 Construction technical specifications for reinforced concrete diaphragm wall	DB 29-103-2004	J10470-2010	天津市地质工程勘察院
55	DB 29-104-2010	天津市管道直饮水工程技术标准 Technical standard of dedicated drinking water engineering in Tianjin	DB 29-104-2004	J10475-2010	天津市供水管理处
56	DB/T 29-106-2011	公路沥青路面基层冷再生设计与施工技术规程 Technical specifications for cold regeneration design & construction of highway	DB 29-106-2004	J10477-2011	天津市公路处

天津市

序号	标准编号	标准名称	被代替标准编号	备案号	主编单位
57	DB 29-110-2010	预应力混凝土管桩技术规程（2013年局部修订） Technical specification for prestressed concreate pipe pile foundation		J11677-2013	天津市勘察院
58	DB 29-111-2004	埋地钢质管道阴极保护技术规程 Technical specifications for cathodic protection of buried steel pipelines		J10496-2005	天津市管道工程集团有限公司
59	DB/T 29-112-2010	钻孔灌注桩成孔、地下连续墙成槽质量检测技术规程 Technical specification for the testing of the drilling hole of cast-in-place pile and the groove of diaphragm wall	DB 29-112-2004	J10497-2010	天津市地质工程勘察院
60	DB 29-113-2011	天津市建设工程施工现场安全管理规程 Tianjin safety management regulations for construction sites	DB 29-113-2004	J10498-2011	天津市建设工程质量安全监督管理总队
61	DB/T 29-114-2016	天津市城市排水设施养护、维修技术规程 Technical specification for maintenance of Tianjin sewerage	DB 29-114-2004	J10499-2016	天津市排水管理处
62	DB/T 29-115-2010	房屋建筑工程施工组织设计编制标准 Standard for desiging a construction management plan of building construction engineering	DB/T 29-115-2004	J10500-2010	天津一建建筑工程有限公司
63	DB/T 29-119-2011	天津市城镇燃气供气设施运行管理标准 Management standard for operation of urban gas supply facility in Tianjin	DB/T 29-119-2005	J10515-2011	天津市燃气管理处 中国市政工程华北设计研究院总院
64	DB/T 29-121-2010	天津市城市供水管网维护管理技术标准 Technical specification for maintenance and management of Tianjin urban water supply networks	DB/T 29-121-2005	J10517-2010	天津市自来水集团有限公司
65	DB/T 29-122-2010	天津市生活垃圾转运站运行管理技术规程 Technical specification for operation of Tianjin municipal domestic waste transfer station	DB/T 29-122-2005	J10518-2010	天津市生活垃圾处理中心
66	DB/T 29-124-2010	天津市生活垃圾卫生填埋场运行管理技术规程 Technical specification for operation of Tianjin municipal domestic waste sanitary landfill	DB/T 29-124-2005	J10520-2010	天津市生活垃圾处理中心
67	DB 29-126-2014	天津市民用建筑节能工程施工质量验收规程 Tianjin city accepting specification for the quality of energy-efficiency engineering of civil building	DB 29-126-2010	J10522-2014	天津市建设工程质量安全监督管理总队
68	DB/T 29-127-2010	天津市仙客来生产栽培技术规程 The technical specifications of cultivating florists cyclamen for Tianjin	DB/T 29-127-2005	J10525-2010	天津市园林绿化研究所
69	DB/T 29-128-2015	天津市蒸压砂加气混凝土制品、应用技术规程 Tianjin technical specification for application of autoclaved aerated concrete	DB 29-128-2008	J10526-2015	天津住宅建设发展集团有限公司 天津市建材业协会

天津市

序号	标准编号	标准名称	被代替标准编号	备案号	主编单位
70	DB/T 29-129-2010	混凝土用矿物掺合料应用技术规程 Technical specification for application of mineral admixture to concrete	DB/T 29-129-2005 DB 29-63-2004	J10527-2010	天津三建建筑工程有限公司
71	DB/T 29-130-2015	预拌砂浆技术规程 Technical specification for ready-mixed mortar	DB/T 29-130-2010	J10528-2015	天津市建筑科学材料研究所 天津市建筑科学研究院
72	DB/T 29-131-2015	天津市建设工程监理规程 The supervisory specifications of construction engineering for Tianjin	DB 29-131-2005	J10529-2015	天津市建设监理协会
73	DB/T 29-132-2010	建筑内外墙涂料应用技术规程 The technical specification for architectural wall coatings	DB/T 29-132-2005	J10537-2010	天津市建筑材料集团（控股）有限公司 天津市建筑材料科学研究所
74	DB/T 29-133-2010	建筑用界面处理剂应用技术规程 The technical specification for building interface treating agent	DB/T 29-133-2005	J10538-2010	天津市建筑材料科学研究所
75	DB/T 29-134-2014	钢结构防火涂料工程施工验收规范 Code for construction and acceptance of fire resistive coating for steel structure	DB 29-134-2005	J10539-2014	天津市公安局消防局
76	DB/T 29-135-2010	脲醛发泡夹芯保温工程施工及验收规程 Code for construction and acceptance of thermal insulation engineering by filling with UF foam	DB/T 29-135-2005	J10540-2010	天津市建筑材料集团（控股）有限公司 天津市建筑材料科学研究所
77	DB/T 29-137-2015	天津市城市快速路设计标准 The design standard of urban expressway in Tianjin	DB 29-137-2005	J10542-2015	天津市市政工程设计研究院
78	DB 29-138-2005	天津市历史风貌建筑保护修缮技术规程 Protect the building of style and features and repair the technical regulation in Tianjin		J10564-2005	天津市保护风貌建筑办公室
79	DB/T 29-139-2015	天津市房屋修缮工程质量验收标准 Standard of inspection of quality that the house repairing project in Tianjin constructs	DB/T 29-139-2005	J10565-2015	天津市房产总公司
80	DB 29-140-2011	天津市空间网格结构技术规程 Technical specification for spatial grid structure of Tianjin	DB 29-140-2005	J10566-2011	天津大学 天津市钢结构学会
81	DB 29-143-2010	天津市地下铁道基坑工程施工技术规程 Technical specification for foundation pit engineering in Tianjin metro construction	DB/T 29-143-2005	J10589-2010	天津市地下铁道集团有限公司
82	DB 29-144-2010	天津市地下铁道盾构法隧道工程施工技术规程 Construction technical specification for shield tunnels in Tianjin metro	DB/T 29-144-2005	J10590-2010	天津市地下铁道集团有限公司
83	DB/T 29-145-2010	天津市地下工程型钢水泥土搅拌墙（SMW）施工技术规程 Construction technical specification for soil mixing wall（SMW）of underground in Tianjin	DB/T 29-145-2005	J10591-2010	天津市地下铁道集团有限公司
84	DB/T 29-146-2010	天津市地下铁道暗挖法隧道工程施工技术规程 Construction technical specification for closed-cut tunnels in Tianjin metro	DB/T 29-146-2005	J10592-2010	天津市地下铁道集团有限公司

天津市

序号	标准编号	标准名称	被代替标准编号	备案号	主编单位
85	DB 29-147-2005	天津市城市桥梁养护操作技术规程 Operational technical specification of maintenance for urban bridges of Tianjin		J10593-2005	天津市道路桥梁管理处
86	DB/T 29-148-2005	结构混凝土实体检测技术规程 Technical specification of testing structure concrete mass		J10594-2005	天津市港湾工程质量检测中心
87	DB/T 29-152-2010	天津市地下管线信息管理技术规程 Technical specification for information management of underground pipelines and cables in Tianjin	DB/T 29-150-2005 DB/T 29-152-2005	J10614-2010	天津市地下空间规划管理信息中心 天津市测绘院
88	DB 29-153-2014	天津市公共建筑节能设计标准 Tianjin design standard for energy efficiency of public buildings	DB 29-153-2010	J10633-2014	天津市建筑设计院
89	DB 29-154-2006	天津市高速公路养护技术规范 Technical specification for maintenance of expressway of Tianjin		J10678-2006	天津市高速公路投资建设发展公司 天津市公路工程设计研究院
90	DB 29-155-2006	塑料排水管道工程闭气检验标准 Standard for the air test method of plastic sewer pipe work		J10679-2006	天津市市政工程研究院
91	DB/T 29-156-2006	天津市居住区绿地设计规范 Code for design of greenland inresident areas for Tianjin urban		J10680-2006	天津市园林规划设计院
92	DB/T 29-157-2006	公路沥青路面裂缝密封施工技术规程 Technical specifications for cracks sealant construction of highway asphalt pavement		J10681-2006	天津市公路管理局
93	DB/T 29-160-2006	高喷插芯组合桩技术规程 Technical code for jg soil-cement-pile strengthened pile（JPP）		J10733-2006	天津市华正岩土工程有限公司 天津市勘察院
94	DB/T 29-161-2006	天津市废轮胎胶粉改性沥青路面技术规程 Technical specification of waste tire rubber powder modified asphalt pavement of in Tianjin		J10732-2006	天津市市政工程局 天津市滨海发展投资控股有限公司 天津海泰环保科技发展有限公司
95	DB/T 29-163-2006	天津市彩色沥青混凝土施工技术规程 Technical specifications for construction of colored asphalt concrete of Tianjin		J10860-2006	天津路桥建设工程有限公司
96	DB 29-164-2013	天津市建筑节能门窗技术标准 Tianjin technique standard for energy efficiency doors and windows of building	DB 29-164-2010	J10861-2013	天津住宅科学研究院有限公司 天津市建材业协会
97	DB/T 29-165-2006	天津市钢筋混凝土桥梁耐久性设计规程 Code for durability design of reinforced concrete bridge in Tianjin		J10862-2006	天津市市政工程局
98	DB/T 29-166-2010	天津市钢桥面浇注式沥青混凝土铺装施工技术规程 Technical specification of stell bridge deck surfacing cast-type asphalt concrete pavement of Tianjin	DB/T 29-166-2006	J10863-2010	天津五市政公路工程有限公司
99	DB 29-167-2007	天津市再生水设计规范 Code for design of reclaimed water system in Tianjin		J10926-2007	中国市政工程华北设计研究院 天津市建筑设计院 天津中水有限公司

天津市

序号	标准编号	标准名称	被代替标准编号	备案号	主编单位
100	DB/T 29-172-2014	天津市非承重烧结页岩（保温）空心砖应用技术规程 Tianjin technical specification for application of non-load-bearing fired shale (insulation) hollow brick	DB 29-172-2007	J10962-2014	天津市建筑设计院、中节能国环新型材料有限公司
101	DB 29-173-2014	天津市叠压供水技术规程 The technical specification for additive pressure water supply in Tianjin	DB/T 29-173-2007	J11016-2014	天津市供水管理处
102	DB 29-174-2007	天津市公路沥青路面微表处施工技术规程 Technical specifications for micro-surfacing construction of highway asphalt pavement of Tianjin		J11044-2007	天津市公路管理局
103	DB/T 29-175-2007	天津市盆栽安祖花和竹芋生产技术规程 Technical specifications of cultivating potted common anthurium and calanthurium for Tianjin		J11045-2007	天津市花卉管理处 天津花卉示范中心
104	DB/T 29-176-2016	天津市预防混凝土碱骨料反应技术规程 Technical specification for prevention alkali-aggregate reaction of concrete in Tianjin	DB/T 29-176-2010	J11063-2016	天津市建筑科学研究院有限公司
105	DB 29-177-2007	天津市建设工程施工现场安全文明施工质量评估标准 Assessment standard of safety and civil engineering quality for construction site in Tianjin		J11062-2007	天津市建设安全监督管理站等
106	DB/T 29-178-2010	天津市地埋管地源热泵系统应用技术规程 The technical specification for ground-coupled heat pump system in Tianjin	DB 29-178-2007	J11109-2010	天津大学
107	DB/T 29-179-2014	天津市玻纤水泥平板复合保温板应用技术规程 Tianjin construction methods and acceptance standards for wall and roofing concrete board insulation panels	DB 29-179-2007	J11110-2014	天津建科建筑节能环境检测有限公司
108	DB 29-181-2008	天津市钢桥面环氧沥青混凝土铺装施工技术规程 Technical specification for stell flooring epoxy asphalt concrete pavement of Tianjin		J11144-2008	天津五市政公路工程有限公司和天津城建滨海路桥有限公司
109	DB 29-182-2008	天津市建设工程施工现场临时用电配电箱安全技术标准 Security technology standard of temporary and site electric switch box of construction in Tianjin		J11154-2008	天津市建设安全监督管理站等
110	DB/T 29-184-2010	天津市预拌混凝土质量管理规程 Specification for quality management of ready-mixed concrete in Tianjin	DB 29-184-2008	J11180-2010	天津市预拌混凝土管理中心 天津市建设工程质量检测中心
111	DB 29-185-2008	天津市建设工程施工现场轻钢结构拼装型临建房屋安全技术规程 Regulations of security and technology for Tianjin light-steel portakabins in construction site		J11187-2008	天津市建设安全监督管理站

天津市

序号	标准编号	标准名称	被代替标准编号	备案号	主编单位
112	DB/T 29-186-2011	天津市矩形钢管混凝土节点技术规程 Technical specification for connection in CFRT structure of Tianjin	DB 29-186-2008	J11202-2011	天津大学 天津市钢结构学会
113	DB 29-187-2008	天津市地热回灌地面工程建设标准 Engineering construction standard of geothermal reinjection in Tianjin		J11241-2008	天津地热勘查开发设计院
114	DB 29-188-2008	天津市路灯设施建设运行技术标准 Technical standard for construction and operation of Tianjin streetlights installation		J11242-2008	天津市路灯管理处
115	DB 29-189-2008	SDE无管网气体灭火系统设计、施工、验收规范 Code for design. installation and deceptamce of SDE gas fire-extinguishing systems.		J11240-2008	天津市公安消防局
116	DB/T 29-190-2010	缓粘结预应力混凝土结构施工技术规程 Construction technical specification for concrete structure prestressed with laterbonded tendons	DB 29-190-2008	J11271-2010	天津市建筑科学研究院
117	DB/T 29-191-2009	天津市地基土层序划分技术规程 Technical specification for division of subsoil sequence in Tianjin		J11414-2009	天津市勘察院
118	DB/T 29-192-2009	中新天津生态城绿色建筑评价标准 Sino-Singapore Tianjin Eco-city green building evaluation standard		J11468-2009	天津城市建设学院 天津市建筑设计院
119	DB/T 29-194-2010	城镇再生水厂运行、维护及安全技术规程 Technical specification for operation, maintenance and safety of city and town recycling water plant		J11518-2010	天津中水有限公司
120	DB 29-195-2010	中新天津生态城绿色建筑设计标准 Green building design standard for Sino-Singapore Tianjin eco-city		J11548-2010	中新天津生态城管理委员会 中国建筑科学研究院
121	DB 29-196-2010	天津市无障碍设计标准 Design standard on accessibility for Tianjin		J11579-2010	天津市建筑标准设计办公室
122	DB 29-197-2010	自密实混凝土应用技术规程 Technical specification for self-compacting concrete application		J11639-2010	天津市预拌混凝土管理中心 天津市质量监督检验站第二十一站
123	DB/T 29-198-2010	中新天津生态城绿色施工技术管理规程 Technical specification for green construction of for Sino-Singapore Tianjin Eco-city		J11645-2010	中新天津生态城管理委员会 中国建筑科学研究院
124	DB/T 29-199-2010	集中供热住宅计量供热热量表应用技术规程 Specification for application of heat meters for residential buildings in the district heating system		J11654-2010	天津市供热办公室
125	DB 29-200-2010	天津市绿色建筑施工管理技术规程 Green building construction management technical specification in Tianjin		J11669-2010	天津市建工集团(控股)有限公司
126	DB 29-201-2010	天津市人行道及人行广场防滑技术标准 Technical standard of antiskid on sidewalk and public plaza in Tianjin		J11671-2010	天津市市政工程设计研究院

天津市

序号	标准编号	标准名称	被代替标准编号	备案号	主编单位
127	DB 29-202-2010	建筑基坑工程技术规程 Technical specification for retaining and protection of building foundation excavation		J11679-2010	天津市勘察院
128	DB 29-203-2010	建筑工程模板支撑系统安全技术规程 Technical regulation for safety of froms and brace in construction		J11710-2010	天津市建设工程质量安全监督管理总队
129	DB/T 29-204-2015	天津市绿色建筑评价标准 Assessment standard of green building for Tianjin	DB/T 29-204-2010	J11715-2015	天津城建大学 天津市建筑设计院
130	DB 29-205-2015	天津市绿色建筑设计标准 Design standard of green building for Tianjin	DB/T 29-205-2010	J11716-2015	天津市建筑设计院 天津城市建设学院
131	DB/T 29-206-2010	天津市污水源热泵系统应用技术规程 The technical specification for sewage source heat pump systems in Tianjin		J11744-2010	天津大学
132	DB/T 29-207-2010	天津市盐碱地园林树木栽植技术规程 Technical regulation for garden trees planting on saline-alkali land in Tianjin		J11745-2010	天津市园林绿化研究所
133	DB/T 29-208-2011	天津市桥梁结构健康监测系统技术规程 Structural health monitoring system technical specification for bridges of Tianjin		J11768-2011	天津市市政工程研究院
134	DB/T 29-209-2011	建筑工程施工质量验收资料管理规程 Management regulations for lnformation about construction quality acceptance		J11850-2011	天津市建设工程质量安全监督管理总队
135	DB/T 29-210-2012	温拌沥青混合料超薄面层技术规程 Technical specifications for warm mix asphalt ultra-thin surface layer technology		J11962-2012	天津市公路工程总公司
136	DB/T 29-211-2012	城市轨道交通工程密闭式污水提升装置技术标准 Technical standard of enclosed lifting device for sewage in urban rail transit project		J12102-2012	天津市地下铁道集团有限公司
137	DB/T 29-212-2012	天津地铁干式消火栓系统技术规程 Technical specification for dry hydrant system of Tianjin metro		J12101-2012	天津市地下铁道集团有限公司
138	DB 29-213-2012	预应力混凝土空心方桩技术规程 Technical specification for prestressed spun concrete square pile foundation		J12113-2012	天津市勘察院
139	DB 29-214-2012	天津市住宅燃气应用技术规程 Technical specification for application of indoor gas of residential buildings i		J12193-2012	中国市政工程华北设计研究总院 天津市燃气管理处
140	DB/T 29-215-2013	现浇泡沫混凝土应用技术规程 Technical specification for application of cast-in-site foamed concrete		J12249-2013	天津市建筑科学研究院有限公司 天津市建设工程质量监督管理总队
141	DB 29-216-2013	天津市民用建筑能耗监测系统设计标准 Tianjin design standard for energy consumption monitoring systems of civil build		J12262-2013	天津市建筑设计院
142	DB/T 29-217-2013	天津市岩棉外墙外保温系统应用技术规程 Tianjin technical specification for spplication of rock material wool external wall insulation syste		J12271-2013	天津建设工程技术研究所 天津市建材业协会

天津市

序号	标准编号	标准名称	被代替标准编号	备案号	主编单位
143	DB/T 29-218-2013	天津市地下建(构)筑物信息管理技术规程 Technical specification for information management of underground buildings in Tianjin		J12417-2013	天津市地下空间规划管理信息中心 天津大学
144	DB/T 29-219-2013	内河沉管法隧道设计、施工及验收规范 Code for design, construction and acceptance of immersed tunnel in the inland river		J12426-2013	天津滨海新区建设投资集团有限公司 中铁隧道勘测设计院有限公司
145	DB 29-220-2013	天津市建筑绿化应用技术规程 The technical specifications of greening buildings for Tianjin	DB/T 29-118-2004	J12464-2013	天津市建筑设计院 天津市市容和园林管理委员会
146	DB 29-221-2013	建筑幕墙工程技术规范 Technical code for building curtain wall engineering		J12494-2013	天津华惠安信装饰工程有限公司 天津市房屋鉴定勘测设计院
147	DB/T 29-222-2014	天津市建设工程施工安全资料管理规程 Management regulations for construction work documentation about construction safety		J12554-2014	天津市建设工程质量安全监督管理总队
148	DB 29-223-2014	建筑地基氡浓度/氡析出率检测技术规程 Technical specification for radon consistence & radon exhalation rate testing of building foundation		J12557-2014	天津市勘察院
149	DB/T 29-224-2014	建筑物室外消防车道、场地及消防设施标识标准 Fire signs using in fire lanes, sites and facilities of outdoor buildings		J12573-2014	天津市公安局消防局 天津市建筑设计院
150	DB 29-225-2014	天津市再生水管网运行、维护及安全技术规程 Technical order for operation, maintenance and safety of reclaimed water pipe network in Tianjin		J12738-2014	天津中水有限公司 国家城市给水排水工程技术研究中心
151	DB/T 29-226-2014	天津市园林绿化土壤质量标准 The standards of soil quality in landscapes and gardens for Tianjin		J12784-2014	天津市园林绿化研究所
152	DB/T 29-227-2014	天津市泡沫塑料板薄抹灰外墙外保温系统应用技术规程 Tianjin technical specification for application of external thermal insulation composite systems based on plastic foam board		J12828-2014	天津建科建筑节能环境检测有限公司 天津市墙体材料革新和建筑节能管理中心
153	DB/T 29-228-2014	现浇桥梁用碗扣式钢管支撑架施工技术规程 Construction technical specifications of cuplok steel tubular scaffolding for cast-in-place bridge		J12843-2014	天津城建集团有限公司 天津第三市政公路工程有限公司
154	DB/T 29-229-2014	建筑基坑降水技术规程 Technical specification for dewatering engineering of building foundation pit		J12895-2014	天津市津勘岩土工程股份有限公司
155	DB 29-230-2015	天津市建设工程检测试验技术管理规程 Tianjin specification for technology management of building engineering inspection and testing		J12907-2015	天津市建设工程质量检测试验行业协会

天津市

序号	标准编号	标准名称	被代替标准编号	备案号	主编单位
156	DB/T 29-231-2015	城市轨道交通自动售检票系统技术标准 Technical standard for automatic fare collection of urban rail transit	DB/T 29-193-2009	J12940-2015	天津市地下铁道集团有限公司
157	DB 29-232-2015	天津市再生水管道工程技术规程 Technical specification for reclaimed water pipe engineering in Tianjin		J13058-2015	天津市市政工程设计研究院 天津中水有限公司
158	DB 29-233-2015	天津市市政公路桥梁减隔震设计规程 Specification for seismic isolation design of urban bridges and highway bridges in Tianjin		J13057-2015	天津城建设计院有限公司 华中科技大学
159	DB/T 29-234-2015	天津市矿物棉喷涂保温应用技术规程 Tianjin technical specification for application of sprayed mineral wool		J13215-2015	天津市建材业协会 天津市墙体材料革新和建筑节能管理中心
160	DB/T 29-235-2015	天津市再生水厂工程设计、施工及验收规范 Code for engineering design, construction and acceptance of water reclaimation plant in Tianjin		J13282-2016	天津中水有限公司
161	DB/T 29-236-2016	天津市雨水径流量计算标准 Storm water runoff calculation standard of Tianjin		J13398-2016	天津城建设计院有限公司
162	DB/T 29-237-2016	后装拔出法检测混凝土强度技术规程 Technical specification for inspection of concrete strength by post-install pullout test		J13403-2016	天津市建筑材料产品质量监督检测中心 天津市建筑材料科学研究院
163	DB/T 29-238-2016	天津市综合管廊工程技术规范 Technical code for the construction of utility tunnel engineering in Tianjin		J13399-2016	天津市市政工程设计研究院

河北省

序号	标准编号	标准名称	被代替标准编号	备案号	主编单位
1	DB13(J)/T 39-2016	水泥土桩复合地基技术规程 Specification for composite foundation of soil-cement piles		J10301-2016	河北省建筑科学研究院
2	DB13(J) 53-2005	市政基础设施工程施工质量验收统一标准 Unified standard for constructional quality acceptance of municipal infrastructure engineering		J10759-2006	唐山市建设工程质量监督检测站
3	DB13(J) 54-2005	市政基础设施工程施工质量验收通用规程 Universal specification for constructional quality acceptance of municipal infrastructure engineering		J10760-2006	唐山市建设工程质量监督检测站
4	DB13(J) 55-2005	市政道路工程施工质量验收规程 Specification for constructional quality acceptance of municipal road engineering		J10761-2006	唐山市建设工程质量监督检测站
5	DB13(J) 56-2005	市政给水管道工程施工质量验收规程 Specification for constructional quality acceptance of municipal water supplying pipeline engineering		J10762-2006	唐山市建设工程质量监督检测站
6	DB13(J) 57-2005	市政排水管渠工程施工质量验收规程 Specification for constructionalq quality acceptance of municipal drainage pipe ditch engineering		J10763-2006	唐山市建设工程质量监督检测站
7	DB13(J) 58-2005	市政给排水构筑物及设备安装工程施工质量验收规程 Specification for constructional quality acceptance of municipal water supply drainage structure and equipment installation engineering		J10764-2006	唐山市建设工程质量监督检测站
8	DB13(J) 59-2006	市政桥梁工程施工质量验收规程 Specification for constructional quality acceptance of municipal bridge engineering		J10932-2007	唐山市建设工程质量监督检测站
9	DB13(J) 60-2006	市政供热管道及设备安装工程施工质量验收规程 Specification for constructional quality acceptance of municipal heating pipeline and facilities installation engineering		J10930-2007	唐山市建设工程质量监督检测站
10	DB13(J) 61-2006	城镇燃气管道及设备安装工程施工质量验收规程 Specification for constructional quality acceptance of city gas pipeline and installation equipment I engineering		J10929-2007	唐山市建设工程质量监督检测站
11	DB13(J) 62-2006	园林工程施工质量验收规程 Specification for constructional quality acceptance of landscape architecture engineering		J10927-2007	石家庄市园林绿化工程质量监督站
12	DB13(J) 63-2011	居住建筑节能设计标准 Design standard for energy efficiency of residential buildings	DB13(J) 63-2007	J10928-2011	河北北方绿野建筑设计有限公司

河北省

序号	标准编号	标准名称	被代替标准编号	备案号	主编单位
13	DB13(J) 67-2007	无障碍设施工程施工质量验收规程 Specification for acceptance of construction quality of barrier-free facilities		J10979-2007	河北建筑设计研究院有限责任公司
14	DB13(J) 68-2007	建设工程安全监理规程 Specification of construction project safety supervision		J11032-2007	河北省建设工程安全生产监督管理办公室
15	DB13(J) 69-2007	混凝土结构无机锚固材料植筋技术规程 Technical specification for construction and acceptance of reinforced bars with inorganic anchoring material inserted in concrete		J11115-2007	河北省建筑科学研究院
16	DB13(J) 70-2007	刚性芯夯实水泥土桩复合地基技术规程 Technical specification for composite foundation of rammed soil-cement piles with rigid core		J11116-2007	河北省建筑科学研究院
17	DB13(J) 71-2007	静载试验组合拉锚内支撑技术规程 Technical specification for static loading of internal bracing combined tie-bar and anchors		J11117-2007	河北省建筑科学研究院
18	DB13(J) 72-2008	建筑同层排水技术规程 Technical specification for same-floor drainage of buliding		J11173-2008	河北建筑设计研究院责任责任公司
19	DB13(J) 73-2008	建筑起重机械报废规程 Specification for construction crane to be scrapped		J11260-2008	河北省建筑业协会材料设备管理分会
20	DB13(J)/74-2008	既有居住建筑节能改造技术标准 Technical standard for energy efficiency renovation of existing residential buildings		J11306-2008	河北北方绿野居住环境发展有限公司
21	DB13(J)/T 75-2008	混凝土结构粘钢加固技术规程 Technical specification for strengthening concrete structures		J11304-2008	河北省建筑科学研究院
22	DB13(J)/T 76-2008	再生骨料混凝土技术规程 Technical specification of recycled aggregate concrete		J11305-2008	河北省建筑科学研究院
23	DB13(J) 77-2009	民用建筑太阳能热水系统一体化技术规程 Technical specification for integrated solar water heating system of civil building		J11369-2009	河北建筑设计研究院有限公司
24	DB13(J)/T 78-2009	城市景观照明技术规范 Technicai code for urban landscape lighting		J11403-2009	河北省照明行业协会
25	DB13(J)/T 79-2009	城市道路照明设施管理维护技术规程 Technical specification for management and maintenance of urban road lighting facilities		J11404-2009	河北省照明行业协会
26	DB13(J)/T 80-2009	建筑物分类与代码标准 Standard for classification and codes of buildings		J11405-2009	石家庄市城乡规划局
27	DB13(J)/T 83-2009	约束混凝土、混凝土和型钢混合结构技术规程 Technical specification for confined concrete structures and hybrid structures of concrete and steel		J11440-2009	河北合创建筑节能科技有限公司

河北省

序号	标准编号	标准名称	被代替标准编号	备案号	主编单位
28	DB13(J)/T 84-2009	钢管混凝土结构技术规程 Technical specification for concrete-filled		J11453-2009	河北大地土木工程有限公司
29	DB13(J)/T 86-2009	回弹法检测预拌混凝土抗压强度技术规程 Technical specification for inspection of ready-mixed concrete compressive strength by rebound method		J11473-2009	石家庄市建筑协会混凝土专业分会
30	DB13(J)/T 87-2009～DB13(J)/T 99-2009	城镇公共厕所管理服务标准、城镇道路清扫保洁服务标准、城镇生活垃圾收集运输服务标准、城建监察管理服务标准、风景名胜区管理服务标准、城镇公园与广场管理服务标准、城镇绿地养护管理服务标准、城镇排水设施养护管理服务标准、城镇道路桥梁设施养护管理服务标准、城镇照明设施管理服务标准、城镇供水服务标准、供热行业服务标准、燃气行业服务标准		J11475-2009	河北省市容环境卫生协会等
31	DB13(J)/T 87-2009	城镇公共厕所管理服务标准 Management and service standard for urban toilets		J11475-2009	河北省市容环境卫生协会
32	DB13(J)/T 88-2009	城镇道路清扫保洁服务标准 Service standard for urban road sweeping and cleaning		J11476-2009	河北省市容环境卫生协会
33	DB13(J)/T 89-2009	城镇生活垃圾收集运输服务标准 Service standard for municipal solid waste collection and transportation		J11477-2009	河北省市容环境卫生协会
34	DB13(J)/T 90-2009	城建监察管理服务标准 Management and service standard for city construction supervision		J11478-2009	河北省市容环境卫生协会
35	DB13(J)/T 91-2009	风景名胜区管理服务标准 Management and service standard for scenic and historic areas		J11479-2009	河北省风景园林协会
36	DB13(J)/T 92-2009	城镇公园与广场管理服务标准 Management and service standard for urban park and plaza		J11480-2009	河北省风景园林协会
37	DB13(J)/T 93-2009	城镇绿地养护管理服务标准 Management and service standard for maintenance of urban green space		J11481-2009	河北省风景园林协会
38	DB13(J)/T 94-2009	城镇排水设施养护管理服务标准 Management and service standard for maintenance of urban drainage facilities		J11482-2009	河北省市政工程协会
39	DB13(J)/T 95-2009	城镇道路桥梁养护管理服务标准 Management and service standard for maintenance of urban road and bridge		J11483-2009	河北省市政工程协会
40	DB13(J)/T 96-2009	城镇照明设施管理服务标准 Management and service standard for urban lighting facilities		J11484-2009	河北省照明行业协会
41	DB13(J)/T 97-2009	城镇供水服务标准 Service standard for municpal water supply		J11485-2009	河北省城镇供水协会
42	DB13(J)/T 98-2009	供热行业服务标准 Service standard for heating industry		J11486-2009	河北省燃气供热管理办公室

河北省

序号	标准编号	标准名称	被代替标准编号	备案号	主编单位
43	DB13(J)/T 99-2009	燃气行业服务标准 Service standard for gas industry		J11487-2009	河北省燃气供热管理办公室
44	DB13(J)/T 100-2009	建设工程安全文明工地标准 Standard for safety civilized site of construction project		J11488-2009	河北省建设工程安全生产监督管理办公室
45	DB13(J)/T 101-2009	建筑施工安全技术资料管理标准 Management standard for construction safety documentation		J11489-2009	河北省建设工程安全生产监督管理办公室
46	DB13(J)/T 103-2009	碳纤维发热线供暖技术规程 Technical specification for heating by carbon fiber heating wire		J11510-2009	河北省建设机械协会暖通空调委员会
47	DB13(J)/T 104-2010	轻骨料混凝土槽型保温砌块砌体技术规程 Technical specification for grooved insulating block masonry of light weight aggregate concrete		J11550-2010	河北省建设机械协会材料构件委员会
48	DB13(J)/T 105-2010	预应力混凝土管桩基础技术规程 Technical specification for prestressed concrete pipe pile foundation		J11551-2010	河北省建筑科学研究院
49	DB13(J)/T 107-2010	热泵系统工程技术规程 Technical specification for heat pump system		J11632-2010	河北冀发地源热泵研究所
50	DB13(J)/T 108-2010	建设工程项目管理规程 Specification of construction project management		J11643-2010	河北中原工程项目管理有限公司
51	DB13(J)/T 110-2010	纤维复合材加固砌体结构技术规程 Technical specification for reinforced masonry structure by fiber composites		J11673-2010	河北省建筑科学研究院
52	DB13(J)/T 111-2010	人民防空工程兼作地震应急避难场所技术标准 Technical standard for using civil air defence works as earthquake emergency shelter		J11709-2010	河北省人民防空办公室 河北省地震局
53	DB13(J)/T 112-2010	混凝土粉煤灰-矿渣粉掺合料应用技术规程 Technical specification for application of fly ash and slag powder in concrete		J11735-2010	石家庄市建设工程质量监督管理站 石家庄建工商品混凝土股份有限公司
54	DB13(J)/T 113-2015	绿色建筑评价标准 Evaluation standard for green building	DB13(J)/T 113-2010	J11753-2015	河北省建筑科学研究院
55	DB13(J) 114-2013	建筑节能门窗工程技术规范 Technical code for energy efficiency door and window engineering	DB13(J) 114-2010	J11741-2013	河北省建筑科学研究院
56	DB13(J)/T 115-2011	柱锤冲扩水泥粒料桩复合地基技术规程 Technical specification for column-hammer compacted cement-aggregate pile composite foundation		J11795-2011	河北工业大学
57	DB13(J)/T 116-2010	外墙外保温技术规程(胶粉聚苯颗粒复合型保温系统) Technical specification for external thermal insulation on walls (composited thermal insulation systems made of rendering with mineral binder and using expanded polystyrene granule as aggregate)		J11796-2011	北京振利节能环保科技股份有限公司 中国建筑科学研究院建筑防火研究所

河北省

序号	标准编号	标准名称	被代替标准编号	备案号	主编单位
58	DB13(J)/T 117-2011	复合墙结构技术规程 Technical specification for compound shear wall structure		J11800-2011	河北合创建筑节能科技有限公司
59	DB13(J)/T 118-2011	建筑物整体移位技术规程 Technical specification for integral moving engineering of buildings		J11827-2011	河北省建筑科学研究院
60	DB13(J)/T 119-2011	高强混凝土强度检测技术规程 Technical specification for strength testing of high strength concrete by NDT		J11828-2011	河北省建筑科学研究院
61	DB13(J)/T 120-2011	施工升降机齿轮锥鼓形渐进式防坠安全器检验技术规程 Test specification for building hoist-pinion and cone progressive type safety device		J11829-2011	河北省建筑科学研究院
62	DB13(J) 121-2011	既有公共建筑节能改造技术标准 Technical standard for energy efficiency retrofitting of existing public buildings		J11870-2011	河北北方绿野居住环境发展有限公司
63	DB13(J)/T 122-2011	复合保温钢筋焊接网架混凝土剪力墙（CL建筑体系）技术规程 Technical specification for concrete shearwall with welded steel frame and composite thermal insulation (CL building system)	DB13(J) 43-2006 DB13(J) 44-2003	J11862-2011	石家庄晶达建筑体系有限公司 北方工程设计研究院有限公司
64	DB13(J)/T 123-2011	长螺旋钻孔泵压混凝土桩复合地基技术规程 Technical specification for composite foundation of long screw drilling cast-in-place pile	DB13(J) 31-2001	J11869-2011	河北大地建设科技有限公司 河北省电力勘测设计研究院
65	DB13(J)/T 127-2011	城市园林绿化评价标准 Evaluation standard for urban landscaping and greening		J11910-2011	河北省风景园林与自然遗产管理中心
66	DB13(J) 128-2011	供热计量技术规程 Technical specification for heat metering of district heating system		J11907-2011	河北建筑设计研究院有限责任公司 河北工业大学 河北大地建设科技有限公司
67	DB13(J)/T 129-2011	外墙保温工程施工防火安全管理规程		J11922-2011	河北省建设工程安全生产监督管理办公室
68	DB13(J) 130-2011	工业场所热气溶胶灭火系统设计、施工及验收规程 Specification of distribution design, installation and acceptance for condensed aerosol automatic fire extinguishing system in industrial site		J11942-2011	河北省公安厅消防局
69	DB13(J)/T 131-2011	林荫停车场绿化标准 Greening standard for shaded parking lot		J11966-2012	河北省风景园林与自然遗产管理中心
70	DB13(J)/T 132-2012	绿色建筑技术标准 Technical sstandard for green building		J12007-2012	河北省建筑科学研究院
71	DB13(J) 133-2012	建筑基坑工程技术规程 Technical specification for building excavation engineering		J11978-2012	河北省建筑科学研究院 中国兵器工业北方勘察设计研究院
72	DB13(J)/T 134-2012	沿海地区造地土地整理工程施工质量检测技术标准		J12038-2012	河北省建筑科学研究院

河北省

序号	标准编号	标准名称	被代替标准编号	备案号	主编单位
73	DB13（J）/T 135-2012	工程建设交易中心建设标准 Establish standard for engineering construction contracting center		J12062-2012	河北省建设工程交易管理中心
74	DB13（J）/T 136-2012	基桩自平衡载试验法检测技术规程 Technical specification for static loading test of self-balanced method of found		J12080-2012	河北大地建设科技有限公司 南京东大自平衡桩基检测有限公司
75	DB13（J）/T137/T138/T 139-2012	旋挖钻机施工技术规程等三项标准 Technical specification for construction of hydraulic rotary rig		J12093-2012	河北建设勘察研究院有限公司
76	DB13（J）/T 140-2012	建设工程施工现场安全管理标准 Safety management standards for construction engineering site		J12122-2012	石家庄市建设工程安全生产监督管理
77	DB13（J）/T 141-2012	柱锤夯实扩底灌注桩技术规程 Technical specification for pile with bearing base rammed		J12123-2012	河北工业大学
78	DB13（J）/T 142-2012	液化石油气一体化设备小型充装站 Construction management code for small filling station of integration liquefied		J12174-2012	保定市住房和城乡建设局
79	DB13（J）/T 143-2012	储罐式氮气灭火系统技术规程 Technical specification for container type nitrogen		J12167-2012	河北省公安厅消防局
80	DB13（J）/T 144-2012	大跨度空间钢结构滑移法施工技术规程 Technical specification for sliding construction of large span spatial steel stucture		J12186-2012	河北建设集团有限公司
81	DB13（J）/T 145-2012	建筑工程资料管理规程 Specification for building engineering document management		J12195-2012	唐山市建设工程质量监督检查站
82	DB13(J) 146-2012	钢网格结构螺栓球节点技术标准 Technical standard for bolted spherical joint of space grid structures		J12189-2012	河北建设集团有限公司
83	DB13（J）/T 147-2012	石家庄市轨道交通工程资料管理规程（土建篇） Specification for rail transportation engineering document management of Shijiazhuang		J12206-2012	石家庄市轨道交通有限责任公司
84	DB13(J)/148-2012	建筑地基基础检测技术规程 Technical specification for testing of building foundation		J12223-2012	河北省电力勘测设计研究院
85	DB13(J) 149-2013	成品住宅装修技术标准 Technical standard for finished residential		J12391-2013	石家庄市建设工程质量监督管理站
86	DB13（J）/T 151-2013	碳纤维布加固低强度混凝土结构技术规程 Technical specification for reinforced low strength concrete structure by carbon fiber sheet		J12433-2013	河北省建筑科学研究院
87	DB13（J）/T 152-2013	岩土工程勘察地层描述技术规程 Technical specification for stratum description of geotechnical investigation		J12463-2013	中国兵器工业北方勘察设计研究院有限公司
88	DB13(J) 153-2013	碳纤维电热地面辐射供暖技术规程 Technical specification for floor radiant heating by carbon fiber electric heating		J11502-2013	河北省建筑科学研究院

河北省

序号	标准编号	标准名称	被代替标准编号	备案号	主编单位
89	DB13（J）/T 154-2013	绿色施工管理规范 Management code for green construction		J12556-2014	河北省建筑科学研究院
90	DB13（J）/T 155-2014	建筑垃圾再生集料路面基层施工技术规程 Technical specification for base pavement construction of recycled aggregate of construction waste		J12565-2014	沧州市市政工程公司
91	DB13（J）/T 156-2014	FS外模板现浇混凝土复合保温系统应用技术规程 Application technical specification for cast-in-place concrete composite thermal insulation system of FS external formwork		J12614-2014	河北省建设机械协会 山东春天建材科技有限公司
92	DB13（J）/T 157-2014	光电缆工程设计、施工及验收规范 Code for design, construction and acceptance of optical power cable engineering		J12629-2014	河北建筑设计研究院有限责任公司
93	DB13（J） 158-2014	高层民用建筑太阳能热水系统应用技术规程 Technical specification for application of solar water heating system of tall civil buildings		J12619-2014	河北九易庄宸工程设计有限公司
94	DB13（J）/T 159-2014	蒸压加气混凝土砌块砌体结构技术规程 Seismic technical of autoclaved aerated concrete blocks masonry structures		J12656-2014	石家庄市建筑设计院
95	DB13（J）/T 160-2014	石家庄市区工程地质地层层序划分标准 Standard for division of geotechnical stratigraphic sequence in Shijiazhuang city		J12692-2014	河北建设勘察研究院有限公司
96	DB13（J）/T 161-2014	建设工程监理工作标准 Working standard of construction project management		J12728-2014	河北工程建设监理有限公司
97	DB14（J）/T 162-2014	HVIP真空绝热板建筑保温系统应用技术规程 Technical specification for application of HVIP vacuum insulation panel external thermal insulation system on building		J12729-2014	河北九易庄宸工程设计有限公司
98	DB13（J）/T 163-2014	保障性住房技术标准 Technical standard for indemnificatory housing		J12770-2014	河北省建筑科学研究院
99	DB13（J）/T 164-2014	铁尾矿骨料混凝土应用技术规程 Technical specification for application of iron tailings aggregate concrete		J12797-2014	承德市住房和城乡建设局
100	DB13（J）/T 165-2014	建筑基坑支护施工图文件编制深度及制图标准 Compile depth and drawing standard for construction document of building foundation excavations retaining and protection		J12800-2014	河北省建筑科学研究院
101	DB13（J）/T 166-2014	倒锥台阶型桩复合地基设计规程 Specification for design of inverted cone-step-pile composite foundation		J12826-2014	河北省建筑科学研究院
102	DB13（J）/T 167-2014	聚苯暖砖现浇混凝土填芯墙体技术规程 Building technical specification for expansive polystyrene granule cast-in-place concrete fill core wall		J12825-2014	石家庄市万成民用建筑设计有限公司

河北省

序号	标准编号	标准名称	被代替标准编号	备案号	主编单位
103	DB13(J)/T 168-2014	空气源热泵直热式辐射供暖技术规程 Technical specification for air-source heat pump direct thermal radiant heating		J12849-2014	河北建筑设计研究院有限责任公司
104	DB13(J)/T 169-2014	建设工程见证取检测管理规程 Testing management specification for witness sampling test of construction engineering		J12868-2014	河北省建筑科学研究院
105	DB13(J)/T 170-2014	建设工程建筑节能检测管理规程 Testing management specification for energy conservation test of construction engineering		J12869-2014	河北省建筑科学研究院
106	DB13(J)/T 171-2014	建设工程使用功能检测管理规程 Testing management specification for service function test of construction engineering		J12870-2014	河北省建筑科学研究院
107	DB13(J)/T 172-2014	双面彩钢板复合风管技术规程		J12866-2014	河北空调工程安装有限公司
108	DB13(J)/T 173-2014	建设工程施工现场质量管理规范 code for quality management of construction site		J12867-2014	河北省工程建设质量监督检测管理总站
109	DB13(J)/T 174-2014	农村居住建筑节能技术标准 Technical standard for energy efficiency of rural residential buildings		J12865-2014	河北北方绿野居住环境发展有限公司
110	DB13(J) 175-2015	雨水控制与利用工程技术规范 Code for design of stormwater management and harvest engineering of Hebei province		J12994-2015	河北农业大学 河北建筑设计研究院有限责任公司
111	DB13(J)/T 176-2015	市政工程资料管理规程 Specification for municipal engineering document management		J13050-2015	唐山市建设工程质量监督检测站
112	DB13(J)/T 177-2015	被动式低能耗居住建筑节能设计标准 Design standard for energy efficiency of passive low-energy residential buildings		J12951-2015	住房和城乡建设部科技发展促进中心
113	DB13(J)/T 178-2015	大树移植技术规程 Technical regulations of big tree transplanting		J12976-2015	河北省风景园林学会
114	DB13(J)/T 179-2015	装配整体式混凝土剪力墙结构设计规程 Specification for design of monolithic precast concrete shear wall structures		J13025-2015	河北建筑设计研究院有限责任公司
115	DB13(J)/T 180-2015	装配式混凝土剪力墙结构建筑与设备设计规程 Technical specification for building and equipment of monolithic precast concrete shear wall structure		J13026-2015	河北建筑设计研究院有限责任公司
116	DB13(J)/T 181-2015	装配式混凝土构件制作与验收标准 Technical specification for manufacture and acceptance of prefabricated concrete components		J13027-2015	河北工程建设监理有限公司
117	DB13(J)/T 182-2015	装配式混凝土剪力墙结构施工及质量验收规程 Specification for construction and quality acceptance of precast concrete structures		J13028-2015	中国二十二冶集团有限公司

河北省

序号	标准编号	标准名称	被代替标准编号	备案号	主编单位
118	DB13(J)/T 183-2015	城市地下综合管廊建设技术规程 Technical specification for construction of municipal tunnel		J13017-2015	河北省建筑科学研究院
119	DB13(J)/T 184-2015	装配整体式混合框架结构技术规程 Technical specification for precast monolithic hybrid frame structures		J13022-2015	河北合创建筑节能科技有限公司
120	DB13(J) 185-2015	居住建筑节能设计标准（节能75％） Design standard for energy efficiency of residential buildings		J12980-2015	河北北方绿野居住环境发展有限公司
121	DB13(J)/T 186-2015	城市管理行政执法装备配备标准 Technical specification for comprehensive administrative law enforcement equipment of urban management		J13036-2015	河北省市容环境卫生协会
122	DB13(J)/T 187-2015	YL无机复合保温板现浇混凝土保温系统应用技术规程 Technical specifications for the application of YL inorganic composite insulation panel cast-in-situ cement insulation system		J13039-2015	石家庄市建筑设计
123	DB13(J)/T 188-2015	HRB500钢筋应用技术规程 Technical specification for application of HRB500 steel bars		J13070-2015	河北建筑设计研究院有限责任公司
124	DB13(J)/T 189-2015	既有建筑地基基础检测技术规程 Technical specification for testing of existing building foundation		J13071-2015	河北省建筑科学研究院
125	DB13(J)/T 190-2015	EPS模块低层现浇混凝土复合墙技术规程 Technical specification for cast-in-place concrete composite wall of low-rise budling with EPS module		J13117-2015	北方工程设计研究院有限公司
126	DB13(J)/T 191-2015	EPS模块现浇混凝土剪力墙保温系统技术规程 Technical specification for thermal insulation system of cast-in-situ concrete shear wall with EPS module		J13118-2015	北方工程设计研究院有限公司
127	DB13(J)/T 192-2015	EPS模块工业建筑围护结构技术规程 Technical specification for enclosure structure of industrial building with EPS module		J13119-2015	北方工程设计研究院有限公司
128	DB13(J)/T 193-2015	酚醛泡沫板外墙外保温应用技术规程 Application technical specifications for phenolic foam panel external thermal insulation on walls		J13120-2015	石家庄市建筑设计院
129	DB13(J)/T 194-2015	LS复合墙体自保温系统应用技术规程 Application technical specification for LS composite wall thermal insulation system		J13175-2015	北方绿野居住环境发展有限公司
130	DB13(J)/T 195-2015	石家庄市轨道交通工程资料管理规程（机电设备篇） Specification for rail transportation engineering document management of Shijiazhuang (part of electromechanical facilities)		J13205-2015	石家庄市轨道交通有限责任公司

河北省

序号	标准编号	标准名称	被代替标准编号	备案号	主编单位
131	DB13(J)/T 196-2015	CL 建筑体系技术规程 Technical specification for CL building system		J11862-2015	河北天艺建筑设计有限公司
132	DB13(J)/T 197-2015	农村住宅建筑抗震设计规程 Seismic design specification for rural residential buildings		J13245-2015	河北建筑设计研究院有限责任公司
133	DB13(J)/T 198-2015	建筑防水工程技术规范 Technical code for waterproof engineering of construction		J13249-2015	河北省建筑防水协会
134	DB13(J)/T 199-2015	金属网岩棉板现浇混凝土复合保温系统应用技术规程 Applied technical specifications of the wire rock wool for external thermal insulation on outer-wall in cast-in-sitn system		J13277-2015	中铁建安工程设计院有限公司
135	DB13(J)/T 200-2015	建筑工程绿色施工示范工程技术标准 Construction engineering green construction technical specification		J13271-2015	河北省建设工程安全生产监督管理办公室
136	DB13(J)/T 201-2015	能量回弹法（Q 值）检测混凝土强度技术规程 Technical specification for inspecting of concrete strength by rebound method in energy (Q value)		J13292-2016	河北大地建设工程检测有限公司
137	DB13(J)/T 202-2016	公共建筑能耗远程监测系统技术标准 Technical standard for the remote monitoring system of public building energy consumption		J13390-2016	河北省建筑科学研究院
138	DB13(J)/T 203-2016	预应力混凝土竹节桩技术规程 Technical specification for pre-stressed concrete knot pile		J13332-2016	河北省建筑科学研究院
139	DB13(J)/T 204-2016	PCB 复合保温板保温系统应用技术规程 Technical specification for application of PCB composite insulation board exterior insulation system		J13389-2016	河北建筑设计研究院有限责任公司
140	DB13(J)/T 205-2016	CIS 复合保温板现浇混凝土保温系统应用技术规程 Technical specification for application of CIS composite insulation board for concrete insulation system		J13395-2016	河北省建筑科学研究院 河北建设集团有限公司
141	DB13(J)/T 206-2016	SK 建筑复合保温板现浇混凝土保温系统应用技术规程 Technical specifications for the application of SK architecture composite insulation panel cast-in-situ concrete insulation system		J13396-2016	北方工程设计研究院有限公司
142	DB13(J)/T 207-2016	CRB600H 高强钢筋应用技术规程 Technical specification for CRB600H high strength steel bar		J13425-2016	河北建筑设计研究院有限责任公司
143	DB13(J)/T 208-2016	古树名木养护管理与复壮技术规程 Technical specification for maintenance management and rejuvenation of ancient and famous trees		J13441-2016	河北省风景园林与自然遗产管理中心

河北省

序号	标准编号	标准名称	被代替标准编号	备案号	主编单位
144	DB13(J)/T 209-2016	OKS复合保温模板系统应用技术规程（OKS温钢模板系统） Technical specification for application of OKS composite thermal insulation formwork system (OKS WenGang formwork system)		J13426-2016	河北建筑设计研究院有限责任公司 奥克森（北京）新材料科技有限公司
145	DB13(J)/T 210-2016	海绵城市建设工程技术规程 Engineering technical specification for the sponge city		J13461-2016	河北农业大学
146	DB13(J)/T 211-2016	海绵城市设施施工及工程质量验收规范 Code for construction and acceptance of project quality of sponge city facility		J13462-2016	河北农业大学 河北建设集团有限公司

山西省

序号	标准编号	标准名称	被代替标准编号	备案号	主编单位
1	DBJ04/T 214-2015	建筑工程施工资料管理规程 Management specification of construction documents of building engineering		J10440-2015	山西省建设工程质量监督管理总站
2	DBJ04/T 226-2015	建筑工程施工质量验收规程 Specification for constructional quality acceptance of building engineering		J10262-2015	山西省建设工程质量监督管理总站
3	DBJ04/T 230-2011	CL 建筑体系技术规程 Technical specification for building system of CL		J10586-2011	山西省建筑设计研究院 石家庄晶达建筑体系有限公司
4	DBJ04/T 235-2014	建筑屋面绝热制品(材料)应用技术规程 Technical specification for application of thermal insulation products for building roofs		J12703-2014	山西省建筑科学研究院
5	DBJ04/T 236-2014	蒸压加气混凝土砌块建筑技术规程 Technical specification for application of autoclaced aerated concrete blocks		J12702-2014	山西省建筑科学研究院
6	DBJ04/T 237-2014	钢丝网架复合夹芯板墙体技术规程 Technical specification for the sandwich board with wire mesh as wallboard		J12701-2014	山西省建筑科学研究院
7	DBJ 04-239-2005	居住建筑节能技术规程 Technical regulations of energy-efficiency for residential building		J10711-2006	山西建筑工程(集团)总公司
8	DBJ04/T 240-2014	无机保温砂浆系统技术规程 Inorganic insulation mortar and system technology procedures		J10710-2014	山西省建筑科学研究院
9	DBJ 04-241-2013	公共建筑节能设计标准 Design standard for energy efficiency of public buildings		J10707-2013	太原市建筑设计研究院
10	DBJ 04-242-2011	居住建筑节能设计标准 Design standard for energy efficiency of residential buildings		J10708-2012	山西省建筑设计研究院 太原理工大学建筑设计研究院
11	DBJ 04-243-2013	既有采暖居住建筑节能改造设计标准 Design standard for energy conservation renovation of existing heating residential buildings		J10709-2013	太原理工大学建筑设计研究院
12	DBJ04/T 245-2014	居住建筑节能检测标准 Energy efficiency inspection standard for residential buildings		J10712-2014	山西省建筑科学研究院
13	DBJ04/T 246-2014	建筑节能门窗应用技术规程 Technical specification for application of high performance concrete		J10713-2014	山西省建筑科学研究院
14	DBJ04/T 248-2014	建筑工程勘察文件编制标准 Standard for geotechnical investigation report in the building fields		J10935-2014	山西勘察设计协会 山西省勘察设计研究院
15	DBJ 04-249-2007	建筑防水工程技术规程 Technical regulations for waterproof of building works		J10976-2007	山西建筑工程(集团)总公司

山西省

序号	标准编号	标准名称	被代替标准编号	备案号	主编单位
16	DBJ 04-250-2007	玻化微珠保温砂浆应用技术规程 Technical specification in application of thermal insulation glazed hollow bead mortar		J10980-2007	太原思科达科技发展有限公司
17	DBJ 04-252-2007	住宅信报箱建设标准 Constructional standard of residential mail-box		J11038-2007	山西省邮政管理局
18	DBJ 04-253-2007	建筑工程施工安全管理标准 Standard of construction safety management of building engineering		J11060-2007	山西省建筑安全监督站
19	DBJ 04-256-2007	城市绿化工程施工及验收规程 Code for construction and acceptance of plant engineering in city and town		J11145-2008	太原市园林绿化工程质量监督站
20	DBJ04/T 258-2016	建筑地基基础勘察设计规范 Code for investigation and design of building foundation		J11182-2016	太原理工大学 山西省建筑科学研究院
21	DBJ 04-259-2008	人工砂生产应用技术规程 Practice code for application of artificial sand		J11215-2008	山西四建集团有限公司
22	DBJ04/T 262-2016	城市道路绿化养护管理标准 Standard for management of greenspace along streets in city		J11274-2016	太原市园林植物研究中心
23	DBJ04/T 263-2016	城市绿化常用苗木标准 Standard for main seedlings of urban greening		J11275-2016	太原市园林植物研究中心
24	DBJ04/T 264-2016	大树移植技术规程 Technical guideline for big trees transplanting		J11276-2016	太原市园林植物研究中心
25	DBJ04/T 265-2016	古树名木保护技术规程 Technical guideline for protection of ancient and famous trees		J11277-2016	太原市园林植物研究中心
26	DBJ 04-266-2008	城镇污水处理厂运行、维护及其安全技术规程 Techinical specification for operation maintenance and safety of municipal wastewater treatment plant		J11278-2008	太原市排水管理处
27	DBJ 04-268-2008	生活垃圾转运站运行维护技术规程 Technical specification for operation and maintenance		J11280-2008	太原市环境卫生科学研究所
28	DBJ 04-269-2008	城市桥梁养护施工质量验收规程 Rules for quality acceptance of bridge maintenance works in city		J11289-2008	太原市市政工程管理处
29	DBJ 04-270-2008	城市园林绿化工程质量验收规程 Standard for quality acceptance of garden engineering in city and town		J11346-2009	太原市园林绿化工程质量监督站
30	DBJ 04-271-2008	城镇道路工程施工技术规范 Construction technical specifications of urban road projects		J11347-2009	太原市市政工程总公司
31	DBJ04/T 273-2012	预应力混凝土管桩建筑桩基技术规程 Specification for prestressed concrete pile foundation		J11381-2012	山西建筑工程(集团)总公司 山西六建集团有限公司

山西省

序号	标准编号	标准名称	被代替标准编号	备案号	主编单位
32	DBJ 04-274-2009	玻化微珠整体式保温隔热建筑应用技术规程 Technical specification in application of integrated thermal insulation buildings made of glazed hollow beads mortar		J11382-2009	太原思科达科技发展有限公司
33	DBJ 04-276-2009	建筑能效测评与标识标准 Standard for energy evaluation and labelling of buildings		J11384-2009	山西省建设科技推广与建筑节能监督管理中心
34	DBJ 04-277-2009	应急避难场所建设标准 Construction standard of emergency shelter		J11385-2009	山西省工程建设标准化协会
35	DBJ 04-279-2009	建筑工程质量检测管理规程 Management regulations of the quality inspections in architectural projects		J11443-2009	山西省建设工程质量监督管理总站
36	DBJ04/T 280-2009	蒸压粉煤灰砖建筑技术规程 Architectural technology regulation for autoclave fly ash brick		J11451-2009	山西省建筑科学研究院 山西省墙改办
37	DBJ04/T 281-2009	烧结煤矸石多孔砖建筑技术规程 Building technical specification of fired coal gangue perforated brick		J11474-2009	山西省建筑科学研究院
38	DBJ04/T 282-2010	行道树栽植技术规程 Code for planting technique of avenue tree		J11567-2010	太原市园林植物研究中心 太原市工程建设标准定额站
39	DBJ04/T 283-2010	城市园林苗圃育苗技术规程 Code for raising seedlings in garden nursery of city		J11568-2010	太原市园林植物研究中心 太原市工程建设标准定额站
40	DBJ04/T 284-2010	建筑幕墙工程质量验收标准 Standard for construction quality acceptance of building curtain wall		J11691-2010	山西省建筑装饰协会 山西国立铝门窗幕墙工程有限公司
41	DBJ04/T 285-2010	预拌砂浆生产与应用技术规程 Technical specification for manufacture and application of ready-mixed mortar		J11726-2010	太原理工大学
42	DBJ04/T 286-2011	液化天然气汽车加气站设计与施工规程 Design and construction regulations of vehicle LNG filling stations		J11801-2011	山西中油新捷天然气有限公司 赛鼎工程有限公司
43	DBJ04/T 287-2011	花坛、花境建植技术规程 Code for planting technique of parterre and flower border		J11933-2011	太原市工程建设标准定额站
44	DBJ04/T 288-2011	地被植物建植技术规程 Code for planting technique of groundcover plants		J11932-2011	太原市工程建设标准定额站
45	DBJ04/T 289-2011	建筑工程施工安全资料管理规程		J11941-2011	山西省建设工程安全监督管理总站
46	DBJ04/T 290-2012	袖阀管注浆加固地基技术规程 Technical code for sleeve-valve pipe grouting reinforced foundation		J12025-2012	山西四建集团有限公司 山西博奥建筑纠偏加固工程有限公司
47	DBJ04/T 291-2012	自限温加热电缆蓄能地面辐射供暖技术规程		J12120-2012	山西耀华电力节能供热有限公司
48	DBJ04/T 292-2012	住宅物业服务标准 The service standard of residential property		J12180-2012	山西省房地产业协会
49	DBJ04/T 293-2012	煤粉工业锅炉系统技术规程 Pulverized coal fired industrial boiler system technical specification		J12179-2012	山西蓝天环保设备有限公司 太原市热力设计院

山西省

序号	标准编号	标准名称	被代替标准编号	备案号	主编单位
50	DBJ04/T 294-2012	块砌式和装配式混凝土检查井技术规程 Technical specification of block inspection chamber and precast concrete inspect		J12181-2012	太原市市政工程总公司
51	DBJ 04-295-2012	消防设备电源监控系统技术规程 Technical specification of monitor and control system for fire equipment power		J12166-2012	山西省公安消防总队
52	DBJ04/T 296-2012	对焊封闭环式箍筋技术规程 Technical specification for butt-welded closed stirrup		J12205-2012	临汾市工程建设标准定额管理站 临汾市尧都区鈜兴建材厂
53	DBJ04/T 297-2013	泡沫陶瓷板外墙外保温工程技术规程		J12287-2013	山西省建筑科学研究院 山西安晟科技发展有限公司
54	DBJ04/T 298-2013	写字楼物业服务标准 The property service standard of office buildings		J12437-2013	山西省房地产业协会
55	DBJ04/T 299-2013	发泡水泥保温板外墙外保温工程技术规程 Technical specification for application of foam concrete board external thermal insulation systems		J12436-2013	潞城市泓钰节能建材有限公司
56	DBJ04/T 300-2013	二次供水系统技术规程 Regulations of technology of secondary water supply system		J12452-2013	太原市节约用水管理中心 北京威派格科技发展有限公司
57	DBJ04/T 301-2013	蒸压粉煤灰加气混凝土砌块墙体自保温技术规程 Specification for wall self-insulation of autoclaved fly-ash aerated concrete block		J12497-2013	太原市新型墙体材料管理中心
58	DBJ04/T 302-2014	岩棉保温装饰一体板外墙外保温系统技术规程		J12600-2014	山西峰岩新型建材股份有限公司
59	DBJ04/T 303-2014	高性能混凝土应用技术规程 Technical specification for application of high performance concrete		J12599-2014	山西省建筑科学研究院
60	DBJ04/T 304-2014	中空内模金属(轻钢肋筋)网水泥内隔墙技术规程 Hollow inner mold metal (light steel yibs) net technical regulations within the wall of cement		J12699-2014	太原市建筑设计研究院
61	DBJ04/T 305-2014	非承重砌块墙体自保温系统应用技术规程 Technical specification for application of non-load-bearing self-insulation system with blocks		J12700-2014	山西省建筑科学研究院
62	DBJ04/T 306-2014	建筑基坑工程技术规范 Technical code for foundation pits excavation for buildings		J12757-2014	山西建筑工程(集团)总公司 太原理工大学
63	DBJ04/T 307-2014	住宅小区配套供电工程技术规程 Construction technical specifications for power supply projects of residential quarters		J12758-2014	国网山西省电力公司
64	DBJ04/T 308-2014	干混砂浆质量管理规程 Quality management specification of dry-mixed mortar		J12791-2014	山西省建筑科学研究院 太原市散装水泥办公室

山西省

序号	标准编号	标准名称	被代替标准编号	备案号	主编单位
65	DBJ04/T 309-2014	蒸压加气混凝土板应用技术规程 Technical regulations of the application of autoclaved aerated concrete slabs		J12790-2014	太钢(集团)粉煤灰综合利用有限公司
66	DBJ04/T 310-2014	工业园区物业服务标准 The property service standard of industrial parks		J12837-2014	山西省房地产业协会
67	DBJ04/T 311-2014	建筑门窗工程质量验收标准 Standard for construction quality acceptance of building doors and windows engineering		J12901-2014	山西国立铝门窗幕墙工程有限公司 山西省宏图建设集团有限公司
68	DBJ04/T 312-2015	湿陷性黄土场地勘察及地基处理技术规范 Code for investigation and ground treatment in collapsible loess regions		J12950-2015	山西省勘察设计研究院
69	DBJ04/T 313-2015	阔叶乔木大规格容器苗培育技术规程 Code for planting technique of large deciduous broadleaf trees container grown plant		J13127-2015	太原市园林植物研究中心
70	DBJ04/T 314-2015	花灌木大规格容器苗培育技术规程 Code for planting technique of large flower shrubs container grown plant		J13128-2015	太原市园林植物研究中心
71	DBJ04/T 315-2015	灌注桩后注浆技术规程 Technical specification for bored pile external grouting		J13173-2015	山西省勘察设计研究院
72	DBJ04/T 316-2015	装配式斜支撑节点钢框架结构技术规程 Technical specifications for prefabricated steel frame structure with diagonal bracing joints		J13174-2015	太原理工大学
73	DBJ04/T 317-2015	无机玻化膨胀珍珠岩保温板外墙外保温工程技术规程 Technical specification for external thermalinsulationon walls with inorganic vitrified expanded perlite insulation board		J13232-2015	山西安顺建材有限公司 山西省建筑节能监管中心
74	DBJ04/T 318-2015	耐热聚乙烯(PE-RT)直埋保温供热管道技术规程 Design and engineering technical code for the high density Polyethylene of raised temperature resistance (PE-RT) of heating supply pipeline		J13272-2015	太原市建筑设计研究院
75	DBJ04/T 319-2016	城市电力电缆电气工程设计规程 Code for design of urban electrical cable engineering		J13357-2016	山西省电力建设工程质量监督中心站
76	DBJ04/T 320-2016	城市电力电缆电气安装工程施工及验收规程 Code for construction and acceptance of urban electrical cable installation projects		J13358-2016	山西省电力建设工程质量监督中心站
77	DBJ04/T 321-2016	城市电力电缆定向钻进拖管工程设计规程 Code for design of urban electrical cable directional drilling pull-in pipeline projects		J13359-2016	山西省电力建设工程质量监督中心站
78	DBJ04/T 322-2016	城市电力电缆定向钻进拖管工程施工及验收规程 Code for construction and acceptance of urban electrical cable directional drilling pull-in pipeline projects		J13360-2016	山西省电力建设工程质量监督中心站

山西省

序号	标准编号	标准名称	被代替标准编号	备案号	主编单位
79	DBJ04/T 323-2016	城市电力电缆排管工程设计规程 Code for design of urban electric cable tubes projects		J13361-2016	山西省电力建设工程质量监督中心站
80	DBJ04/T 324-2016	城市电力电缆排管工程施工及验收规程 Code for construction and acceptance of urban electrical cable tubes projects		J13362-2016	山西省电力建设工程质量监督中心站
81	DBJ04/T 325-2016	城市电力电缆隧道工程设计规程 Code for design of urban electric cable tunnel projects		J13363-2016	山西省电力建设工程质量监督中心站
82	DBJ04/T 326-2016	城市电力电缆隧道工程施工及验收规程 Code for construction and acceptance of urban electrical cable tunnel projects		J13364-2016	山西省电力建设工程质量监督中心站
83	DBJ04/T 327-2016	城市地下管线工程文件归档规范 Town underground pipeline engineering filling specifications		J13365-2016	山西省城市建设档案馆
84	DBJ04/T 328-2016	钢板桩支护技术规程 Technical code for retaining and protection of steel sheet pile		J13400-2016	山西钢铁建设(集团)有限公司
85	DBJ04/T 329-2016	建筑施工键插接式钢管支架安全技术规程 Technical code for safety of steel tubular support with plug in construction		J13401-2016	山西省宏图建设集团有限公司
86	DBJ04/T 330-2016	场馆物业服务标准 The property service standard of venue		J13402-2016	山西赛欧物业管理有限公司
87	DBJ04/T 331-2016	玻璃幕墙安全性能鉴定技术标准 Technical standard for safety property appraisal of glass curtain wall		J13487-2016	山西科建工程检测研究院
88	DBJ04/T 332-2016	钢结构检测技术规程 Technical regulation for inspection and appraiser of steel structures		J13486-2016	太原理工大学

内蒙古自治区

序号	标准编号	标准名称	被代替标准编号	备案号	主编单位
1	DBJ 03-24-2011	复合保温钢筋焊接网架混凝土剪力墙(CL建筑体系)技术规程 The technical specification for concrete shearwall with composite heat insulation welded steel frame (CL building system)	DBJ 03-24-2006	J10896-2011	内蒙古盛威结构设计咨询事务所
2	DBJ 03-25-2007	农村牧区户用沼气 The residential use of methane in rural pastoral areas		J11001-2007	赤峰富龙燃气热力设计有限公司
3	DBJ 03-26-2007	民用建筑太阳能热水器保温及就位桥架 The insulated and suppositing bridge of solar-powered water heater in the civil construction		J11002-2007	内蒙古工大建筑设计有限公司
4	DBJ 03-27-2011	公共建筑节能设计标准 The standard for energy efficiency design of public buildings	DBJ 03-27-2007	J11066-2011	内蒙古建筑科学研究所有限责任公司 内蒙古工业大学
5	DBJ 03-28-2008	钢管混凝土结构技术规程 The technical specification for concrete-filled steel tubular structure		J11207-2009	中冶东方工程设计有限公司
6	DBJ 03-34-2008	混凝土多孔砖建筑技术规程 The technical specification for perforated concrete brick buildings		J11392-2009	包头市建筑设计院研究院有限责任公司
7	DBJ 03-35-2011	居住建筑节能设计标准 The standard for energy efficiency design of residential buildings	DBJ 03-35-2008	J11332-2011	内蒙古建筑科学研究所有限责任公司 内蒙古工业大学
8	DBJ 03-36-2008	住宅信报箱建设标准 The constructional standard for residential private mailboxes		J11293-2008	内蒙古自治区邮政管理局
9	DBJ 03-41-2011	ZGT稀土保温材料应用技术规程 The applied technological specification for ZGT rare-earth insulation materials		J11781-2011	内蒙古包头市建筑设计研究院有限责任公司
10	DBJ 03-46-2012	EPS模块框(刚)架结构工业节能建筑工程技术规程 The technical specifcation for industrial energy-saving constuctional engineering of(steel)frame structures with EPS modules		J12171-2012	中城建北方建筑勘察设计研究院
11	DBJ 03-47-2012	EPS模块外保温工程技术规程 The technical specification for EPS modular external insulation		J12170-2012	中城建北方建筑勘察设计研究院
12	DBJ 03-48-2012	EPS模块混凝土剪力墙结构工程技术规程 The technical specification for EPS modular concrete shearwall structural engineering		J12173-2012	中城建北方建筑勘察设计研究院
13	DBJ 03-49-2012	ICF外墙外保温工程技术规程 The technical specification for ICF outer wall insulation engineering		J12172-2012	中城建北方建筑勘察设计研究院
14	DBJ 03-50-2013	预拌混凝土技术管理规程 The specification for technology management of ready-mixed concrete		J12264-2013	内蒙古自治区建设工程质量监督总站

内蒙古自治区

序号	标准编号	标准名称	被代替标准编号	备案号	主编单位
15	DBJ 03-51-2013	成品住房装修技术标准 The technical standard for interior finishing of residential buildings		J12419-2013	内蒙古建校建筑勘察设计有限公司
16	DBJ 03-52-2013	住宅小区开发建设标准 The standard for exploitation and construction of residence community		J12481-2013	内蒙古城市规划市政设计研究院有限公司
17	DBJ 03-53-2014	FS外模板现浇混凝土复合保温系统应用技术规程 The applied technological specification for integrated insulation system of cast-in-place concrete with FS outer template		J12641-2014	内蒙古智汇工程设计咨询有限责任公司 内蒙古自治区建设科技开发推广中心
18	DBJ 03-55-2014	建设工程质量检测管理规程 The management regulation for construction engineering quality test		J12665-2014	内蒙古自治区建设工程质量检测中心
19	DBJ 03-61-2014	绿色建筑评价标准 The evaluation standard for green buildings		J12730-2014	内蒙古城市规划市政设计研究院
20	DBJ 03-64-2015	JBY混凝土剪力墙自保温系统技术规程 The technical specification for JBY concrete shearwall integrated insulation system		J12938-2015	乌兰浩特市铭龙建筑设计有限责任公司
21	DBJ 03-66-2015	绿色建筑设计标准 The standard for green building design		J13113-2015	内蒙古建校建筑勘察设计有限公司
22	DBJ 03-72-2016	公共建筑节能改造技术规范 The technical code for energy-efficienct retrofitting of public buildings		J13310-2016	内蒙古建筑科学研究所有限责任公司
23	DBJ 03-73-2016	房屋建筑工程技术资料管理规程 The management specification for technological data of constructional engineering		J13330-2016	内蒙古自治区建设工程质量监督总站

辽宁省

序号	标准编号	标准名称	被代替标准编号	备案号	主编单位
1	DB21/T 907-2015	建筑地基基础技术规范 Technical code for building foundation		J10615-2015	辽宁省建筑设计研究院
2	DB21/T 1274-2009	含空气层现浇复合材料保温屋面工程技术规程 Technical code for insulating roof with air layer using insulating polymer placed in site		J11505-2009	东北大学
3	DB21/T 1304-2012	预拌混凝土技术规程（修订） Technical specification for ready-mixed concrete		J12215-2012	辽宁省混凝土协会
4	DB21/T 1450-2015	建筑基桩及复合地基检测技术规程 Technical specification for testing of building foundation piles and composite ground		J10854-2015	辽宁省建设科学研究院
5	DB21/T 1456-2006	埋地钢肋复合缠绕排水管道工程技术规程 Technical specification for buried PE pipeline with steel-ribbed wrapped of sewer engineering		J10918-2007	辽宁省建设科技发展促进中心
6	DB21/T 1461-2006	封底混凝土小型空心砌块建筑技术规程 Technical specification for sub-sealed concrete small hollow block building		J10920-2007	辽宁省建筑设计研究院
7	DB21/T 1462-2006	民用建筑节能保温工程施工质量验收规程 Specification for insulation constructional quality acceptance of civil building		J10921-2007	辽宁省建设科学研究院
8	DB21/T 1463-2006	硬泡聚氨酯外保温工程技术规程 Technical specification for rigid polyurethane foam to external insulation engineering of buildings		J10919-2007	辽宁省建设科技发展促进中心
9	DB21/T 1476-2011	居住建筑节能设计标准 Design standard for energy efficiency of residential buildings	DB21/T 1476-2006	J10922-2011	辽宁省建设科学研究院
10	DB21/T 1477-2006	公共建筑节能设计标准 Design standard for energy efficiency of public buildings		J10923-2007	沈阳建筑大学建筑设计研究院
11	DB21/T 1479-2007	矿渣混凝土砖建筑技术规程 Technical specification for slag concrete brick building		J10933-2007	沈阳建筑大学
12	DB21/T 1488-2007	辽宁省民用建筑太阳能热水系统一体化技术规程 Technical specification for integrated solar hot water systems of civil buildings in Liaoning province		J10945-2007	大连理工大学
13	DB21/T 1512-2007	工业废渣混凝土多孔砖建筑技术规程 Technical specification for buildings of concrete perforated brick with industrial wasted materials		J10992-2007	中国建筑东北设计研究院
14	DB21/T 1520-2007	拼装棚模现浇混凝土密肋楼盖 Technical specification for multi-ribbed concrete floor		J11021-2007	沈阳建筑大学

辽宁省

序号	标准编号	标准名称	被代替标准编号	备案号	主编单位
15	DB/T 1521-2007	自燃煤矸石混凝土结构技术规程 Technical specification for self-combusted coal gangues		J11020-2007	辽宁省建设科学研究院
16	DB21/T 1558-2007	融雪剂质量与使用技术规程 Technical specification of quality and use for snow melt agent		J11097-2007	沈阳市环境卫生工程设计研究院
17	DB/T 1559-2007	回弹法检测泵送混凝土抗压强度技术规程 Technical specification for inspection of pumped concrete compressive strength		J11098-2007	辽宁省建设科学研究院
18	DB21/T 1561-2007	防水透汽膜和隔汽膜建筑工程技术规程 Technical specification for waterpro of weather and vapor barriers of construction engineering		J11099-2007	辽宁省建设科技发展促进中心
19	DB21/T 1562-2007	燕尾槽XPS板外保温工程技术规程 Technical specification for external thermal insulation engineering of XPS board with swallow tail groove		J11100-2007	辽宁省建设科技发展促进中心
20	DB21/T 1563-2007	住宅小区及商住楼通信配套设施技术规程 Technical regulations of communications facility in residential district and buildings		J11111-2007	沈阳市电信规划设计院
21	DB21/T 1564.1～14-2007	岩土工程勘察技术规程 Technical specification for investigation of geotechnical engineering		J11119-2007	辽宁有色勘察研究院
22	DB21/T 1565-2015	预应力混凝土管桩基础技术规程 Technical specification for prestressed concrete tube pile foundation		J11131-2015	辽宁省建筑设计研究院
23	DB21/T 1568-2008	商品砂浆生产应用技术规程 Practice code for manufacture and application of factory-manufactured mortar		J11161-2008	辽宁省建筑材料监督检验院
24	DB21/T 1571-2013	CL板保温夹心混凝土墙建筑技术规程 Technical specification for CL board insulation sandwich concrete wall construction	DB21/T 1571-2008	J12500-2013	辽宁省住房和城乡建设厅建筑节能与建设科技发展中心
25	DB21/T 1572-2008	聚苯板混凝土复合保温砌块填充墙技术规程 Technical specification for the infilled wall of heat insulating building blocks with polystyrene foam and concrete		J11163-2008	辽宁省建设科技发展促进中心
26	DB21/T 1622-2008	铜包铝芯电线电缆技术规程 Technical specification for coppor-clad aluminium cable		J11218-2008	东北大学设计研究院(有限公司)
27	DB21/T 1626-2008	埋地钢塑复合螺旋缠绕排水管道工程技术规程 Technical specification of buried steel reinforced plastic compositepipe from helical ribbed process for sewerage engineering		J11243-2008	沈阳市市政工程设计研究院 辽宁远东新型管业发展有限公司
28	DB21/T 1643-2008	地源热泵系统工程技术规程 Specification code for ground-source heat pump system		J11261-2008	沈阳建筑法学建筑设计研究院

辽宁省

序号	标准编号	标准名称	被代替标准编号	备案号	主编单位
29	DB21/T 1684-2008	地面辐射采暖泡沫混凝土绝热层技术规程 Technical specification for foaming concrete heat-insulating layer in ground irradiation heating engineering		J11298-2008	辽宁省建筑节能环保协会
30	DB21/T 1685-2008	太阳能光伏照明应用技术规程 Solar PV lighting applications technology specification		J11299-2008	辽宁省建筑节能环保协会
31	DB21/T 1686-2008	地面辐射供暖技术规程 Technical specification for floor radiant heating		J11300-2008	辽宁省建筑节能环保协会
32	DB21/T 1692-2008	建筑工程冬期施工技术规程 Technical specification for winter construction of building engineering		J11303-2008	沈阳建筑大学
33	DB21/T 1705-2008	辽宁省石材幕墙工程技术规程 Technical specification for stone curtain wall engineering		J11312-2008	辽宁省装饰协会
34	DB21/T 1720-2009	海水源热泵系统工程技术规程 Technical specification for seawater-source heat pump system		J11373-2009	大连理工大学
35	DB21/T 1722-2009	居住建筑供暖热计量系统技术规程 Technical specification for heat metering system in residential building		J11397-2009	沈阳建筑大学
36	DB21/T 1725-2009	水泥基渗透结晶型防水材料施工技术规程 Construction technical specification for cementitious capillary crystalline waterproofing materials		J11431-2009	辽宁省建设科技发展促进中心
37	DB21/T 1729-2009	钢筋混凝土异形柱结构技术规程 Technical specification for reinforced concrete structures with specially shaped columns		J11432-2009	辽宁省建筑设计研究院
38	DB21/T 1745-2009	地下混凝土结构防裂技术规程 Technical specification for crack resistance of underground concrete structure		J11513-2009	辽宁省建筑设计研究院
39	DB21/T 1746-2009	钢管混凝土结构技术规程 Technical specification for concrete-filled steel tubular structures		J11507-2009	沈阳建筑大学
40	DB21/T 1748-2009	XPS板外墙外保温技术规程 Technical specifcation for external thermal insulation system		J11506-2009	沈阳市建筑设计院
41	DB21/T 1752-2009	混凝土结构耐久性修复与防护技术规程 Technical specification for repair and protection of durability of concrete structure		J11519-2009	辽宁省建设科学研究院
42	DB21/T 1778-2010	混凝土界面处理剂应用技术规程 Technical specification for interface treating agent for concrete		J11586-2010	辽宁省建设科学研究院
43	DB21/T 1779-2009	混凝土结构砌体填充墙技术规程 Technical specification for masonry infilled wall of concrete structures		J11585-2010	辽宁省建筑设计研究院

辽宁省

序号	标准编号	标准名称	被代替标准编号	备案号	主编单位
44	DB21/T 1792-2010	太阳能光伏与建筑一体化技术规程 Building integrated photovoltaic technical specification		J11588-2010	辽宁省建筑节能环保协会 辽宁太阳能研究应用有限公司
45	DB21/T 1794-2010	现浇轻质复合墙体应用技术规程 Technical specification applied for cast-in-place light-weight composite wall		J11589-2010	大连市建筑科学研究设计院股份有限公司
46	DB21/T 1795-2010	污水源热泵系统工程技术规程 Technical specification for sewage source heat pump system		J11626-2010	大连理工大学
47	DB21/T 1823-2010	既有居住建筑节能改造技术规程 Technical specification for the retrofitting of residential buildings		J11702-2010	沈阳建筑大学
48	DB21/T 1824-2010	既有公共建筑节能改造技术规程 Technical specification for the retrofitting of public buildings		J11700-2010	沈阳建筑大学
49	DB21/T 1825-2010	自动跟踪定位射流灭火系统技术规程 Technical specification of auto tracking and targeting jet suppression system		J11701-2010	沈阳建筑大学建筑设计研究院
50	DB21/T 1837-2010	蒸压粉煤灰砖建筑技术规程 Technical code for autoclaved fly ash brick masonry buildings		J11766-2010	中国建筑东北设计研究院有限公司
51	DB21/T 1843-2010	混凝土养护剂应用技术规程 Technical specification for curing compounds for concrete		J11768-2010	辽宁省建设科学研究院
52	DB21/T 1844-2010	保温装饰复合板外墙外保温工程技术规程 Technical regulations of external thermal insulation and decoration composite board engineering for external walls		J11769-2010	辽宁省建筑材料工业协会
53	DB21/T 1845-2010	XPS保温装饰复合板外墙外保温工程技术规程 Technical regulations of XPS external thermal insulation and architectural composite board engineering for external walls		J11767-2010	辽宁省住房和城乡建设厅建筑节能与建设科技发展中心
54	DB21/T 1868-2010	装配整体式混凝土结构技术规程（暂行） Technical specifications for precast concrete structures (trial)		J11792-2011	沈阳建筑大学
55	DB21/T 1869-2010	水泥聚苯模壳格构式混凝土墙体建筑技术规程 Technical specification for EPSC heat-preservation form latticed concrete wall buildings		J11793-2011	辽宁省住房和城乡建设厅建筑节能与建设科技发展中心
56	DB21/T 1872-2011	预制混凝土构件制作与验收规程（暂行） Technical specification for manufacture and acceptance of precast concrete components (trial)		J11791-2011	沈阳兆寰现代建筑产业园有限公司
57	DB21/T 1891-2011	混凝土矿物掺合料应用技术规程 Technical specification for application of concrete mineral admixtures		J11835-2011	沈阳北方建设股份有限公司

辽宁省

序号	标准编号	标准名称	被代替标准编号	备案号	主编单位
58	DB21/T 1893-2011	装配式建筑全装修技术规程（暂行） Technical specification for full decoration of fabricated buildings（trial）		J11896-2011	沈阳建筑大学
59	DB21/T 1896-2011	发泡浆料楼（地）面、屋面保温技术规程 Technical specification for foam slurry floor and roof thermal insulation		J12016-2012	辽宁省建筑设计研究院
60	DB21/T 1899-2011	公共建筑节能（65%）设计标准 Design standard for（65%）energy saving of public buildings		J11895-2011	辽宁省建设科学研究院
61	DB21/T 1900-2011	玻璃钢供热管道工程技术规程 Technical specification for GRP heating pipe		J12014-2012	辽宁省建筑节能环保协会
62	DB21/T 1914-2011	建筑中水回用技术规程 Technical specification for building reclaimed water system		J12013-2012	辽宁省建设科学研究院
63	DB21/T 1917-2011	卫生设备同层排水系统技术规程 Technology regulation of same-floor drainage system in sanitation		J12015-2012	辽宁省建筑标准设计研究院
64	DB21/T 1925-2011	装配整体式建筑设备与电气技术规程（暂行） Technical specifications for equipment and electrical of assembled monolithic buildings（trial）		J12011-2012	中国建筑东北设计研究院有限公司
65	DB21/T 1947-2012	建材产品废渣掺量检验技术规程 Specification for inspection of addition amount of waste residue in building material products		J12010-2012	辽宁省建设科学研究院
66	DB21/T 1954-2012	园林绿化养护管理标准 Standard for urban greening maintenance		J12055-2012	沈阳市城市建设管理局 沈阳市园林科学研究院
67	DB21/T 1991-2012	现浇发泡浆料外保温夹心墙技术规程 Technical specification for cast foam slurry in placeof external thermal insulation		J12140-2012	辽宁省建筑设计研究院
68	DB21/T 2000-2012	装配整体式剪力墙结构设计规程（暂行） Specification for design of assembled monolithic shear wall structures（trial）		J12141-2012	沈阳建筑大学
69	DB21/T 2001-2012	建筑外墙仿花岗岩涂料应用技术规程 Technical specification for external wall granitelike coatings		J12142-2012	辽宁省建设科学研究院
70	DB21/T 2005-2013	LED照明工程安装与质量验收规程 Specification for installation and quality acceptance of LED lighting		J12602-2014	辽宁省建筑节能环保协会
71	DB21/T 2017-2012	绿色建筑评价标准 Evaluation standard for green building		J12188-2012	辽宁省住房和城乡建设厅建筑节能与建设科技发展中心
72	DB21/T 2018-2012	电热储能炉工程应用技术规程 Technical specification for engineering application of the electric storage furnace		J12187-2012	辽宁省建筑节能环保协会
73	DB21/T 2025-2012	膨胀珍珠岩保温板薄抹灰外墙外保温工程技术规程 Technical specification for external thermal insulation on walls with thin-plast		J12216-2012	辽宁省建筑节能环保协会

辽宁省

序号	标准编号	标准名称	被代替标准编号	备案号	主编单位
74	DB21/T 2041-2012	寒区温拌沥青混凝土路面工程技术规程 Technical specification for warm mix asphalt concrete pavement engineering in cold region		J12272-2013	沈阳市政集团有限公司
75	DB21/T 2043-2012	复合粉状防水剂应用技术规程 Technical specification for composite powder water-repellent admixture		J12273-2013	辽宁省建筑材料工业协会
76	DB21/T 2096-2013	相变无机材料复合保温技术规程 Technical specification for composite insulation of phase change and inorganic materials		J12333-2013	辽宁省建设科学院研究院
77	DB21/T 2109-2013	园林绿化养护管理技术规程 Technical regulations for urban greening maintenance		J12354-2013	沈阳市园林科学研究院
78	DB21/T 2117-2013	泡沫混凝土外墙外保温工程技术规程 Technical specification for external thermal insulation systems of foamed concrete		J12402-2013	辽宁省建筑节能环保协会
79	DB21/T 2124-2013	建筑外保温防火隔离带技术规程 Technical specification for external thermal insulation fireproof isolation belt construction		J12421-2013	沈阳市建筑研究院
80	DB21/T 2156-2013	防火保温块外墙外保温技术规程 Technical specification for fire-resistant insulation block external thermal insulation on external		J12506-2013	沈阳市建筑研究院有限公司
81	DB21/T 2157-2013	夹芯墙脲醛树脂现场发泡保温技术规程 Thermal insulation technical specification for on-the-spot foaming urea-formaldehyde resin used in cavity wall		J12448-2013	辽宁省建设科学研究院
82	DB21/T 2158-2013	夹芯墙聚氨酯硬泡浇注保温技术规程 Thermal insulation technical specification for the on-the-spot foaming PU used in sandwich wall		J12447-2013	辽宁省建设科学研究院
83	DB21/T 2159-2013	岩棉板或玻璃棉板保温幕墙工程技术规程 Rock wool board, glass curtain wall exterior insulation engineering and technical regulations		J12446-2013	辽宁省建设科学研究院
84	DB21/T 2170-2013	采暖塑料管管道工程技术规程 Engineering technical specification of heating supply plastic pipeline for buildings		J12499-2013	辽宁省建筑标准设计研究院
85	DB21/T 2171-2013	酚醛泡沫板外墙外保温技术规程 Technical specification for external thermal insulation system of rigid phenolic foam board		J12458-2013	辽宁省建筑材料工业协会 辽宁省建设科学研究院
86	DB21/T 2204-2013	塑料排水检查井应用技术规程 Technical specification for application of plastic inspection chambers for drainage system		J12603-2014	辽宁省建筑节能环保协会
87	DB21/T 2206-2013	岩棉薄抹灰外墙外保温技术规程 Technical specification for rock wool external thermal insulation with thin-plastered coat		J12605-2014	辽宁工业大学

辽宁省

序号	标准编号	标准名称	被代替标准编号	备案号	主编单位
88	DB21/T 2225-2014	高性能混凝土应用技术规程 Technical specification for application of high performance concrete		J12604-2014	辽宁省建设科学研究院
89	DB21/T 2237-2014	建筑外保温工程施工防火安全技术规程 Technical specification to fire safety of exterior insulation construction for building		J12606-2014	沈阳市建筑研究院有限公司
90	DB21/T 2241-2014	城市雨水利用系统技术规程 Technical specification of urban rainwater utilization system		J12601-2014	沈阳建筑大学建筑设计研究院
91	DB21/T 2259-2014	废钢渣预拌建筑砂浆技术规程 Technical specification for building ready-mixed mortar with wasted steel slag		J12678-2014	本溪荣鑫建筑环保节能材料有限公司
92	DB21/T 2295-2014	市政工程施工质量验收实施细则 Implementation regulations for construction quality acceptance of public works		J12661-2014	辽宁省建设工程质量监督总站
93	DB21/T 2316-2014	建筑节能门窗工程技术规程 Technical code for energy efficiency doors and windows engineering		J12829-2014	辽宁省建筑节能环保协会
94	DB21/T 2353-2014	泡沫混凝土板隔墙系统技术规程 Technical specification for foam concrete slab partition wall system		J12781-2014	沈阳建筑大学
95	DB21/T 2358-2014	EPS模块外保温工程技术规程 Technical specification for external thermal insulation energy-saving building with EPS module		J12831-2014	辽宁省建筑材料工业协会
96	DB21/T 2359-2014	膨胀玻化微珠保温砂浆墙体保温工程技术规程 Technical specification for wall insulation engineering of expansive glazed hollow beads thermal mortar		J12832-2014	辽宁省建筑材料工业协会
97	DB21/T 2360-2014	硬泡聚氨酯板外墙外保温技术规程 Technical specification for exterior thermal insulation of rigid polyurethane foam board		J12830-2014	辽宁省建设科学研究院
98	DB21/T 2399-2015	园林植物栽植技术规程 Technical regulations for landscape vegetation planting		J12946-2015	沈阳市园林科学研究院
99	DB21/T 2419-2015	装配式混凝土结构工程检测技术规程 Technical specification for inspection of precast concrete structure		J12953-2015	辽宁省建设科学研究院
100	DB21/T 2445-2015	民用建筑新风系统工程技术规程 Technical specification for civil architecture fresh air systems		J13066-2015	辽宁省建筑节能环保协会
101	DB21/T 2446-2015	机喷RQW轻质复合保温材料外保温工程技术规程 RQW lightweight composite thermal insulation material		J13064-2015	辽宁省建筑材料工业协会
102	DB21/T 2447-2015	无卤阻燃木塑地板工程技术规程 Technical specification of non-halogen flame retardant wood-plastic flooring		J13065-2015	辽宁省建筑节能与建设科技发展中心

辽宁省

序号	标准编号	标准名称	被代替标准编号	备案号	主编单位
103	DB21/T 2480-2015	发泡陶瓷保温板应用技术规程 Technical specification for application of foamed ceramic thermal insulation board		J13270-2015	辽宁省建筑设计研究院
104	DB21/T 2519-2015	现浇轻质保温复合墙体应用技术规程 Technical specification for cast-in-situ lightweight insulation composite wall		J13221-2015	辽宁省住房和城乡建设厅建筑节能与建设科技发展中心
105	DB21/T 2520-2015	断桥混凝土保温砌块外填充墙技术规程 Technical specification for outer filling walls of thermal-bridge-broken concrete insulation block		J13222-2015	辽宁省住房和城乡建设厅建筑节能与建设科技发展中心
106	DB21/T 2545-2015	XPS复合模板外保温系统技术规程 Technical specification for exterior thermal insulation of XPS composite template		J13405-2016	辽宁省住房和城乡建设厅建筑节能与建设科技发展中心
107	DB21/T 2568-2016	装配式混凝土结构构件制作、施工与验收规程 Specifications for manufacture construction and acceptance of precast concrete structures		J13406-2016	沈阳建筑大学
108	DB21/T 2572-2016	装配式混凝土结构设计规程 Specification for design of precast concrete structures		J13407-2016	沈阳建筑大学
109	DB21/T 2575-2016	太阳能供热供暖工程技术规程 Technical regulations for solar heating system		J13408-2016	辽宁省建筑节能环保协会
110	DB21/T 2585-2016	装配式住宅全装修技术规程 Technical specification for full decoration of assembled housing		J11896-2016	沈阳建筑大学
111	DB21/T 2586-2016	聚氨酯-聚脲防水涂料应用技术规程 Technical specification for application of polyurethane-polyurea waterproofing coating		J13409-2016	辽宁省住房和城乡建设厅建筑节能与建设科技发展中心
112	DB21/T 2601-2016	后锚固法检测混凝土抗压强度技术规程 Technical specification for inspection of concrete compressive strength by post-installed adhesive anchorage method		J13410-2016	辽宁省建设科学研究院

吉林省

序号	标准编号	标准名称	被代替标准编号	备案号	主编单位
1	DB22/JT 13-2014	城市桥梁养护技术规程 Technical specification of maintenance for city bridge		J12735-2014	中庆建设有限责任公司
2	DB22/T 112-2012	灌注桩基础技术规程 Technical specification for building cast-in-place pile foundations		J12213-2012	吉林省建筑设计院有限责任公司
3	DB22/JT 125-2014	建筑物灾后应急鉴定技术规程（震灾、洪灾部分） Technical specification for building's post-disaster emergency appraisal (earthquake and waterflood part)		J12664-2014	吉林省建筑科学研究设计院
4	DB22/JT 126-2014	可再生能源工程施工质量验收规程 Specification for construction quality acceptance of renewable energy engineering		J12663-2014	吉林省建筑科学研究设计院
5	DB22/JT 127-2014	建筑工程资料管理标准 Architectural engineering document management standard		J11730-2014	吉林省金科工程监理有限公司
6	DB22/JT 128-2014	成品住宅室内装修标准 Decoration standard for finished housing indoors		J12680-2014	长春万科房地产开发有限公司
7	DB22/JT 129-2014	保温装饰板外墙外保温工程技术规程 Technical specification for insulation decorative plates external thermal insulation on walls		J12736-2014	吉林省工程建设标准化协会 吉林省方泰建筑材料有限公司
8	DB22/JT 131-2014	建筑基坑支护工程检测标准 Standard for building foundation pit engineering test		J12960-2015	吉林省建筑科学研究设计院 吉林省建筑工程质量检测中心
9	DB22/JT 132-2014	石墨EPS板外墙外保温工程技术规程 Technical specification for graphite EPS board external thermal insulation on walls		J12961-2015	吉林科龙建筑节能科技股份有限公司
10	DB22/JT 133-2014	长螺旋钻孔压灌混凝土桩基础技术规程 Technical specification for long auger bored cast-with-pressure concrete pile foundations		J12962-2015	吉林省建苑设计集团有限公司
11	DB22/JT 134-2014	建筑工程绿色施工规程 Specification for green construction of building		J12963-2015	吉林建工集团有限公司
12	DB22/JT 135-2015	预制装配式道路基层工程技术规程 Technical specification for prefabricated roadbases engineering		J12964-2015	长春市市政工程设计研究院
13	DB22/JT 136-2015	城镇道路养护技术规程 Technical specification of maintenance for urban road		J12965-2015	长春市城市快速路管理维护有限责任公司
14	DB22/JT 138-2015	建设工程施工现场安全管理内业标准 Standard pratice for safety management of construction project on construction site	DB22/T 479-2010	J11636-2015	吉林省建设安全协会
15	DB22/JT 139-2015	建筑基坑工程监测技术规程 Technical specification for monitoring of building excavation engineering		J13031-2015	中铁现代勘察设计院有限公司 中国铁建大桥工程局集团有限公司

吉林省

序号	标准编号	标准名称	被代替标准编号	备案号	主编单位
16	DB22/JT 140-2015	城镇供热系统节能技术规程 Technical specification for energy efficiency of city heating system		J13032-2015	长春工程学院设计研究院
17	DB22/JT 141-2015	内击沉管压灌桩技术规程 Technical specification for hammer-bottom-driven cast-in-place pile with pumping pressure concrete		J13034-2015	吉林建东科技开发有限公司
18	DB22/JT 142-2015	EPS模块外墙外保温工程技术规程 Technical specification for EPS modular external thermal insulation on walls		J11682-2015	吉林省工程建设标准化协会 哈尔滨鸿盛房屋节能体系研发中心
19	DB22/JT 144-2015	建筑太阳能光伏系统技术规程 Technical specification for building solar photovoltaic systems		J13255-2015	吉林省建筑科学研究设计院
20	DB22/JT 145-2015	建筑基坑支护技术规程 Technical specification for retaining and protecting of building foundation excavations		J11947-2015	吉林省中鼎建筑设计有限公司
21	DB22/JT 146-2015	房屋结构安全性与抗震鉴定标准 Home structure seismic evaluation and safety standards		J13256-2015	吉林省阳光建设工程咨询有限公司
22	DB22/JT 147-2015	岩土工程勘察技术规程 Specification for investigation of geotechnical engineering		J10582-2016	吉林建筑大学 吉林建筑大学勘测公司
23	DB22/T 164-2007	居住建筑节能设计标准（节能50%） Design standard for energy efficiency of residential buildings(energy efficiency 50%)	DB 22/164-1998	J11120-2007	吉林省建设标准化管理办公室
24	DB22/T 285-2013	二次供水工程技术规程 Technical specification for secondary water supply engineering		J10114-2013	吉林省建苑设计集团有限公司
25	DB22/T 436-2006	吉林省公共建筑节能设计标准 Design standard for energy efficiency		J10942-2007	吉林省建筑设计院有限责任公司
26	DB22/T 437-2006	住宅厨房卫生间变压式排气道 Technical specification of variable pressure		J11088-2007	吉林省建筑标准化管理所
27	DB22/T 438-2007	建设工程质量检测数据 Technical standard for data information		J11105-2007	延边州建设工程质量检测中心
28	DB22/T 438-2013	建设工程质量检测文件标准 Arrangement standard for document of construction quality test		J11105-2013	吉林省建筑科学设计院
29	DB22/T 439-2007	住宅小区通信配套设施建设标准 Standard for construction of communication		J11089-2007	吉林省通信管理局
30	DB22/T 442-2007	混凝土多孔砖砌体结构技术规程		J11091-2007	新星宇建设有限责任公司 吉林省墙材革新与建筑节能办公室
31	DB22/T 444-2007	超薄天然石材外墙装饰工程 Decorative technical specification for natural super-thin stone for outer-wall		J11112-2007	吉林省万泉石业有限公司
32	DB22/T 445-2007	混凝土小型空心砌块砌体工程施工及验收规程 Code for construction and acceptance of concrete		J11090-2007	新星宇建设有限责任公司 吉林省墙材革新与建筑节能办公室

吉林省

序号	标准编号	标准名称	被代替标准编号	备案号	主编单位
33	DB22/T 448-2007	住宅小区有线数字电视工程技术规范(暂行) Technical code for regunation of cable digital television in residential district		J11106-2007	吉林省广电网络集团公司
34	DB22/T 448-2013	居住区有线数字电视系统工程技术规程 Technical specification for residential areas of digital cable television system engineering		J11106-2013	吉视传媒股份有限公司
35	DB22/T 450-2007	居住建筑节能设计标准(节能65%) Design standard for energy efficiency of residential buildings(energy efficiency 65%)		J11121-2007	吉林省建设标准化管理办公室
36	DB22/T 452-2008	配筋混凝土小型空心砌块砌体工程施工及验收规程 Code for construction and acceptance of reinforced concrete small-sized hollow block masonry engineering		J11201-2008	新星宇建设有限责任公司
37	DB22/T 457-2008	低合金双钢筋混凝土构件技术规程 Technical specification for concrete structures with low-alloy coupled steel bars		J11287-2008	吉林建筑工程学院
38	DB22/T 464-2009	《独立式感烟火灾探测报警无线远程监控系统设计、安装及验收规程》 Code for design and installation and acceptance of wirless remot monitoring equipments for self-contained fire smoke alarm systems		J11449-2009	吉林省公安厅消防局
39	DB22/T 477-2010	胶粉聚苯颗粒复合外墙外保温工程施工及验收规程 Construction and acceptance specification for colloid powder polyphenylene grain composite external thermal insulation engineering		J11552-2010	吉林省东佳建筑节能材料有限公司
40	DB22/T 478-2010	岩土工程勘察大纲及详勘报告编制标准 The preparation of standard for the outline and the detailed exploration of geotechnical investigation engineering		J11553-2010	吉林省阳光建设工程咨询有限公司
41	DB22/T 479-2010	建设工程施工现场安全管理内业标准 Standard pratice for safety management of constructi on project on construction site		J11636-2010	吉林省建设安全协会
42	DB22/T 480-2012	建设工程抗震超限界定标准(建筑工程部分) Definition standard for seismic out of codes of building constructions（building）	DB22/T 480-2010	J11066-2012	吉林省阳光建设工程咨询有限公司
43	DB22/T 481-2010	低温辐射电热膜采暖技术规程 Technical specification for low temperature electric radiant heating film		J11657-2010	吉林省建筑科学研究设计院
44	DB22/T 483-2010	ICF外墙外保温工程技术规程 Technical specification for ICF external thermal insulation on walls		J11681-2010	新星宇建设集团有限公司
45	DB22/T 496-2012	民用建筑外保温工程防火技术规程 Technical specification to fire safety of exterior insulation construction for civil building	DB22/T 496-2010	J11680-2012	吉林省建筑设计院有限责任公司
46	DB22/T 497-2010	静压预应力混凝土管桩基础技术规程 Technical specification for staric pressing-in prestressed concrete pipe pile foundation		J11719-2010	吉林省阳光建设工程咨询有限公司

吉林省

序号	标准编号	标准名称	被代替标准编号	备案号	主编单位
47	DB22/T 499-2010	烟囱拆除工程升降架法施工规程 Specification for chimney removal of lifting		J11732-2010	吉林市云天工程设备技术开发有限公司 新星宇建设集团有限公司
48	DB22/T 1021-2010	建设工程施工图设计文件审查文本编制标准 The standard text of the construction shop drawing design documents		J11731-2010	吉林省同德市政工程设计审查有限公司 吉林省阳光建设工程咨询有限公司
49	DB22/T 1022-2010	蒸压粉煤灰多孔砖砌体结构技术规程 Technical specification for autoclaved fly ash brick masonry structures		J11779-2011	吉林省建筑材料工业设计研究院 吉林省墙材革新与建筑节能办公室
50	DB22/T 1023-2010	烧结注孔保温砌块（砖）砌体工程技术规程 Technical specification for sintered note hole insulation block（brick）masonry works		J11778-2011	吉林省墙材革新和建筑节能办公室 吉林省光大实业集团有限责任公司
51	DB22/T 1024-2011	预拌砂浆技术规程 Technical specification for ready-mixed mortar		J11808-2011	吉林省住房和城乡建设厅散装水泥办公室
52	DB22/T 1026-2011	模塑聚苯乙烯泡沫塑料板外墙外保温工程技术规程 Technical specification for moulded polystyrene foam board external thermal insulation		J10092-2012	吉林省建筑科学研究设计院
53	DB22/T 1027-2011	民用建筑太阳能热水系统应用技术规程 Technical specification for solar water heating system of civil buildings		J11872-2011	吉林省土木建筑学会热能动力分会 吉林省建筑设计院有限责任公司
54	DB22/T 1028-2011	EPS模块混凝土剪力墙结构技术规程 Technical specification for concrete wall structure system with EPS module		J11893-2011	吉林建筑工程学院 哈尔滨鸿盛房屋节能体系研发中心
55	DB22/T 1029-2011	EPS模块钢结构工业建筑节能体系技术规程 Technical specification for energy conservation system of steel structure industrial building with EPS module		J11892-2011	吉林建筑工程学院 哈尔滨鸿盛房屋节能体系研发中心
56	DB22/T 1030-2011	EPS模块混凝土芯柱结构体系技术规程 Technical specification for EPS module concrete core column structure system		J11891-2011	新星宇建设集团有限公司 哈尔滨鸿盛房屋节能体系研发中心
57	DB22/T 1031-2011	EPS模块混凝土框架结构体系技术规程 Technical specification for EPS module concrete frame structure system		J11890-2011	新星宇建设集团有限公司 哈尔滨鸿盛房屋节能体系研发中心
58	DB22/T 1032-2011	EPS模块轻钢结构体系技术规程 Technical specification for EPS module light steel structure system		J11889-2011	新星宇建设集团有限公司 哈尔滨鸿盛房屋节能体系研发中心
59	DB22/T 1035-2011	建筑材料见证取样检测标准 Buidling materials evidential testing criterion		J12214-2012	吉林省建筑科学研究设计院
60	DB22/T 1036-2011	超薄石材硬泡聚氨酯复合板外墙外保温工程技术规程 Technical specification for super-thin stone rigid polyurethane foam composite		J11900-2011	吉林省万泉石业有限公司 新星宇建设集团有限公司
61	DB22/T 1037-2011	塑料排水检查井应用技术规程		J11919-2011	吉林省建筑设计院有限责任公司
62	DB22/1038-2011	地源与低温余热水源热泵系统工程技术规程		J12032-2012	吉林省土木建筑学会热能动力分会 长春工程学院设计研究院 吉林省安装工程股份有限公司

吉林省

序号	标准编号	标准名称	被代替标准编号	备案号	主编单位
63	DB22/T 1039-2011	预制钢筋混凝土复合保温外墙挂板技术规程 Technical specification for thermal insulation precast reinforced concrete composite exterior wall cladding		J11948-2011	长春工程学院 吉林佳泓新型节能建材有限公司
64	DB22/T 1052-2012	CL复合墙体建筑体系技术规程 Technical specification for CL composite wall building system		J12061-2012	吉林省建筑设计院有限责任公司 新星宇建设集团有限公司
65	DB22/T 1548-2012	岩土工程勘察原位测试规程 Technical specification for in-situ tests of geotechnical engineering investigation		J12088-2012	吉林建筑工程学院勘测公司
66	DB22/T 1568-2012	低温热水地面辐射供暖技术规程 Technical specification for low temperature hot water floor radiant heating		J12125-2012	吉林省建筑科学研究设计院
67	DB22/T 1569-2012	离网型风光互补室外照明系统应用技术规程 Technical specification for applications of off-grid outdoor wind & solar hybrid		J12124-2012	中科恒源科技股份有限公司
68	DB22/T 1591-2012	绿色建筑评价标准 Evaluation standard for green building		J12212-2012	长春工程学院设计研究院
69	DB22/T 1596-2012	硬泡聚氨酯外墙外保温工程技术规程 Technical specification for rigid polystyrene foam to external thermal insulation		J10317-2012	吉林省建筑科学研究设计院
70	DB22/T 1631-2012	建筑消防救援破拆标识设置技术规程 Technical specification for building fire and rescue rescue logo set		J12238-2013	吉林省公安消防总队 长春市公安消防支队 辽源市公安消防支队
71	DB22/T 1752-2012	预拌混凝土技术规程 Technical specification for ready-mixed concrete		J12237-2013	吉林省建设工程质量监督检测协会
72	DB22/T 1779-2013	装配整体式混凝土剪力墙结构体系居住建筑技术规程 Technical specification for technical specification for assembled monolithic concrete shear wall structure		J12326-2013	吉林亚泰(集团)股份有限公司
73	DB22/T 1780-2013	混凝土模块砌体结构市政工程技术规程 Technical specification for concrete the modules masonry structure municipal engineering		J12331-2013	吉林省工程建设标准化协会 吉林市北盟新型墙体材料有限公司
74	DB22/T 1781-2013	QT无机保温材料复合墙体保温工程技术规程 Technical specification for QT inorganic insulation materials composite wall insulation works		J12330-2013	吉林省工程建设标准化协会 吉林省瑞茂升建筑保温材料有限公司
75	DB22/T 1887-2013	居住建筑节能设计标准(节能75%) Design standard for energy efficiency of residential buildings(Energy efficiency 75%)		J12486-2013	吉林省工程建设标准化协会
76	DB22/T 1888-2013	民用建筑设计防火统一技术措施 Unified technical measures for fire protection design of civil buildings		J12485-2013	吉林省建苑设计集团有限公司
77	DB22/T 1957-2013	公共建筑能耗监测系统技术规程 Technical specification for metering system construction of energy consumption of public building		J12520-2013	吉林省建苑设计集团有限公司
78	DB22/T 2038-2014	农村居住建筑节能设计标准 Design standard for energy efficiency of rural residential buildings		J12636-2014	吉林省工程建设标准化协会

黑龙江省

序号	标准编号	标准名称	被代替标准编号	备案号	主编单位
1	DB23/1131-2007	HS-ICF 建筑节能体系技术规程 Technical specification for HS-ICF architectural energy-conservation system	Q/HHS 002-2006	J11029-2007	哈尔滨鸿盛房屋节能体系研发中心
2	DB23/T 1167-2013	HS-ICF 外墙外保温建筑节能体系技术规程 Technical specification for external wall thermal insulation system eith insulating concrete form	DB23/1167-2007	J11055-2013	哈尔滨鸿盛房屋节能体系研发中心
3	DB23/T 1203-2007	低温辐射电热地膜采暖技术规程 Technical specification for low temperature radiant heating of floor film with electric energy		J11151-2008	哈尔滨市专家咨询顾问委员会
4	DB23/T 1314-2009	硬泡聚氨酯外墙外保温工程技术规程 Techical specification for rigid polyuret hanefoam to external		J11536-2010	哈尔滨天硕建材工业有限公司
5	DB23/1328-2013	预拌混凝土质量管理规程 Quality management specification for ready-mixed concrete	DB23/T 1328-2009	J11408-2013	哈尔滨市建设工程质量监督总站
6	DB23/T 1340-2009	外墙复合轻质饰面砖工程技术规程 Technical specification for exterior wall compound light-weight wall tile		J11452-2009	黑龙江省建工集团有限责任公司
7	DB23/1350-2011	EPS 模块外保温工程技术规程 Technical specifications of exterior insulation construction with EPS modular	DB23/T 1350-2009	J11533-2011	哈尔滨鸿盛房屋节能体系研发中心
8	DB23/T 1355-2009	HS-EPS 模块轻钢结构建筑节能体系技术规程 Technical specification for building energy-saving system of HS-EPS module with light steel structure		J11534-2010	哈尔滨鸿盛房屋节能体系研发中心
9	DB23/T 1356-2009	HS-EPS 模块钢筋混凝土框架结构建筑节能体系技术规程 Technical specification for building energy-saving system of HS-EPS module with reinforced concrete frame structure		J11535-2010	哈尔滨鸿盛房屋节能体系研发中心
10	DB23/T 1357-2010	预拌砂浆生产应用技术规程 Technical specification for manufacture and application of ready-mixed mortar		J11537-2010	哈尔滨市散装水泥办公室
11	DB23/T 1389-2010	钻孔压灌超流态混凝土桩基础技术规程 Technical code of the affused superfluid concrete pile foundation		J11613-2010	黑龙江龙华岩土工程有限公司
12	DB23/T 1390-2010	EPS 模块钢结构工业建筑节能体系技术规程 Technical specification for industrial building energy-saving system of HS-EPS module with steel structure		J11615-2010	哈尔滨鸿盛房屋节能体系研发中心 黑龙江省冶金设计规划院
13	DB23/T 1391-2010	EPS 模块钢筋混凝土芯柱结构建筑节能体系技术规程 Technical specification for building energy-saving system of EPS module with reinforced concrete core column structure		J11614-2010	哈尔滨工业大学建筑设计研究院 哈尔滨鸿盛房屋节能体系研发中心

黑龙江省

序号	标准编号	标准名称	被代替标准编号	备案号	主编单位
14	DB23/T 1398-2010	民用建筑供热计量收费技术规程 Technical specification for heat metering and charge of civil buildings		J11668-2010	哈尔滨工业大学
15	DB/T 1400-2010	预制装配整体式房屋混凝土剪力墙结构技术规范 Technical specification for concrete shear wall structure assembled with precast components		J11693-2010	哈尔滨工业大学 黑龙江宇辉新型建筑材料有限公司
16	DB23/T 1403-2010	住宅工程质量通病防控规范 Technical specification for prevention and control of common defects of residential buildings		J11770-2010	黑龙江省建设工程质量监督管理总站
17	DB23/T 1411-2010	严寒地区居民小汽车库节能设计规程 Design standard of resident garage energy efficiency in frigid region		J11780-2011	哈尔滨盛源百利科技开发有限公司
18	DB23/T 1437-2011	扩底搅拌复合地基混凝土灌注桩技术规程 Technical specification for gast-in-place concrete pile with underreamed mixed composite foundation		J11822-2011	黑龙江省东煤建筑基础工程有限公司
19	DB23/T 1453-2011	黑龙江省住宅小区有线数字电视工程技术规程 Technical code for regulation of cable digital television in residential district		J11847-2011	黑龙江广播电视网络股份有限公司
20	DB23/T 1454-2011	EPS模块混凝土剪力墙结构体系技术规程 Technical specifications for concrete shear wall structure system		J11874-2011	哈尔滨鸿盛房屋节能体系研发中心 黑龙江建筑职业技术学院
21	DB23/T 1491-2012	地下水源热泵技术规程 Technical specification for groundwater heat pump		J12345-2013	黑龙江省寒地建筑科学研究院
22	DB23/T 1492-2012	民用建筑太阳能热水系统应用技术规程 Technical specification for solar water heating system of civil buildings		J12344-2013	哈尔滨工业大学
23	DB23/T 1493-2012	污水源热泵技术规程 Technical specification for sewage source heat pump		J12353-2013	黑龙江省寒地建筑科学研究院
24	DB23/T 1502-2013	建筑工程抗震性态设计规范 Seismic code for performance-based design of buildings		J12343-2013	中国地震局工程力学研究所
25	DB23/T 1537-2013	黑龙江省农村居住建筑节能设计标准 Design standard for energy efficiency of rural residential buildings in Heilongjiang Province		J12551-2014	哈尔滨工业大学
26	DB23/T 1538-2013	水泥聚苯颗粒模壳格构式混凝土墙体建筑技术规程 Technical specification for CEPG form latticed concrete wall buildings		J12552-2014	黑龙江省寒地建筑科学研究院
27	DB23/T 1539-2014	钢筋混凝土空间网格结构空腹夹层板楼盖技术规程 Technical specification for reinforced concrete space griding structure with open-web sandwich plate		J12582-2014	黑龙江省建设集团有限公司

黑龙江省

序号	标准编号	标准名称	被代替标准编号	备案号	主编单位
28	DB23/T 1586-2015	钢筋工厂化集中加工及配送技术规程 Reinforced factory centralized processing and distribution of technical specification		J13176-2015	黑龙江省建设集团有限公司
29	DB23/T 1621.1-2015	黑龙江省建设工程施工操作技术规程 土方与基坑支护工程 Technical specification for construction engineering operation in Heilongjiang Province the engineering of earthwork and foundation pit support		J12995-2015	黑龙江省东煤建筑基础工程有限公司
30	DB23/T 1621.2-2015	黑龙江省建设工程施工操作技术规程 地基与基础工程 Technical specification for construction engineering operation in Heilongjiang Province ground and foundation engineering		J12996-2015	黑龙江省桩基础工程公司
31	DB23/T 1621.3-2015	黑龙江省建设工程施工操作技术规程 地下防水工程 Echnical specification for construction engineering operation in Heilongjiang Province underground waterproof		J12997-2015	哈尔滨大东集团股份有限公司
32	DB23/T 1621.4-2015	黑龙江省建设工程施工操作技术规程 装饰装修工程 Technical specification for construction engineering operation in Heilongjiang Province decoration engineering		J12998-2015	黑龙江国光建筑装饰工程有限公司
33	DB23/T 1621.5-2015	黑龙江省建设工程施工操作技术规程 混凝土结构工程 Technical specification for construction engineering operation in Heilongjiang Province concrete structural engineering		J12999-2015	黑龙江省寒地建筑科学研究院
34	DB23/T 1621.6-2015	黑龙江省建设工程施工操作技术规程 砌体结构工程 Technical specification for construction engineering operation in Heilongjiang Province masonry structural engineering		J13000-2015	黑龙江省建工集团有限责任公司
35	DB23/T 1621.7-2015	黑龙江省建设工程施工操作技术规程 钢结构工程 Technical specification for construction engineering operation in Heilongjiang Province steel structures		J13001-2015	黑龙江省安装工程公司
36	DB23/T 1621.8-2015	黑龙江省建设工程施工操作技术规程 建筑屋面工程 Technical specification for construction engineering operation in Heilongjiang Province building roof engineering		J13002-2015	黑龙江东辉建筑工程有限公司
37	DB23/T 1621.9-2015	黑龙江省建设工程施工操作技术规程 建筑围护结构节能工程 Technical specification for construction engineering operation in Heilongjiang Province energy conservation of building envelope		J13003-2015	齐翔建工集团
38	DB23/T 1621.10-2015	黑龙江省建设工程施工操作技术规程 建筑给排水及采暖工程 Technical specification for construction engineering operation in Heilongjiang Province building water supply & drainage and heating project		J13004-2015	黑龙江东辉建筑工程有限公司

黑龙江省

序号	标准编号	标准名称	被代替标准编号	备案号	主编单位
39	DB23/T 1621.11-2015	黑龙江省建设工程施工操作技术规程 通风与空调工程 Technical specification for construction engineering operation in Heilongjiang Province ventilation and air conditioning engineering		J13005-2015	哈尔滨三建工程有限责任公司
40	DB23/T 1621.12-2015	黑龙江省建设工程施工操作技术规程 建筑电气工程 Technical specification for construction engineering operation in Heilongjiang Province electrical installation in building		J13006-2015	黑龙江省安装工程公司
41	DB23/T 1621.13-2015	黑龙江省建设工程施工操作技术规程 电梯安装工程 Technical specification for construction engineering operation in Heilongjiang Province elevator installation project		J13007-2015	黑龙江电梯厂
42	DB23/T 1621.14-2015	黑龙江省建设工程施工操作技术规程 建筑燃气工程 Technical specification for construction engineering operation in Heilongjiang Province indoor gas engineering		J13008-2015	哈尔滨市气化工程质量监督站
43	DB23/T 1621.15-2015	黑龙江省建设工程施工操作技术规程 城镇道路工程 Technical specification for construction engineering operation in Heilongjiang Province urban road engineering		J13009-2015	哈尔滨市市政工程质量监督站
44	DB23/T 1621.16-2015	黑龙江省建设工程施工操作技术规程 市政桥梁工程 Technical specification for construction engineering operation in Heilongjiang Province municipal bridge project		J13010-2015	中铁二十二局哈尔滨铁路建设集团有限责任公司
45	DB23/T 1621.17-2015	黑龙江省建设工程施工操作技术规程 市政给排水管道工程 Technical specification for construction engineering operation in Heilongjiang Province municipal water supply and drainage pipeline project		J13011-2015	哈尔滨市供水集团公司市政建设有限公司
46	DB23/T 1621.18-2015	黑龙江省建设工程施工操作技术规程 市政给排水构筑物工程 Technical specification for construction engineering operation in Heilongjiang Province construction of municipal water and sewerage structures		J13012-2015	哈尔滨市供水集团公司市政建设有限公司
47	DB23/T 1621.19-2015	黑龙江省建设工程施工操作技术规程 市政燃气工程 Technical specification for construction engineering operation in Heilongjiang Province municipal gas engineering		J13013-2015	哈尔滨市气化工程质量监督站
48	DB23/T 1640-2015	村镇建筑热湿环境检测及空气品质治理规程 Specification for thermal environment measurement and indoor air quality treatment of rural buildings		J12987-2015	哈尔滨工业大学
49	DB23/T 1641-2015	小型集中供热系统节能设计及运行管理规程 Specification for energy efficient design and operation management of small centralized heating system		J12988-2015	哈尔滨工业大学

黑龙江省

序号	标准编号	标准名称	被代替标准编号	备案号	主编单位
50	DB23/T 1642-2015	黑龙江省绿色建筑评价标准 Assessment standard for green building of Heilongjiang Province		J13073-2015	黑龙江省寒地建筑科学研究院 哈尔滨市城乡建设委员会
51	DB23/T 1664-2015	轻集料混凝土复合保温砌块（侧腔气阻型）墙体应用技术规程 Technical specification for application of lightweight aggregate insulation concrete compound blocked (side chamber air-lock type) walls		J13218-2015	黑龙江省寒地建筑科学研究院
52	DB23/1673-2015	园林工程施工及质量验收规范 Code for construction and acceptance of landscape engineering		J13246-2015	黑龙江国华环境规划设计咨询有限公司

上海市

序号	标准编号	标准名称	被代替标准编号	备案号	主编单位
1	DG/TJ 08-01-2014	上海市工程建设标准体系表 Technical specification of maintenance for city bridges		J12871-2014	上海市建筑建材业市场管理总站 上海现代建筑设计（集团）有限公司
2	DG/TJ 08-003-2013	建筑锚栓抗拉拔、抗剪性能试验方法 Tension and shear test methods for building anchors		J12474-2013	上海市建筑科学研究院（集团）有限公司
3	DG/TJ 08-7-2014	建筑工程交通设计及停车库（场）设置标准 Standards for design of traffic and setting up parking garages (lots) in architectural engineering		J10716-2014	上海市交通委员会 上海市公安局交通警察总队 同济大学
4	DG/TJ 08-008-2000	建筑钢结构防火技术规程 Technical code on fire safety of steel building structures		J10041-2000	同济大学 上海市消防协会
5	DGJ 08-9-2013	建筑抗震设计规程 Code for seismic design of buildings		J10284-2013	同济大学
6	DGJ 08-10-2004	城市煤气、天然气管道工程技术规程 Technicat specification of manufactured gas and natunel gas pipeline engineering		J10472-2005	上海燃气工程设计研究有限公司 上海燃气（集团）有限公司 华东建筑设计研究院有限公司
7	DG/TJ 08-010-2001	轻型钢结构制作及安装验收规程 Specification procedures for fabrication and erection of light steel structure		J10125-2002	上海市金属结构协会
8	DGJ 08-11-2010	地基基础设计规范 Foundation design code		J11595-2010	上海现代设计（集团）有限公司
9	DG/TJ 08-12-2004	普通中小学校建设标准 Standard for schoolconstruction of general and primary secondary and high schools		J10355-2004	上海市教育委员会
10	DG/TJ 08-012-2002	纤维增强复合材料加固混凝土结构技术规程 Technical code for strengthening concrete structure with fiber rcinforced polymer		J10158-2002	上海市建筑科学研究院
11	DG/TJ 08-15-2009	绿地设计规程 Code for the design of green space	DBJ 08-15-89	J11525-2009	上海市绿化和市容管理局 上海市园林设计院
12	DG/TJ 08-015-2004	高层建筑钢—混凝土混合结构设计规程 Code for design of steel-concrete hybrid structure for high-rise building		J10285-2003	同济大学 上海金属结构行业协会
13	DG/TJ 08-16-2011	钢管扣件式模板垂直支撑系统安全技术规程 Technology code for safety of fastener type steel pipe formwork vertical support		J10347-2012	上海市建工设计研究院有限公司
14	DG/TJ 08-18-2011	园林绿化植物栽植技术规程 Planting technical specification for garden plants		J11913-2011	上海市绿化和市容管理局
15	DG/TJ 08-19-2011	园林绿化养护技术规程 Technical specification of landscape gardening	DBJ 08-19-91	J11852-2011	上海市绿化和市容管理局
16	DG/TJ 08-019-2005	建筑结构用索应用技术规程 Technical specification for cable applications in building structures		J10553-2005	上海建筑设计研究院有限公司

上海市

序号	标准编号	标准名称	被代替标准编号	备案号	主编单位
17	DGJ 08-20-2013	住宅设计标准 Design standard for residential buildings		J10090-2014	上海建筑设计研究院有限公司 上海市建筑建材业市场管理总站
18	DG/TJ 08-020-2005	混凝土结构工程施工技术 Specification for construction of concrete structures		J10618-2005	上海建工（集团）总公司
19	DGJ 08-20-2013	住宅设计标准（2016年局部修订） Design standard for residential buildings		J10090-2014	上海建筑设计研究院有限公司 上海市建筑建材业市场管理总站
20	DG/TJ 08-21-2013	砌体工程施工规程 Specification for construction of masonry engineering		J12324-2013	上海建工集团股份有限公司
21	DG/TJ 08-22-2013	屋面工程施工规程 Construction regulation for roof engineering		J12320-2013	上海建工集团股份有限公司
22	DGJ 08-22-2003	城市排水泵站设计规程 Specification for design of drainage pumping station in municipallity		J10316-2003	上海市政工程设计研究院
23	DG/TJ 08-023-2006	多层住宅平屋面改坡屋面工程技术规程 Code of applied technique for roof renovation		J10817-2006	上海市房屋土地资源管理局
24	DG/TJ 08-32-2008	高层建筑钢结构设计规程（修编） Specification for steel structure design of tall buildings		J11195-2008	上海市机电设计研究院 同济大学 中船第九设计研究院工程有限公司
25	DG/TJ 08-35-2014	绿化植物保护技术规程 Technical regulations of plant protection for urban green space		J12723-2014	上海市绿化和市容管理局 上海市园林科学研究所 上海市绿化管理指导站
26	DGJ 08-37-2012	岩土工程勘察规范 Code for investigation of geotechnical engineering		J12034-2012	上海岩土工程勘察设计研究院有限公司
27	DG/TJ 08-40-2010	地基处理技术规范 Ground treatment code	DBJ 08-40-1994	J11631-2010	同济大学
28	DG/TJ 08-45-2005	普通幼儿园建设标准 The construction standards of general kindergarten			上海市教育委员会
29	DG/TJ 08-50-2012	隧道工程防水技术规程 Technical specification for tunnel waterproofing		J12117-2012	上海市隧道工程轨道交通设计研究院
30	DG/TJ 08-52-2004	空间格构结构设计规程 Specification for design of reticulated structures		J10508-2005	同济大学 上海建筑设计研究院有限公司
31	DG/TJ 08-53-2016	行道树栽植技术规程 Technical regulations of transplanting for street tree		J13350-2016	上海市绿化管理指导站
32	DGJ 08-55-2006	城市居住地区和居住区公共服务设施设置标准 Standard for public facilites of urban residential area and district		J10059-2006	上海市城市规划管理局 上海市城市规划设计研究院
33	DGJ 08-56-2012	建筑幕墙工程技术规范 Technical code for building curtain wall		J12028-2012	上海市金属结构行业协会 上海信安幕墙建筑装饰有限公司 沈阳远大铝业工程有限公司

上海市

序号	标准编号	标准名称	被代替标准编号	备案号	主编单位
34	DGJ 08-59-2006	钢锭铣削型钢纤维混凝土应用技术规程 Technical specification for application of mill-cut steel fiber reinforced concrete	DBJ 08-59-97	J10949-2006	上海市建筑科学研究院有限公司
35	DGJ 08-60-2006	机械式停车库（场）设计规程 Specification for design of mechanical parking garages (lots)	DBJ 08-60-97	J10924-2007	上海自动化车库研究所 上海市公安局交警总队
36	DG/TJ 08-61-2010	基坑工程技术规范 Technical code for excavation engineering	DG/TJ 08-2001-2006	J11577-2010	上海市勘察设计行业协会 上海现代设计（集团）有限公司 上海建工（集团）有限公司
37	DG/TJ 08-66-2016	花坛、花境技术规程 Technical regulation for flower beds and borders		J13351-2016	上海市绿化和市容管理局
38	DG/TJ 08-67-2015	园林绿化草坪建植和养护技术规程 Technical guidelines for landscaping turf establishment and maintenance		J12989-2015	上海市园林科学研究院
39	DGJ 08-69-2015	预应力混凝土结构设计规程 Code for design of prestressed concrete structures	DGJ 08-69-2007	J13156-2015	同济大学 上海市建筑科学研究院（集团）有限公司 上海市勘察设计行业协会
40	DGJ 08-70-2013	建筑物、构筑物拆除技术规程 Regulation for demolishing construction		J12367-2013	上海市住房保障和房屋管理局
41	DG/TJ 08-72-2012	岩土工程勘察文件编制深度规定 Specification of geotechnical investigation report compiling depth		J12190-2012	上海市勘察设计行业协会 上海岩土工程勘察设计研究院有限公司
42	DGJ 08-74-2004	燃气直燃型吸收式冷热水机组工程技术规程 Technical specification for gas direct-fired absorption chiller and heater engineering		J10430-2004	上海燃气（集团）有限公司 上海燃气工程设计研究有限公司
43	DG/TJ 08-75-2014	立体绿化技术规程 Technical specification for green building planting		J12714-2014	上海市绿化和市容管理局 上海市规划和国土资源管理局
44	DG/TJ 08-79-2008	房屋质量检测规程 The code of building quality inspection	DGJ 08-79-99	J11208-2008	上海市房地产科学研究院 上海市房屋检测中心
45	DGJ 08-81-2015	现有建筑抗震鉴定与加固规程 Code for seismic appraisement and strengthening of existing buildings		J10016-2014	同济大学
46	DGJ 08-82-2000	养老设施建筑设计标准 Architectural design standard of facilities for the elderly			上海市工程建设标准化办公室 上海市民政局
47	DG/TJ 08-83-2009	防静电工程技术规程 Technical regulation for protection of electrostatic discharge	DGJ 08-83-2000	J10011-2009	上海市建设工程安全质量监督总站上海现代建筑设计（集团）有限公司
48	DG/TJ 08-85-2010	地下管线测绘规范 Code for surveying and mapping underground pipelines and cables	DGJ 08-85-2000	J10046-2010	上海市测绘院
49	DG/TJ 08-86-2010	1∶500 1∶1000 1∶2000 数字地形测量规范 Surveying code for 1∶500 1∶1000 1∶2000 topographic maps	DGJ 08-86-2000	J11696-2010	上海市测绘院

上海市

序号	标准编号	标准名称	被代替标准编号	备案号	主编单位
50	DG/TJ 08-87-2009	道路、排水管道成品与半成品施工及验收规程 The specification for road's and drainage pipe's product and semi-product construction and acceptance	DGJ 08-87-2000	J11540-2010	上海建设机场道路工程有限公司 上海市市政规划设计研究院
51	DGJ 08-88-2006	建筑防排烟技术规程 Technical specification for building smoke control		J10035-2006	公安部上海消防科研所 上海市消防局
52	DGJ 08-89-2000	网架与网架工程质量检验及评定标准 Specification for examination acteptancl and quality qualification of reticulated structures		J10048-2000	同济大学 上海建筑设计研究院
53	DG/TJ 08-90-2014	水利工程施工质量检验与评定标准 Inspection and assessment standard for construction quality of hydraulic engineering		J10053-2014	上海市水务局 上海市水务建设工程安全质量监督中心站
54	DG/TJ 08-92-2013	城市道路养护技术规程 Technical regulations for urban road maintenance		J12361-2013	上海市路政局
55	DGJ 08-93-2002	民用建筑电线电缆防火设计规程 Code for fireproofing design of wires and cables used in civil buildings		J10166-2002	华东建筑设计研究院有限公司
56	DGJ 08-94-2007	民用建筑水灭火系统设计规程		J11056-2007	上海现代建筑设计（集团）有限公司 华东建筑设计研究院有限公司
57	DGJ 08-95-2001	铝合金格构结构技术规程（试行） Technical specification for design and construction of aluminum-alloys reticulated structures		J10084-2001	同济大学 上海建筑设计研究院
58	DGJ 08-96-2013	城市道路平面交叉口规划与设计规程 Design regulations for at-grade intersections on urban street		J10099-2013	同济大学 上海市公安局交通警察总队
59	DGJ 08-97-2002	膜结构技术规程 Technical specification of membrane structures		J10209-002	上海现代建筑设计（集团）有限公司
60	DGJ 08-98-2014	机动车停车场（库）环境保护设计规程 Instruction for the environmental protection designing of vehicular parking		J12859-2014	上海市环境科学研究院有限公司 上海市机电设计研究院
61	DGJ 08-99-2003	10kV预装式变电站应用设计规程 Installation desion code for 10kV prefabricated substations		J10239-2003	上海市电力公司 上海电力设计研究院有限公司
62	DGJ 08-100-2003	低压用户电气装置规程 Regulations governing electrical installations supplied with low voltage		J10247-2003	上海市电力公司
63	DGJ 08-101-2003	城市轨道交通信号系统技术规程 Technical specification for signal system of urban rail transit		J10258-2003	上海市城市交通管理局
64	DGJ 08-102-2003	城镇高压、超高压天然气管道工程技术规程 Technical specification of city high pressure and super hiigh perssure natural gas distribution pipeline engineering		J10263-2003	上海燃气工程设计研究有限公司

上海市

序号	标准编号	标准名称	被代替标准编号	备案号	主编单位
65	DGJ 08-103-2003	无障碍设施设计标准 Standard for design on accessibility of buildings and facilities		J10264-2003	上海市建设工程标准定额管理总站 上海现代建筑设计（集团）有限公司
66	DG/TJ 08-104-2014	城市轨道交通专用无线通信系统技术规范 Technical specification for operation exclusive radio system of urban rail transit		J10343-2014	上海市交通委员会 上海市申通轨道交通研究咨询有限公司
67	DGJ 08-106-2015	城市轨道交通工程车辆选型技术规范 Technical code of vehicle selecting for urban rail transit engineering		J10269-2015	上海市交通委员会 上海申通地铁集团有限公司
68	DGJ 08-107-2015	公共建筑节能设计标准 Design standard for energy efficiency in public buildings	DGJ 08-107-2012	J12068-2015	上海现代建筑设计（集团）有限公司 同济大学
69	DG/TJ 08-108-2014	优秀历史建筑保护修缮技术规程 Technical specifications for preservation and restoration of historic buildings		J10319-2014	上海市住房保障和房屋管理局 上海市房地产科学研究院 上海市历史建筑保护事务中心
70	DGJ 08-109-2004	城市轨道交通设计规范 Urban rail transit design standard		J10325-2004	上海市隧道工程轨道交通设计研究院 上海轨道交通学科（专项技术）研究发展中心
71	DGJ 08-110-2004	饮食行业环境保护设计规程 Specification for the environmental protection design of catering trade		J10473-2004	上海市环境科学研究院
72	DG/TJ 08-111-2014	城市轨道交通信息传输系统技术规程 Technical specification for information transmission system of urban rail transit		J12840-2014	上海市交通委员会 上海申通轨道交通研究咨询有限公司
73	DGJ 08-112-2005	机动车隧道机电设备安装工程施工质量验收规范 Code for acceptance of construction quality of mechanical and electrical equipment installation of vehicle tunnel		J10531-2005	上海市市政工程质量监督站
74	DGJ 08-113-2009	建筑节能工程施工质量验收规程 Code for acceptance of energy efficient building construction	DGJ 08-113-2005	J11594-2010	上海市建筑科学研究院（集团）有限公司 上海市建筑建材业市场管理总站 上海市建设工程安全质量监督总站
75	DGJ 08-114-2016	临时性建（构）筑物应用技术规程 Technical regulations for temporary work operation		J13370-2016	上海市建设工程安全质量监督总站
76	DG/TJ 08-115-2016	燃气分布式供能系统工程技术规程 Technical specification of engineering for gas distributed energy system		J10602-2016	国网上海市电力公司 上海燃气（集团）有限公司
77	DGJ 08-119-2005	公路工程施工质量验收规范 Code for acceptance of construction quality of highway engineering		J10632-2005	上海市市政工程管理局 上海市公路学会
78	DGJ 08-120-2006	城市轨道交通安全运营技术规范 Technical specifications for safety operation of mass rail transit		J10766-2006	上海市城市交通管理局

上海市

序号	标准编号	标准名称	被代替标准编号	备案号	主编单位
79	DGJ 08-121-2006	建设工程扬尘污染防治规范 Dust pollution control code on construction		J10767-2006	上海市建设工程安全质量监督总站
80	DG/TJ 08-202-2007	钻孔灌注桩施工规程 Specification for construction of bored cast-in-place pile	DBJ 08-202-92	J11042-2007	上海现代建筑设计（集团）有限公司 上海建工（集团）总公司
81	DGJ 08-205-2015	居住建筑节能设计标准 Design standard for energy efficiency of residential buildings	DG/TJ 08-205-2011	J10044-2015	上海市建筑科学研究院（集团）有限公司 上海市建筑建材业市场管理总站
82	DG/TJ 08-206-2002	住宅建筑围护结构节能应用技术规程 Code of applied technique for energy conservation on the envelope of residential building		J10186-2002	上海市建筑科学研究院 上海市建筑节能办公室
83	DG/TJ 08-207-2008	房屋修缮工程技术规程 Technical specification for repair construction of buildings	DBJ 08-207-92	J11165-2008	上海市房屋土地资源管理局
84	DG/TJ 08-211-2014	假山叠石工程施工规程 Specification for rockery laying		J12724-2014	上海市绿化和市容（林业）工程管理站 上海市园林管理局
85	DG/TJ 08-216-2016	钢结构制作与安装规程 Specification for fabrication and installation of steel structures		J11103-2016	上海市金属结构行业协会 上海建工（集团）总公司
86	DGJ 08-218-2003	建筑基桩检测技术规程 Technical specification for testing of building foundation piles		J10287-2003	上海市建筑科学研究院
87	DG/TJ 08-227-2014	预拌混凝土生产技术规程 Technical specification for ready-mixed concrete		J11462-2014	上海市建筑科学研究院（集团）有限公司 上海市建设工程安全质量监督总站 上海市混凝土行业协会
88	DG/TJ 08-230-2006	粉煤灰混凝土应用技术规程 Practice code for application of fly ash concrete		J10868-2006	上海市粉煤灰综合利用办公室
89	DG/TJ 08-231-2013	园林绿化栽植土质量标准 Quality standards of landscape gardening soil		J12562-2014	上海市园林科学研究所
90	DBJ 08-232-98	道路交通管理设施施工验收规程 Regulations for road traffic control equipment installation and acceptance			上海市公安局交通警察总队
91	DGJ 08-234-2001	玻璃纤维增强塑料夹砂排水管道施工及验收规程 The code for installation and acceptance of glass fiber reinforced plastic mortar pipe for sewerage			上海市市政工程管理局
92	DG/TJ 08-235-2012	后张预应力施工规程 Specification for post-tensioned construction		J12145-2012	上海市建筑科学研究院（集团）有限公司 上海建科预应力技术工程有限公司 上海建工集团股份有限公司
93	DG/TJ 08-236-2013	市政地下工程施工质量验收规范 Code for acceptance of construction quality of municipal underground engineering		J10819-2013	上海市市政工程管理局 上海城建（集团）公司 上海市隧道工程轨道交通设计研究院 上海市土木工程学会

上海市

序号	标准编号	标准名称	被代替标准编号	备案号	主编单位
94	DG/TJ 08-308-2002	埋地塑料排水管道工程技术规程 Technical specification of buried plastic pipeline for sewer engineering		J10185-2002	上海市市政工程研究院 上海现代建筑设计（集团）有限公司
95	DG/TJ 08-309-2005	建筑给水塑料管道工程技术规程 Engineering technical specification of plastic pipeline for building water supply		J10507-2005	上海现代建筑设计（集团）有限公司 上海沪标工程建设咨询有限公司
96	DG/TJ 08-401-2007	城市公共厕所规划和设计标准（修编） Standards for planning design of public toilet in city code	DBJ 08-401-99	J11049-2007	上海市环境工程设计科学研究院
97	DG/TJ 08-402-2000	小型压缩式生活垃圾收集站设置标准 Setting standard for small domestic garbage collection stationwith compaction		J10015-2000	上海市环境工程设计科学研究院
98	DG/TJ 08-501-2016	粒化高炉矿渣粉在水泥混凝土中应用技术规程 Code for utility technique of ground granulated blast-furnace slag used in concrete	DG/TJ 08-501-2008	J11239-2016	上海市建筑科学研究院（集团）有限公司 上海宝钢新型建材科技有限公司
99	DG/TJ 08-502-2012	预拌砂浆应用技术规程 Technical specification for application of ready-mixed mortar		J10012-2012	上海市建筑科学研究院（集团）有限公司
100	DG/TJ 08-503-2000	高强泵送混凝土生产和施工规程 Production and construction code for pumped high-strength concrete		J10014-2000	上海建工（集团）总公司
101	DG/TJ 08-504-2014	外墙涂料工程应用技术规程 Technical specification of application for exterior wall coating		J10023-2014	上海市建筑科学研究院（集团）有限公司
102	DG/TJ 08-506-2002	机制砂在混凝土中应用技术规程 Practice code for application of concrete with mechanical sand		J10137-2002	上海建工材料工程有限公司
103	DG/TJ 08-507-2003	高强混凝土抗压强度非破损检测技术规程 The specification code for non-destructively testing the compressive strength of high-strength concrete		J10246-2003	上海市建筑科学研究院
104	DG/TJ 08-601-2009	智能建筑施工及验收规范 Construction & acceptance specification of intelligent building	DG/TJ 08-601-2001	J10111-2010	上海市安装行业协会 上海市安装工程有限公司
105	DG/TJ 08-604-2013	住宅小区智能化应用技术规程 Technical code for intelligent systems in residential district		J10138-2013	上海市建筑科学研究院（集团）有限公司 上海市智能建筑建设协会
106	DG/TJ 08-606-2011	住宅建筑通信配套工程技术规范 Technical specifications for communication accessory project of residential buildings	DG/TJ 08-606-2004	J10334-2011	上海市通信管理局
107	DG/TJ 08-701-2008	园林绿化工程施工质量验收标准 Code for acceptance of construction quality of gardening and landscapes engineering	DG/TJ 08-701-2000	J10042-2009	上海市绿化和市容管理局
108	DG/TJ 08-702-2011	园林绿化养护技术等级标准 Grade technical standards of landscape gardening maintenance		J10603-2011	上海市绿化和市容管理局
109	DG/TJ 08-803-2013	建筑幕墙安全性能检测评估技术规程 Testing and evaluating technical code for building curtain walls safety performance		J12438-2013	上海市建筑科学研究院（集团）有限公司 上海建科检验有限公司

上海市

序号	标准编号	标准名称	被代替标准编号	备案号	主编单位
110	DG/TJ 08-804-2005	既有建筑物结构检测与评定标准 Standard of structural inspection and assessment for existing buildings		J10616-2005	同济大学 上海市房屋检测中心
111	DG/TJ 08-901-2014	城市轨道交通站台屏蔽门技术规程 Technical specification for platform screen		J12896-2014	上海市交通委员会 上海申通地铁集团有限公司
112	DG/TJ 08-902-2006	旁通道冻结法技术规程 Technical code for crosspassage freezing method		J10851-2006	上海申通轨道交通研究咨询有限公司
113	DGJ 08-903-2010	现场施工安全生产管理规范 Site construction safety management specification		J11761-2010	上海市建设工程安全质量监督总站 上海市建筑施工行业协会工程质量安全专业委员会
114	DG/TJ 08-1001-2013	岩土工程勘察外业操作规程 Site operating specification for geotechnical investigation		J12363-2013	上海岩土工程勘察设计研究院
115	DGJ 08-1101-2007	城市轨道交通自动售检票系统技术规程 General technical specification for automatic fare collection system of urban rail transit		J10510-2007	上海市城市交通管理局
116	DGJ 08-1102-2005	城市轨道交通单程票非接触集成电路（IC）卡通用技术规范 General technical specification for contactless integrated circuit (IC) card of single journey ticket for urban rail transit	DGJ 08-1102-2005	J10511-2006	上海市标准化研究所 上海城市发展信息研究中心 上海公共交通卡股份有限公司 上海地铁运营有限公司 上海申通集团有限公司
117	DGJ 08-1103-2005	城市公共交通非接触集成电路（IC）卡通用技术规范 General technical specification for contactless integrated circuit (IC) card of urban public transport	DGJ 08-1103-2005	J10512-2006	上海市标准化研究所 上海城市发展信息研究中心 上海公共交通卡股份有限公司 上海地铁运营有限公司 上海申通集团有限公司
118	DG/TJ 08-1104-2005	公共建筑电磁兼容设计规范 The code of electromagnetic compatibility design for public buildings		J10532-2005	上海建筑设计研究院有限公司
119	DG/TJ 08-1105-2010	移动通信室内信号覆盖系统设计与验收规范 Standard for design & acceptance of mobile communication indoor signal covering system	DG/TJ 08-1105-2006	J10689-2010	上海市信息系统质量技术协会
120	DG/TJ 08-1201-2005	施工现场安全质量保证体系 Engineering quality assurance system on construction		J10563-2005	上海市建设工程安全质量监督总站
121	DG/TJ 08-1202-2011	建设工程造价咨询规范 Code of practice for construction cost consultants	DG/TJ 08-1202-2006	J10810-2011	上海市建筑建材业市场管理总站 上海市建设工程咨询行业协会
122	DG/TJ 08-1401-2006	车用压缩天然气加气站设备安全技术规程 Technical code for equipment safety of automobile CNG filling station		J10794-2006	上海市城市交通管理局
123	DG/TJ 08-2001-2016	基坑工程施工监测规程 Specification for foundation excavation monitoring		J13459-2016	上海岩土工程勘察设计研究院有限公司
124	DG/TJ 08-2002-2006	悬挑式脚手架安全技术规程 Technical code for safety of steel tubular scaffold		J10885-2006	上海市建工设计研究院有限公司

上海市

序号	标准编号	标准名称	被代替标准编号	备案号	主编单位
125	DG/TJ 08-2003-2006	建筑外立面附加设施设置安全技术规程 Technical specification for adjunction on external elevation of building		J10894-2006	上海市工程建设质量管理协会
126	DG/TJ 08-2004A-2014	太阳能热水系统应用技术规程 Specification of applied technique for solar hot water system		J12916-2015	上海交通大学 华东建筑设计研究院有限公司 上海市建筑科学研究院（集团）有限公司
127	DG/TJ 08-2004B-2008	民用建筑太阳能应用技术规程（光伏发电系统分册） Technical specification for solar energy application in civil buildings (fascicule: photovoltaic system)		J11326-2008	上海交通大学 上海市电力公司 上海市电力设计院有限公司
128	DG/TJ 08-2005-2006	城市轨道交通机电设备安装工程质量验收规范 Quality inspection & acceptance code of M&E installation of urban rail transit		J10913-2006	上海城建（集团）公司
129	DG/TJ 08-2006-2006	配筋混凝土小型空心砌块砌体建筑技术规程 Technical specification for reinforced concrete small-sized hollow block masonry building		J10898-2006	同济大学 上海中房建筑设计有限公司
130	DGJ 08-2007-2016	地质灾害危险性评估技术规程 Technical code for risk assessment of geological disaster	DGJ 08-2007-2006	J10897-2015	上海市规划和国土资源管理局 上海市地质调查研究院
131	DG/TJ 08-2008-2006	建筑地面工程施工规程 Specification for construction of building ground		J10963-2007	上海建工（集团）总公司
132	DG/TJ 08-2009-2016	有线网络建设技术规范 Technical standard for cable network construction		J13341-2016	东方有线网络有限公司
133	DG/TJ 08-2011-2007	钢结构检测与鉴定技术规程 Technical specification for inspection and appraisal of steel structures		J10973-2007	同济大学 上海市金属结构行业协会
134	DG/TJ 08-2012-2007	燃气管道设施标识应用规范 Application specification of marker for gas pipeline and facility		J10975-2007	上海市燃气管理处
135	DG/TJ 08-2013-2007	钢渣粉混凝土应用技术规程 Practice code for application of steel slag powder concrete		J10974-2007	上海市市政规划设计研究院 上海市建筑科学研究院有限公司
136	DG/TJ 08-2014-2007	液化天然气事故备用调峰站设计规程 Design code for LNG emergency peak-shaving station		J10938-2007	上海市燃气（集团）有限公司 上海燃气工程设计研究有限公司
137	DGJ 08-2015-2007	郊区中心村住宅设计标准 Design standard of housing for central villages		J11003-2007	上海市房屋土地资源管理局
138	DG/TJ 08-2016-2007	上海市郊区新市镇与中心村规划编制技术标准（试行） Technical standard of plans for new towns and central villages in Shanghai suburbs		J10990-2007	上海市城市规划管理局
139	DGJ 08-2017-2014	综合管廊工程技术规范 Technical code for municipal tunnel engineering	DG/TJ 08-2017-2007	J10982-2014	上海市政工程设计研究总院（集团）有限公司 同济大学

上海市

序号	标准编号	标准名称	被代替标准编号	备案号	主编单位
140	DG/TJ 08-2018-2007	再生混凝土应用技术规程 Technical codeon the application of recycled concrete		J10995-2007	同济大学 上海市建筑科学研究院有限公司 上海市市政工程研究院
141	DG/TJ 08-2019-2007	膜结构检测技术规程 Technical specification for inspection		J11015-2007	同济大学 上海建筑设计研究院有限公司
142	DG/TJ 08-2020-2007	结构混凝土抗压强度检测技术规程 Technical specification for inspection of structure concrete		J11027-2007	上海市建设工程检测行业协会
143	DG/TJ 08-2021-2007	商品砌筑砂浆现场检测技术规程 The specification code for site testing of manufactured masonry mortar		J11014-2007	上海市建设工程检测行业协会
144	DG/TJ 08-2022-2007	油浸式电力变压器火灾报警与灭火系统技术规程 Technical specification of fire-alarm and extinguishing system for oil-immersed power transformers		J11039-2007	上海市电力公司
145	DG/TJ 08-2023-2007	集约化通信局房设计规范 Design code for multi-operator integrated communication station		J11022-2007	上海市信息系统质量技术协会 上海现代建筑设计（集团）有限公司 上海建瓴建设科技发展有限公司
146	DG/TJ 08-2024-2007	用户高压电气装置规范 Code of high voltage electrical installations		J11051-2007	上海市电力公司
147	DG/TJ 08-2025-2007	建设工程远程监控系统应用技术规程 Application technology specification for remote supervisory system in construction project		J11050-2007	上海市建设工程安全质量监督总站
148	DG/TJ 08-2027-2007	城市居住区交通组织规划与设计规程 Specification for traffic organizing planning and design of urban residential district		J11040-2007	同济大学
149	DG/TJ 08-2029-2007	多高层钢结构住宅技术规程 Technical specification for multi-story and high-rise steel residential buildings		J11102-2007	同济大学 上海建筑设计研究院有限公司 上海宝钢工程技术有限公司
150	DG/TJ 08-2030-2007	建设工程规划检测规范 Specification for inspecting of building planning by surveying and mapping		J11125-2008	上海市测绘院
151	DG/TJ 08-2031-2007	城镇燃气管道工程施工质量验收标准 Standard for construction quality acceptance of city gas pipeline engineering		J11126-2008	上海市建设工程质量监督站公用事业分站 上海燃气（集团）有限公司
152	DG/TJ 08-2033-2008	道路隧道设计规范 Design code for road tunnel		J11197-2008	上海市隧道工程轨道交通设计研究院
153	DG/TJ 08-2034-2008	预拌混凝土和预制混凝土构件生产质量质量管理规程 Code for quality management of ready-mixed concrete and precast concrete component		J11198-2008	上海市建设工程安全质量监督总站 上海市建筑科学研究院有限公司
154	DG/TJ 08-2035-2014	建设工程监理施工安全监督规程 Safety supervision code for engineering construction		J12649-2014	上海市建设工程咨询行业协会 上海市建设工程安全质量监督总站

上海市

序号	标准编号	标准名称	被代替标准编号	备案号	主编单位
155	DG/TJ 08-2036-2008	既有民用建筑能效评估标准		J11200-2008	上海市房地产科学研究院
156	DG/TJ 08-2037-2016	城市轨道交通自动售检票系统（AFC）检测规程 Inspecting specification for automatic fare collection system of urban rail transit	DG/TJ 08-2037-2008	J11177-2016	上海市城市交通管理局
157	DG/TJ 08-2038-2008	建筑围护结构节能现场检测技术规程 Code of applied techniques for in-situ inspection on energy efficiency for building envelope		J11209-2008	上海市建设工程检测行业协会
158	DG/TJ 08-2039-2008	上海市交通规划编制技术标准 Code for Shanghai transport planning		J11210-2008	上海市城市综合交通规划研究所 上海市城市规划管理局
159	DG/TJ 08-2040-2008	公共建筑节能工程智能化技术规程 Intelligent technical specification		J11199-2008	上海现代建筑设计（集团）有限公司 上海建筑科学研究院（集团）有限公司
160	DG/TJ 08-2041-2008	地铁隧道工程盾构施工技术规程 Technical standard for metro shield tunneling construction		J11317-2008	上海申通地铁集团有限公司 上海隧道工程股份有限公司
161	DG/TJ 08-2042-2008	建设工程检测管理规程 Management specification of construction inspection and testing		J11318-2008	上海市建设工程安全质量监督总站 上海市建设工程检测行业协会
162	DG/TJ 08-2043-2008	绿化林业信息获取及分类编码标准 Standard for information acquisition and classification codes of green and forest		J11238-2008	上海市绿化管理局 华东师范大学
163	DG/TJ 08-2044-2008	建设工程总承包管理规程 Code for management of engineering procurement construction		J11319-2008	上海现代建筑设计（集团）有限公司
164	DG/TJ 08-2045-2008	大型泵站设备设施运行规程 Running specification of requipments and construction of the large pumping station		J11320-2008	上海市城市排水市南运营有限公司
165	DG/TJ 08-2046-2008	建设项目（工程）竣工档案编制技术规范 Technical code for construction project archive filling and arrangement		J11321-2008	上海市城市建设档案馆
166	DG/TJ 08-2047-2013	公共建筑通信配套设施设计规范 Code of communication infrastructures designing for public buildings		J11322-2013	上海市通信管理局 上海建筑设计研究院有限公司 上海信息化发展研究协会
167	DG/TJ 08-2048-2008	民用建筑电气防火设计规程 Code for fireproofing design of electric in civil buildings		J11323-2008	上海现代建筑设计（集团）有限公司 华东建筑设计研究院有限公司 上海市消防局
168	DG/TJ 08-2049-2008	顶管工程施工规程 Specification for construction of pipe jacking		J11324-2008	上海建工（集团）总公司
169	DG/TJ 08-2050-2008	智能建筑工程应用技术规程 Technological specification for intelligent building		J11325-2008	上海市安装行业协会 上海市安装工程有限公司
170	DG/TJ 08-2051-2008	地面沉降监测与防治技术规程 Technical specification for land subsidence monitor and control		J11371-2009	上海市地质调查研究院

上海市

序号	标准编号	标准名称	被代替标准编号	备案号	主编单位
171	DG/TJ 08-2052-2009	公共汽（电）车中途站候车设施配套规范 Code for disposition of bus stop facilities		J11372-2009	上海市交通运输和港口管理局
172	DG/TJ 08-2053-2009	电梯安装作业平台技术规程 Technical specification of scaffold and scaffoldless method in elevator installation procedure		J11398-2009	上海市建设工程安全质量监督总站
173	DG/TJ 08-2054-2013	涉外建设项目视频安防监控系统设计规范 Design regulations of video monitoring system for foreign associated construction projects	DG/TJ 08-2054-2009	J11399-2013	上海市城乡建设和交通委员会安保设计审批处
174	DGJ 08-2055-2009	燃料电池汽车加氢站技术规程 Technical specification of hydrogen refuelling		J11330-2009	上海新奥九环车用能源股份有限公司 同济大学
175	DG/TJ 08-2056-2009	重型结构（设备）整体提升技术规程 Technical specification for lifting of heavy structure（equipment）		J11400-2009	同济大学 上海建工（集团）总公司
176	DG/TJ 08-2057-2009	公共汽车和电车首末站、枢纽站建设标准 Standards for bus and trolley bus terminal and interchange construction		J11467-2009	上海市交通运输和港口管理局 上海城市交通设计院
177	DG/TJ 08-2058-2009	生态公益林建设技术规程 Technical specification for non-commercial forest construction		J11460-2009	上海市绿化和市容管理局 上海市林业总站
178	DG/TJ 08-2059-2009	轻型木结构建筑技术规程 Technical specification for wood frame construction		J11461-2009	上海现代建筑设计（集团）有限公司
179	DG/TJ 08-2060-2009	公共汽（电）车车载信息系统技术规范 Code for technical requirement of bus information system		J11492-2009	上海市交通运输和港口管理局
180	DGJ 08-2061-2009	建设工程班组安全管理标准 The safety management standard of construction term		J11543-2010	上海市建设工程安全质量监督总站 上海市施工现场安全生产保证体系第一审核认证中心
181	DG/TJ 08-2062-2009	住宅工程套内质量验收规范 Code for acceptance of construction quality of house dwelling space		J11493-2009	上海市建设工程安全质量监督总站 上海建科建设监理咨询有限公司
182	DG/TJ 08-2063-2009	地铁土压平衡盾构机技术规程 Technical specification of metro shield machine		J11526-2009	上海隧道工程股份有限公司
183	DG/TJ 08-2064-2009	地下铁道建筑结构抗震设计规范 Code for seismic design of subway structures		J11527-2009	同济大学 上海申通轨道交通研究咨询有限公司
184	DG/TJ 08-2065-2009	住宅二次供水设计规程 Design specification for residential secondary water supply		J11528-2009	上海市水务局 上海市住房保障和房屋管理局 上海市政工程设计研究总院
185	DG/TJ 08-2066-2009	农村生活污水土地处理系统技术规范 Technical specification of rural domestic wastewater land treatment system		J11541-2010	上海交通大学 上海城投污水处理有限公司
186	DG/TJ 08-2067-2009	农村公路建设与养护技术规范 Technical standard for rural road construction and maintenance		J11539-2010	上海市公路管理处

上海市

序号	标准编号	标准名称	被代替标准编号	备案号	主编单位
187	DGJ 08-2068-2012	公共建筑用能监测系统工程技术规范 Technical code for energy consumption monitoring systems of public buildings		J11542-2012	上海市建筑科学研究院（集团）有限公司
188	DGJ 08-2069-2016	装配整体式混凝土结构预制构件制作与质量检验规程 Specification for manufacture and quality inspection of precast component for assembled monolithic concrete structure	DG/TJ 08-2069-2010	J11578-2016	上海建工集团股份有限公司 上海市建筑科学研究院（集团）有限公司
189	DGJ 08-2070-2010	房屋白蚁防治设计规程 Technical specification for termite prevention and control in buildings		J11599-2010	上海市房地产科学研究院
190	DG/TJ 08-2071-2010	装配整体式混凝土住宅体系设计规程 Code for design of assembled monolithic concrete buildings		J11660-2010	同济大学
191	DG/TJ 08-2072-2016	建设工程招标代理规范 Specification of construction project tendering agency	DG/TJ 08-2072-2010	J11659-2016	上海市建筑建材业市场管理总站 上海市建设工程咨询行业协会
192	DG/TJ 08-2073-2010	地下连续墙施工规程 Specification for construction of diaphragm wall		J11658-2010	上海建工（集团）总公司
193	DG/TJ 08-2074-2016	道路排水性沥青路面技术规范 Technical specification for drainage asphalt pavement of road		J11695-2016	上海市政工程设计研究总院
194	DG/TJ 08-2075-2010	管线定向钻进技术规范 Technical specifications for pipelines laying with directional drilling		J11722-2010	上海市非开挖技术协会
195	DG/TJ 08-2076-2010	户外发光二极管（LED）显示屏安装技术规程 Technical specification for installation outdoor LED display screen		J11746-2010	上海市信息系统质量技术协会
196	DGJ 08-2077-2010	危险性较大的分部分项工程安全管理规范 Safety management specification for divisional work & subdivisional work with higher risk		J11755-2010	上海市建设工程安全质量监督总站 上海市建筑施工行业协会工程质量安全专业委员会
197	DG/TJ 08-2078-2014	建筑能效标识技术标准 Standard for building energy performance certification	DG/TJ 08-2078-2010	J11788-2015	上海市建筑科学研究院 上海市建筑建材业市场管理总站
198	DG/TJ 08-2079-2010	土地开发整理工程建设技术标准 Standard for land consolidation engineering		J11790-2011	上海市规划和国土资源管理局 上海市地质调查研究院
199	DG/TJ 08-2080-2010	建筑起重机械安全检验与评估规程 Inspection and assessment regulation for building crane safety capability		J11789-2011	上海市建筑科学研究院（集团）有限公司
200	DG/TJ 08-2081-2011	空调水系统化学处理设计规程 Design standard of chemical water treatment for HVAC water systems		J11830-2011	华东建筑设计研究院有限公司
201	DG/TJ 08-2082-2011	脱硫石膏土体增强剂应用技术规程 Technical specification for application of desulpho-gypsum soil consolidator		J11831-2011	上海市建筑科学研究院（集团）有限公司

上海市

序号	标准编号	标准名称	被代替标准编号	备案号	主编单位
202	DG/TJ 08-2083-2011	温拌沥青混合料路面技术规程 Technical regulations for pavements using warm mix asphalt		J11851-2011	上海市公路管理处 上海市市政工程管理处
203	DG/TJ 08-2084-2011	沉井与气压沉箱施工技术规程 Specification for open caisson and pneumatic caisson		J11875-2011	上海市基础工程有限公司
204	DG/TJ 08-2085-2011	脱硫石膏粉刷砂浆应用技术规程 echnical specification for application of gypsum plaster from flue gas desulfurization		J11876-2011	上海市建筑科学研究院（集团）有限公司
205	DG/TJ 08-2086-2011	道路声屏障结构技术规范 Technical specification for noise barrier of urban roads		J11877-2011	上海市市政工程管理处 上海市公路管理处 上海建设钢结构安全检测有限公司
206	DG/TJ 08-2087-2011	混凝土模卡砌块应用技术规程 Technical specification for application of concrete moka block		J11915-2011	上海市房屋建设设计研究院有限公司
207	DG/TJ 08-2088-2011	无机保温砂浆系统应用技术规程 Technical specification for thermal insulation systems based on inorganic insulation		J11914-2011	同济大学 上海市建筑科学研究院（集团）有限公司
208	DG/TJ 08-2089-2012	轻型钢结构技术规程 Technical specification for light weight steel building structures		J12002-2012	同济大学 上海市金属结构行业协会
209	DG/TJ 08-2090-2012	绿色建筑评价标准 Evaluation standard for green building		J12001-2012	上海市建筑科学研究院（集团）有限公司 上海市建筑建材业市场管理总站
210	DG/TJ 08-2091-2012	多联式空调（热泵）工程施工技术规程 Technological specification of multi-split air conditioning (heat pump) constructure		J12000-2012	上海市安装行业协会
211	DG/TJ 08-2092-2012	居住区信息系统网络互联技术规程 Residential area networks interconnecting specification		J12004-2012	上海市信息家电行业协会
212	DG/TJ 08-2093-2012	电动汽车充电基础设施建设技术规范 Construction specifications for electric vehicles' charging infrastructure		J12104-2012	上海市电力公司 上海交通大学 上海电力设计院有限公司
213	DG/TJ 08-2094-2012	内河航道信息化设施设置规范 Setting code for information construction of inland waterway		J12003-2012	上海市交通运输和港口管理局
214	DG/TJ 08-2095-2012	公路技术状况评定规程 Highway performance assessment specification		J12046-2012	上海市公路管理处
215	DG/TJ 08-2096-2012	生态公益林养护技术规程 Technical specification for tending of non-commercial forest		J12047-2012	上海市林业局
216	DGJ 08-2097-2012	地下管线探测技术规程 Technical specification for detecting and surveying underground pipeline sand cables		J12043-2012	上海市测绘院
217	DG/TJ 08-2098-2012	轻钢龙骨石膏板隔墙、吊顶应用技术规程 Technical specification for application of light steel keel gypsum partition wa		J12083-2012	上海市房地产科学研究院

上海市

序号	标准编号	标准名称	被代替标准编号	备案号	主编单位
218	DG/TJ 08-2099-2012	脱硫石膏轻质砌块、条板应用技术规程 Technical specification for application of lightweight FGD gypsum block and pane		J12082-2012	上海市房地产科学研究院
219	DG/TJ 08-2100-2012	人工湿地污水处理技术规程 Technical specification for wastewater treatment with constructed wetland		J12086-2012	上海市政工程设计研究总院（集团）有限公司
220	DG/TJ 08-2101-2012	上海市区（县）、镇（乡）土地利用总体规划编制标准 Standard for general land use planning of district and town level		J12087-2012	上海市规划和国土资源管理局 上海市地质调查研究院
221	DGJ 08-2102-2012	文明施工规范 Code for construction of HSE（health, safety and environment）		J12069-2012	上海市建设安全协会 上海市建筑业管理办公室
222	DG/TJ 08-2103-2012	城镇天然气站内工程施工质量验收标准 Standard for construction quality acceptance of engineering in city gas station		J12084-2012	上海市建设工程安全质量监督总站 上海燃气（集团）有限公司
223	DG/TJ 08-2104-2012	城市地面公共交通基础设施维护规程 Specification of urban bus infrastructure maintenance		J12085-2012	上海市交通运输和港口管理局
224	DG/TJ 08-2105-2012	行道树养护技术规程 Technical regulations for street tree maintenance		J12103-2012	上海市绿化和市容管理局
225	DGJ 08-2106-2012	城市道路设计规程 Specification for design of urban road engineering		J12100-2012	上海市政工程设计研究总院（集团）有限公司 上海市城市建设设计研究总院
226	DG/TJ 08-2107-2012	住宅小区移动通信系统设计和验收规范 Standard for design and acceptance of mobile telecom system in residential block		J12115-2012	上海信息化发展研究协会
227	DG/TJ 08-2108-2012	出租汽车站点设置规范 Standards for setting up taxi rank		J12105-2012	上海市交通运输和港口管理局
228	DG/TJ 08-2109-2012	橡胶沥青路面技术规范 Technical code for asphalt rubber pavement		J12116-2012	上海市政工程设计研究总院（集团）有限公司
229	DG/TJ 08-2110-2012	城镇排水工程施工质量验收规范 Code for construction quality acceptance of municipal sewerage engineering		J12114-2012	上海市排水管理处 上海市城市排水有限公司 上海市政工程设计研究总院（集团）有限公司
230	DG/TJ 08-2111-2012	滩涂促淤圈围造地工程设计规范 Design code for promoting sedimentation and land reclamation project		J12144-2012	上海市水利工程设计研究院有限公司 中船第九设计研究院工程有限公司
231	DG/TJ 08-2112-2012	环城绿带工程设计规范 Code for the design of greenbelt		J12143-2012	上海市绿化和市容管理局 上海市规划和国土资源管理局
232	DG/TJ 08-2113-2012	逆作法施工技术规程 Technical specification for construction of top-down method		J12191-2012	上海建工二建集团有限公司 华东建筑设计研究院有限公司
233	DG/TJ 08-2114-2012	公共建筑能源审计标准 Standard for energy audit of public buildings		J12192-2012	上海市建筑科学研究院（集团）有限公司 上海市建筑建材业市场管理总站

上海市

序号	标准编号	标准名称	被代替标准编号	备案号	主编单位
234	DG/TJ 08-2115-2012	城市网格化管理信息系统技术规范 Technical code for city grid management information system		J12228-2012	上海市数字化城市管理中心
235	DG/TJ 08-2116-2012	内河航道工程设计规范 Inland channel engineering design code		J12229-2012	上海市交通运输和港口管理局 中交上海航道勘察设计研究院有限公司
236	DGJ 08-2117-2012	装配整体式混凝土结构施工及质量验收规范 Code for construction and quality acceptance inspection of precast concrete structure		J12259-2013	上海市建设工程安全质量监督总站 上海建工集团股份有限公司 上海城建（集团）公司
237	DG/TJ 08-2118-2013	既有住宅建筑光纤到户改造工程技术规范 FTTH renovation project for existing residential building technical specification		J12321-2013	上海市通信管理局
238	DG/TJ 08-2119-2013	地源热泵系统工程技术规范 Technical specification for ground-source heat pump system		J12325-2013	上海市地矿工程勘察院 上海现代建筑设计（集团）有限公司
239	DG/TJ 08-2120-2013	集体土地所有权调查技术规范 Technical code for investigation of collective land ownership		J12368-2013	上海市规划和国土资源管理局 上海市地籍事务中心
240	DG/TJ 08-2121-2013	卫星定位测量技术规范 Technical code for surveying using satellite positioning system		J12362-2013	上海市测绘院 上海市城市建设设计研究总院
241	DG/TJ 08-2122-2013	保温装饰复合板墙体保温系统应用技术规程 Technical specification for application of wall thermal insulation system of insulation decorative plywood		J12364-2013	上海申城建筑设计有限公司
242	DG/TJ 08-2123-2013	盾构法隧道结构服役性能鉴定规范 Code for structural appraisal of shield tunnel		J12360-2013	同济大学
243	DG/TJ 08-2124-2013	排水泵站自动化系统设计规程 Design specification of automation system for drainage pump station		J12396-2013	上海市城市建设设计研究总院
244	DG/TJ 08-2125-2013	数据中心基础设施设计规程 Design specification for data center infrastructure		J12443-2013	上海现代建筑设计（集团）有限公司 上海信息化发展研究协会 上海电信科技发展有限公司
245	DG/TJ 08-2126-2013	岩棉板（带）薄抹灰外墙外保温系统应用技术规程 Application technology code of rock wool external thermal insulation systems		J12395-2013	同济大学 上海市建筑科学研究院（集团）有限公司 上海中房建筑设计有限公司
246	DG/TJ 08-2127-2013	机关办公建筑用能监测系统工程技术规范 Technical code for energy consumption monitoring systems of organ office buildings		J12394-2013	上海市机关事务管理局 上海市建筑科学研究总院（集团）有限公司 上海市建筑建材业市场管理总站
247	DG/TJ 08-2128-2013	轨道交通及隧道工程混凝土结构耐久性设计施工技术规程 Specification for durability design and construction of concrete structures in rail transport		J12444-2013	上海市建筑科学研究院（集团）有限公司

上海市

序号	标准编号	标准名称	被代替标准编号	备案号	主编单位
248	DG/TJ 08-2129-2013	建设工程绿色施工管理规范 Green construction management specification for construction project		J12440-2013	上海市建设工程安全质量监督总站 上海市建筑施工行业协会
249	DG/TJ 08-2130-2013	城市轨道交通基于通信的列车控制系统（CBTC）列车自动监控（ATS）技术规程 Technical specification for communication-based train control (CBTC) system's automatic train superv		J12439-2013	上海市交通运输和港口管理局 上海申通地铁集团有限公司
250	DG/TJ 08-2131-2013	路面设计规范 Code for pavement design of road		J12442-2013	上海市政工程设计研究总院（集团）有限公司
251	DG/TJ 08-2132-2013	地下防水工程橡胶防水材料成品检测规程 Test specification of rubber waterproof product for underground waterproof projects		J12475-2013	上海建科检验有限公司 上海申通地铁集团有限公司
252	DGJ 08-2133-2013	精细化工企业设计防火规范 Fire protection specification for the design of fine chemical enterprise		J12465-2013	上海华谊工程有限公司 上海市消防局
253	DG/TJ 08-2134-2013	建筑装饰工程石材应用技术规范 Technical code for application of stone in building decoration engineering		J12508-2013	上海石材行业协会
254	DGJ 08-2135-2013	建筑装饰装修工程施工规程 Specification for building renovation and decoration engineering		J12572-2014	上海建工集团股份有限公司 上海市装饰装修行业协会
255	DG/TJ 08-2136-2014	既有居住建筑节能改造技术规程 Technical specification for energy efficiency retrofit of existing residential buildings	DG/TJ 08-2010-2006	J10952-2014	上海市房地产科学研究院 上海市建筑科学研究院（集团）有限公司
256	DG/TJ 08-2137-2014	既有公共建筑节能改造技术规程 Technical specification for energy efficiency retrofitting of existing public buildings		J12587-2014	上海市房地产科学研究院 上海市建筑科学研究院（集团）有限公司
257	DG/TJ 08-2138-2014	发泡水泥板保温系统应用技术规程 Application technology code of the formed cement board thermal insulation system		J12650-2014	上海市建筑科学研究院 同济大学
258	DGJ 08-2139-2014	住宅建筑绿色设计标准 Green design standard for residential building		J12621-2014	同济大学建筑设计研究院（集团）有限公司上海建筑设计研究院有限公司
259	DG/TJ 08-2140-2014	城镇污水处理厂分类技术规范 Technical specification for classification of municipal wastewater treatment plants		J12648-2014	上海城投污水处理有限公司
260	DG/TJ 08-2141-2014	隧道LED照明应用技术规范 Application code for the LED tunnel lighting technology		J12715-2014	上海市隧道工程轨道交通设计研究院
261	DG/TJ 08-2142-2014	非承重蒸压灰砂多孔砖应用技术规程 Technical specification for application of non load bearing autoclaved sand-lime perforated brick		J12716-2014	上海市建筑科学研究院（集团）有限公司
262	DGJ 08-2143-2014	公共建筑绿色设计标准 Green design standard for public building		J12671-2014	同济大学建筑设计研究院（集团）有限公司 华东建筑设计研究院有限公司 华东建筑设计研究总院
263	DG/TJ 08-2144-2014	公路养护工程质量检验评定标准（土建工程） Quality inspection and evaluation standards for highway maintenance engineering		J12722-2014	上海市路政局

上海市

序号	标准编号	标准名称	被代替标准编号	备案号	主编单位
264	DG/TJ 08-2145-2014	城市桥梁养护技术规程 Technical specification of maintenance for city bridges		J12721-2014	上海市路政局
265	DG/TJ 08-2146-2014	既有建筑外立面整治设计规范 Code for design of existing building elevations renovation		J10600-2014	上海市住房保障和房屋管理局 上海市房地产科学研究院
266	DG/TJ 08-2147-2014	既有建筑幕墙维修工程技术规程 Technical specification for maintenance and repair of building curtain wall		J12717-2014	上海市装饰装修行业协会 上海市建筑科学研究院（集团）有限公司
267	DG/TJ 08-2149-2014	桥梁结构检测技术规程 Shanghai technical specification for bridge structure inspection		J12783-2014	上海市路政局
268	DG/TJ 08-2150-2014	电气火灾监控系统工程技术规程 Technical code of alarm and control system for electric fire prevention		J12768-2014	上海建筑设计研究院有限公司
269	DG/TJ 08-2151-2014	区（县）、镇（乡）土地利用总体规划实施评估标准 Standard for implementation evaluation of general land use planning of district and town level		J12819-2014	上海市规划和国土资源管理局 上海市地质调查院
270	DG/TJ 08-2152-2014	城市道路桥梁工程施工质量验收规范 Code for acceptance of construction quality of city road and bridge engineering		J12823-2014	上海市建设工程安全质量监督总站 上海公路桥梁（集团）有限公司
271	DG/TJ 08-2153-2014	住宅装饰装修工程施工技术规程 Technical specification for construction in residential building decoration engineering		J12820-2014	上海市装饰装修行业协会
272	DGTJ 08-2154-2014	装配整体式混凝土公共建筑设计规程 Code for design of assembled monolithic concretepublic buildings		J12874-2014	同济大学 上海市建工设计研究院有限公司 上海市城市建设设计研究总院
273	DG/TJ 08-2155-2014	全螺纹压灌桩技术规程 Technical specification for continuous flight auger-pressure grouting pile		J12906-2015	上海现代建筑设计集团工程建设咨询有限公司
274	DG/TJ 08-2156-2014	地下空间规划编制规范 Code for underground space planning		J12905-2015	上海市城市规划设计研究院
275	DG/TJ 08-2157-2015	排水性沥青路面养护技术规程 Technical regulation for maintenance of draining asphalt pavement		J13029-2015	上海市路政局
276	DG/TJ 08-2158-2015	预制混凝土夹心保温外墙板应用技术规程 Technical specification for precast concrete external wall panel with sandwich insulation		J13019-2015	上海市建筑科学研究院（集团）有限公司 同济大学
277	DG/TJ 08-2159-2015	高速公路网电子不停车收费系统（ETC）技术规程 Technical specification for electric toll collection system of expressway network		J12948-2015	上海市路政局
278	DG/TJ 08-2160-2015	预制拼装桥墩技术规程 Technical specification for prefabricated bridge piers		J12992-2015	上海公路投资建设发展有限公司 上海市城市建设设计研究总院

上海市

序号	标准编号	标准名称	被代替标准编号	备案号	主编单位
279	DGJ 08-2161-2015	地面辐射供暖技术规程 Technical specification for floor radiant heating		J13121-2015	上海市建筑材料行业协会 华东建筑设计研究院有限公司
280	DG/TJ 08-2162-2015	可再生能源建筑应用测试评价标准 Evaluation standard for application of renewable energy in buildings		J12991-2015	上海市建筑科学研究院（集团）有限公司 上海市建筑建材业市场管理总站
281	DG/TJ 08-2163-2015	公共场所无线局域网信号覆盖系统集约化技术规范 Code for design & acceptance of integrated WLAN signal covering system in public place		J12990-2015	上海市无线电管理局
282	DGJ 08-2164-2015	民用建筑外保温材料防火技术规程 Technical specification for external insulation materials ofcivil construction on fire-protection		J13122-2015	上海市建筑科学研究院（集团）有限公司 上海建科检验有限公司
283	DG/TJ 08-2165-2015	建设项目交通影响评价技术标准 Technical standards of traffic impact analysis of Shanghai construction project		J13030-2015	上海市城市规划设计研究院 上海市公安局交通警察总队
284	DG/TJ 08-2166-2015	城市地下综合体设计规范 Code for design of urban underground complex		J13035-2015	上海市政工程设计研究总院（集团）有限公司
285	DG/TJ 08-2167-2015	公路绿化养护技术规程 Technical specification for maintenance of highway planting		J13041-2015	上海市路政局
286	DG/TJ 08-2168-2015	城市综合管廊维护技术规程 Technical specification of maintenance for urban municipal tunnel		J13053-2015	上海市路政局
287	DG/TJ 08-2169-2015	轨道交通地下车站与周边地下空间的连通工程设计规程 Technical standard for connection engineering between rail transit underground station and surrounding underground space		J13068-2015	上海申通地铁集团有限公司 上海市民防科学研究所 上海市消防局
288	DG/TJ 08-2170-2015	城市轨道交通结构监护测量规范 Code for maintenance monitoring measurement of urban rail transit structure		J13206-2015	上海岩土工程勘察设计研究院有限公司 上海申通地铁集团有限公司
289	DG/TJ 08-2171-2015	市政道路机电系统维护技术规程 Technical specification for maintaining electrical and mechanical system of municipal road engineering		J13116-2015	上海市路政局
290	DG/TJ 08-2172-2015	公交专用道系统设计规范 Code for design of bus exclusive lanes system		J13115-2015	上海市交通委员会 上海市城市建设设计研究总院
291	DGJ 08-2173-2016	展览建筑及布展设计防火规程 Code of design for fire protection and prevention of exhibition building and exhibits arrangement		J13147-2015	华东建筑设计研究总院 上海市消防局
292	DG/TJ 08-2174-2015	高速公路改扩建设计规范 Code for design of expressway reconstruction andexpansion		J13185-2015	上海市路政局 上海市政工程设计研究总院（集团）有限公司

上海市

序号	标准编号	标准名称	被代替标准编号	备案号	主编单位
293	DG/TJ 08-2175-2015	隧道养护技术规程 Technical specification of maintenance for tunnel		J13224-2015	上海市路政局
294	DG/TJ 08-2176-2015	沥青路面预防性养护技术规程 Technical specifications for preventive maintenance of asphalt pavement		J13186-2015	上海市路政局
295	DG/TJ 08-2177-2015	建筑工程消防施工质量验收规范 Code of acceptance for construction quality of fire protection in building engineering		J13342-2016	上海市消防局 上上海市建筑科学研究院（集团）有限公司
296	DG/TJ 08-2178-2015	全装修住宅室内装修设计标准 Design standard of fully-fit-out residential buildings		J13187-2015	上海市住房保障和房屋管理局 上海天华建筑设计有限公司
297	DGJ 08-2180-2015	体外预应力加固技术规程 Technical specification for strengthening reinforced concrete structures with external prestressing		J13152-2015	同济大学 上海建筑设计研究院有限公司
298	DG/TJ 08-2181-2015	混凝土生产回收水应用技术规程 Technical specification for application of concrete recycled water		J13208-2015	上海建工材料工程有限公司 上海建科工程咨询有限公司
299	DG/TJ 08-2182-2015	道路LED照明应用技术规范 Application code for the LED road lighting technology		J13217-2015	上海市路政局
300	DG/TJ 08-2183-2015	城市道路养护维修作业安全技术规程 Technical regulation for safety operation of urban road maintenance		J13225-2015	上海市路政局
301	DG/TJ 08-2185-2015	沥青路面冷再生技术规程 Technical specification for asphalt pavement cold recycling		J13265-2015	上海市路政局 上海市公路学会
302	DG/TJ 08-2186-2015	软土地层降水工程施工作业规程 Construction operational code for engineering dewatering in soft soil stratum		J13266-2015	上海隧道工程有限公司 上海广联建设发展有限公司 上海建科工程咨询有限公司
303	DG/TJ 08-2187-2015	钢管扣件式木模板支撑系统施工作业规程 Construction and operation code for steel tubular supporting system with couplers of timber formwork		J13267-2015	上海建工七建集团有限公司 上海建科工程咨询有限公司
304	DG/TJ 08-2188-2015	应急避难场所设计规范 Design standard for emergency shelter		J13268-2015	上海市园林设计院有限公司 上海现代建筑设计（集团）有限公司
305	DG/TJ 08-2189-2015	静力触探技术规程 Technical specification for cone penetration test		J13316-2016	上海岩土工程勘察设计研究院有限公司
306	DG/TJ 08-2190-2015	平板膜生物反应器法污水处理工程技术规范 Technical specification for wastewater treatment by flat-sheet membrane bioreactor		J13317-2016	上海城投污水处理有限公司
307	DG/TJ 08-2191-2015	公路大中修工程设计规范 Code of the design for highway heavy and intermediate maintenance engineering		J13318-2016	上海市路政局
308	DG/TJ 08-2192-2016	工程木结构设计规范 Design code for engineered wood structure		J13336-2016	上海现代建筑设计（集团）有限公司

上海市

序号	标准编号	标准名称	被代替标准编号	备案号	主编单位
309	DG/TJ 08-2193-2016	泡沫玻璃板保温系统应用技术规程 Technical specification for application of thermal insulating system based on cellular glass board		J13339-2016	上海市建筑科学研究院 同济大学
310	DG/TJ 08-2195-2016	电力黄线规划编制技术规范 Technical code for electric control line planning		J13340-2016	上海市城市规划设计研究院
311	DG/TJ 08-2196-2016	城市有轨电车线网规划编制标准 Code for compilation of urban tramway network planning		J13348-2016	上海市城市建设设计研究总院 上海市交通委员会
312	DG/TJ 08-2197-2016	迪士尼度假区场地形成工程技术规范 Technical specification for disney resort site formation		J13371-2016	上海国际旅游度假区管理委员会 中国建筑西南勘察设计研究院有限公司
313	DG/TJ 08-2198-2016	工业化住宅建筑评价标准 Evaluation standard for industrialized housing		J13369-2016	上海市房地产科学研究院
314	DG/TJ 08-2199-2016	绿色建筑检测技术标准 Green building performance evaluation standard		J13460-2016	上海市建筑科学研究院 中国建筑科学研究院上海分院
315	DG/TJ 08-2200-2016	建筑反射隔热涂料应用技术规程 Technical specification for application of solar reflective coating on building surface		J13430-2016	上海市建筑科学研究院 上海市绿色建筑协会
316	DG/TJ 08-2201-2016	建筑信息模型应用标准 Standard for building information model application		J13453-2016	华东建筑设计研究院有限公司 上海建科工程咨询有限公司
317	DG/TJ 08-2202-2016	城市轨道交通信息模型技术标准 Technical standard for information model of urban rail transit		J13454-2016	上海申通地铁集团有限公司 上海市隧道工程轨道交通设计研究院
318	DG/TJ 08-2203-2016	城市轨道交通信息模型交付标准 Delivery standard for information model of urban rail transit		J13455-2016	上海申通地铁集团有限公司 上海市隧道工程轨道交通设计研究院
319	DG/TJ 08-2204-2016	市政道路桥梁信息模型应用标准 Application standard for municipal road and bridge information model		J13456-2016	上海市城市建设设计研究总院
320	DG/TJ 08-2205-2016	市政给排水信息模型应用标准 Application standard for municipal plumbing information model		J13457-2016	上海市城市建设设计研究总院
321	DG/TJ 08-2206-2016	人防工程设计信息模型交付标准 Information model delivery standard for civil air defense works		J13472-2016	上海市地下空间设计研究总院有限公司
322	DG/TJ 08-2207-2016	城市供水管网泵站远程监控系统技术规程 Technical specification for remote supervisory control system of pumping station in urban water supply distribution system		J13434-2016	上海市供水行业协会 上海市供水调度监测中心
323	DG/TJ 08-2208-2016	住宅建筑电能计量技术规范 Technical code for electric energy metering of residential buildings		J13458-2016	国网上海市电力公司 上海现代建筑设计（集团）有限公司
324	DG/TJ 08-2211-2016	建筑施工现场应急预案编制规程 Specification for building construction site to develop emergency response plan		J13515-2016	上海市建工集团股份有限公司 上海市建设安全协会

江苏省

序号	标准编号	标准名称	被代替标准编号	备案号	主编单位
1	DGJ32/TC 01-2015	江苏省城市环境卫生作业质量标准 Standard for city sanitation quality in Jiangsu Province		J10624-2015	常州市环境卫生管理处 江苏省住房和城乡建设厅城市管理局
2	DGJ32/TC 03-2015	江苏省城市供水服务质量标准 Standard for service quality of urban water supply in Jiangsu Province	DGJ32/C 03-2007	J11033-2015	江苏省住房和城乡建设厅城市建设处 南京市供水节水管理处
3	DGJ32/TJ 04-2010	优质建筑工程施工质量验收评定标准 Acceptance and evaluating standard for constructional quality of excellent quality building engineering	DGJ32/TJ 04-2004	J11725-2010	江苏省建设工程质量监督总站
4	DGJ32/TC 04-2015	江苏省城镇燃气服务质量标准 Quality standard for urban and town gas service in Jiangsu Province	DGJ32/C 04-2007	J11034-2015	江苏省住房和城乡建设厅城市建设处
5	DGJ32/C 05-2008	江苏省城建档案馆业务工作规程 Operational specifications of Jiangsu urban construction archives		J11264-2008	江苏省建设厅建设档案办公室
6	DGJ32/TC 06-2011	城市道路照明技术规范 Technical code for lighting of urban road		J11821-2011	江苏省住房和城乡建设厅城市建设与管理处 南京市路灯管理处
7	DGJ32/C 07-2016	江苏省城市容貌标准 Standard for urban appearance in Jiangsu Province		J13251-2015	江苏省住房和城乡建设厅城市管理局 苏州市市容市政管理局
8	DGJ32/J 08-2015	建筑太阳能热水系统应用技术规范 Technical code for application of solar water heating system of buildings	DGJ32/J 08-2008	J13123-2015	江苏省住房和城乡建设厅科技发展中心 江苏省建设工程质量监督总站
9	DGJ32/J 12-2005	南京地区建筑地基基础设计规范 Code for design of building foundation in Nanjing area		J10648-2005	南京市建设委员会
10	DGJ32/J 14-2007	35kV 及以下客户端变电所建设标准 The construction standard of client substation at 35kV and below		J10685-2008	江苏省电力公司
11	DGJ32/J 16-2014	住宅工程质量通病控制标准 Standard for residential engineering common quality defect control		J10687-2014	江苏省建设工程质量监督总站
12	DGJ32/TJ 18-2012	建筑物沉降、垂直度检测技术规程 Technical regulations of detection of building settlement and verticality		J10771-2012	江苏方建工程质量鉴定检测有限公司 南京建正建设工程质量检测中心
13	DGJ32/J 19-2015	绿色建筑工程施工质量验收规范 Accepting code for the quality of green building	DGJ32/J 19-2007	J13158-2015	江苏省住房和城乡建设厅科技发展中心 江苏省建设工程质量监督总站
14	DGJ32/TJ 20-2015	建筑装饰装修制图标准 Drawing standard for building decoration		J10776-2015	东南大学建筑学院
15	DGJ32/J 21-2009	建设工程质量检测规程 Inspection and testing specification of construction engineering quality	DGJ32/J 21-2006	J10777-2009	江苏省建设工程质量监督总站
16	DGJ32/J 23-2006	民用建筑节能工程现场热工性能检测标准 In-situ inspection on energy conservation for civil buildings		J10779-2006	江苏省建筑科学研究院有限公司 江苏省建设厅新技术推广站

江苏省

序号	标准编号	标准名称	被代替标准编号	备案号	主编单位
17	DGJ32/J 54-2016	施工现场装配式轻钢结构活动板房技术规程 Technical specification for prefabricated light steel movable plank building in construction site	DGJ32/J 54-2006	J13159-2016	江苏金土木建设集团有限公司 常熟市建筑管理处
18	DGJ32/J 55-2011	建筑施工安全生产条件评价规范 Code for construction condition of safety work evaluation	DGJ32/J 55-2006	J11977-2012	江苏省建筑安全与设备管理协会
19	DGJ32/J 56-2006	六氟丙烷灭火系统设计、施工及验收规程 Code of design, installation and acceptance for HFC 236fa fire extinguishing systems		J10876-2006	江苏省公安厅消防局防火部
20	DGJ32/TJ 57-2015	建（构）筑物整体移位技术规程 Technical specification for building monolithic moving engineering	DGJ32/J 57-2007	J12956-2015	东南大学
21	DGJ32/TJ 59-2007	混凝土用矿物掺合料（粒化高炉矿渣粉、粉煤灰）应用技术规程 Specification for application of mineral admixture for concrete (ground granulated blast furnace slag, fly ash)		J10988-2007	江苏省建筑钢结构混凝土协会 江苏博特新材料有限公司
22	DGJ32/TJ 60-2007	灌注桩钢筋笼长度检测技术规程 Technical code for testing of reinforcement length in cast-in-situ piles		J11118-2007	江苏省建设工程质量监督总站 南京南大工程检测有限公司
23	DGJ32/TJ 61-2015	透水水泥混凝土路面应用技术规程 Technical specification for application of pervious cement concrete pavement	DGJ32/J 61-2007	J11133-2015	江苏省建工集团有限公司 南京标美彩石建材有限公司
24	DGJ32/J 63-2008	既有玻璃幕墙可靠性能检验评估技术规程 Technical code for checking and evaluating the reliability of existed glass curtain wall		J11140-2008	江苏省建筑科学研究院有限公司
25	DGJ32/J 64-2008	室内无线电信号覆盖系统建设规范 The construction code for indoor radio signal coverage system		J11141-2008	江苏省工程建设标准站
26	DGJ32/J 65-2015	建筑工程施工机械安装质量检验规程 Testing specification for installation quality of constructional engineering machinery	DGJ32/J 65-2008	J11178-2015	江苏省建筑安全与设备管理协会
27	DGJ32/J 66-2008	江苏省建筑施工安全质量标准化管理标准 Standard of safety stanardization management for construction company and site in Jiangsu Province		J11179-2008	江苏省建筑安全与设备管理协会
28	DGJ32/TJ 69-2008	建筑施工承插型盘扣式钢管支架安全技术规程 Technical code for safety of plate-plug-pin steel tubular scaffold with socket and spigot in building construction		J11281-2008	无锡市锡山三建实业有限公司 南通新华建筑集团有限公司 东南大学
29	DGJ32/J 71-2014	江苏省居住建筑热环境和节能设计标准 Design standard of thermo-environment & energy conservation for residential buildings in Jiangsu Province	DGJ32/J 71-2008	J11266-2014	江苏省建筑科学研究院有限公司 南京工业大学 江苏省住房和城乡建设厅科技发展中心
30	DGJ32/J 75-2009	力值检测数据采集与传输规程 Technological regulations for data acquisition and transmission of force value		J11362-2009	江苏省建设工程质量监督总站

江苏省

序号	标准编号	标准名称	被代替标准编号	备案号	主编单位
31	DGJ32/TJ 77-2009	基桩自平衡法静载试验技术规程 Technical specification for static loading test of self-balanced method of foundation pile	DB32/T 291-1999	J11364-2009	南京东大自平衡桩基检测有限公司 东南大学土木工程学院
32	DGJ32/TJ 79-2009	雷达法检测建设工程质量技术规程 Technical specification for testing construction quality by radar method		J11424-2009	南京工业大学 南京工大建设工程技术有限公司
33	DGJ32/TJ 80-2009	超细干粉灭火系统设计、安装与验收规程 Code of design, installation and acceptance for super fine powder extinguishing systems		J11425-2009	江苏省公安厅消防局防火部
34	DGJ32/TJ 81-2009	建筑工程红外热成像法检测技术规程 Technical specification for inspection of construction engineering using infrared thermography		J11426-2009	江苏省建筑科学研究院有限公司 江苏省建筑工程质量检测中心有限公司
35	DGJ32/TJ 83-2009	轻型木结构检测技术规程 Technical specification for inspection of light wood-frame construction		J11428-2009	江苏省建筑科学研究院有限公司 江苏省建筑工程质量检测中心有限公司 江苏东方建筑设计有限公司
36	DGJ32/TJ 85-2009	混凝土复合保温砌块（砖）非承重自保温系统应用技术规程 Technical specification for application of non-load-bearing self-isolation system with compound insulation concrete blocks (bricks)		J11491-2009	江苏省建筑节能技术中心 苏州市世好建材新技术工程有限公司
37	DGJ32/TJ 86-2013	保温装饰板外墙外保温系统技术规程 Technical specification for external thermal insulation systems based on insulated decorative panel	DGJ32/TJ 86-2009	J11508-2014	江苏丰彩保温装饰板有限公司 江苏省建筑科学研究院有限公司
38	DGJ32/J 87-2009	太阳能光伏与建筑一体化应用技术规程 The technology code in integration of building and photovoltaic		J11496-2009	无锡尚德太阳能电力有限公司 华仁建设集团有限公司
39	DGJ32/TJ 88-2009	建筑外遮阳工程质量验收规程 Accepting specification for engineering quality of external blinds and shutters-resistance		J11529-2009	江苏省工程质量监督总站 南京二十六度建筑节能工程有限公司
40	DGJ32/TJ 89-2009	地源热泵系统工程技术规程 Technical specification for ground-source heat pump system		J11522-2009	南京市建筑设计研究院有限责任公司 南京工业大学
41	DGJ32/TJ 90-2009	建筑太阳能热水系统检测与评定标准 Guide for engineering testing of construction solar water heating systems		J11523-2009	江苏方建工程质量鉴定检测有限公司 江苏省产品质量监督检验研究院 江苏省工程建设标准站
42	DGJ32/TJ 91-2009	强化液自动灭火装置设计、安装与验收规程 Code of distribution design, installation and acceptance for automatic fire extinguishing device of strengthen fluid		J11524-2009	江苏省公安厅消防局防火部
43	DGJ32/TJ 93-2016	建设工程电子招投标专业工具软件数据交换标准 Standard for data exchange between e-bidding application of construction projects	DGJ32/TJ 93-2009	J11569-2016	江苏省建设工程招标投标办公室
44	DGJ32/TJ 94-2010	住宅信报箱建设标准 Constructional standard of residential private mailbox		J11575-2010	江苏省邮政管理局 江苏省城市规划设计研究院

江苏省

序号	标准编号	标准名称	被代替标准编号	备案号	主编单位
45	DGJ32/TJ 95-2010	聚氨酯硬泡体防水保温工程技术规程 Technical specification rigid polyurethane foam waterproof and insulation engineering		J11574-2010	江苏省建筑节能技术中心 江苏省建筑节能协会 江苏久久防水保温隔热工程有限公司
46	DGJ32/J 96-2010	公共建筑节能设计标准 Design standard for energy efficiency of public buildings		J11544-2010	江苏省建筑设计研究院有限公司
47	DGJ32/TJ 98-2010	钻芯法检测外墙外保温构造技术规程 Technical specification for inspection with drilled-core of wall-thermal-insulation-layer		J11611-2010	江苏省建设工程质量监督总站
48	DGJ32/J 99-2010	成品住房装修技术标准 The technical standard for finished housing decoration		J11592-2010	江苏省住房和城乡建设厅住宅与房地产业促进中心
49	DGJ32/TJ 101-2010	采选矿废渣页岩模数多孔砖建筑技术规程 Technical specification for buildings with waste slag of mining beneficiation and shale module perforated bricks		J11623-2010	江苏省建筑科学研究院有限公司 南京鑫翔新型建筑材料有限责任公司
50	DGJ32/TJ 102-2010	城市隧道竖井型自然通风设计与验收规范 Code of design and acceptance on natural ventilation for city tunnel with vertical shafts		J11624-2010	南京市住房和城乡建设委员会 南京城建项目建设管理有限公司
51	DGJ32/TJ 103-2010	住宅工程质量分户验收规程 Acceptance specification for unit quality of housing engineering		J11638-2010	江苏省建设工程质量监督总站
52	DGJ32/TJ 104-2010	现浇轻质泡沫混凝土应用技术规程 Technical specification for application of lightweight foam concrete		J11644-2010	江苏省建工集团有限公司 常州市武进东方人防实业有限公司
53	DGJ32/TJ 107-2010	蒸压加气混凝土砌块自保温系统应用技术规程 Technical specification for application of self-insulation system with autoclaved aerated concrete blocks		J11711-2010	江苏省建筑节能技术中心 南通通佳新型建筑材料有限公司
54	DGJ32/TJ 108-2010	玻璃纤维增强树脂土钉基坑支护技术规程 Technical specification for grass fiber reinforced polymer soil nailing in foundation excavations		J11721-2010	河海大学 南京锋晖复合材料有限公司
55	DGJ32/TJ 109-2010	预应力混凝土管桩基础技术规程 Technical specification for pretensioned concrete pipe pile foundation		J11750-2010	江苏省住房和城乡建设厅科技发展中心 连云港市建筑设计研究院有限责任公司
56	DGJ32/TJ 110-2010	工程结构动力特性及动力响应检测技术规程 Technical specification for inspecting dynamic characteristic and dynamic response of engineering structures		J11748-2010	南京工业大学 南京工大建设工程技术有限公司
57	DGJ32/TJ 111-2010	公共建筑能耗监测系统技术规程 Technical code of energy consumed monitoring systems for public buildings		J11747-2010	江苏省建筑节能技术中心 江苏省住房和城乡建设厅科技发展中心
58	DGJ32/TJ 112-2010	人工湿地污水处理技术规程 Technical specification for wastewater treatment on constructed wetland		J11749-2010	南京工业大学

江苏省

序号	标准编号	标准名称	被代替标准编号	备案号	主编单位
59	DGJ32/TJ 113-2011	雨水利用工程技术规范 Technical code for engineering of rain utilization		J11817-2011	南京工业大学 江苏省工程建设标准站
60	DGJ32/TJ 114-2011	回弹法检测砌体中砖抗压强度技术规程 Specification for inspection of brick conpressive strength by rebound method in masonry structures		J11816-2011	江苏方建工程质量鉴定检测有限公司 扬州市建伟建设工程检测中心有限公司
61	DGJ32/TJ 115-2011	房屋白蚁预防工程技术规程 Technical specification for termite prevention treatment for buildings		J11815-2011	南京市白蚁防治研究所 江苏省白蚁防治协会
62	DGJ32/TJ 116-2011	里氏硬度计现场检测建筑钢结构钢材抗拉强度技术规程 Technical specification for testing tensile strength of steel construction in site by leeb-hardness tester		J11814-2011	江苏省建筑科学研究院有限公司 江苏省建筑工程质量检测中心有限公司
63	DGJ32/TJ 117-2011	钻孔灌注桩成孔、地下连续墙成槽质量检测技术规程 Technical specification for the testing of the drilling hole of cast-in-place pile and the groove of diaphragm wall		J11813-2011	江苏省建筑科学研究院有限公司 江苏省建筑工程质量检测中心有限公司
64	DGJ32/J 118-2011	住宅小区光纤到户通信配套设施建设标准 Construction standard of communication ancillary facilities of fiber to the home in residential district		J11809-2011	江苏省通信管理局 江苏省邮电规划设计院有限责任公司
65	DGJ32/TJ 119-2011	预拌混凝土绿色生产管理规程 Management specification of green production of ready-mixed concrete		J11853-2011	南京市预拌混凝土管理处
66	DGJ32/TJ 120-2011	城市居住区人防工程规划设计规范 Code of civil air defence works for urban residential district planning and construction		J11856-2011	江苏省民防局 南京市人民防空办公室 中国人民解放军理工大学工程兵工程学院
67	DGJ32/J 121-2011	建筑施工悬挑式钢管脚手架安全技术规程 Technical specification for safety of cantilever steel tubular scaffolding in construction		J11837-2011	江苏金土木建设集团有限公司 常熟市建筑管理处
68	DGJ32/J 122-2011	城市应急避难场所建设技术标准 Technical standard of construction of urban emergency shelter		J11842-2011	江苏省住房和城乡建设厅抗震办公室 中国人民解放军理工大学工程兵工程学院
69	DGJ32/J 123-2011	建筑外遮阳工程技术规程 Technical regulations for solar shading engineering of buildings		J11843-2011	南京市建筑设计研究院有限责任公司 江苏省建筑科学研究院有限公司 江苏省住房和城乡建设厅科技发展中心
70	DGJ32/J 124-2011	建筑幕墙工程质量验收规程 code for construction quality acceptance of curtain wall		J11879-2011	江苏省建设工程质量监督总站 江苏省建筑工程质量检测中心有限公司
71	DGJ32/TJ 125-2011	预制装配整体式剪力墙结构体系技术规程 Technical specification for precast and monolithic shear wall structure system		J11911-2011	南通建筑工程总承包有限公司 东南大学

江苏省

序号	标准编号	标准名称	被代替标准编号	备案号	主编单位
72	DGJ32/TJ 126-2011	太阳能光伏与建筑一体化工程检测规程 Guide for testing of building integrated photovoltaic systems		J11938-2011	国家太阳能光伏产品质量监督检验中心 江苏省建设工程质量监督总站
73	DGJ32/TJ 127-2011	既有建筑节能改造技术规程 Technical specification for energy efficiency retrofit of existing buildings		J11939-2011	江苏丰彩节能科技有限公司 江苏省建筑科学研究院有限公司 江苏省住房和城乡建设厅科技发展中心
74	DGJ32/J 128-2011	装饰装修木制品应用技术标准 The technical standard for decoration woodwork application		J11920-2011	江苏省装饰装修发展中心 苏州金螳螂建筑装饰股份有限公司 江苏省华夏天成建设股份有限公司
75	DGJ32/TJ 129-2011	轻型木结构建筑技术规程 Technical specification for wood-frame construction		J11949-2011	江苏省建筑科学研究院有限公司 江苏东方建筑设计有限公司
76	DGJ32/TJ 130-2011	地源热泵系统检测技术规程 Technical specification for test of ground-source heat pump system		J11955-2011	南京工业大学 南京工大建设工程技术有限公司
77	DGJ32/TJ 131-2011	房屋面积测算技术规程 Construction technical specification of building area surveying and mapping		J11973-2012	苏州市房地产测绘队 无锡市曦晨测绘有限公司
78	DGJ32/TJ 132-2011	城市轨道交通能源管理系统技术规程 Technical code for energy management system of urban rail transit		J11963-2012	南京地下铁道有限责任公司 江苏联宏自动化系统工程有限公司
79	DGJ32/TJ 133-2011	装配整体式自保温混凝土建筑技术规程 Technical specification for assembled monolithic concrete structures with self-insulation system		J11972-2011	南京华韵建筑科技发展有限公司 东南大学
80	DGJ32/TJ 134-2012	常开防火门控制装置应用技术规程 Technical specification for often open fire door control device application		J12024-2012	江苏省公安厅消防局
81	DGJ32/TJ 135-2012	民用建筑能效测评标识标准 Standard for energy efficiency evaluation labeling of civil buildings		J12023-2012	江苏省住房和城乡建设厅科技发展中心 昆山市建设工程质量检测中心
82	DGJ32/TJ 136-2012	既有建筑结构加固工程现场检测技术规程 Technical standard for inspection of strengthening building structures		J12031-2012	江苏省建筑科学研究院有限公司 江苏省建筑工程质量检测中心有限公司 江苏九鼎环球建设科技集团有限公司
83	DGJ32/TJ 137-2012	预应力混凝土U型梁施工与验收规程 Specification for construction and acceptance of prestressed concrete u-beam		J12057-2012	南京地下铁道有限责任公司
84	DGJ32/TJ 138-2012	公共建筑能源审计标准 Standard for energy consumption auditing of public buildings		J12131-2012	江苏省住房和城乡建设厅科技发展中心
85	DJG32/TJ 139-2012	太阳能热水系统运行管理规程 Operation and management specification of solar water heating system		J12199-2012	东南大学

江苏省

序号	标准编号	标准名称	被代替标准编号	备案号	主编单位
86	DGJ32/J 140-2012	民用建筑室内装修工程环境质量验收规程 Code for indoor environmental quality acceptance of civil building decoration		J12091-2012	江苏省装饰装修发展中心 江苏省建筑工程质量检测中心有限公司
87	DGJ32/TJ 141-2012	地源热泵系统运行管理规程 Operation and management specification for ground source heat pump system		J12198-2012	南京工业大学
88	DGJ32/TJ 142-2012	建筑地基基础检测规程 Technical code for testing of foundation soil and building foundation		J12210-2012	江苏省建设工程质量监督总站
89	DGJ32/TJ 143-2012	房屋建筑和市政基础设施工程档案资料管理规范 Housing construction and municipal infrastructure engineering archives management specification		J12209-2012	江苏省建设工程质量监督总站 江苏省住房和城乡建设厅建设档案办公室
90	DGJ32/TJ 144-2012	超声回弹综合法检测混凝土抗压强度技术规程 Technical specification for inspecting of concrete compressive strength by ultrasonic-rebound combines method		J12208-2012	江苏省建筑科学研究院有限公司
91	DGJ32/TJ 145-2012	回弹法检测混凝土抗压强度技术规程 Technical specification for inspecting of concrete compressive strength by rebound method		J12207-2012	江苏省建筑科学研究院有限公司
92	DGJ32/J 146-2013	城镇户外广告和店招标牌设施设置技术规范 Technical code for setting out door advertising facilities and shop signs		J12232-2012	江苏省住房和城乡建设厅城市建设与管理处 苏州大学
93	DGJ32/TJ 147-2013	固化粉煤灰应用技术规程 Technical application specification for hydrated fly ash		J12357-2013	江苏镇江建筑科学研究院集团有限公司 镇江市城市干道工程建设办公室
94	DGJ32/TJ 148-2013	城镇道路开挖、回填、恢复快速施工及验收规程 Code for construction and acceptance of urban road excavation and quick recovery		J12275-2013	英达热再生有限公司
95	DGJ32/TJ 149-2013	城镇道路沥青路面就地热再生施工及验收规程 Code for construction and acceptance of urban road asphalt pavement hot-in-place recycling		J12274-2013	英达热再生有限公司
96	DGJ32/J 150-2013	建筑施工中心提升式附着升降脚手架安全技术规程 Technical specification for safety of center attached lifting scaffold in construction		J12242-2013	江苏省建筑安全与设备管理协会 江苏天任建设装备有限公司
97	DGJ32/TJ 151-2013	劲性复合桩技术规程 Technical specification for strength composite piles		J12291-2013	江苏兴鹏基础工程有限公司 江苏省建筑科学研究院有限公司
98	DGJ32/TJ 152-2013	可再生能源建筑应用数据监测系统技术规程 Technical specification of data monitoring systems for application of renewable energy in buildings		J12455-2013	江苏省建筑节能技术中心 江苏省住房和城乡建设厅科技发展中心

江苏省

序号	标准编号	标准名称	被代替标准编号	备案号	主编单位
99	DGJ32/TJ 153-2013	民用建筑能耗统计标准 Standard for civil buildings energy consumption statistics		J12451-2013	江苏省住房和城乡建设厅科技发展中心
100	DGJ32/TJ 154-2013	水泥土试验方法 Test methods for cement soil		J12492-2013	苏州市建设工程质量检测中心有限公司
101	DGJ32/TJ 155-2013	中小学校舍抗震鉴定与加固技术规程 Technical specification for seismic appraisal and strengthening for primary and secondary school buildings		J12491-2013	江苏省建筑科学研究院有限公司
102	DGJ32/TJ 156-2013	中小学校舍抗震加固工程施工质量验收规程 Code for acceptance of constructional quality in seismic strengthening of school buildings		J12490-2013	江苏省住房和城乡建设厅抗震办公室
103	DGJ32/J 157-2013	居住建筑标准化外窗系统应用技术规程 Technical specification for application of standard external window system of residential buildings		J12488-2013	江苏省建筑科学研究院有限公司 南京市建筑设计研究院有限公司
104	DGJ32/TJ 158-2013	地源热泵系统工程勘察规程 Specification for investigation of ground-source heat pump system		J12545-2014	江苏省地质工程勘察院
105	DGJ32/TJ 159-2013	建筑电气工程绝缘电阻、接地电阻检测规程 Technical specification for inspection and test methods for ground resistance & insulation resistance in buildings		J12550-2014	江苏方建工程质量鉴定检测有限公司 南京建正建设工程质量检测有限公司
106	DGJ32/TJ 160-2014	钠基膨润土防水毯施工技术规程 Technical specification for construction of sodium bentonite waterproof blanket		J12618-2014	通州建总集团有限公司
107	DGJ32/J 161-2014	居民住宅二次供水工程技术规程 Technical specification for secondary water supply engineering of residential buildings		J12579-2014	江苏省住房和城乡建设厅城市建设与管理处
108	DGJ32/TJ 162-2014	玻璃纤维增强复合材料筋基坑工程应用技术规程 Technical specification for excavation engineering application of GFRP bar		J12675-2014	江苏华东工程设计有限公司 南京锋晖复合材料有限公司
109	DGJ32/TJ 164-2014	工程建设监理企业质量管理规范 Code of quality management for construction project management enterprise		J12776-2014	江苏省建设监理协会 江苏建科建设监理有限公司
110	DGJ32/TJ 165-2014	建筑反射隔热涂料保温系统应用技术规程 Technical specification for application of thermal insulation system of reflective coating on buildings		J12705-2014	江苏省建筑科学研究院有限公司 江苏晨光涂料有限公司
111	DGJ32/J 166-2014	建筑施工机械设备维护保养技术规程 Technical specification for machanery and equipment maintenance in construction		J12654-2014	江苏省建筑安全与设备管理协会
112	DGJ32/TJ 167-2014	烧结保温砖（砌块）自保温系统应用技术规程 Technical specification for application of self-insulation wall system with fired thermal insulation bricks and blocks		J12704-2014	江苏省建筑节能技术中心 江苏省住房和城乡建设厅科技发展中心
113	DGJ32/TJ 168-2014	有机填料型人工湿地生活污水处理技术规程 Technical specification for domestic sewage treatment of constructed wetland of organic filler		J12743-2014	江苏省住房和城乡建设厅科技发展中心

江苏省

序号	标准编号	标准名称	被代替标准编号	备案号	主编单位
114	DGJ32/TJ 169-2014	江苏省城市居住区和单位绿化标准 Standard for greening of residential district and companies in Jiangsu Province		J12761-2014	江苏省住房和城乡建设厅风景园林处
115	DGJ32/TJ 170-2014	太阳能热水系统建筑应用能效测评技术规程 Technical specification for solar heating water system energy efficiency evaluation used in buildings		J12760-2014	江苏省住房和城乡建设厅科技发展中心 南京工业大学
116	DGJ32/TJ 171-2014	地源热泵系统建筑应用能效测评技术规程 Technical specification for ground-source heat pump system energy efficiency evaluation used in buildings		J12759-2014	南京工业大学 南京工大建设工程技术有限公司
117	DGJ32/TJ 172-2014	城市道路环卫机械化作业质量标准 Quality standard for environmental sanitation mechanized operation of city roads		J12778-2014	苏州市环境卫生管理处 江苏省住房和城乡建设厅城市建设与管理处
118	DGJ32/J 173-2014	江苏省绿色建筑设计标准 Jiangsu design standard for green building		J12777-2014	江苏省住房和城乡建设厅科技发展中心
119	DGJ32/TJ 174-2014	复合发泡水泥板外墙外保温系统应用技术规程 Technical specification for external thermal insulation systems of composite foam cement panel		J12834-2014	东南大学 江苏春迈建筑科技有限公司
120	DGJ32/TJ 175-2014	江苏省城市总体规划成果数据标准 Data standard for Jiangsu urban master planning		J12850-2014	江苏省住房和城乡建设厅城乡规划处 江苏省城市规划设计研究院 江苏省建设信息中心
121	DGJ32/TJ 176-2014	勘察设计企业质量管理规范 Code of quality management of geotechnical investigation and design enterprises		J12915-2015	江苏省勘察设计协会 江苏省住房和城乡建设厅建筑节能与科研设计处
122	DGJ32/TJ 177-2014	智能建筑工程质量检测规范 Code for checking and measuring the quality of intelligent building engineering		J12914-2015	南京工业大学 江苏省计量科学研究院
123	DGJ32/TJ 178-2014	智能建筑工程施工质量验收规范 Code for construction quality acceptance of intelligent building engineering		J12922-2015	南京工业大学 南京熊猫信息产业有限公司
124	DGJ32/J 179-2014	通信用户驻地网室内无线信号覆盖系统技术标准 Technical standard for indoor radio signal coverage system of communication customer premises network		J12876-2014	江苏省通信管理局 江苏省邮电规划设计院有限责任公司
125	DGJ32/J 180-2014	生活垃圾卫生填埋场岩土工程勘察规程 Technical specification for municipal solid waste sanitary landfill		J12873-2014	江苏苏州地质工程勘察院 江苏省水文地质工程地质勘察院
126	DGJ32/TJ 181-2015	地铁保护区内岩石爆破施工技术规程 Technical specification for rock blasting of subway protection areas		J12947-2015	南京市城市建设投资控股（集团）有限责任公司 南京市地铁建设有限责任公司
127	DGJ32/TJ 182-2015	孔压静力触探技术规程 Technical specification for CPTU		J13021-2015	东南大学岩土工程研究所
128	DGJ32/TJ 183-2015	墙体饰面砂浆应用技术规程 Application technical specification for wall decorative render and plaster		J12955-2015	江苏尼高科技有限公司 东南大学

江苏省

序号	标准编号	标准名称	被代替标准编号	备案号	主编单位
129	DGJ32/J 184-2016	装配式结构工程施工质量验收规程 Specification for construction quality acceptance of prefabricated structural engineering		J13276-2016	江苏省建设工程质量监督总站
130	DGJ32/TJ 185-2015	排水用塑料检查井应用技术规程 Technical specification for application of plastic inspection chamber for drainage		J13045-2015	江苏省建筑设计研究院有限公司 江苏河马井股份有限公司
131	DGJ32/TJ 186-2015	江苏省城市地下管线探测技术规程 Technical specification of detection and survey for Jiangsu urban underground pipelines		J13054-2015	江苏省住房和城乡建设厅城乡规划处 南京市测绘勘察研究院有限公司
132	DGJ32/TJ 187-2015	江苏省城市地下管线数据标准 Data standard for Jiangsu urban underground pipelines		J13055-2015	江苏省住房和城乡建设厅城乡规划处 南京市测绘勘察研究院有限公司
133	DGJ32/TJ 188-2015	立体绿化技术规程 Technical specification for green building planting		J13072-2015	江苏省住房和城乡建设厅科技发展中心
134	DGJ32/J 189-2015	南京地区建筑基坑工程监测技术规程 Technical specification for monitoring of building excavation engineering in Nanjing area		J13079-2015	南京市建筑安装工程质量监督站 南京工业大学测绘学院
135	DGJ32/TJ 190-2015	公共建筑节能运行管理规程 Operation management specification for energy efficency of public buildings		J13213-2015	江苏省住房和城乡建设厅科技发展中心
136	DGJ32/TJ 191-2015	供暖通风与空气调节系统检测技术规程 Technical specification for testing of heating ventilation and air-conditioning system		J13214-2015	昆山市建设工程质量检测中心 南京工业大学
137	DGJ32/TJ 193-2015	回弹法检测泵送混凝土抗压强度技术规程 Technical specification for inspecting of pimped concrete compressive strength by rebound method		J13212-2015	江苏省建设工程质量监督总站 江苏省建筑工程质量检测中心有限公司
138	DGJ32/TJ 194-2015	绿色建筑室内环境检测技术标准 Technical standard for green building indoor environmental testing		J13261-2015	南京工业大学 南京工大建设工程技术有限公司
139	DGJ32/J 195-2015	江苏省城市轨道交通工程监测规程 Technical specification for monitoring measurement of urban rail transit engineering in Jiangsu Province		J13160-2015	南京市轨道交通建设工程质量安全监督站 南京市测绘勘察研究院有限公司
140	DGJ32/TJ 196-2015	预拌砂浆技术规程 Technical specification of ready-mixed mortar		J13283-2016	东南大学 江苏省建筑科学研究院有限公司
141	DGJ32/TJ 197-2015	建筑外窗工程检测与评定规程 Specification for inspection and evaluation of building exterior windows		J13284-2016	江苏方建工程质量鉴定检测有限公司 江苏方正工程技术开发检测有限公司
142	DGJ32/TJ 198-2015	城市轨道交通接触网系统工程质量验收规范 Accepting code for constructional quality of urban rail transit contact line system engineering		J13300-2016	江苏省产品质量监督检验研究院 南京地铁建设有限责任公司
143	DGJ32/TJ 199-2016	预制预应力混凝土装配整体式结构技术规程 Technical specification for structures comprised of precast prestressed concrete components		J13397-2016	南京大地建设集团有限责任公司 东南大学土木工程学院

江苏省

序号	标准编号	标准名称	被代替标准编号	备案号	主编单位
144	DGJ32/TJ 200-2016	江苏省房地产经纪服务标准 Standard for real estate brokerage services in Jiangsu Province		J13415-2016	江苏省住房和城乡建设厅房地产市场监管处 南京工业大学
145	DGJ32/TJ 201-2016	园林绿化工程施工及验收规范 Code for construction and acceptance of landscape engineering		J13413-2016	南京市绿化园林局 南京工业大学
146	DGJ32/TJ 202-2016	热处理带肋高强钢筋混凝土结构技术规程 Technical specification for application of heat-treatment high-strength ribbed bar in concrete structures		J13424-2016	江苏天舜金属材料集团有限公司 东南大学土木工程学院
147	DGJ32/TJ 203-2016	建筑工地扬尘防治标准 Standard of construction site dust control		J13349-2016	江苏省建筑安全与设备管理协会 扬州市建筑安全监察站
148	DGJ32/TJ 204-2016	复合材料保温板外墙外保温系统应用技术规程 Technical specification for external thermal insulation systems of composite material panel		J13445-2016	江苏省住房和城乡建设厅科技发展中心 江苏省建筑工程质量检测中心有限公司
149	DGJ32/TJ 205-2016	江苏省游泳场馆建筑节能设计技术规程 Technical specification for energy efficiency design of natatoria in Jiangsu		J13450-2016	南京城镇建筑设计咨询有限公司 南京市建筑设计研究院有限责任公司
150	DGJ32/TJ 206-2016	城市轨道交通工程高性能混凝土质量控制技术规程 Technical specification for quality control of high performance concrete in urban rail transit construction		J13516-2016	南京市轨道交通建设工程质量安全监督站 江苏省建筑科学研究院有限公司
151	DGJ32/TJ 207-2016	生活垃圾填埋场恶臭控制技术规范 Technical code for odor control of municipal solid waste landfill		J13527-2016	苏州市环境卫生管理处 江苏省住房和城乡建设厅城市管理局
152	DGJ32/TJ 208-2016	岩土工程勘察规范 Code for investigation of geotechnical engineering		J13574-2016	江苏省地质工程勘察院 江苏省建苑岩土工程勘测有限公司 江苏省水文地质工程地质勘察院
153	DGJ32/TJ 209-2016	既有医疗建筑抗震鉴定与加固技术规程 Technical specification for seismic appraisal and strengthening of existing hospital buildings		J13572-2016	江苏省建筑科学研究院有限公司
154	DGJ32/TJ 210-2016	江苏省民用建筑信息模型设计应用标准 Application standard for civil building information model in Jiangsu		J13573-2016	江苏省勘察设计行业协会 江苏省邮电规划设计院有限责任公司
155	DGJ32/TJ 211-2016	电磁法填土密度检测技术规程 Technical specification for electromagnetic method to detect earth fill density		J13609-2016	苏州科技大学 常熟市住房和城乡建设局
156	DGJ32/TJ 212-2016	江苏省建筑防水工程技术规程 Technical specification for waterproof engineering of construction in Jiangsu		J13666-2016	苏州中材非金属矿工业设计研究院有限公司 苏州建筑科学研究院集团股份有限公司
157	DGJ32/TJ 213-2016	江苏省城市地下管线信息管理系统技术规范 Technical code for urban underground pipeline information management system in Jiangsu Province		J13668-2016	江苏省住房和城乡建设厅城乡规划处 南京市测绘勘察研究院有限公司
158	DGJ32/TJ 214-2016	江苏省既有房屋鉴定标准 Standard for existing building appraisal in Jiangsu Province		J13667-2016	南京市房屋安全鉴定处

福建省

序号	标准编号	标准名称	被代替标准编号	备案号	主编单位
1	DBJ 13-15-2008	建筑隔墙用轻质条板应用技术规程 Technical specification for application of lightweight panels used as partition wall		J10076-2008	厦门市建筑科学研究院集团股份有限公司
2	DBJ 13-22-2007	建筑电气工程施工技术规程 Technical specification for the construction of electrical installation in building	DBJ 13-22-99	J10967-2007	福建省安装技术情报网等
3	DBJ/T 13-23-2015	建筑排水硬聚氯乙烯管道安装工程技术规程 Technical specification for unplasticized polyvinyl chloride pipeline of building drainage		J12984-2015	福建六建集团有限公司 福建巨岸建设工程有限公司
4	DBJ 13-25-2008	城市隧道工程质量验收标准 Standard of acceptance for construction quality of city tunnel	DBJ 13-25-1999	J11263-2008	厦门市建设工程质量安全监督站
5	DBJ/T 13-27-2015	福建省建筑内外墙涂料涂饰工程施工及验收规程 Specification for construction and acceptance of internal and external wall surface decoration in Fujian		J11337-2015	中建海峡建设发展有限公司 宏峰集团（福建）有限公司
6	DBJ/T 13-28-2016	福建省基础工程钻芯法检测技术规程 Technical specification of core drilling method for foundation engineering in Fujian		J13442-2016	福建省建筑设计研究院
7	DBJ/T 13-29-2016	福建省蒸压加气混凝土砌块应用技术规程 Technical specification for application of autoclaved aerated concrete block of Fujian		J10757-2016	厦门市建筑科学研究院集团股份有限公司 常青树建材（福建）开发有限公司
8	DBJ/T 13-42-2012	预拌混凝土生产施工技术规程 Technical specification for production and construction of ready-mixed concrete	DBJ 13-42-2008	J10141-2012	福建省建筑科学研究院 福建省建筑业协会混凝土分会
9	DBJ/T 13-46-2013	建筑装修工程施工质量验收规程 Specification for construction quality acceptance of building decoration engineering		J10197-2013	福州铁建建筑有限公司
10	DBJ/T 13-51-2010	钢管混凝土结构技术规程 Technical specification for concrete-filled steel tubular structures	DBJ 13-51-2003	J10279-2010	福州大学 福建省建筑科学研究院
11	DBJ/T 13-54-2013	改性沥青玛蹄脂碎石混合料（SMA）施工技术规程 Specifications of stone matrix mixtune of modified asphalt for construction		J10327-2013	福州市市政工程质量监督站 福建博海工程技术有限公司
12	DBJ/T 13-56-2011	福建省建筑工程施工文件管理规程 Specification for building engineering construction document management of Fujian Province		J10352-2012	福建省建设工程质量安全监督总站
13	DBJ 13-62-2014	福建省居住建筑节能设计标准 Design standard for energy efficiency of residential buildings in Fujian		J10441-2014	福建省建筑科学研究院 福建省工程建设科学技术标准化协会
14	DBJ/T 13-65-2015	福建省智能建筑工程质量检测技术规程 Technical specification for check and measure of quality for intelligent building in Fujian		J10549-2015	福建省建筑科学研究院 福建省工程建设科学技术标准化协会智能化分会

福建省

序号	标准编号	标准名称	被代替标准编号	备案号	主编单位
15	DBJ/T 13-66-2015	福建省粒化高炉矿渣粉在水泥混凝土中应用技术规程 Technical specification for ground granulated blast furnace slag applied in concrete of Fujian		J10601-2015	厦门市建筑科学研究院集团股份有限公司
16	DBJ/T 13-67-2010	建筑施工起重机械安全检测规程 Test specification for safety of construction heavy-lifting machinery	DBJ/T 13-67-2005	J11699-2010	福建省工程建设质量安全协会建筑机械分会
17	DBJ/T 13-69-2013	沥青混合料配合比设计规程 Specification for mix proportion design of bituminous mixtures		J10657-2013	福州市市政工程质量监督站 福建博海工程技术有限公司
18	DBJ/T 13-71-2015	回弹法检测混凝土抗压强度技术规程 Technical specification for inspection of concrete compressive strength by rebound method		J10717-2015	福建省建筑科学研究院
19	DBJ/T 13-76-2016	福建省预拌砂浆生产与应用技术规程 Technical specification for production and application of ready-mixed mortar in Fujian	DBJ 13-76-2006	J10756-2016	厦门市建筑科学研究院集团股份有限公司 中建四局第一建筑工程有限公司
20	DBJ 13-77-2006	混凝土外加剂应用技术规程 Technical specification for application of admixtures in concrete		J10892-2006	福建省建筑科学研究院等
21	DBJ/T 13-78-2016	扣压式和紧定式钢导管电线管路施工及验收规程 Specification for construction and acceptance of wire pipelines with extruding connection and fastening connection steel conduit	DBJ 13-78-2007	J10925-2016	福建省建设工程质量安全监督总站 福建省闽南建筑工程有限公司
22	DBJ 13-79-2007	后装拔出法检测混凝土强度技术规程 Technical specification for inspection of concrete strength by pull out post-insert method		J10971-2007	福建省建筑科学研究院
23	DBJ/T 13-83-2013	福建省建筑节能工程施工质量验收规程 Specification for acceptance of energy efficient building construction of Fujian		J10791-2013	福建建工集团总公司
24	DBJ 13-84-2006	岩土工程勘察规范 Code for investigation of geotechnical engineering		J10857-2006	福建省建筑设计研究院
25	DBJ 13-85-2016	福建省LED夜景照明工程安装与质量验收规程 Specification for installation and quality acceptance of LED nightscape lighting in Fujian	DBJ 13-85-2007	J10994-2016	厦门市土木建筑学会 厦门市建设工程质量安全监督站
26	DBJ 13-86-2007	先张法预应力混凝土管桩基础技术规程 Technical specification for pretensioned spun concrete pile foundation		J11017-2007	福建省建筑设计研究院 福州市建筑设计院
27	DBJ 13-87-2007	福建省城市公共交通运营服务规范 Code for Fujian Province urban public transport operate and service		J11028-2007	福建省城市建设协会
28	DBJ/T 13-88-2010	福建省城镇污水处理厂运行管理标准 Operation & management of standard for municipal sewage treatment plant in Fujian Province	DBJ 13-88-2007	J11054-2010	福建省城市建设协会 厦门水务集团有限公司

福建省

序号	标准编号	标准名称	被代替标准编号	备案号	主编单位
29	DBJ 13-89-2007	建筑施工塔式起重机、施工升降机报废规程 Specification for construction tower crane & builder's hoist to be scrapped		J11065-2007	福建省工程建设质量安全协会建筑机械分会
30	DBJ 13-90-2007	城市绿化工程质量验收规程 Accepting specification of quality for landscape project in city		J11075-2007	厦门市园林绿化工程质量站
31	DBJ 13-91-2007	建设工程施工重大危险源辨识与监控技术规程 Specification for identification, monitoring and controltechnologyof major hazard installations in construction engineering		J11064-2007	厦门市建设工程质量安全监督站
32	DBJ 13-92-2007	埋地塑料排水管道工程技术规程 Technical specification for buried plasticpipeline of sewer engineering		J11107-2007	福州市规划设计研究院
33	DBJ/T 13-93-2015	福建省生活垃圾焚烧厂运行维护、检测监管及考核评价标准 Standard of operation, monitoring and assessment for municipal solid waste incineration	DBJ 13-93-2010	J11108-2015	住房和城乡建设部环境卫生工程技术研究中心、福建省城市建设协会
34	DBJ 13-94-2007	燃气用衬塑（PE）、衬不锈钢铝合金管道工程技术规程		J11113-2007	福建省城市建设协会
35	DBJ 1313-95-2007	住宅建筑供水"一户一表"设计、施工及验收规程 Design, construction and acceptance specification of one water-meter per householder in residential building		J11114-2007	福州市自来水总公司 福建省城市建设协会
36	DBJ 13-96-2008	混凝土结构工程后张预应力施工技术规程 Technical specification of post-tension prestressed construction		J11158-2008	福建省建筑科学研究院
37	DBJ 13-97-2008	市政工程施工安全技术标准 Technology standard for construction security of municipal engineering		J11159-2008	厦门市建设工程质量安全监督站
38	DBJ 13-98-2008	城镇沥青路面施工技术规程 Technical specification for construction of road asphalt pavement		J11171-2008	福建省建设工程质量安全监督总站
39	DBJ 13-99-2008	建设工程质量检测数据信息管理技术规程 Technical standard for data information management of construction quality test		J11184-2008	福建省建设工程质量安全监督总站
40	DBJ 13-100-2008	建设工程质量检测信息监管系统技术规程 Technical standard for supervise system of construction quality test information		J11185-2008	福建省建设工程质量安全监督总站
41	DBJ 13-101-2008	水泥土配合比设计规程 The regulations of mixed proportion design for soil mixed with cement		J11186-2008	福建省建筑科学研究院
42	DBJ 13-102-2008	水平定向钻进管线铺设工程技术规程 Standard specification for installing pipeline by horizontal directional drilling		J11214-2008	福建东辰市政工程有限公司
43	DBJ 13-103-2008	福州市温泉供应技术规程 Technical specification for hot spring supply in Fuzhou		J11237-2008	福州市地热管理处

福建省

序号	标准编号	标准名称	被代替标准编号	备案号	主编单位
44	DBJ 13-104-2008	透水砖路面（地面）设计与施工技术规程 The regulations of design and construction for roads and grounds with permeable bricks		J11252-2008	福建省建筑科学研究院 福建省建设科技发展促进中心
45	DBJ 13-105-2008	住宅小区通信配套设施建设标准 Standard for construction of communication ancillary facilities in residential district		J11294-2008	福建省邮电规划设计院有限公司 福建省建筑设计研究院
46	DBJ 13-106-2008	钢-混凝土混合刚（排）架单层房屋结构技术规程 Technical specification for steel-concrete mixed gabled frames (bent) structure		J11328-2009	福建省建筑工程施工图审查中心等
47	DBJ/T 13-107-2015	福建省建筑工程常见质量问题控制规程 Specification for control the popular defect of quality of building engineering	DBJ 13-107-2008	J11338-2015	中建海峡建设发展有限公司 福建省华荣建设集团有限公司
48	DBJ/T 13-108-2015	建筑抹灰工程金属网护角技术规程 Technical code for metallic mesh protected corners of building decoration engineering	DBJ 13-108-2008	J11344-2015	中建海峡建设发展有限公司 福州建工（集团）总公司
49	DBJ 13-109-2009	大树移植技术规程 Technical specification for transplanting large trees		J11350-2009	福州市园林科研所
50	DBJ 13-110-2009	古树名木评估鉴定标准 Code for identification & assessment of ancient and famous trees		J11351-2009	福州市园林科研所
51	DBJ/T 13-111-2009	建筑用薄钢板焊缝超声检测及质量分级法 Method of ultrasonic testing and classification of welds for building sheets		J11357-2009	福建省建筑科学研究院
52	DBJ/T 13-112-2009	福建省建筑节能工程施工文件管理规程 Management specification for energy efficient building construction document of Fujian Province		J11358-2009	福州市建筑工程质量监督站 福建省南安市第一建设有限公司
53	DBJ/T 13-113-2009	回弹法检测高强混凝土抗压强度技术规程 Technical specification for inspection of high strength concrete compressive strength by rebounded method		J11359-2009	福建省建筑科学研究院
54	DBJ/T 13-114-2009	城镇水泥路面施工技术规程 Technical specifications for construction of road cement pavements		J11374-2009	福建省建设工程质量安全监督总站
55	DBJ/T 13-115-2009	城市生活垃圾分类标准 Classification standard of municipal solid waste in Fujian		J11430-2009	厦门市市容环境卫生管理处
56	DBJ/T 13-116-2015	预拌机制砂混凝土生产及施工技术规程 Technical specifications for application of manufactured sand in concrete	DBJ/T 13-116-2009	J11490-2015	福建省建设工程质量安全监督总站
57	DBJ/T 13-117-2016	福建省钢丝网架水泥岩棉夹芯板（GSY板）墙体应用技术规程 Technical specification for the application of cement rockwool sandwich board with wire mesh as wallboard	DBJ/T 13-117-2009	J11517-2009	福建二建建设集团公司 福建省抗震防灾技术中心
58	DBJ/T 13-118-2014	福建省绿色建筑评价标准 Evaluation standard for green building of Fujian	DBJ/T 13-118-2010	J11573-2014	厦门市建筑科学研究院集团股份有限公司 厦门市合道工程设计集团有限公司

福建省

序号	标准编号	标准名称	被代替标准编号	备案号	主编单位
59	DBJ/T 13-119-2010	福建省住宅工程质量分户验收规程 Specifications for individual household acceptance of quality of Fujian residential project		J11570-2010	泉州市建设工程质量监督站 福建省南安市第一建设有限公司
60	DBJ/T 13-120-2010	古树名木管理与养护技术标准 Technical standard of management & maintenance of ancient and famous trees		J11576-2010	福州市园林科学研究院
61	DBJ/T 13-121-2010	建筑工程施工技术管理规程 Specifications for building construction technology management		J11698-2010	中国建筑第七工程局第三建筑公司 福建省九龙建设集团有限公司
62	DBJ/T 13-122-2010	轻型种植屋面工程技术规程 Technical specification for lightweight rooftop planting systems		J11597-2010	福建省建设科技发展促进中心 福建工程学院
63	DBJ/T 13-123-2010	建筑装饰装修工程设计制图标准 Standard for building decoration engineering design drawing		J11618-2010	厦门市建筑装饰协会
64	DBJ/T 13-124-2010	城市垂直绿化技术规范 Technical code for vertical greening in city		J11620-2010	厦门市筼筜湖管理中心
65	DBJ/T 13-125-2010	福建省城市供水企业安全运行管理标准 Security, operations and management standard for urban water supply enterprise in Fujian Province		J11633-2010	厦门水务集团有限公司
66	DBJ/T 13-126-2010	酚醛保温板外墙保温工程应用技术规程 Technical specification for the application of phenolic insulation board in exteral wall insulation systems		J11634-2010	福建六建建设集团有限公司 福建省建设科技发展促进中心
67	DBJ/T 13-127-2010	福建省城市用水量标准 Standard for urban water consumption of Fujian Province		J11697-2010	福建省城乡规划设计研究院 福建省城市建设协会
68	DBJ/T 13-128-2010	水泥粉煤灰碎石桩复合地基技术规程 Specification for composite foundation of cement-flyash-gravel piles		J11728-2010	厦门市建设工程质量安全监督站 福建省建筑科学研究院
69	DBJ/T 13-129-2010	后锚固填充墙拉结钢筋施工及验收规程 Constructional and acceptance specification of post-installed fastenings tie bar for filledin wall		J11738-2010	福州市建设工程质量监督站
70	DBJ/T 13-130-2010	管道液化石油气供气系统天然气转换规程 Technical specification for piping gas system of LPG converting to NG		J11729-2010	福州市液化石油气管理处
71	DBJ/T 13-131-2010	城市行道树栽植技术规程 Technical specification for street tree planting in city		J11739-2010	厦门市园林绿化工程质量监督站
72	DBJ/T 13-132-2010	园林绿化种植土质量标准 Quality standard for garden soil		J11765-2010	福州市园林科学研究院
73	DBJ/T 13-133-2011	液化天然气（LNG）汽车加气站设计与施工规范 Code for design and construction of automotive liquified natural gas (LNG) filling station		J11785-2011	福建中闽物流有限公司 福建省轻工业安装公司

福建省

序号	标准编号	标准名称	被代替标准编号	备案号	主编单位
74	DBJ/T 13-134-2011	市政工程施工质量评价标准（城镇道路、城市桥梁及给排水管道工程） Standard for quality evaluation of municipal engineering (road works in city and town, bridge works in city, water and sewerage pipeline works)		J11804-2011	福建省建设工程质量安全监督总站 中建七局第三建筑有限公司
75	DBJ/T 13-135-2011	市政工程施工技术文件管理规程（城镇道路、城市桥梁、给排水构筑物及管道工程） Specification for construction documentation management of municipal engineering (road works in city and town, bridge works in city, water and sewerage structures and pipeline works)		J11804-2011	福建省建设工程质量安全监督总站 中建七局第三建筑有限公司
76	DBJ/T 13-136-2011	钢管混凝土拱桥技术规程 Technical specification for concrete filled steel tubular arch bridges		J11833-2011	福州大学 中建七局第三建筑有限公司
77	DBJ/T 13-137-2011	水泥基耐磨地面应用技术规程 Technical code for cement-based abrasion-proof floor		J11844-2011	厦门市建筑科学研究院集团股份有限公司
78	DBJ/T 13-138-2011	福建省居住建筑节能检测技术规程 Technical specification for energy efficiency test of residential buildings of Fujian		J11846	福建省建筑科学研究院 福建省工程建设科学技术标准化协会
79	DBJ/T 13-139-2011	生活垃圾卫生填埋场运行维护及考核评价标准		J11917-2011	厦门市市容环境卫生管理处
80	DBJ 13-140-2011	探火管感温自启动灭火装置设计、施工及验收规程 Code for design, installation and acceptance of self-actuated fire extinguishing equipment for temperature used fire detect tub		J11873-2011	福建省公安消防总队 福建省建筑设计研究院
81	DBJ/T 141-2011	福建建筑结构风压规程 Specification for wind pressure of building structure in Fujian		J11897-2011	福建省建筑工程施工图审查中心
82	DBJ/T 13-142-2011	细水雾灭火系统技术规程 Technical standard for water mist fire protection systems		J11926-2011	福建省公安消防总队 福建省建筑设计研究院
83	DBJ/T 13-143-2011	福建省城市住宅小区物业服务规范 Code of urban residential property services of Fujian Province		J11925-2011	福建省物业管理协会
84	DBJ/T 13-144-2011	福建省建筑工程监理文件管理规程 Specification for building engineering supervision document management of Fujian Province		J11943-2011	福建省建设工程质量安全监督总站
85	DBJ/T 13-145-2012	后锚固法检测混凝土抗压强度技术规程 Technical specification for inspection of concrete compressive strength by post-installed adhesive anchorage method		J11958-2012	福建省建筑科学研究院
86	DBJ/T 13-146-2012	建筑地基检测技术规程 Technical specification for testing of building foundation subgrade		J12009-2012	福建省建筑科学研究院
87	DBJ/T 13-147-2012	稳定型橡胶改性沥青路面施工技术规程 Technical specification for construction of stabilized rubber modified asphalt		J12051-2012	福建省金泉建设集团有限公司 福建省鑫海湾建材科技有限公司

福建省

序号	标准编号	标准名称	被代替标准编号	备案号	主编单位
88	DBJ/T 13-148-2012	城市园林植物种植技术规程 Technical specification for planting of landscape plant in cities		J12052-2012	厦门市园林绿化工程质量监督站
89	DBJ/T 13-149-2012	城市道路养护作业安全设施设置技术规程 Technical specification of setting for maintenance work safety devices of city		J12056-2012	福州市市政工程管理处 福州大学
90	DBJ/T 13-150-2012	自密实混凝土加固工程结构技术规程		J12109-2012	福州大学 福建省中嘉建设工程有限公司
91	DBJ/T 13-151-2012	预拌混凝土绿色生产管理规程 Specification of green production management for ready-mixed concrete		J12107-2012	厦门市建筑科学研究院集团股份有限公司
92	DBJ/T 13-153-2012	城市桥梁养护管理机构设置及专业技术人员、设备配置标准 Standard for structuring of administrative organization of municipal bridge maintenance and configuration of technicians and equipment		J12132-2012	福州市政工程管理处
93	DBJ/T 13-154-2012	城市园林绿地养护质量标准 Quality standard for garden soil		J12157-2012	三明市园林管理局
94	DBJ/T 13-155-2012	福建省既有居住建筑节能改造技术规程		J12159-2012	福建省建筑科学研究院 福建省工程建设科学技术标准化协会
95	DBJ/T 13-156-2012	福建省地源热泵系统应用技术规程 Technical specification for ground-source heat pump system of Fujian		J12158-2012	福建省建筑科学研究院 福建建工集团总公司
96	DBJ/T 13-157-2012	建筑太阳能光伏系统应用技术规程 The technology code in building photovoltaic system		J12160-2012	福建省建筑科学研究院 福建省鼎日光电科技有限公司
97	DBJ/T 13-158-2012	福建省公共建筑能耗监测系统技术规程 Technical regulation of energy consumed monitoring systems for public buildings		J12165-2012	福建省建筑设计研究院 福建建工集团总公司
98	DBJ/T 13-160-2012	稳定型橡胶改性沥青应力吸收层施工技术规程		J12200-2012	福建省金泉建设集团有限公司 福建省鑫海湾建材科技有限公司
99	DBJ/T 13-161-2012	建筑用净化海砂应用技术规程 Technical specification for purified sea sand for building		J12109-2012	福建省建筑科学研究院 福州市城乡建设发展总公司
100	DBJ/T 13-162-2012	城市道路占用与挖掘管理标准 Directorial standard of impropriating and digging urban roads		J12201-2012	福州市政工程管理处
101	DBJ/T 13-163-2012	福建省城市桥梁限载标准 Standard of limiting load for the municipal bridge		J12246-2013	福州市城乡建设委员会
102	DBJ/T 13-164-2012	城市桥梁检测评估标准 Standard of detection and evaluation for the municipal bridge		J12247-2013	福州市城乡建设委员会
103	DBJ/T 13-165-2013	城市桥梁养护维修管理标准 Standard of maintenance and management for municipal bridge		J12261-2013	泉州市市政工程管理处
104	DBJ/T 13-166-2013	剪压法检测混凝土抗压强度技术规程 Technical specification for testing of concrete compressive strength by shear-press		J12284-2013	福建省建筑科学研究院

福建省

序号	标准编号	标准名称	被代替标准编号	备案号	主编单位
105	DBJ/T 13-167-2013	福建省城市道路雨水排水设计标准 Standard for urban road drainage of Fujian Province		J12282-2013	福州市规划设计研究院
106	DBJ/T 13-168-2013	轻集料混凝土多孔砖应用技术规程		J12283-2013	福建省建筑科学研究院 福建省二建建设集团有限公司
107	DBJ/T 13-169-2013	城市道路LED照明设计标准 Standard for LED lighting design of urban road		J12303-2013	福州市规划设计研究院
108	DBJ/T 13-170-2013	木塑地板铺装技术规程 Technical specification for installation works of wood-plastic composites flooring		J12302-2013	福建省建筑科学研究院 福建弘景木塑科技股份有限公司
109	DBJ/T 13-171-2013	园林植物保护技术规程 Technical regulation of landscape plant protection		J12332-2013	泉州市园林管理局
110	DBJ/T 13-172-2013	城市园林绿化工程用木本苗木标准 Standard for urban landscape greening tree seedling		J12338-2013	福州市园林科学研究院
111	DBJ/T 13-173-2013	树池透水彩石应用技术规程 Applied technique rules of porous color stone in tree pool		J12339-2013	福州市园林科学研究院
112	DBJ/T 13-174-2013	福建省古建筑屋面施工及验收规程 Specification for roof construction and acceptance of historic buildings in Fujian Province		J12401-2013	福建省南安市第一建设有限公司
113	DBJ/T 13-175-2013	热拌沥青混合料生产技术规程 Technical specification for production of hot-mix asphalt mixtures		J12405-2013	福建省建筑科学研究院
114	DBJ/T 13-176-2013	铝合金电缆工程设计、施工及验收规程 Specification for design, installation and acceptance of aluminum alloy cable engineering		J12404-2013	福建省建筑设计研究院
115	DBJ/T 13-177-2013	预应力混凝土折线形屋架施工技术规程 Construction technical specification of broken line roof truss of prestressed reinforced concrete		J12478-2013	福建省建筑科学研究院
116	DBJ/T 13-178-2013	6m后张法预应力混凝土吊车梁施工技术规程 Construction technical specification for 6m crane girder of post-tension prestressed reinforced concrete		J12479-2013	福建省建筑科学研究院
117	DBJ/T 13-179-2013	福建省村庄整治技术规程 Technical specification for village rehabilitation of Fujian Province		J12477-2013	厦门理工学院
118	DBJ/T 13-180-2013	建筑工程绿色施工技术规程 Technical specification for green construction of building		J12476-2013	中建海峡建设发展有限公司
119	DBJ/T 13-181-2013	扣件式钢管支撑高大模板工程安全技术规程 Code for safety technical of fastening steel pipe support formwork engineering		J12529-2014	福建工程学院 福建省建设工程质量安全监督总站

福建省

序号	标准编号	标准名称	被代替标准编号	备案号	主编单位
120	DBJ/T 13-182-2013	彩色路面应用技术规程 Technical specification for application of colored pavement		J12567-2014	福建省建筑科学研究院 中建海峡建设发展有限公司
121	DBJ/T 13-183-2014	基桩竖向承载力自平衡法静载试验技术规程 Technical specification for static loading test of self-balanced method of vertical bearing capacity of foundation pile		J12566-2014	福建省建筑科学研究院
122	DBJ/T 13-184-2014	福建省城镇排水管道检查井防坠落安全网标准 Fall arrest safety nets standard for manhole of urban drainage pipeline in Fujian		J12574-2014	福建省城市建设协会 厦门水务中环污水处理有限公司
123	DBJ/T 13-185-2014	建筑腻子施工及验收规程 Specification for construction and acceptance of building putty		J12575-2014	福建省建筑科学研究院 中建海峡建设发展有限公司
124	DBJ/T 13-186-2014	建筑排水聚丙烯静音管道工程技术规程 Engineering technical specification of polypropylene sound insulating pipeline for building drainage		J12576-2014	福建省建筑设计研究院 福州大学土木工程学院
125	DBJ/T 13-187-2014	住宅区和住宅建筑内有线广播电视设施工程设计、施工和验收规程 Specification for design, installation and acceptance of cable television facilities in residential district		J12633-2014	福建省建筑设计研究院 福建广电网络集团股份有限公司
126	DBJ/T 13-188-2014	城镇道路水泥稳定粒料基层施工技术规程 Technical specification for construction of urban cement stabilized aggregate		J12616-2014	福建省建设工程质量安全监督总站 福建省榕圣市政工程股份有限公司
127	DBJ/T 13-189-2014	福建省房屋建筑 工程材料检测试验文件管理规程 Specification for ducument management of building engineering material inspection and testing of Fujian Province		J12617-2014	福建省建设工程质量安全监督总站 厦门市工程检测中心有限公司
128	DBJ/T 13-190-2014	混凝土结构 耐久性现场检测与评定技术规程 Technical specification for in-situ inspection and assessment of concrete structure durability		J12615-2014	福建省建设工程质量安全监督总站 福建博海工程技术有限公司
129	DBJ/T 13-191-2014	温拌沥青混合料路面施工技术规程 Specification for warm mix asphalt for pavement construction		J12639-2014	福建省金泉建设集团有限公司 福建省鑫海湾建材科技有限公司
130	DBJ/T 13-192-2014	电气火灾监控系统设计、施工及验收规程 Code for design, installation and acceptance of alarm and control system for electric fire prevention		J12640-2014	福建省公安消防总队 福建省建筑设计研究院
131	DBJ/T 13-193-2014	福建省省级企业技术中心（建筑施工企业）管理与评价标准 Standard for Fujian provincial enterprise technology center (construction enterprises) management and evaluation		J12813-2014	福建建工集团总公司 福建省顺安建筑工程有限公司
132	DBJ/T 13-194-2014	福建省绿道规划建设标准 Standard for greenway planning and construction of Fujian Province		J12814-2014	福建省城乡规划设计研究院 福建省城市规划学会

福建省

序号	标准编号	标准名称	被代替标准编号	备案号	主编单位
133	DBJ/T 13-195-2014	烧结煤矸石多孔砖（砌块）应用技术规程 Technical specification for application of fired coal gangue perforated brick and block		J12815-2014	福建省建筑科学研究院 福建成森建设集团有限公司
134	DBJ/T 13-196-2014	水泥净浆材料配合比设计与试验规程 Specification for mix proportion design and test of cement paste		J12821-2014	福建省建筑科学研究院 中建海峡建设发展有限公司
135	DBJ/T 13-197-2014	福建省绿色建筑设计规范 Code for green buildings design of Fujian		J12822-2014	福建省建筑科学研究院 福建省城乡规划设计研究院和 福建省建筑设计研究院
136	DBJ/T 13-199-2014	废弃混凝土拌合物回收利用技术规程 Specification of reclamation for abandoned fresh concrete		J12844-2014	厦门市建筑科学研究院集团股份有限公司 中泰（福建）混凝土发展有限公司
137	DBJ/T 13-200-2014	桩基础与地下结构防腐蚀技术规程 Technical specification for anticorrosive of pile foundations and underground structure		J12845-2014	厦门理工学院 厦门市建设工程质量安全监督站
138	DBJ/T 13-201-2014	福建省全装修住宅工程技术规程 Technical specification of full decorated housing of Fujian		J12846-2014	厦门市建设工程质量安全监督站 厦门特房建设工程集团有限公司
139	DBJ/T 13-202-2014	福建省建设工程造价咨询规程 Code of practice for construction cost consultants in Fujian		J12897-2014	厦门市建设工程造价管理站
140	DBJ/T 13-203-2014	既有建筑幕墙可靠性鉴定及加固规程 Specification for reliability appraisal and strengthening of curtain wall		J12898-2014	福建省建筑科学研究院 中建海峡建设发展有限公司
141	DBJ/T 13-204-2014	福建省城市地下管线探测及信息化技术规程 Technical specification for detecting and surveying underground pipelines and cables of informatization in Fujian Province		J12899-2014	厦门市城市建设档案馆
142	DBJ/T 13-205-2014	福建省城市地下管线信息数据库建库规范 Code for information database building of underground pipelines and cables in Fujian Province		J12900-2014	厦门市城市建设档案馆
143	DBJ/T 13-206-2014	福建省混凝土用机制砂质量及检验规程 Quality and inspection specification of manufactured sand for concrete of Fujian Province		J12911-2015	福建省建筑科学研究院和福建源鑫集团
144	DBJ/T 13-207-2014	脱硫石膏砂浆应用技术规程 Technical specification for application of gypsum mortar from flue gas desulfurization		J12913-2015	厦门市建筑科学研究院集团股份有限公司 漳州正霸建材科技有限公司
145	DBJ/T 13-208-2014	冲击回波法检测混凝土厚度和内部缺陷技术规程 Technical specification for testing on thickness and inner flaw of concrete by impact echo method		J12912-2015	福建省抗震防灾技术中心和福建省建筑科学研究院
146	DBJ/T 13-209-2015	桥梁结构动力特性检测技术规程 Technical specification for inspecting dynamic characteristic of bridge engineering structures		J12932-2015	福建省建筑科学研究院 福建省抗震防灾技术中心

福建省

序号	标准编号	标准名称	被代替标准编号	备案号	主编单位
147	DBJ/T 13-210-2015	福建省建筑基桩检测试验文件管理规程 Specification for document management of building foundation piles testing of Fujian Province		J12931-2015	建省建设工程质量安全监督总站 厦门市翔安区建设工程质量安全监督站
148	DBJ/T 13-211-2015	纤维水泥夹芯复合墙板应用技术规程 Technical specification for application of composite sandwich panel with fiber cement in building		J12933-2015	福建省建筑科学研究院和福建鑫晟钢业有限公司
149	DBJ/T 13-212-2015	既有建筑结构加固工程现场检测技术规程 Technical specification for on site test of strengthening building structures		J12941-2015	厦门市工程检测中心有限公司 厦门市建设工程质量安全监督站
150	DBJ/T 13-213-2015	园林绿化工程监理规程 Specification for construction supervision of landscape greening		J12942-2015	福州市园林科学研究院 晋江市市政园林局
151	DBJ/T 13-214-2015	桥面绿化种植养护技术规程 Specification for greening and maintenance techenic of bridge deck		J12943-2015	福州市园林科学研究院和晋江市市政园林局
152	DBJ/T 13-215-2015	福建省建筑节能工程质量检测试验文件管理规程 Specification for ducument management of the quality of energy efficient building test of Fujian Province		J12971-2015	福建省建设工程质量安全监督总站 福建九鼎建设集团有限公司
153	DBJ 13-216-2015	福建省预制装配式混凝土结构技术规程 Technical specification for precast concrete structures of Fujian		J12934-2015	福建省建筑设计研究院 润铸建筑工程（上海）有限公司 厦门合道工程设计集团有限公司
154	DBJ/T 13-217-2015	福建省城镇道路检查井技术规范 Technical specification for urban road inspection wells of Fujian Province		J12975-2015	福州市规划设计研究院
155	DBJ/T 13-218-2015	福建省城市桥梁检测试验文件管理规程 Specification for ducument management of inspection and testing for city bridge of Fujian Province		J12983-2015	福建省建设工程质量安全监督总站 福建九鼎建设集团有限公司
156	DBJ/T 13-219-2015	城市隧道管理养护技术规程 Technical specification of management and maintenance for municipal tunnel		J12982-2015	福建省建设工程质量安全监督总站 厦门路桥建设集团有限公司
157	DBJ/T 13-220-2015	泡沫混凝土应用技术规程 Technical specification for application of foamed concrete		J12986-2015	福建省建筑科学研究院 福建厚德节能科技发展有限公司
158	DBJ/T 13-221-2015	刚性桩桩网路基设计与施工技术规程 Design and construction specification for the roadbed using rigid piles		J12985-2015	福建省建筑科学研究院
159	DBJ/T 13-222-2015	福建省花坛布置技术规程 The technical specification for flower bed decorate in Fujian		J13023-2015	福州市园林科学研究院 龙岩市园林管理局
160	DBJ/T 13-223-2015	福建省绿地草坪建植及养护技术规程 Techincal specification for establishment and maintenance on lawn in Fujian		J13024-2015	福州市园林科学研究院 龙岩市园林管理局
161	DBJ/T 13-224-2015	福建省地下连续墙检测技术规程 Technical specification for testing of diaphragm wall in Fujian		J13080-2015	福建省建筑科学研究院

福建省

序号	标准编号	标准名称	被代替标准编号	备案号	主编单位
162	DBJ/T 13-225-2015	福建省建设工程材料综合价格编制规程 The compilation rules of materials comprehensive price in construction engineering		J13211-2015	厦门市建设工程造价管理站
163	DBJ/T 13-226-2015	福建省塑料排水检查井应用技术规程 Technical specification for application of plastic inspection chambers for sewerages in Fujian		J13258-2015	福州市规划勘测设计研究总院 浙江天井塑业有限公司
164	DBJ/T 13-227-2016	福建省城镇污水处理厂污水污泥监测技术规程 Technical specification for wastewater and sludge monitoring of municipal wastewater treatment plant in Fujian		J13414-2016	厦门市排水监测站 福建省城市建设协会
165	DBJ/T 13-228-2016	福建省城市道路管理与养护考核标准 Assessment criteria for management and maintenance of urban roads in Fujian		J13374-2016	福州市政工程管理处
166	DBJ/T 13-229-2016	福建省城市桥梁管理与养护考核标准 Evaluation standard of management and maintenance for municipal bridge in Fujian		J13375-2016	泉州市市政工程管理处
167	DBJ/T 13-230-2016	福建省城市道路照明管理与养护考核标准 Standard for lighting management and maintenance of urban road in Fujian		J13376-2016	厦门市市政工程管理处（厦门市城市照明管理中心）
168	DBJ/T 13-231-2016	福建省建筑智能照明系统工程技术规程 Technical specification for engineering of building intelligent lighting system in Fujian		J13381-2016	福建省建筑设计研究院 安明斯智能股份有限公司
169	DBJ/T 13-232-2016	福建省聚苯颗粒轻集料混凝土砌块墙体应用技术规程 Technical specifications for application of expanded polystyrene granule lightweight aggregate concrete block wall in Fujian		J13391-2016	福建省建筑科学研究院 程远节能建材制造（福建）有限公司
170	DBJ/T 13-233-2016	福建省混凝土结构加固修复用聚合物水泥砂浆施工及验收规程 Specification for construction and acceptance of concrete structure strengthened with polymer cement mortar in Fujian		J13412-2016	福建省建筑科学研究院 中建海峡建设发展有限公司
171	DBJ/T 13-234-2016	福建省不发火建筑地面应用技术规程 Technical specification for application of misfiring building ground of Fujian		J13439-2016	厦门市建筑科学研究院集团股份有限公司 中建名城集团有限公司
172	DBJ/T 13-235-2016	福建省磁测井法测试基桩钢筋笼长度技术规程 Technical specification for testing of reinforcement cage length of foundation piles by magnetic logging building in Fujian		J13428-2016	厦门市工程检测中心有限公司
173	DBJ/T 13-236-2016	福建省铝合金模板体系技术规程 Fujian provincial technical specification for aluminum alloy formwork system		J13392-2016	福建工程学院 福建方鼎建筑材料有限公司 福建金正丰金属工业有限公司
174	DBJ/T 13-237-2016	福建省城市公共自行车规划建设与管理技术规程 Technical specification for planning & construction & management of urban public bicycle system in Fujian		J13423-2016	福州市规划设计研究院

福建省

序号	标准编号	标准名称	被代替标准编号	备案号	主编单位
175	DBJ/T 13-238-2016	福建省房屋建筑工程质量监督检测与管理规程 Specification for supervision, inspection and management of building engineering quality in Fujian		J13437-2016	福建省建设工程质量安全监督总站 厦门市建设工程质量安全监督站
176	DBJ/T 13-239-2016	福建省钢铁渣粉混凝土应用技术规程 Technical specification for application of ground iron and steel slag concrete in Fujian		J13443-2016	福建省建筑材料工业科学研究所 福建省钢源粉体材料有限公司
177	DBJ/T 13-241-2016	福建省钻芯法检测混凝土强度技术规程 Technical specification of Fujian for testing concrete strength by drilled core method		J13482-2016	福建省建筑科学研究院
178	DBJ/T 13-242-2016	福建省可控刚度桩筏基础技术规程 Technical code for piled raft foundation of controlled stiffness in Fujian		J13488-2016	中建海峡建设发展有限公司 南京工业大学
179	DBJ/T 13-243-2016	福建省石粉在混凝土中应用技术规程 Technical specification for application of stone powder in concrete in Fujian		J13508-2016	厦门市建筑科学研究院集团股份有限公司 中交一公局厦门工程有限公司
180	DBJ/T 13-244-2016	福建省预制混凝土衬砌管片质量验收规程 Standard for acceptance of quality of prefabricated concrete segments in Fujian		J13509-2016	福州市地铁建设工程质量安全监督站 福建省建筑科学研究院
181	DBJ/T 13-245-2016	福建省园林植物修剪技术规程 Technical regulation for landscape plant pruning in Fujian Province		J13536-2016	泉州市园林管理局 泉州市市政公用管理局总工程师办公室

浙江省

序号	标准编号	标准名称	被代替标准编号	备案号	主编单位
1	DB33/1001-2003	建筑地基基础设计规范		J10252-2003	浙江省建筑设计研究院
2	DB33/1003-2006	住宅建筑通信设施设计规范		J0798-2006	浙江省通信管理局
3	DB33/1012-2003	挤扩支盘混凝土灌注桩技术规程		J10270-2004	浙江工业大学建筑工程学院
4	DB33/T 1013-2016	混凝土矿物掺合料应用技术规程		J13586-2016	浙江大学建筑工程学院
5	DB33/1014-2003	混凝土多空砖建筑技术规程		J10295-2003	浙江省建筑设计研究院
6	DB33/1015-2015	居住建筑节能设计标准 Design standard for energy efficiency of residential buildings		J10310-2015	浙江大学建筑设计研究院有限公司
7	DB33/1016-2004	先张法预应力混凝土管桩基础技术规程		J10348-2003	浙江大学建筑设计研究院
8	DB33/1017-2004	房屋白蚁预防技术规程		J10432-2004	浙江省白蚁防治所
9	DB33/1019-2005	看守所建筑设计规范		J10585-2005	浙江省公安厅
10	DB33/1020-2005	建筑工程地质钻探安全技术操作规程		J10584-2005	杭州市勘测设计研究院
11	DB33/1021-2013	城市建筑工程停车场（库）设置规则和配建标准 Standards for planning parking lots (garages) in urban architectural engineering		J12424-2013	浙江省城乡规划设计研究院 浙江省标准设计站
12	DB33/T 1022-2005	蒸压砂加气混凝土砌块应用技术规程		J10653-2005	浙江省新型墙体材料改革办公室
13	DB33/T 1024-2005	大体积混凝土工程施工技术规程		J10655-2005	浙江省建筑业管理局
14	DB33/T 1027-2006	蒸压粉煤灰加气混凝土应用技术规程		J10799-2006	浙江省新型墙体材料办公室
15	DB33/T 1028-2006	岩体结构面抗剪强度综合评价应用技术规程		J10800-2006	浙江建设职业技术学院
16	DB33/T 1029-2006	地面辐射供暖及供冷应用技术规程 Technical specification for floor radiant heating and cooling		J10858-2006	浙江建设职业技术学院
17	DB33/T 1030-2006	传染病区（房）建筑设计标准		J10801-2006	浙江省现代建筑设计有限公司
18	DB33/T 1032-2006	饭店建筑节能管理标准 Energy conservation management standard of hotel construction and facility		J10910-2006	浙江大学旅游学院
19	DB33/1033-2006	天然气联合循环电厂设计防火规范 Code of fire protection design for natural gas-nutural combined cycle power plant		J10883-2006	浙江省公安厅
20	DB33/1034-2007	居住建筑太阳能热水系统设计、安装及验收规范 Code for design, installation and acceptance of residential building solar water heating system		J10906-2006	浙江大学建筑设计研究院
21	DB33/1035-2007	建筑施工扣件式钢管模板支架技术规程 Technical rule for steel tubular formwork support with couplers in building construction		J10905-2006	浙江大学建工学院
22	DB33/1036-2007	公共建筑节能设计标准 Design standard for energy efficiency of public buildings		J11071-2007	浙江大学建筑设计研究院 浙江省建筑设计研究院 浙江省气象科学研究所
23	DB33/1037-2007	金属网建筑阳角技术规程 Technical code for metallic building cornris		J10964-2007	浙江大学建筑设计研究院
24	DB33/1038-2007	河道生态建设技术规程 Technical specification for construction of river couse		J11148-2007	浙江省河道管理总站 浙江省水利水电专科学校
25	DB33/T 1039-2007	绿色建筑评价标准 Evaluation standard for green building		J11149-2008	温州市联合建筑设计有限公司

浙江省

序号	标准编号	标准名称	被代替标准编号	备案号	主编单位
26	DB33/T 1042-2007	城镇排水设施养护作业安全技术规程 Safety technical specification for operation and maintenance of urban drainage		J11167-2008	杭州高新（滨江）水务有限公司
27	DB33/T 1044-2007	大直径现浇混凝土薄壁管桩技术规程 Technical code for cast-in-situ concrete large-diameter tubular pile		J11169-2008	浙江大学建工学院
28	DB33/1047-2008	混凝土小型空心砌块建筑技术规程 Technical specification for concrete small hollow block masonry building		J11221-2008	浙江大学建筑设计研究院
29	DB33/T 1048-2010	刚-柔性复合桩基技术规程 Technical code for rigid-flexible composite pile foundation		J11690-2010	温州市建筑设计研究院
30	DB33/T 1049-2016	回弹法检测泵送混凝土抗压强度技术规程 Technical specification for inspection of pumped concrete compressive strength by rebound method		J13498-2016	浙江省建筑科学设计研究院有限公司
31	DB33/1050-2016	城市建筑工程日照分析技术规程 The technical regulation of daylight analysis for urban construction project		J11223-2016	杭州市规划局 杭州市城市规划信息中心
32	DB33/T 1051-2008	复合地基技术规程 Technical code for composite foundation		J11244-2008	浙江大学土木工程学系
33	DB33/T 1052-2008	土壤固化剂加固道路路基应用技术规程 Application technical specification for reinforce with soil stabilization of roadbed		J11245-2008	杭州广播电视大学城市建设系
34	DB33/T 1053-2008	固定式塔式起重机基础技术规程 Technical specification for immovable foundation of tower crane		J11284-2008	浙江宝业建设集团有限公司
35	DB33/T 1054-2016	无机轻集料砂浆保温砂浆及系统技术规程 Technical specification for thermal insulating systems of inorganic lightweight aggregate mortar		J13585-2016	浙江大学建筑工程学院等
36	DB33/1055-2008	环境照明工程技术规范 Technical specification of environment lighting engineering		J11333-2009	浙江省照明学会 杭州市建筑设计研究院有限公司
37	DB33/1056-2008	城市道路平面交叉口规划与设计规范 Planning and design criteria of at-grade intersections on urban streets		J11334-2009	杭州市综合交通研究中心
38	DB33/1057-2008	城市道路机动车道宽度设计规范 Design criteria of vehicle lane width on urban streets		J11335-2009	杭州市综合交通研究中心
39	DB33/1058-2008	城市道路人行过街设施规划与设计规范 Planning and design criteria of pedestrian facilities on urban streets		J11336-2009	杭州市综合交通研究中心
40	DB33/1059-2008	数字化城市管理部件和事件分类与立案结案标准 Registration and conclusion and classification standards for the component and event of digitalized city management		J11354-2009	浙江省建设厅 杭州市城市管理信息中心 绍兴县建设局

浙江省

序号	标准编号	标准名称	被代替标准编号	备案号	主编单位
41	DB33/1060-2008	数字化城市管理信息系统绩效评价规范 Performance appraisal norms for digitalized city management information system		J11355-2009	绍兴县建设局 杭州市城管信息中心
42	DB33/T 1061-2009	城镇广场工程质量验收规范 Code for acceptance of construction quality of Zhejiang Province city and town public square		J11415-2009	浙江省市政行业协会 杭州市建设工程质量安全监督总站
43	DB33/1062-2009	城镇景观河道养护技术规程 Technical specification for maintenance of urban landscape river		J11445-2009	杭州高新（滨江）水务有限公司
44	DB33/1064-2009	建筑门窗应用技术规程 Doors and windows application architecture point of order		J11566-2010	浙江省建筑科学设计研究院有限公司
45	DB33/T 1065-2009	工程建设岩土工程勘察规范 Engineering construction code for investigation of geotechnical engineering		J11637-2010	浙江大学建筑设计研究院 杭州市勘察设计研究院 浙江工程勘察院 浙江省综合勘察研究院有限公司
46	DB33/1066-2010	村镇避灾场所建设技术规程 Technical specification for construction of (natural) disaster evacuation shelter		J11565-2010	浙江省建筑厅
47	DB33/1067-2010	预应力混凝土结构技术规程 Code for fire acceptance of building engineering		J11630-2010	浙江大学建筑工程学院
48	DB33/1068-2010	园林绿化工程施工质量验收规范 Code for construction quality acceptance of landscape engineering of Zhejiang Province		J11603-2010	温州市市政园林局
49	DB33/1069-2010	聚氨酯硬泡保温装饰一体化板外墙外保温系统技术规程 Technical specification for external thermal insulation systems of insulation decorative integration board made of polyurethane rigid foam		J11604-2010	浙江省建筑科学设计研究院 浙江科达新型建材有限公司 浙江省嵊州市建筑工程有限公司
50	DB33/1070-2010	大型公共建筑能耗测评标准 Standard of energy use measurement and evaluation for large scale public buildings		J11605-2010	浙江清华长三角研究院 清华大学 杭州华电华源环境工程有限公司
51	DB33/1071-2010	建筑工程消防验收规范 Code for fire acceptance of building engineering		J11593-2010	浙江省公安厅消防局
52	DB33/1072-2010	泡沫玻璃建筑外墙外保温系统技术规程 Technical specification for external foam glass thermal insulating system on walls		J11617-2010	嘉兴学院土木工程研究所 浙江省建筑科学设计研究院有限公司
53	DB33/T 1073-2010	混凝土企业质量管理规范 Code for quality management of concrete enterprises		J11713-2010	浙江省天和建材集团有限公司
54	DB33/T 1074-2010	城镇道路工程施工质量评价标准 Evaluating standard for construction quality of road works in city and town		J11714-2010	浙江省市政行业协会 绍兴市市政公用工程质量监督站
55	DB33/T 1075-2010	城镇道路工程施工安全操作规程 Construction safety operation specification of road works in city and town		J11727-2010	恒基建设集团有限公司 杭州恒泰建设工程有限公司 杭州明华市政工程有限公司
56	DB33/T 1076-2011	翻转式原位固化法排水管道修复技术规程 Technical specification for rehabilitation of existing drainage pipelines by the inversion and curing of a resin-impregnated tube		J11776-2011	杭州市排水有限公司 杭州市建设工程质量安全监督总站 杭州管丽管道工程有限公司

浙江省

序号	标准编号	标准名称	被代替标准编号	备案号	主编单位
57	DB33/T 1077-2011	建筑装饰装修工程质量评价标准 Zhejiang evaluating standard for quality of building decoration engineering		J11783-2011	浙江省建筑装饰行业协会 浙江省一建建设集团有限公司
58	DB33/T 1078-2011	风景区绿色施工管理规范 Code for green construction management of scenic area		J11799-2011	杭州市京杭运河（杭州段）综合保护委员会 杭州西湖风景名胜区湖滨管理处 中天建设集团有限公司
59	DB33/T 1079-2011	控制性详细规划人民防空设施编制标准 Planning standard of civil air defence facilities of regulatory planning		J11807-2011	杭州市人民防空办公室 解放军理工大学 杭州市城市规划设计研究院
60	DB33/T 1080-2011	城镇道路特种沥青路面施工与质量验收规范 Code for construction and quality acceptance of special type asphalt pavements in city and town		J11826-2011	杭州市建设工程质量安全监督总站
61	DB33/T 1081-2011	既有居住建筑节能改造技术规程 Technical specification for the retrofitting of existing residential buildings on energy efficiency		J11839-2011	浙江省建筑科学设计研究院有限公司
62	DB33/T 1082-2011	型钢水泥土搅拌墙技术规程 Technical specification for soil mixed wall		J11861-2011	浙江省建筑设计研究院 杭州大通建筑工程有限公司 浙江新盛建设集团有限公司
63	DB33/T 1083-2011	净水厂生产自动控制系统质量验收规范 Acceptance specification for quality of waterworks production automatism control		J11918-2011	浙江浙大中控信息技术有限公司 杭州天健流体控制设备有限公司 杭州恒泰建设工程有限公司
64	DB33/T 1084-2011	民用建筑装饰装修工程室内环境检测与验收规范 Code for indoor environmental test and acceptances		J11945-2011	浙江大学环境科学研究所 浙江省东阳市天亿建设有限公司 浙江省建筑科学设计研究院有限公司
65	DB33/T 1085-2011	建设工程竣工规划核实测量技术规程 Technical specification for finished construction survey of planning to verify		J11944-2011	杭州市勘察设计研究院
66	DB33/T 1086-2012	渠式切割水泥土连续墙技术规程 Technical specification for trench cutting re-mixing deep wall		J12071-2012	浙江省建筑设计研究院 东杭大通岩土科技（杭州）有限公司 浙江广诚建设有限公司
67	DB33/T 1087-2012	基桩承载力自平衡检测技术规程 Technical specification for self-balanced bearing capacity testing of foundation		J12089-2012	浙江大合建设工程检测有限公司
68	DB33/1088-2013	高层建筑结构设计技术规程 Technical specification for structures of tall building		J12314-2013	浙江省建筑设计研究院
69	DB33/T 1089-2012	公共建筑空气调节系统节能运行管理标准 Standard for energy efficiency of operation and management in air conditioning system		J12250-2013	浙江清华长三角研究院 浙江大学建筑设计研究院
70	DB33/1090-2013	国家机关办公建筑和大型公共建筑用电分项计量系统设计标准 Design standard for sub-metering system of large scale public buildings		J12323-2013	浙江清华长三角研究院
71	DB33/T 1091-2013	基坑工程钢管支撑施工技术规程 Technical specification for steel pipe support of foundation excavation engineering		J12466-2013	浙江华铁建筑安全科技股份有限公司

浙江省

序号	标准编号	标准名称	被代替标准编号	备案号	主编单位
72	DB33/1092-2016	绿色建筑设计标准 Design standard for green building		J13379-2016	浙江大学建筑设计研究院有限公司 浙江大学建筑学系 浙江大学绿色建筑研究中心
73	DB33/T 1093-2013	村镇房屋防灾技术规程 Technical specification for disaster precaution in town and village		J12487-2013	浙江省建筑设计研究院 浙江工业大学建筑工程学院
74	DB33/T 1094-2013	基桩钢筋笼长度磁测井法探测技术规程 Technical code for magnetic logging prospecting and testing of reinforcement cage length in foundation		J12495-2013	浙江有色地球物理技术应用研究院 浙江省建设工程质量检验站有限公司 绍兴县建设工程安全质量监督站
75	DB33/T 1095-2013	预拌砂浆应用技术规程 Technical specification for application of ready-mixed mortar		J12519-2013	浙江省建筑科学设计研究院有限公司 浙江新盛建设集团有限公司 浙江天华建设集团有限公司
76	DB33/T 1096-2014	建筑基坑工程技术规程 Code for technique of building foundation excavation engineering		J12533-2014	浙江省建筑设计研究院 浙江大学
77	DB33/T 1097-2014	城市桥梁检测技术规程 Technical specification for inspection of urban bridge		J12563-2014	浙江华东工程安全技术有限公司等
78	DB33/T 1098-2014	城市桥梁与隧道运行管理规范 Technical code of operation management for city bridge and tunnel		J12564-2014	杭州市市政设施监管中心
79	DB/T 1099-2014	浙江省园林工程施工规范 Code for landscape architecture construction engineering of Zhejiang Province		J12637-2014	杭州大通市政园林工程有限公司 浙江农林大学风景园林与建筑学院
80	DB33/1100-2014	城镇居家养老服务设施规划配建标准 Standard for planning of city and town accessory service facilities for the home-based care for the aged		J12739-2014	浙江省建筑设计研究院 杭州市规划局 杭州通达集团有限公司
81	DB33/T 1101-2014	浙江省保障性住房建设标准 Indemnificatory residential construction standard of Zhejiang Province		J12682-2014	浙江省标准设计站 浙江大学建筑设计研究院有限公司 广厦建设集团
82	DB33/T 1102-2014	墙体自保温系统应用技术规程 Technical specification for self-insulation system of wall		J12681-2014	浙江大学建筑工程学院 华汇工程设计集团股份有限公司 浙江开元新型墙体材料有限公司
83	DB33/T 1103-2014	浙江省建设工程计价成果文件数据标准 Data standards for outcome documents of construction project valuation in Zhejiang Province		J12686-2014	浙江省建设工程造价管理总站 杭州擎洲软件有限公司
84	DB33/T 1104-2014	建设工程监理工作标准 Standards of management for construction project		J12749-2014	浙江省建设工程监理管理协会 浙江工程建设监理公司 浙江江南工程管理股份有限公司
85	DB33/1105-2014	民用建筑可再生能源应用核算标准 Application standard for renewable energy of civil building		J12847-2014	浙江大学建筑设计研究院有限公司 杭州市地源空调研究所 浙江省能源与核技术应用研究院

浙江省

序号	标准编号	标准名称	被代替标准编号	备案号	主编单位
86	DB33/T 1106-2015	建筑太阳能光伏系统应用技术规程 Technical specification for application of solar photovoltaic system of buildings		J13043-2015	浙江省建筑设计研究院 浙江正泰新能源开发有限公司 浙江合大太阳能科技有限公司
87	DB33/T 1107-2014	建筑工程施工安全隐患防治管理规范 Specification for construction safety risk management		J12838-2014	浙江省建筑科学设计研究院有限公司 长业建设集团有限公司 浙江勤业建工集团有限公司
88	DB33/T 1108-2014	房屋白蚁监测控制技术规程 Technical specifications for monitoring and controlling termites in buildings and surroundings		J12833-2014	浙江省白蚁防治中心
89	DB33/1109-2015	城镇防涝规划标准 Standards for local flooding prevention and control system planning		J12981-2015	浙江省城乡规划设计研究院
90	DB33/T 1110-2015	一体化预制泵站应用技术规程 Technical code for integrated prefabricated pumping station		J12936-2015	杭州市市政设施监管中心 格兰富水泵（上海）有限公司 杭州市城乡建设设计院有限公司
91	DB33/1111-2015	居住建筑风环境和热环境设计标准 Design standard for wind and thermal environmental design of residential buildings		J13018-2015	浙江大学建筑设计研究院有限公司
92	DB33/T 1112-2015	建筑基坑工程逆作法技术规程 Technical specification for building foundation excavation engineering constructed by top-down method		J12959-2015	浙江省建筑设计研究院
93	DB33/T 1113-2015	屋面保温隔热工程技术规程 Technical specification of thermal preservation and heat insulation for roof		J12966-2015	浙江大学建筑工程学院
94	DB33/T 1114-2015	建设工程塔机安全监控系统应用技术规程 Technical specification of tower crane safety monitoring system application for construction project		J13056-2015	中国电子科技集团公司第五十二研究所 杭州市建设工程质量安全监督总站 浙江科正电子信息产品检验有限公司
95	DB33/T 1115-2015	塑料排水检查井应用技术规程 Description of plastic drainage inspection well application of technical regulations and provisions		J13172-2015	浙江省城乡规划设计研究院 浙江天井塑业有限公司 浙江工业大学
96	DB33/1116-2015	建筑施工安全管理规范 Code for construction safety management		J13155-2015	中天建设集团有限公司 浙江欣捷建设有限公司 浙江省长城建设集团有限公司
97	DB33/T 1117-2015	建筑施工承插型插槽式钢管支架安全技术规程 Technical specification for safety of bracket steel tubular scaffold in construction		J13188-2015	杭州二建设有限公司 永康市高磊五金工贸有限公司 浙江杭州湾建筑集团有限公司
98	DB33/T 1118-2015	纳米二氧化硅保温毡应用技术规程 Technical specification for application of nano silica insulation felt		J13299-2016	纳诺科技有限公司
99	DB33/T 1119-2016	住宅厨房和卫生间排气道系统应用技术规程 Technical specification for application of ventilating ductssystem for kitchen and bathroom		J13347-2016	杭州市建设工程质量安全监督总站 浙江省绿色建筑与建筑节能行业协会 温州诚博建设工程有限公司

浙江省

序号	标准编号	标准名称	被代替标准编号	备案号	主编单位
100	DB33/T 1120-2016	叠合板式混凝土剪力墙结构技术规程 Technical specification for composite slab concrete shear wall structure		J13383-2016	浙江宝业住宅产业化有限公司 中国联合工程公司 浙江省建筑设计研究院
101	DB33/1121-2016	民用建筑电动汽车充电设施配置与设计规范 Code for allocation and design of electric vehicle charging facilities of civil buildings		J13380-2016	浙江大学建筑设计研究院有限公司 浙江省城乡规划设计研究院 国网浙江省电力公司
102	DB33/T 1122-2016	污水泵站运行质量评价标准 Assessment standard of the operation quality for wastewater pumping station		J13467-2016	绍兴柯桥排水有限公司 萧山水务投资发展有限公司 杭州市市政设施监管中心

安徽省

序号	标准编号	标准名称	被代替标准编号	备案号	主编单位
1	DB34/178-2012	烧结多孔砖砌体工程施工及质量验收规程 Acceptance secification for construction quality of fired perforated brick maso	DB34/178-1999	J12313-2013	安徽省建筑科学研究设计院
2	DB34/179-2012	混凝土小型空心砌块砌体工程施工及质量验收规程 Acceptance specification for construction quality of concrete small-sized hollow	DB34/179-1999	J12312-2013	安徽省建筑科学研究设计院
3	DBJ 34/T 206-2005	安徽省建设工程工程量清单计价规范			安徽省建设厅
4	DB34/T 233-2002	贯入法检测砌筑砂浆抗压强度技术规程			安徽省建筑科学研究设计院
5	DB34/T 234-2002	回弹法检测砌体中普通粘土砖抗压强度技术规程			安徽省建筑科学研究设计院
6	DB34/462-2004	安徽省城市抗震防灾规划编制技术标准			安徽省建设厅 北京工业大学抗震减灾研究所
7	DB34/T 463-2004	钢筋滚轧直螺纹连接技术规程			安徽省建筑科学研究设计院
8	DB34/T 465-2004	混凝土多孔砖砌体工程施工及验收规程			安徽省建筑科学研究设计院
9	DB34/T 490-2005	住宅小区安全防范系统设计规范			安徽省公安厅科技处 合肥市公安局科技处
10	DB34/T 547-2005	城市控制性详细规划编制规范			安徽省建设厅 安徽建筑工业学院
11	DB34/T 569-2005	剪压法检测混凝土抗压强度技术规程			安徽省建筑科学研究设计院 安徽省建筑工程质量第二监督检测站
12	DB34/T 579-2005	住宅小区智能化系统工程设计、验收规范			安徽省智能建筑学会
13	DB34/T 647-2006	无比钢建筑技术规程			安徽省建筑科学研究设计院
14	DB34/T 648-2006	桩承载力自平衡法深层平板载荷测试技术规程			安徽省建筑科学研究设计院
15	DB34/T 733-2009	企口型铝合金聚氨酯复合装饰板外墙外保温工程施工质量验收规程 Tenon slot polyurethane composite decorative plate ofaluminum alloy exterior insulation works for construction quality acceptance of order		J11402-2009	安徽省建筑科学研究设计院 安徽罗宝建筑节能材料有限公司
16	DB34/T 734-2009	住宅工程防渗漏技术规程 Technical specification for antiseep of residential buildings		J11401-2009	合肥市建筑质量安全监督站 安徽建工集团技术开发中心
17	DB34/745-2009	城镇桥梁安全鉴定技术规程 Technical specification for safety appraisal of bridge in city and town		J11368-2009	安徽省建设工程质量安全监督总站
18	DB34/746-2009	住宅小区和商住楼通信设施技术标准 Technology standard for construction of telecommunications facility in residential district and buildings		J11370-2009	合肥工业大学建筑设计研究院 安徽省电信规划设计有限责任公司
19	DB34/T 751-2007	安徽省建设工程质量检测规程			安徽省建设工程质量安全监督总站
20	DB34/810-2008	叠合板式混凝土剪力墙结构技术规程 Technical specification for superimposed slab concrete shear wall structure		J11235-2008	安徽建筑工业学院

安徽省

序号	标准编号	标准名称	被代替标准编号	备案号	主编单位
21	DB34/T 856-2008	光纤光栅感温火灾报警系统设计、施工及验收规范 Code for design, installation & acceptance of the fiber grating temperature sensing fire alarm system		J11296-2008	安徽省公安消防总队
22	DB34/T 867-2008	办公楼物业管理服务规范 Office building property service criterion		J11302-2008	安徽省物业管理专业委员会 安徽辰光物业管理有限公司
23	DB34/T 921-2009	安徽省城市市政基础设施抗震设计审查规范 Code for seismic design review of municipal infrastructure in Anhui Province		J11434-2009	安徽省建设工程勘察设计院
24	DB34/T 978-2009	高桩码头施工安全检查标准 Standard of high-piled wharf safety inspection		J11472-2009	安徽建工集团有限公司
25	DB34/T 1116-2009	汽车库照明节能设计规范 Code for energy saving design of underground garage lighting		J11651-2010	合肥三川自控工程有限责任公司 中铁合肥建筑市政工程设计研究院有限公司
26	DB34/T 1118-2010	城镇检查井盖技术规范 Technical standard for city and town manhole cover		J11675-2010	安徽省产品质量监督检验研究院 合肥市城乡建设委员会 合肥市重点工程建设管理局 合肥市市政工程管理处 合肥市排水管理办公室 合肥市市政设计院 合肥市建筑工程质量监督站 合肥市道路窨井盖管理应急处置中心
27	DB34/T 1232-2010	安徽省房屋和市政工程施工招标文件（标准） Housing & municipal engineering construction bidding document (Standard) of Anhui Province		J11685-2010	安徽省住房和城乡建设厅标准定额处
28	DB34/T 1263-2010	非承重蒸压粉煤灰多孔砖砌体工程施工及质量验收规程 Specification for construction and quality acceptance of non bearing autoclaved fly ash perforated brick masonry		J11775-2011	安徽省建筑科学研究设计院
29	DB34/T 1264-2010	住宅装饰装修验收标准 The checking and accepting standards of the housing-decoration		J11771-2011	安徽省建筑科学研究设计院 安徽省住宅产业化促进中心 合肥经济技术开发区住宅产业化促进中心 合肥建工集团有限公司
30	DB34/1466-2011	安徽省居住建筑节能设计标准 Design standard for energy efficiency of residentia buildings	DB34/T 754-2007	J11810-2011	安徽省建筑设计研究院 合肥市建设委员会
31	DB34/1467-2011	安徽省公共建筑节能设计标准 Design standard for energy efficiency of public buildings	DB34/T 753-2007	J11887-2011	安徽省建筑设计研究院 合肥市建设委员会
32	DB34/T 1468-2011	叠合板式混凝土剪力墙结构施工及验收规程 Specification for construction and acceptanceof superimposed slab concrete wall		J11904-2011	安徽省建筑科学研究设计院 合肥经济技术开发区住宅产业化促进中心 安徽建工集团有限公司 西伟德混凝土预制件（合肥）有限公司

安徽省

序号	标准编号	标准名称	被代替标准编号	备案号	主编单位
33	DB34/T 1469-2011	居住区供配电系统技术规范 Technical code for power supply and distribution systems of urban residential areas		J11903-2011	合肥工业大学建筑设计研究院
34	DB34/T 1470-2011	金融建筑智能化系统技术规范 Technical code for the intelligent systems of financial buildings		J11902-2011	合肥工业大学建筑设计研究院
35	DB34/1503-2011	无机保温砂浆墙体保温系统应用技术规程 Technical specification applied for dry-mixed thermal insulating composition for buildings interior-exterior composition wall exterior insulation system		J11950-2011	安徽省建筑科学研究设计院
36	DB34/T 1504-2011	无水型粉刷石膏应用技术规程 Technical specification for application of anhydrous gypsum plaster		J11951-2011	安徽建筑工业学院 安徽省皖北煤电集团有限责任公司含山恒泰非金属材料分公司 安徽建工集团有限公司
37	DB34/T 1505-2011	建筑反射隔热涂料应用技术规程 Technical application specification of reflective thermal insulating coatings on buildings		J11952-2011	安徽省建筑科学研究设计院 安徽天锦云漆业有限公司
38	DB34/T 1586-2012	非承重混凝土空心砖砌体工程施工及验收规程 Specification for construction and acceptance of masonry engineering with non-load		J12075-2012	滁州市建筑勘察设计院 安徽德森建材科技发展有限公司
39	DB34/T 1587-2012	非承重混凝土复合保温砖砌体自保温应用技术规程 Technical specification for application of self-isolation system of masonry engineering		J12074-2012	合肥工业大学建筑设计研究院 安徽德森建材科技发展有限公司
40	DB34/T 1588-2012	安徽省民用建筑节能工程现场检测技术规程 Code of applied techniques for in-situ inspection on energy conservation engineering		J12077-2012	安徽省建设工程勘察设计院 安徽省建设工程抗震测试研究所有限责任公司
41	DB34/T 1589-2012	建筑节能门窗应用技术规程 Technical specification for application of energy saving doors and windows		J12076-2012	安徽省建筑科学研究设计院 安徽省住宅产业化促进中心 安徽国风塑料建材有限公司 安徽新视野门窗幕墙工程有限公司 合肥经济技术开发区住宅产业化促进中心
42	DB34/1659-2012	住宅工程质量通病防治技术规程 Technical specification for general defect provent in quality of residential building		J12176-2012	安徽省建设工程质量安全监督总站
43	DB34/T 1660-2012	钢结构建筑维护技术规程 Technical specification for maintenance of steel structures buildings		J12169-2012	安徽省建筑科学研究设计院 安徽富煌钢构股份有限公司
44	DB34/T 1786-2012	预制装配式钢筋混凝土检查井技术规程 Technical specification for prefabricated reinforced concrete manhole		J12276-2013	合肥市城乡建设委员会 合肥市市政设计院有限公司 合肥市公路桥梁工程有限责任公司 合肥经济技术开发区住宅产业化促进中心
45	DB34/T 1787-2012	长螺旋钻孔压灌桩技术规程 Technical specification for long screw drilling cast-in-place pile foundation		J12277-2013	合肥工业大学建筑设计研究院

安徽省

序号	标准编号	标准名称	被代替标准编号	备案号	主编单位
46	DB34/T 1788-2012	公路隧道防火涂料喷涂施工及验收规程 Road tunnel fire-resistive coating spray and acceptance standard		J12278-2013	安徽省工业设备安装公司 安徽省高等级公路工程监理有限公司 安徽三建工程有限公司
47	DB34/T 1789-2012	给排水工程顶管技术规程 Specifications of pipe jacking for water and sewer pipeline construction		J12279-2013	合肥市重点工程建设管理局 中国地质大学（武汉）
48	DB34/1800-2012	安徽省地源热泵系统工程技术规程 Technical standard for Anhui ground-source heat pump systems engineering		J12231-2012	安徽省住房和城乡建设厅节能科技处 合肥市建筑业协会建筑节能与勘察设计分会 合肥工业大学
49	DB34/1801-2012	太阳能热水系统与建筑一体化技术规程 Specification for solar water heating system integrated with building	DB34/854-2008	J12294-2013	合肥市城乡建设委员会 安徽省建筑设计研究院有限责任公司 合肥经济技术开发区住宅产业化促进中心
50	DB34/T 1859-2013	岩棉板外墙外保温系统应用技术规程 Specification for application of external thermal insulation systems based on rock wool boards		J12347-2013	安徽建工建设科技有限公司 合肥市城乡建设委员会 安徽威耐得新型建材有限公司
51	DB34/T 1860-2013	建设工程勘察技术资料归档整理规程 Specification for geotechnical investigation technical information filling and arrangement		J12348-2013	安徽省城建设计研究院
52	DB34/T 1874-2013	装配整体式剪力墙结构技术规程（试行） Technical specification for monolithic precast concrete shear wall structure		J12346-2013	安徽省建筑设计研究院有限责任公司 中国十七冶集团有限公司
53	DB34/T 1921-2013	35kV及以下铝合金电力电缆工程设计规范 Code for design of 35kV and below aluminum alloy power cables		J12421-2013	安徽省建筑科学研究设计院 安徽省城乡规划设计研究院
54	DB34/T 1922-2013	公共建筑能耗监测系统技术规范 Technical code for the energy monitor metering systems of public buildings		J12420-2013	合肥工业大学建筑设计研究院 安徽省住房城乡建设厅建筑节能与科技处 安徽省省直机关事务管理局公共机构节能工作处
55	DB34/T 1923-2013	医疗建筑智能化系统技术规范 Technical code for the intelligent systems of medical buildings		J12422-2013	合肥工业大学建筑设计研究院
56	DB34/T 1924-2013	民用建筑能效标识技术标准 Technical standard for energy performance certification of civil buildings		J12423-2013	安徽省城建设计研究院 安徽省绿色建筑协会 安徽建筑大学 安徽建工建设科技有限公司
57	DB34/T 1948-2013	建设工程造价咨询档案立卷标准 The filing standard province construction cost consultancy		J12482-2013	安徽省住房城乡建设厅标准定额处 安徽省建设工程造价管理总站 合肥市建设工程造价管理站
58	DB34/T 1949-2013	挤塑聚苯板薄抹灰外墙外保温系统应用技术规程 Technical specification applied for XPS exterior wall exterior thermal insulation system		J12483-2013	安徽省建筑科学研究设计院
59	DB34/T 1950-2013	安徽省县城规划编制标准 Standard for compiling of county town planning in Anhui Province		J12484-2013	安徽省住房和城乡建设厅 安徽建筑大学

安徽省

序号	标准编号	标准名称	被代替标准编号	备案号	主编单位
60	DB34/T 1991-2013	安徽省建筑工程项目信息编码标准 Technical standard for solar shading engineering of buildings		J12524-2013	安徽省住房和城乡建设厅
61	DB34/5000-2013	住宅区物业服务标准 Service standard of residential community	DB34/T 334-2003	J12525-2013	安徽省住房和城乡建设厅房地产市场监管处 安徽省物价局服务价格处 安徽辰元物业管理有限公司
62	DB34/T 5001-2014	高层钢结构住宅技术规程 Technical specification for high-rise steel structure residential building		J12623-2014	安徽省住宅产业化促进中心 安徽省钢结构协会 合肥工业大学
63	DB34/T 5002-2014	钢管桁架结构技术规程 Technical specification for steel tube truss structures		J12624-2014	合肥工业大学 中铁四局集团钢结构有限公司
64	DB34/T 5003-2014	工程勘察现场作业人员职业标准 Occupational standards for site operating personnel of project prospecting		J12625-2014	安徽省工程勘察设计协会
65	DB34/T 5004-2014	县管省级开发区多层标准化厂房规划设计导则（试行）		J12591-2014	安徽省住房城乡建设厅
66	DB34/5005-2014	先张法预应力混凝土管桩基础技术规程 Technical specification for pretensioned prestressed spun concrete pile foundation	DB34/T 1198-2010	J11684-2014	安徽省建筑科学研究设计院 合肥工业大学建筑设计研究院
67	DB34/5006-2014	太阳能光伏与建筑一体化技术规程 Technical specification for building integrated solar photovoltaic		J12706-2014	合肥市城乡建设委员会 合肥经济技术开发区住宅产业化促进中心 合肥工业大学教育部光伏系统工程研究中心 安徽省建设设计研究院有限责任公司 安徽省住宅产业化促进中心
68	DB34/T 5007-2014	屋面和楼地面泡沫混凝土保温工程技术规程 Technical specification applied for foamed concrete thermal insulating engineering for roofing and flooring		J12767-2014	安徽省建筑科学研究设计院 安徽天筑建设（集团）有限公司
69	DB34/5008-2014	工程建设场地抗震性能评价标准 Standard for earthquake resistant performance assessment of engineering sites	DB34/144-2005	J12794-2014	安徽省城建设计研究院
70	DB34/T 5009-2014	绿色建筑检测技术标准 Technical standard for inspection of green building		J12894-2014	安徽省建筑科学研究设计院 安徽省建筑节能与科技协会
71	DB34/T 5010-2014	螺杆桩基础技术规程 Technical specification for screw pile foundation		J12893-2014	安徽省金田建筑设计咨询有限责任公司 安徽卓典建筑工程有限公司
72	DB34/T 5011-2015	安徽省村庄规划编制标准 Specification on construction tendering made by agency		J12928-2015	安徽省住房和城乡建设厅村镇规划与建设处 安徽省城建设计研究院
73	DB34/T 5012-2015	回弹法检测泵送混凝土抗压强度技术规程 Technical specification for inspection of pumped concretecompressive strength by rebound method		J12930-2015	合肥市建筑质量安全监督站 合肥工业大学 芜湖市建设工程质量监督站

安徽省

序号	标准编号	标准名称	被代替标准编号	备案号	主编单位
74	DB34/T 5013-2015	建设工程招标代理规程 Specification on construction tendering made by agency		J12929-2015	安徽省建设工程招标投标办公室 安徽省建筑工程招标投标协会
75	DB34/T 5014-2015	先张法预应力混凝土竹节桩基础技术规程 Technical specification for pretensioned pre-stressed spun concrete knot pile foundation		J13133-2015	安徽省金田建筑设计咨询有限责任公司 安徽省建筑科学研究设计院 江苏天海建材有限公司
76	DB34/T 5015-2015	先张法预应力混凝土空心方桩基础技术规程 Technical specification for pretensioned pre-stressed spun concrete squared pile foundation		J13134-2015	安徽省金田建筑设计咨询有限责任公司 安徽省建筑科学研究设计院 江苏天海建材有限公司
77	DB34/T 5016-2015	安徽省园林绿化养护管理标准 Conservation management standards of garden afforestationin Anhui Province		J13135-2015	安徽省住房和城乡建设厅城市建设处 安徽经典景观工程（集团）有限公司
78	DB34/T 5017-2015	安徽省城镇燃气服务规范 Code for the city of Anhui Province gas service		J13136-2015	安徽省燃气协会 合肥燃气集团有限公司
79	DB34/T 5018-2015	双向螺旋挤土灌注桩技术规程 Technical specification of soil displacement screw pile		J13137-2015	安徽省建筑科学研究设计院 中冶建筑研究总院有限公司 安徽宏图伟业基础工程有限公司
80	DB34/T 5019-2015	细水雾灭火系统设计、施工与验收规范 Code for design, installation and acceptance of water mist fire extinguishing system	DB34/T 493-2005	J10570-2015	安徽省公安消防总队
81	DB34/T 5020-2015	超细干粉灭火装置设计、施工及验收规范 Code for design, installation and acceptance of ultrafine powder fire extinguishing devices	DB34/T 494-2005	J10571-2015	安徽省公安消防总队
82	DB34/T 5021-2015	电气火灾监控系统设计、施工及验收规范 Code for design, installation and acceptance of alarm and control system for electric fire prevention	DB34/T 855-2008	J11295-2015	安徽省公安消防总队
83	DB34/T 5022-2015	喷射型自动射流灭火系统设计、施工与验收规范 Code of design, installation and acceptance for auto tracking and targeting jet suppression system	DB34/T 608-2006	J13190-2015	安徽省公安消防总队
84	DB34/T 5023-2015	蒸压加气混凝土砌块砌体和抹灰工程施工及质量验收规程 Construction and quality acceptance procedures for masonry& plastering structures of autoclaved aerated concrete blocks		J13191-2015	安徽省建筑科学研究设计院 安徽省墙体材料革新和建筑材料节能协会
85	DB34/T 5024-2015	二次供水工程技术规程 Technical specification for secondary water supply engineering	DB34/T 752-2007	J13192-2015	安徽省城镇供水协会 合肥供水集团有限公司
86	DB34/T 5025-2015	城镇供水服务标准 Standard of urban water supply service		J13193-2015	安徽省城镇供水协会 合肥供水集团有限公司
87	DB34/T 5026-2015	住宅厨卫烟气集中排放系统施工与质量验收规程 Specification for construction and acceptance of gas fumes centralized drainage system for kitchen and bathroom of civil		J13194-2015	安徽建工集团有限公司 合肥学院 安徽建筑大学

安徽省

序号	标准编号	标准名称	被代替标准编号	备案号	主编单位
88	DB34/T 5027-2015	水基渗透型无机防水剂防水工程施工与验收规程 Specification for construction and acceptance of waterproofering of waterbased capillary inorganic waterproofer		J13195-2015	安徽建工集团有限公司
89	DB34/T 5028-2015	工程建设标准员职业标准（试行） Occupational standard for standardization supervisors in construction projects (Trial)		J13196-2015	安徽省住房和城乡建设厅标准定额处 东华工程科技股份有限公司
90	DB34/T 5029-2015	建筑遮阳工程技术规程 Technical standard for solar shading engineering of buildings		J13197-2015	安徽省建筑科学研究设计院 安徽省建筑节能与科技协会 合肥市建筑节能科技与勘察设计协会
91	DB34/T 5031-2015	安徽省海绵城市规划技术导则——低影响开发雨水系统构建（试行） Planning technical guidance of sponge city in Anhui Province build of low impact development stromwater system (pilot draft)		J13241-2015	安徽省城乡规划设计研究院
92	DB34/5032-2015	城镇污水处理厂运行、维护及安全技术规程 Technical specification for operation, maintenance and safety of municipal wastewater treatment plant		J13141-2015	安徽省住房城乡建设厅城市建设处 安徽国祯环保节能科技股份有限公司
93	DB34/T 5033-2015	装配整体式建筑预制混凝土构件制作与验收规程 Technical specification for manufacture and acceptance of precast concrete components of assembled monolithic buildings		J13242-2015	合肥市城乡建设委员会 合肥经济技术开发区住宅产业化促进中心 中建国际投资（中国）有限公司 长沙远大住宅工业安徽有限公司
94	DB34/T 5034-2015	岩沥青改性沥青路面技术规程 Technical code for asphalt pavements using natural rock asphalt		J13243-2015	同济大学、合肥市重点工程建设管理局
95	DB34/T 5035-2015	预拌-增强型温拌沥青路面技术规程 Technical code for pavements using warm mix asphalt based premixed and reinforced technology		J13244-2015	安徽省交通勘察设计院 合肥市重点工程建设管理局 同济大学
96	DB34/T 5036-2015	蒸压砂加气混凝土砌块非承重墙体自保温工程施工及质量验收规程 Construction and quality acceptance procedures for block masonry engineering of sand autoclaved aerated concrete block nonbearing wall heat preservation	DB34/T 1233-2010	J11686-2016	安徽省建筑科学研究设计院 安徽省墙体材料革新和建筑材料节能协会
97	DB34/T 5037-2015	船闸工程施工安全检查标准 Standard for safety inspecton of construction on ship lock wharf		J13297-2016	安徽省路港工程有限责任公司 安徽建工集团有限公司
98	DB34/T 5038-2015	安徽省城市地下空间暨人防工程综合利用规划编制导则 Guide for comprehensive utilization planning about urban underground space and civil air defense project in Anhui Province		J13298-2016	安徽省住房和城乡建设厅城市规划处 安徽省人民防空办公室工程处 合肥市规划设计研究院
99	DB34/T 5039-2016	城市地下空间兼顾人民防空工程设计标准 Design standard of city underground space give consideration to civil air defense works		J13353-2016	安徽省人防建筑设计研究院

安徽省

序号	标准编号	标准名称	被代替标准编号	备案号	主编单位
100	DB34/T 5040-2016	建筑安全生产标准化示范工地评价标准 Assessment standard of construction work safety standardization demonstration sites		J13354-2016	安徽省建设行业安全协会
101	DB34/T 5041-2016	建筑工程质量安全信用评分标准 Construction quality and safety standard for evaluation of credit		J13355-2016	合肥市建筑质量安全监督站 安徽三建工程有限公司
102	DB34/T 5042-2016	地下管线探测与数据标准 Underground pipeline detection and data of surveying and mapping standard		J13393-2016	安徽省城建档案学会 合肥市测绘设计研究院
103	DB34/T 5043-2016	装配整体式混凝土结构工程施工及验收规程 Specification for construction and acceptance of monolithic precast concrete structure		J13394-2016	合肥市城乡建设委员会 合肥经济技术开发区住宅产业化促进中心 中建七局第二建筑有限公司 安徽宝业住宅产业化有限公司
104	DB34/T 5044-2016	养老服务设施规划建设导则 The planning and construction guidelines for elderly care facilities			安徽省住房和城乡建设厅标准定额处 安徽省城乡规划设计研究院 安徽省建筑设计研究院有限责任公司
105	DB34/T 5045-2016	城建档案信息化建设标准 Standard for informationization construction of urban construction archive	DB34/T 1014-2009	J11501-2016	安徽省住房和城乡建设厅城市建设处 芜湖市城建档案馆
106	DB34/T 5046-2016	徽州传统聚落适应性改造与提升规划技术导则 Guideline of Huizhou traditional settlement adaptability reformation and improvement plan technique			安徽建筑大学 安徽省城乡规划设计研究院
107	DB34/T 5047-2016	徽州传统建筑修缮技术规程 Technical specification of Huizhou traditional building reinforcement			安徽建筑大学 安徽省城乡规划设计研究院 同济大学 安徽省徽州古典园林建设有限公司
108	DB34/T 5048-2016	徽州传统建筑改造利用技术导则 Technic guideline of Huizhou traditional architecture recycling			安徽建筑大学
109	DB34/T 5049-2016	保持徽州传统建筑风貌的现代住区规划设计导则 Guidelines for the planning and design of modern residential areas to maintain the traditional architectural features of Huizhou			安徽建筑大学
110	DB34/T 5050-2016	徽派建筑可再生能源利用设计导则 Design guidelines of renewable energy source for Huizhou buildings			安徽建筑大学
111	DB34/T 5051-2016	徽州传统聚落景观水体保护与修复技术导则 Technical specifications for scenic water remediation and protective technics of Huizhou traditional settlements			中国科学技术大学
112	DB34/T 5052-2016	徽州传统建筑木雕构件保护修复技术标准 Technical code for conservation and restoration of woodcarvings of Huizhou traditional architecture			安徽建筑大学

安徽省

序号	标准编号	标准名称	被代替标准编号	备案号	主编单位
113	DB34/T 5053-2016	徽州新聚落生活污水生态处理技术导则 Technical guidelines of domestic wastewater ecological treatment in new settlement of Huizhou			安徽建筑大学 安徽省城建设计研究院
114	DB34/T 5054-2016	徽州传统聚落火灾预警与安全防范技术导则 Technical guide for fire early warming and security protection of traditional settlements in Huizhou			安徽建筑大学
115	DB34/T 5055-2016	城镇污水处理厂运行管理和评价标准 Operation management and evaluation standard for municipal wastewater treatment plant		J13621-2016	安徽省住房和城乡建设厅城市建设处 安徽国祯环保节能科技股份有限公司
116	DB34/T 5056-2016	市政道路桥梁工程施工安全检查标准 Municipal road and bridge construction safety inspection standard		J13627-2016	安徽省公路桥梁工程有限公司 安徽建工集团有限公司 安徽省建设行业安全协会
117	DB34/T 5057-2016	既有公共建筑节能改造技术规程 Technical specification for the retrofitting of existing public building on energy efficiency		J13624-2016	安徽省建筑设计研究院有限责任公司
118	DB34/T 5058-2016	既有居住建筑节能改造技术规程 Technical specification for the retrofitting of existing public building on energy efficiency		J13625-2016	安徽省建筑设计研究院有限责任公司
119	DB34/T 5059-2016	合肥市居住建筑节能设计标准 Design standard for energy efficiency of residential buildings in Hefei City		J13623-2016	合肥市建筑节能科技与勘察设计协会 安徽省建筑设计研究院有限责任公司 合肥神舟建筑集团有限公司
120	DB34/T 5060-2016	合肥市公共建筑节能设计标准 Design criterion for energy efficiency of public buildings in Hefei City		J13622-2016	合肥市建筑节能科技与勘察设计协会 安徽省建筑设计研究院有限责任公司 合肥神舟建筑集团有限公司
121	DB34/T 5061-2016	城镇燃气场站经营企业安全生产标准化评分标准 Safety standardization standard for evaluation of city gas station business		J13620-2016	合肥市燃气行业协会 安徽省燃气协会
122	DB34/T 5062-2016	加气站安全运行与管理检查规程 Safety operation and management of gas station		J13626-2016	合肥市燃气行业协会 安徽省燃气协会
123	DB34/T 5063-2016	非承重烧结保温砌块墙体自保温工程应用技术规程 Rules for construction and acceptance of self-insulation engineering based on non-load-bearing fired heat preservation block wall		J13678-2016	安徽建筑大学 安徽建工集团有限公司建筑设计研究院
124	DB34/T 5064-2016	民用建筑设计信息模型（D-BIM）交付标准 Delivery standard of civilian building design-information modeling (D-BIM)		J13679-2016	安徽地平线建筑设计有限公司
125	DB34/T 5065-2016	玻璃纤维增强复合材料（GFRP）筋地下工程应用技术规程 Technical specification of underground engineering applied of glass fiber reinforced polymer bars (GFRP)		J13677-2016	安徽省建筑科学研究设计院 淮北市城乡建设委员会 淮北宇鑫新型材料有限公司

江西省

序号	标准编号	标准名称	被代替标准编号	备案号	主编单位
1	DB36/J 001-2007	钢管混凝土结构技术规程 Technical specification for concrete-filled steel tubular structure		J11092-2007	华东交通大学土木工程学院
2	DB36/J 001-2009	民用建筑围护结构保温工程施工技术规程（一） Technical specification for civil construction surrounding protection structure heat preservation project（1）		J11438-2009	南昌市建筑工程集团有限公司
3	DB36/J 001-2011/T	江西省新建住宅供配电设施建设标准 New residential construction standards of power supply facilities		J11858-2011	江西省电力公司
4	DB36/J 002-2007	建筑工程资料编制规程 Specification for compiling engineering documents of construction		J11093-2007	江西省建设工程质量监督管理总站
5	DB/T 36/J002-2009	安全控制与报警逃生门锁系统设计、施工及验收规程 Code for design installation and acceptance of security control and exit device with alarm system		J11512-2009	江西省公安消防总队
6	DB36/J 002-2010/T	住宅装饰装修工程技术规程 Technical specification for decoration engineering of housings		J11672-2010	江西省建筑科学研究院
7	DB36/J 003-2006	抱压式桩端自引孔静压入岩 PHC 管桩技术规程 Technical specification of holding type PHC tube pile with tip entering into rock by augering technique		J10818-2006	江西省建设工程质量监督管理总站 江西建筑设计研究总院
8	DB36/J 004-2006	江西省居住建筑节能设计标准 Design standard for energy efficiency of residential buildings in Jiangxi Province		J10908-2006	江西省建筑材料工业科学研究设计院
9	DB36/J 009-2012/T	江西省建筑用反射隔热涂料应用技术规程 Technical specification for application of reflective thermal insulating coating		J12281-2013	江西省建筑科学研究院
10	DB36-J 010-2012/T	江西省住宅小区及商住楼通信管线及配套设施技术应用规范 The communication pipeline and facilities construction standard for residential district and commercial building		J12280-2013	江西省通信管理局 江西省邮电规划设计院有限公司
11	DBJ/T 36-013-2014	烧结保温砌块外墙自保温应用技术规程 Technical specification for exterior wall self-insulation of fired heat-insulation block		J12679-2014	江西省建筑材料工业科学研究设计院
12	DBJ/T 36-024-2014	江西省居住建筑节能设计标准 Design standard for energy efficiency of residential buildings in Jiangxi Province		J12183-2015	江西省建筑材料工业科学研究设计院
13	DBJ/T 36-025-2014	江西省城镇生活饮用水二次供水工程技术规程 Technical specifications for secondary water supply engineering of city and town drinking water in Jiangxi		J12917-2015	江西省城市建设管理协会城市供水专业委员会

江西省

序号	标准编号	标准名称	被代替标准编号	备案号	主编单位
14	DBJ/T 36-026-2015	灌芯复合保温砌块自保温墙体应用技术规程 Technical specification applicated to self-insulation wall of core filling compound insulation block		J12974-2015	江西省建筑材料工业科学研究设计院
15	DBJ/T 36-027-2015	江西省绿色医院建筑评价标准 Evaluation standard for green hospital building of Jiangxi Province		J13138-2015	江西省建筑设计研究总院
16	DBJ/T 36-028-2015	蒸压加气混凝土砌块外墙自保温墙体应用技术规程 Technical specification for application of external self-insulation wall with autoclaved aerated concrete block		J13207-2015	江西省建筑材料工业科学研究设计院
17	DBJ/T 36-029-2016	江西省绿色建筑评价标准 Evaluation standard for green building of Jiangxi Province		J11591-2016	江西省建筑设计研究总院
18	DBJ/T 36-030-2016	遮阳技术应用规程 Solar shading technical specification for application		J13463-2016	江西省建筑科学研究院

山东省

序号	标准编号	标准名称	被代替标准编号	备案号	主编单位
1	DBJ 14-BJ12-2001	建筑内外墙涂料应用技术规程		J10129-2002	山东省建设发展研究院
2	DBJ 14-BJ13-2001	现场喷涂硬质发泡聚氨酯屋面防水保温技术规程		J10130-2002	山东省建设发展研究院
3	DBJ 14-015-2002	建筑施工物料提升机安全技术规程		J10178-2002	山东省建筑科学研究院
4	DBJ 14-018-2002	工程建设管理服务规范		J10184-2002	山东省建设厅
5	DBJ 14-019-2002	挤扩灌注桩技术规程		J10231-2003	山东省建筑设计研究院
6	DBJ 14-021-2002	煤矸石多孔砖砌体结构技术规程		J10233-2003	淄博市墙体材料革新与建筑节能办公室
7	DBJ 14-024-2004	建筑基坑工程监测技术规范		J10372-2004	济南大学
8	DBJ 14-032-2004	建筑工程施工工艺规程		J10495-2005	山东省建筑工程管理局、山东建筑学会
9	DBJ 14-033-2005	建筑施工现场管理标准		J10560-2005	青岛市建设委员会
10	DBJ 14-035-2008	外墙外保温应用技术规程（修订版） Exposed wall outside keep warm application technology regulation		J11135-2008	山东省建设发展研究院
11	DBJ 14-036-2006	公共建筑节能设计标准		J10786-2006	山东省墙体革新与建筑节能办公室
12	DBJ 14-038-2006	混凝土小型空心砌块建筑技术规程		J10793-2006	山东省建筑科学研究院
13	DBJ 14-039-2006	建筑工程施工现场装配式轻钢结构临建房屋技术规程		J10792-2006	青岛建委
14	DBJ 14-040-2006	预应力混凝土管桩基础技术规程 Prestressed concrete tubular pile foundation technical specifcation		J11160-2008	山东省建设发展研究院
15	DBJ/T 14-043-2012	CL建筑体系技术规程 Technical specification for CL building system		J11956-2011	山东省建设发展研究院
16	DBJ 14-044.1～11-2007	建设工程质量专项检测操作规程 Operating regulations for construction engineering quality		J10959-2007	山东省建筑科学研究院
17	DBJ 14-045.1～8-2007	建设工程质量见证取样检测操作规程 Operating regulations for construction engineering quality verificetion		J10960-2009	山东省建筑科学研究院
18	DBJ 14-046-2007	民用建筑室内环境污染物检测操作规程 Test operating regulations for indoor environmental		J10961-2007	山东省建筑科学研究院
19	DBJ 14-047-2007	复合土钉墙施工及验收规范 Code for construction and acceptance of composite soil nailing wall		J11004-2007	济南大学
20	DBJ 14-050-2007	建筑节能检测技术规范 Specifications for architectural energy saving examination		J11078-2007	山东省建设发展研究院
21	DBJ 14-051-2007	聚苯板薄抹灰外墙外保温系统质量控制技术标准 Technical standard for quality control of polyphenyl board plastered external wall outer thermal insulation system		J11079-2007	山东省建设发展研究院
22	DBJ 14-053-2008	蒸压灰砂砖砌体技术规程 Technical specification for autoclave sand-line brick masonry		J11188-2008	山东省建筑科学研究院

山东省

序号	标准编号	标准名称	被代替标准编号	备案号	主编单位
23	DBJ/T 14-054-2009	棚模密肋楼盖结构技术规程 Technical regulation on formwork and decking unitized rib floor		J11343-2009	山东省建设发展研究院
24	DBJ/T 14-055-2009	基桩承载力自平衡检测技术规程 Technical code for self-balanced testing of foundation pile bearing capacity		J11342-2009	山东省建筑科学研究院
25	DBJ/T 14-057-2009	保温隔热材料性能检测操作规程		J11455-2009	山东省建筑科学研究院
26	DBJ/T 14-058-2009	建筑玻璃幕墙与外窗性能检测操作规程		J11456-2009	山东省建筑科学研究院
27	DBJ/T 14-059-2009	建筑节能工程现场实体检测操作规程		J11457-2009	山东省建筑科学研究院
28	DBJ/T 14-060-2009	建筑设备节能工程材料及设备性能检测操作规程		J11458-2009	山东省建筑科学研究院
29	DBJ/T 14-061-2009	粘结与增强材料性能检测操作规程		J11459-2009	山东省建筑科学研究院
30	DBJ 14-063-2010	建筑施工现场施工升降机安全性能评估技术规程 Technical rules for inspection and appraisal to safety capability of builder's hoist used in construction locale		J11640-2010	山东省建筑科学研究院
31	DBJ 14-064-2010	建筑施工现场塔式起重机安全性能评估技术规程 Technical rules for inspection and appraisal to safety capability of tower crane used in construction locale		J11641-2010	山东省建筑科学研究院
32	DBJ 14-065-2010	建筑施工现场塔式起重机安装拆卸安全技术规程 Technical rules for installing and dismantling operation of tower crane used in construction locale		J11642-2010	山东省建筑施工安全监督站
33	DBJ/T 14-066-2010	燃气用衬塑（PE）铝合金管道工程技术规程 Applying technical specification of plastic (PE) liner aluminum alloy composite pipeline for gas supply		J11648-2010	青岛市燃气协会
34	DBJ/T 14-067-2010	旋挖成孔灌注桩施工技术规程 Construction technical specification for rotary drilled and bored pile		J11655-2010	山东省地矿工程勘察院
35	DBJ 14-068-2010	地源热泵系统工程技术规程 Technical specification for ground-source heat pump system		J11674-2010	山东省建设发展研究院
36	DBJ/T 14-069-2010	散热器热水采暖系统塑料管道工程技术规程 Technical regulation for plastic pipe of water heating radiators		J11662-2010	山东省建设发展研究院
37	DBJ 14-070-2010	城市建设项目配建停车位规范 Code for building accessorial parking		J11734-2010	山东省城乡规划设计研究院
38	DBJ/T 14-071-2010	公共建筑节能监测系统技术规范 Technical specifications of energy conservation monitoring system for public buildings		J11733-2010	山东省建设发展研究院
39	DBJ/T 14-072-2010	保温装饰板外墙外保温系统应用技术规程 Application technical specification for external thermal insulation systems based on insulated decorative panel		J11763-2010	山东省建筑科学研究院

山东省

序号	标准编号	标准名称	被代替标准编号	备案号	主编单位
40	DBJ/T 14-073-2010	岩棉板外墙外保温系统应用技术规程 Application technical specification of rock wool board external thermal insulation and finish system		J11762-2010	山东省建筑科学研究院
41	DBJ/T 14-076-2011	柔性饰面砖建筑装饰工程技术规程 Engineering technical specification of flexible tile for building decoration		J11836-2011	山东省建设发展研究院
42	DBJ 14-077-2011	居住建筑太阳能热水系统一体化应用技术规程 Technical specification for integrated solar water heating system of residential buildings		J11859-2011	山东省建设发展研究院
43	DBJ 14-078-2011	太阳能-地源热泵复合系统技术规程 Technical specification for solar-ground source heat pump hybrid system		J11860-2011	山东省建设发展研究院
44	DBJ/T 14-079-2011	非承重砌块自保温体系应用技术规程 Technical specification for application of non-load-bearing self-insulation system with blocks		J11881-2011	山东省建设发展研究院
45	DBJ 14-080-2011	管桩水泥土复合基桩技术规程 Technical specification for composite pile made up of jet-mixing cement and PHC		J11880-2011	山东省建筑科学研究院
46	DBJ/T 14-081-2011	建筑边坡与基坑工程设计文件编制标准 Design documentation standard for building slope and foundation excavations		J11953-2011	山东省建设工程勘察质量监督站
47	DBJ/T 14-082-2012	绿色建筑评价标准 Evaluation standard for green building		J11957-2011	山东省建筑科学研究院
48	DBJ/T 14-083-2012	建筑地基安全性鉴定技术规程 Technical code for safety appraisal of building subgrade		J11984-2012	山东省建筑科学研究院
49	DBJ 14-084-2012	建筑用 NFPP-RCT 管道工程技术规程 Technical specification for building NFPP-RCT pipeline engineering		J12048-2012	山东省建设发展研究院
50	DBJ/T 14-085-2012	外墙外保温应用技术规程（发泡水泥保温板外墙外保温系统） Application technical specification of external wall thermal insulation		J12049-2012	山东省建设发展研究院
51	DBJ/T 14-086-2012	组合塑料模盒混凝土空心楼盖结构技术规程 Technical specification for combination plastic mold box concrete hollow floor		J12050-2012	山东省建设发展研究院
52	DBJ/T 14-087-2012	智能建筑工程技术标准 Technical standard for engineering construction of intelligent building systems		J12059-2012	山东省智能建筑技术专家委员会
53	DBJ/T 14-088-2012	IPS现浇混凝土剪力墙自保温体系应用技术规程 Technical specification for IPS cast-in-place concrete shearwall self-thermal in		J12072-2012	山东省建设发展研究院
54	DBJ 14-089-2012	城镇综合管廊工程施工及验收规范 Code for construction and acceptance of urban municipal tunnel engineering		J12139-2012	济南城建集团有限公司

山东省

序号	标准编号	标准名称	被代替标准编号	备案号	主编单位
55	DBJ 14-090-2012	城镇道路高模量沥青混合料设计与施工技术规范 Technical code for design and construction of high modulus asphalt mixture in urban road		J12146-2012	济南城建集团有限公司
56	DBJ 14-091-2012	螺旋挤土灌注桩技术规程 Technical code for construction of soil displacement screw pile		J12203-2012	威海建设集团股份有限公司
57	DBJ/T 14-092-2012	建筑基桩竖向抗压承载力快速检测技术规程 Technical code for fast testing of foundation pile vertical bearing capacity		J12227-2012	山东省建筑科学研究院
58	DBJ/T 14-094-2012	工程勘察岩土层序列划分方法标准 Standard for division method of rock and soil layers sequence in geotechnical investigation		J12239-2013	青岛市勘察测绘研究院
59	DBJ/T 14-095-2013	铝合金电缆应用技术规范 Technical application specification for aluminum alloy power cable		J12258-2013	山东省智能建筑专家技术委员会
60	DBJ/T 14-096-2013	既有玻璃幕墙检验评估技术规程 Technical code for checking and evaluating of existing glass curtain wall		J12257-2013	山东省建筑科学研究院
61	DBJ/T 14-097-2013	建筑施工现场齿轮齿条式施工升降机安装质量检验技术规程 Technical rules for installation quality inspection of rack and pinion buiders		J12256-2013	山东省建筑科学研究院
62	DBJ/T 14-098-2013	建筑施工现场塔式起重机安装质量检验技术规程 Technical rules for installation quality inspection of tower crane used in const		J12255-2013	山东省建筑科学研究院
63	DBJ/T 14-099-2013	外墙外保温应用技术规程（胶粉聚苯颗粒浆料复合型外墙外保温系统） Application technical specification of external wall		J12350-2013	山东省建设发展研究院
64	DBJ/T 14-100-2013	外墙外保温应用技术规程（改性酚醛泡沫板薄抹灰外墙外保温系统） Application technical specification of external wall		J12351-2013	山东省建设发展研究院
65	DBJ/T 14-101-2013	膨胀玻化微珠浆料复合保温板外墙外保温系统 Application technical specification of composite thermal insulation systems made of thermal insulation		J12349-2013	山东省建筑科学研究院
66	DBJ 14-102-2013	供热计量与系统调控工程技术标准 Technical standard for engineering of heating metering and system controlling		J12378-2013	山东省热力管理办公室
67	DB37/T 5000-2013	建筑工程优质结构评价标准 Evaluation standard for high quality of construction engineering		J12471-2013	山东省建设工程质量监督总站
68	DB37/T 5001-2013	住宅外窗工程水密性现场检测技术规程 On-site inspection technology procedures of watertightness for residential building external windows		J12470-2013	青岛市建筑工程质量监督站

山东省

序号	标准编号	标准名称	被代替标准编号	备案号	主编单位
69	DB37/T 5002-2013	市政公用设施分类与编码规范 Classification and coding specification for municipal public facilities		J12469-2013	济南市市政公用事业局
70	DB37/T 5003-2013	市政公用设施普查数据规范 Census data specification for municipal public facilities		J12468-2013	济南市市政公用事业局
71	DB37/T 5004-2013	硅丙聚合物水泥防水涂料应用技术规程 Application technical specification of silicone-acrylate copolymer and composite modified cement		J12496-2013	山东省建设发展研究院
72	DB37/T 5005-2013	水泥聚苯模壳格构式混凝土墙体保温结构一体化应用技术规程 Application technical specification of structural and thermal insulating integration building based		J12518-2013	山东省建筑科学研究院
73	DB37/T 5006-2013	发热电缆地面低温辐射供暖技术规程 Technical specification for heating cable floor low-temperature radiant heating		J12537-2014	山东省建筑科学研究院
74	DB37/5007-2014	太阳能光伏建筑一体化应用技术规程 Technical code for solar photovoltaic application of construction		J12655-2014	山东省建设发展研究院 山东力诺太阳能电力工程有限公司
75	DB37/5008-2014	建筑施工直插盘销式模板支架安全技术规范 Technical code for safety of straightly inserted disk and pin-joined formwork support in construction		J12588-2014	山东天齐置业集团股份有限公司 山东省建设发展研究院
76	DB37/T 5009-2014	建设工程监理文件资料管理规程 Specification for building engineering management document		J12628-2014	山东省建设监理协会
77	DB37/T 5010-2014	房屋建筑和市政基础设施工程质量检测技术管理规程 Testing technology management code for building and municipal infrastructure engineering quality		J12676-2014	山东省建设工程质量监督总站
78	DB37/T 5011-2014	城市环境照明工程规范 Specification for the urban environment lighting engineering		J12677-2014	山东照明学会
79	DB37/T 5012-2014	热轧带肋高强钢筋应用技术规程 Technical specification for application of high strength hot rolled ribbed bars		J12742-2014	山东省建设发展研究院
80	DB37/T 5013-2014	EPS模块保温系统技术规程 Thermal insulation system with EPS module		J12771-2014	山东省建设发展研究院
81	DB37/T 5014-2014	EPS模块现浇混凝土剪力墙结构技术规程 Technical specification for concrete shear wall structure with EPS cavity module		J12773-2014	山东省建设发展研究院
82	DB37/T 5015-2014	EPS模块工业建筑围护结构技术规程 Technical specification for industrial building envelope with EPS hollow module		J12772-2014	山东省建设发展研究院
83	DB37/T 5016-2014	民用建筑外窗工程技术规范 Technical code for external windows of civil building		J12741-2014	山东省建筑科学研究院

山东省

序号	标准编号	标准名称	被代替标准编号	备案号	主编单位
84	DB37/T 5017-2014	燃气管道环压连接技术规程 Technical specification for ring compression connection for gas pipes		J12796-2014	山东省建设发展研究院
85	DB37/T 5018-2014	装配整体式混凝土结构设计规程 Specification for design of monolithic precast concrete structures		J12812-2014	山东建筑大学
86	DB37/T 5019-2014	装配整体式混凝土结构工程施工与质量验收规程 Specification for construction and quality acceptance of precast monolithic concrete structures		J12811-2014	山东省建筑科学研究院
87	DB37/T 5020-2014	装配整体式混凝土结构工程预制构件制作与验收规程 Specification for component manufacture and acceptance of precast monolithic concrete structures		J12810-2014	山东省建设发展研究院
88	DB37/T 5021-2014	Ⅱ型耐热聚乙烯（PE-RTⅡ）低温直埋供热管道设计与施工规范 Design and construction code for buried polyethylene of raised temperature (PE-RT Ⅱ) pipeline for low temperature-heating distribution		J12809-2014	青岛能源泰能热电有限公司
89	DB37/T 5022-2014	温拌沥青混合料施工技术规程 Technical specification for construction of warm asphalt mixtures		J12818-2014	济南黄河路桥工程公司
90	DB37/T 5023-2014	非透明幕墙建筑外墙保温工程技术规程 Technical specification of nontransparent curtain wall building external wall thermal insulation		J12857-2014	山东省建设发展研究院
91	DB37/T 5024-2014	建筑施工操作平台与运送设备组合系统技术规范 Technical code for system of construction operating platform and transporting equipment		J12858-2014	山东省建筑科学研究院
92	DB37/T 5025-2014	建筑结构动力性能检测技术规程 Technical specification for testing dynamic performance of building structure		J12891-2014	青岛理工大学
93	DB37/5026-2014	居住建筑节能设计标准 Design standard for energy efficiency of residential buildings	DBJ 14-037-2012	J12036-2015	山东省建设发展研究院
94	DB37/T 5027-2014	聚合物水泥防水材料工程应用技术规程 Technical specification for engineering application of polymer modified cementitous waterproofing materials		J12892-2014	山东省建筑科学研究院
95	DB37/T 5028-2015	建设工程监理工作规程 Tasks code for construction project management		J12939-2015	山东省建设监理协会
96	DB37/5029-2015	镶嵌式太阳能热水系统建筑一体化应用技术规程 Technical specification for building integrated solar water heating system of mosaictype collector		J12979-2015	山东省建设发展研究院

山东省

序号	标准编号	标准名称	被代替标准编号	备案号	主编单位
97	DB37/T 5030-2015	柔性石材墙体饰面工程应用技术规程 Technical specification for application of flexible stone on wall decoration		J13037-2015	山东省建设发展研究院
98	DB37/T 5031-2015	SMC玻璃钢检查井应用技术规程 Technical specification for application of SMC glass fiber reinforced plastics inspection chamber		J13076-2015	山东省建设发展研究院
99	DB37/T 5032-2015	建设工程质量检测信息监管系统技术规程 Technical code of construction engineering quality inspection information supervision system		J13077-2015	山东省建设工程质量监督总站
100	DB37/T 5033-2015	彩色防滑涂层施工技术规程（丙烯酸类） Technical specification for colored anti-skid coating construction (Acrylic acid series)		J13078-2015	济南城建集团有限公司
101	DB37/T 5034-2015	行道树栽植及养护技术规程 Technical specification for street tree planting and maintenance		J13179-2015	济南市园林绿化工程质量监督站
102	DB37/T 5035-2015	城镇道路绿地养护管理标准 Standard for management of road greening in city and town		J13180-2015	济南市园林绿化工程质量监督站
103	DB37/T 5036-2015	城镇河道绿化技术规程 Technical specification for greening of urban river		J13181-2015	济南市园林绿化工程质量监督站
104	DB37/T 5037-2015	城市湿地公园园林工程技术规程 Technical specification for garden engineering of urban wetland park		J13182-2015	济南市园林绿化工程质量监督站
105	DB37/T 5038-2015	建筑基底静水压力控制技术规程 Technical specification for building basement hydrostatic pressure control		J13183-2015	山东省建设发展研究院
106	DB37/T 5039-2015	城镇供水水质现场快速检测技术规程 Technical specification of on-site rapid test for urban water supply system		J13200-2015	山东省城市供排水水质监测中心
107	DB37/T 5040-2015	城镇给水厂二氧化氯消毒技术规范 Disinfection technical specification of chlorine dioxide		J13201-2015	山东省城市供排水水质监测中心
108	DB37/T 5041-2015	城镇供水水质应急监测技术规范 Technical specifications for emergency monitoring of city and town waterworks		J13202-2015	山东省城市供排水水质监测中心
109	DB37/T 5042-2015	城镇供水水质在线监测系统技术规范 Urban water supply technical specification for on-line monitoring system		J13203-2015	山东省城市供排水水质监测中心
110	DB37/T 5043-2015	绿色建筑设计规范 Code for green building design		J13204-2015	山东省建筑科学研究院
111	DB37/T 5044-2015	建筑桩基检测技术规范 Technical code for testing of building foundation piles	DBJ 14-020-2002	J10232-2016	山东省建筑科学研究院
112	DB37/T 5045-2015	房屋建筑结构安全评估技术规程 Technical specification for safety assessment of building construction		J13288-2016	山东省建筑科学研究院

山东省

序号	标准编号	标准名称	被代替标准编号	备案号	主编单位
113	DB37/T 5046-2015	里氏硬度法现场检测建筑钢材抗拉强度技术规程 Technical specification for testing tensile strength of steel construction in site by leeb-hardness method		J13289-2016	山东省建筑科学研究院
114	DB37/T 5047-2015	低温热水地面辐射供暖技术规程 Technical specificationfor low temperature hot water floor radiant heating	DBJ/T 14-056-2009	J10156-2016	山东省建设发展研究院
115	DB37/T 5048-2015	塑料排水检查井应用技术规程 Technical specification for application of plastics manholes and inspection chambers for sewerage	DBJ/T 14-074-2010	J11764-2016	山东省建设发展研究院
116	DB37/T 5049-2015	预拌混凝土绿色生产及管理技术规程 Technical specification for green production and management of ready-mixed concrete		J13290-2016	山东省建筑科学研究院
117	DB37/T 5050-2015	PVC复合塑料模板技术规程 Technical specification for PVC composite plastic template		J13291-2016	青岛理工大学
118	DB37/T 5051-2015	聚苯模块现浇混凝土墙体保温系统技术规程 Technical specification for cast-in-place concrete wall thermal insulation system with polystyrene module		J13312-2016	山东省建设发展研究院
119	DB37/5052-2015	建筑岩土工程勘察设计规范 Code for geotechnical engineering investigation and design of buildings		J13146-2015	山东省城乡建设勘察设计研究院
120	DB37/T 5053-2016	装配式结构独立钢支柱临时支撑系统应用技术规程 Technical specification for application of independent steel temporary support in precast concrete structures		J13345-2016	山东天齐置业集团股份有限公司
121	DB37/T 5054-2016	建筑用双层共挤绝缘辐照交联电线电缆应用技术规程 Technical application specification for double-layer extruded irradiation crosslinking insulation wires and cables for buildings		J13346-2016	山东省建设发展研究院
122	DB37/T 5055-2016	建筑工程抗震性态设计规范 Seismic code for performance-based design of buildings		J13444-2016	哈尔滨工业大学（威海）
123	DB37/5056-2016	民用建筑电线电缆防火设计规范 Code for fireproofing design of wiresand cables used in civil buildings		J13286-2016	山东省建筑科学研究院
124	DB37/5057-2016	建筑物移动通信基础设施建设规范 Code for mobile communication infrastructure of buildings		J13287-2016	山东省通信管理局
125	DB37/T 5058-2016	微膨胀防水混凝土应用技术规程 Technical specification for waterproof concrete with micro-expansion	DBJ 14-017-2002	J10180-2016	山东省建筑材料工业设计研究院
126	DB37/T 5059-2016	工程建设地下水控制技术规范 Technical code for groundwater control of engineering buildings		J13468-2016	山东设协勘察设计审查咨询中心

山东省

序号	标准编号	标准名称	被代替标准编号	备案号	主编单位
127	DB37/T 5060-2016	海绵城市设计规程 Specification for design of sponge city		J13469-2016	济南市市政工程设计研究院（集团）有限责任公司
128	DB37/T 5061-2016	住宅小区供配电设施建设标准 Code for the construction of power supply and distribution facilities in residential area		J13470-2016	国网山东省电力公司
129	DB37/T 5062-2016	农村一体式无害化卫生厕所施工及验收规范 Specification for construction and acceptance of rural harmless integrated sanitary toilet		J13510-2016	山东省标准化研究院
130	DB37/T 5063-2016	建筑施工现场安全管理资料规程 Management specification of construction site safety management documents		J13512-2016	山东省建筑施工安全监督站
131	DB37/T 5064-2016	STP真空绝热板建筑保温系统应用技术规程 Technical specification of building application for thermal insulation systems of STP vacuum insulation panels		J13513-2016	山东省建设发展研究院
132	DB37/T 5065-2016	建筑外遮阳工程应用技术规程 Technical regulations for solar shading engineering of buildings		J13514-2016	山东省建设发展研究院
133	DB37/T 5066-2016	干混砂浆应用技术规程 Technical specification for application of dry-mixed mortar	DBJ/T 14-056-2009	J11454-2016	山东省建筑科学研究院
134	DB37/T 5067-2016	FS外模板现浇混凝土复合保温系统应用技术规程 Application technical specification for cast-in-place concrete composite thermal insulation system of FS external formwork	DBJ/T 14-075-2011	J11811-2016	山东省建设发展研究院
135	DB37/T 5068-2016	CRB600H钢筋应用技术规程 Technical specification for CRB600H steel wires and bars		J13510-2016	山东省建筑设计研究院
136	DB37/T 5070-2016	低能耗建筑外墙隔离式防火保温体系应用技术规程 Application technical specification of low energy consumption building external thermal insulation system using partition type of structural fireproof thermal board	DBJ/T 14-093-2012	J12240-2016	山东省建筑科学研究院
137	DB37/T 5071-2016	低能耗建筑外墙粘贴复合防火保温体系应用技术规程 Application technical specification of low energy consumption building external pasting composite fireproof and thermal insulation system		J13546-2016	山东省建筑科学研究院
138	DB37/T 5072-2016	建筑工程（建筑与结构工程）施工资料管理规程 Specifications for construction engineering document management (construction & structural engineering)	DBJ 14-023-2004	J10350-2016	山东省建设工程质量监督总站
139	DB37/T 5073-2016	建筑工程（建筑设备、安装与节能工程）施工资料管理规程 Specifications for construction engineering document management (equipment, installation、energy-saving engineering)		J13551-2016	山东省建设工程质量监督总站

山东省

序号	标准编号	标准名称	被代替标准编号	备案号	主编单位
140	DB37/T 5074-2016	被动式超低能耗居住建筑节能设计标准 Design standard for energy efficiency of passive ultra-low energy residential buildings		J13589-2016	山东省建筑科学研究院
141	DB37/T 5075-2016	生态一体式厕所净化处理技术标准 Ecological integrated toilet purification treatment technical standard		J13592-2016	山东省建设发展研究院
142	DB37/T 5076-2016	宾馆酒店建筑能耗限额标准 Standard for energy consumption quota of hotel buildings		J13633-2016	山东建筑大学
143	DB37/T 5077-2016	机关办公建筑能耗限额标准 Standard for energy consumption quota of government office buildings		J13635-2016	山东建筑大学
144	DB37/T 5078-2016	商务办公建筑能耗限额标准 Standard for energy consumption quota of commercial office buildings		J13636-2016	山东建筑大学
145	DB37/T 5079-2016	医院建筑能耗限额标准 Standard for energy consumption quota of hospital buildings		J13638-2016	山东建筑大学
146	DB37/T 5080-2016	一体化预制泵站技术规程 Technical specification of integrated prefabricated pumping station		J13637-2016	山东省建设发展研究院
147	DB37/T 5081-2016	住宅厨房卫生间排烟气系统应用技术规程 Application technical specification for exhaust flue gas system of residential kitchen and bathroom		J13639-2016	山东省建设发展研究院
148	DB37/T 5082-2016	沉管压灌桩技术规程 Technical specification for tube-sinking cast-in-situ pile with pressure grouting concrete		J13631-2016	山东高速科技发展集团有限公司
149	DB37/T 5083-2016	海绵城市城镇道路雨水控制利用系统施工与验收规程 Specification for construction and acceptance of urban road stormwater management and utilization system		J13634-2016	济南城建集团有限公司
150	DB37/T 5069-2016	太阳能热水系统安装及验收技术规程 Technical specification for installation and acceptance of solar waterheating system		J13602-2016	山东同圆设计集团有限公司
151	DB37/T 5084-2016	立体绿化技术规程 The technical specifications of stereoscopic greening		J13723-2017	山东省建设发展研究院
152	DB37/T 5085-2016	组合铝合金模板工程技术规程 Technical specification for combined aluminum alloy formwork engineering		J13724-2017	山东省建设发展研究院
153	DB37/T 5086-2016	建筑与市政工程绿色施工管理标准 Standard for green construction management of building and municipal engineering		J13721-2017	山东省建筑科学研究院
154	DB37/T 5087-2016	建筑与市政工程绿色施工评价标准 Standard for green construction evaluation of building and municipal engineering		J13722-2017	山东省建筑科学研究院

河南省

序号	标准编号	标准名称	被代替标准编号	备案号	主编单位
1	DBJ41/T 062-2012	河南省居住建筑节能设计标准（寒冷地区）Henan Province design standard for energy efficiency of residential building (cold region)	DBJ 41/062-2005	J12164-2012	河南省建筑科学研究院
2	DBJ41/T 070-2014	住宅工程质量常见问题防治技术规程 Technical specification for prevention & cure of common quality faults of residential buildings	DBJ41/T 070-2005	J12856-2014	河南省建设工程质量监督总站
3	DBJ 41/071-2012	河南省居住建筑节能设计标准（夏热冬冷地区）	DBJ 41/071-2006	J12163-2012	河南省建筑科学研究院
4	DBJ41/T 072-2006	安全控制与报警逃生门锁设计、施工及验收规范		J10797-2006	河南省公安消防总队
5	DBJ41/T 073-2008	复合灌注聚氨酯硬泡外墙外保温技术规程 Technical specification of external thermal insulation on walls by compound panel being primed with polyurethane foam	DBJ41/T 073-2006	J10915-2008	郑州大学综合设计研究院
6	DBJ41/T 074-2013	高压细水雾灭火系统设计、施工及验收规范 Code for design, installation and acceptance of high pressure water mist fire protection systems	DBJ41/T 074-2006	J12532-2014	河南省公安消防总队
7	DBJ41/T 075-2016	河南省公共建筑节能设计标准 Henan Province energy efficiency design standard for public buildings	DBJ41/T 075-2006	J13554-2016	河南省建筑科学研究院
8	DBJ41/T 077-2009	蒸压粉煤灰砖建筑技术规程 Technical specification of autoclaved flyash-lime brick building	DBJ41/T 077-2007	J10946-2009	河南省建筑科学研究院
9	DBJ41/T 078-2015	预拌砂浆生产与技术规程 Technical specification for manufacture and application of ready-mixed mortar	DBJ41/T 078-2007	J13132-2015	河南省建筑科学研究院
10	DBJ41/T 080-2012	复合保温钢筋焊接网架混凝土墙（CL建筑体系）技术规程 Design specification for CL structure	DBJ41/T 080-2008	J12155-2012	河南省建筑科学研究院
11	DBJ41/T 083-2008	模块化同层排水节水系统应用技术规程 The design and construction acceptance technical specification of water-saving system on modular same floor drainage		J11226-2008	中国石化集团中原石油勘探局勘察设计研究院
12	DBJ41/T 085-2008	大型商业建筑设计防火规范 Code for fire protection design of large-scale commercial buildings		J11234-2008	河南省公安消防总队
13	DBJ41/T 086-2008	集中空调计量收费及应用技术规程 Technical specification of device for central air conditioning measurement charge and its application		J11256-2008	河南省建设科技协会建筑节能专业委员会
14	DBJ41/T 087-2008	建设工程造价软件数据交换标准 Standard for construction cost software data exchange		J11259-2008	河南省建筑工程标准定额站
15	DBJ41/T 088-2008	建筑玻璃贴膜工程技术规程 Technical specification for window film		J11286-2008	河南省建筑科学研究院

河南省

序号	标准编号	标准名称	被代替标准编号	备案号	主编单位
16	DBJ41/T 089-2008	河南省既有建筑节能改造技术规程 Henan Province technical specification of renovation of energy efficiency for existing building		J11307-2008	河南省建筑科学研究院
17	DBJ 41/090-2014	房屋建筑快带网络设施建设标准 Standard for construction of communication ancillary facilities in residential district and business-living building	DBJ41/T 090-2008	J12854-2014	河南省通信管理局
18	DBJ41/T 091-2009	现浇泡沫混凝土墙体技术规程 Technical specification of cast-in-stiu bubble concrete wall		J11376-2009	河南省建筑科学研究院
19	DBJ41/T 094-2009	城镇供水"一户一表"管道工程水力检测及验收技术规程 Technical specification for hydraulic test and acceptance in city and town water supply of ONE HOUSE ONE METER pipe system		J11447-2009	河南省城镇供水协会
20	DBJ41/T 097-2009	既有建筑幕墙安全性鉴定标准 Standard for appraiser of safety performance of existing curtain wall		J11471-2009	河南省建筑科学研究院
21	DBJ41/T 098-2009	河南省建设工程施工现场管理标准 Henan Province standard for administration		J11509-2009	焦作市标准定额管理站
22	DBJ41/T 099-2010	河南省附属绿地绿化规划设计规范 The code for greening planning and designing of the attached green space of Henan		J11590-2010	平顶山园林绿化管理处
23	DBJ41/T 100-2015	砌块墙体自保温体系技术规程 Echnical specification for self-thermal insulation block wall	DBJ41/T 100-2010	J11663-2015	河南省建筑科学研究院
24	DBJ41/T 101-2010	河南省太阳能光热建筑应用示范项目评价标准 Henan Province technical code for solar water heating system of civil buildings		J11688-2010	河南省建筑科学研究院
25	DBJ41/T 102-2010	河南省地源热泵系统建筑应用示范项目评价标准 Henan Province technical code for ground-source heat pump system of civil buildings		J11687-2010	河南省建筑科学研究院
26	DBJ41/T 103-2010	干挂法施工保温装饰板外墙外保温技术规程 Technical specification for external thermal insulation on walls		J11689-2010	河南省第一建筑工程集团有限责任公司
27	DBJ41/T 104-2010	河南省民用建筑太阳能热水系统应用技术规程 Henan Province technical code for solar water		J11740-2010	河南省建筑科学研究院
28	DBJ41/T 105-2010	河南省城镇公共供水行业服务规范 Customer service standards for public of Henan city water supply		J11718-2010	河南省城镇供水协会
29	DBJ41/T 106-2010	中小学校舍工程安全管理规范 Safe management standard for primary and secondary schoolhouse		J11818-2011	河南省城乡建筑设计院
30	DBJ41/T 107-2010	绿色施工管理规程 Management specification of green construction		J11819-2011	河南省第一建筑工程集团有限责任公司

河南省

序号	标准编号	标准名称	被代替标准编号	备案号	主编单位
31	DBJ41/T 108-2011	钢丝网架水泥膨胀珍珠岩夹芯板隔墙应用技术规程 Technical specification for application of partition with stell mesh cement expanded pearlite sandwich panel		J11882-2011	河南省建筑设计院有限公司
32	DBJ41/T 109-2015	河南省绿色建筑评价标准 Henan Province evaluation standard for green building	DBJ41/T 109-2011	J11960-2015	河南省建筑科学研究院
33	DBJ41/T 110-2011	合成树脂柔性饰面砖工程技术规程 Specification for construction and acceptance		J11959-2012	河南省建筑科学研究院
34	DBJ41/T 111-2011	消防控制室管理技术规程 Safety regulations of fire control room		J11983-2012	河南省公安消防总队
35	DBJ41/T 112-2016	混凝土保温幕墙工程技术规程 Technical specification for concrete curtain walls with insulation layer engineering	DBJ41/T 112-2012	J12053-2016	河南省第一建筑工程集团有限责任公司
36	DBJ41/T 113-2012	民用建筑工程室内装饰材料污染物限量技术规程 Technical specification for indoor decorating materials pollution limit of civil		J12155-2012	河南省建筑科学研究院
37	DBJ41/T 114-2012	电气火灾监控系统设计、施工及验收规范 Code for design, installation and acceptance		J12217-2012	河南省公安消防总队
38	DBJ41/T 115-2012	市政排水管道半开槽顶管施工技术规程		J12218-2012	郑州市市政工程总公司
39	DBJ41/T 118-2013	医院建筑智能化系统设计标准 Hospital building intelligent systems design specifications		J12292-2013	河南丹枫科技有限公司
40	DBJ41/T 119-2013	河南省地源热泵建筑应用检测及验收技术规程 Henan Province testing and acceptance technical		J12316-2013	河南省建筑科学研究院
41	DBJ41/T 120-2013	砌筑保温砂浆应用技术规程 Technical specification for application of thermal insulation masonry mortar		J12319-2013	河南省建筑科学研究院
42	DBJ41/T 121-2013	污水源热泵系统应用技术规程 Technical code for sewage source heat pump system		J12315-2013	河南省建筑科学研究院
43	DBJ41/T 122-2013	无机保温砂浆墙体保温系统应用技术规程 Technical specification for thermal insulating systems of inorganic aggregate mortar		J12318-2013	河南省建筑科学研究院
44	DBJ41/T 123-2013	河南省太阳能热水建筑应用检测及验收技术规程 Henan Province inspection and acceptance standards of solar water heating system on buildings		J12317-2013	河南省建筑科学研究院
45	DBJ41/T 124-2013	高强钢筋混凝土结构应用技术规程 Technical specification for concrete structures		J12335-2013	郑州大学综合设计研究院
46	DBJ41/T 125-2013	建筑装饰装修工程质量验收标准 Technical standard for construction quality acceptance		J12377-2013	河南省建筑科学研究院
47	DBJ41/T 126-2013	发泡陶瓷保温板保温系统应用技术规程		J12406-2013	河南省建筑科学研究院有限公司
48	DBJ41/T 127-2013	城市桥梁检测技术规程 Technical specification for inpection of the municipal bridge		J12408-2013	河南省建筑科学研究院有限公司

河南省

序号	标准编号	标准名称	被代替标准编号	备案号	主编单位
49	DBJ41/T 128-2013	钻芯检测抹灰砂浆粘结强度方法标准 Testing standard for adhesive strength of hardened rendering and plastering mortars on substrates		J12407-2013	河南省建筑科学研究院有限公司
50	DBJ41/T 129-2013	外墙外保温用聚合物砂浆质量检验标准 Quality inspection standard of polymer mortar for exterior wall thermal insulation		J12425-2013	河南省建筑科学研究院有限公司
51	DBJ41/T 130-2013	预应力混凝土空心方桩技术规程 Technical specification for prestressed spun		J12473-2013	河南省建筑设计研究院有限公司
52	DBJ41/T 131-2014	水泥稳定碎石基层道路施工及质量验收规程 Specification for construction and quality		J12651-2014	郑州市市政工程总公司
53	DBJ41/T 132-2014	双向螺旋挤土灌注桩技术规程 Technical specification of soil displacement screw pile		J12652-2014	河南省有色工程勘察有限公司 中冶建筑研究总院有限公司
54	DBJ41/T 133-2014	建筑及市政园林建筑标志牌设置技术规程 Technical specification of signs for engineering		J12740-2014	河南省建筑科学研究院有限公司
55	DBJ41/T 134-2014	膨胀珍珠岩板薄抹灰外墙外保温工程技术规程 Technical specification for external thermal		J12684-2014	河南省建筑设计院有限公司
56	DBJ41/T 135-2014	河南省公共建筑能耗监测系统技术规程 Henan Province technical code of energy consumed monitoring systems of public buildings		J12731-2014	河南省建筑科学研究院有限公司
57	DBJ41/T 136-2014	河南省民用建筑太阳能光伏系统应用技术规程 Henan Province technical code for application of solar		J12754-2014	河南省建筑科学研究院有限公司
58	DBJ41/T 137-2014	防渗墙质量无损检测技术规程 Specification of quality inspection for cut-off wall		J12766-2014	河南黄科工程技术检测有限公司
59	DBJ41/T 138-2014	河南省建筑地基基础勘察设计规范 Henan code for investigation and design on geotechnical engineering of building foundation		J12756-2014	河南省建筑设计研究院有限公司
60	DBJ41/T 139-2014	河南省基坑工程技术规范 Henan technical code for foundation excavations		J12755-2014	郑州大学综合设计研究院有限公司
61	DBJ41/T 140-2014	石膏秸秆复合隔墙技术规程 Technical specification for gypsum-crop waste composite wall		J12779-2014	河南省建筑科学研究院有限公司
62	DBJ41/T 141-2014	河南省建筑与市政工程施工现场从业人员管理标准 Management standards for construction site staff		J12788-2014	河南省第一建筑集团工程有限公司
63	DBJ41/T 142-2014	居住区建筑智能化系统设计标准 Standards for design of building intelligent system of residential districts		J12787-2014	河南丹枫科技有限公司
64	DBJ41/T 143-2014	城市综合体建筑智能化系统设计标准		J12786-2014	河南丹枫科技有限公司
65	DBJ41/T 144-2014	浆料嵌缝敷面EPS板外墙外保温技术规程 Technical specification of external thermal insulation on wall		J12785-2014	河南省建筑科学研究院有限公司

河南省

序号	标准编号	标准名称	被代替标准编号	备案号	主编单位
66	DBJ41/T 145-2015	三轴水泥土搅拌桩帷幕技术规程 The curtain of soil-cement pile mixed by three shafts		J12910-2015	河南省建筑科学研究院有限公司
67	DBJ41/T 146-2015	免拆复合保温模板（FS）应用技术规程 Applied technical specification for free-removed formwork of composite thermal insulation system (FS)		J12909-2015	河南省建筑科学研究院有限公司
68	DBJ41/T 147-2015	泡沫玻璃保温板保温系统应用技术规程 Technical specification for application of thermal		J12923-2015	河南省建筑科学研究院有限公司
69	DBJ41/T 148-2015	建筑工程钢筋专业化加工应用技术规程 Technical specification for professional processing and application of steel bars in construction engineering		J12937-2015	郑州大学综合设计研究院
70	DBJ41/T 149-2015	高校智能化系统设计标准 The standard of design for intelligent systems of colleges and universities in Henan		J12967-2015	郑州大学综合设计研究院
71	DBJ41/T 150-2015	聚氨酯基透水路面技术规程 Technical specification for polyurethane pervious pavement		J13254-2015	河南省建筑科学研究院有限公司
72	DBJ41/T 151-2015	河南省成品住房装修工程技术规程 Henan Province technical code for decoration of residential buildings		J13260-2015	河南省建筑科学研究院有限公司
73	DBJ41/T 152-2015	建筑施工悬挑式脚手架安全技术规程 Technical specification for construction safety of cantilever beam supported scaffolding		J13328-2016	河南五建建设集团有限公司
74	DBJ41/T 153-2016	建设工程造价咨询档案立卷标准 Filing standard of engineering cost consultancy		J13343-2016	河南省鑫诚工程管理有限公司
75	DBJ41/T 154-2016	装配整体式混凝土结构技术规程 Technical specification for monolithic precast		J13416-2016	河南省建筑科学研究院有限公司
76	DBJ41/T 155-2016	装配式混凝土构件制作与验收技术规程 Technical code for precast concrete structure production and acceptance		J13417-2016	河南省建筑科学研究院有限公司
77	DBJ41/T 156-2015	保障性住房装修工程质量控制规程 Technical code for decoration of indemnificatory housing		J13474-2016	河南省建筑科学研究院有限公司
78	DBJ41/T 157-2016	城市轨道交通路基工程施工质量验收规范 Code for acceptance of construction quality of subgrade engineering of urban rail transit		J13473-2016	郑州轨道交通有限公司

湖南省

序号	标准编号	标准名称	被代替标准编号	备案号	主编单位
1	DBJ 43/001-2004	湖南省居住建筑节能设计标准 Design standard of residential buildings for energy efficiency in Hunan Province		J10410-2004	湖南省建设厅建筑节能办公室 湖南大学
2	DBJ43/T 001-2005	预拌混凝土生产与施工技术规程 Technical specification for production and construction of ready-mixed concrete		J10663-2006	长沙理工大学
3	DBJ43/T 001-2007	蒸压灰砂砖建筑技术规程 Technical specification for autoclaved sand-lime brick masonry building		J10951-2007	长沙理工大学
4	DBJ43/T 001-2009	陶粒混凝土小型空心砌块与陶粒混凝土砖建筑技术规程 Technical specification for ceramsite concrete small hollow block and ceramsite concrete brick masonry building		J11500-2009	长沙理工大学
5	DBJ43/T 001-2012	蒸压加气混凝土砌块建筑技术规程 Technical specification for antoclaved aerated concrete of block building		J12154-2012	长沙理工大学
6	DBJ 43/002-2005	混凝土多孔砖建筑技术规程 Technical specification for concrete perforated brick masonry building		J10750-2006	长沙理工大学
7	DBJ43/T 002-2007	N式混凝土小型空心砌块建筑技术规程 Technical specification for N type concrete small hollow block masonry building		J11101-2007	二十三冶建设集团有限公司 湘西州建设局 湖南省建设工程质量安全督管理总站 湖南大学 湘西州荣昌建设集团公司
8	DBJ 43/002-2009	城市二次供水设施技术规程 Technical code for urban secondary water supply facilities		J11499-2009	长沙市城市用水管理办公室
9	DBJ43/T 002-2010	预拌砂浆生产与应用技术规程 Technical specification for production and application of ready-mixed mortar		J11723-2010	长沙理工大学 长沙市散装水泥办公室
10	DBJ43/T 003-2009	轻骨料混凝土多孔砖建筑技术规程 Technical specification for lightweight aggregate concrete perforated brick masonry building		J11516-2009	长沙理工大学
11	DBJ 43/003-2010	湖南省公共建筑节能设计标准 Design standard for energy efficiency of public buildings in Hunan Province		J11742-2010	湖南大学
12	DBJ 43/003-2012	住宅小区及商住楼通信设施建设标准 Standard for construction of communication facilities in residential district & business residential building	DBJ 43/003-2007	J12133-2012	湖南省通信工程质量监督中心 湖南省邮电规划设计院有限公司
13	DBJ 43/004-2006	房屋白蚁预防技术规程 Technical specification for termite control in buildings	DBJ 43/004-2006	J10878-2006	湖南省房协白蚁防治专业委员会 长沙市白蚁防治站

湖南省

序号	标准编号	标准名称	被代替标准编号	备案号	主编单位
14	DBJ43/T 135-2016	现浇混凝土保温免拆模板复合体系应用技术规程 Technical specification for application of cast-in-place concrete composite system with thermal insulation and non-dismantle formwork		J13555-2016	长沙理工大学
15	DB43/155-2001	城市电信服务设施设计规范 Design standards of telecom service facilities in city		J10282-2003	湖南省邮电规划设计院
16	DBJ43/T 301-2013	混凝土叠合楼盖装配整体式建筑技术规程 Technical guidelines for precast concrete building using composite floor system		J12528-2013	湖南大学 长沙远大住宅工业有限公司
17	DBJ43/T 301-2015	混凝土装配-现浇式剪力墙结构技术规程 Technical specification for assembled monolithic concrete shear wall structure		J12957-2015	湖南大学 长沙远大住宅工业集团有限公司
18	DBJ43/T 302-2014	保温装饰板外墙外保温系统应用技术规程 Technical specification for application of external thermal insulation systems based on insulated de		J12578-2014	长沙理工大学
19	DBJ43/T 303-2014	建筑反射/保温隔热涂料应用技术规程 Technical specification for application of building reflective/thermal insulation coating		J12643-2014	湖南大学
20	DBJ43/T 304-2014	多层房屋钢筋沥青基础隔震技术规程 Technical specification for reinforced asphalt base isolation technology in multi-story building		J12710-2014	湖南大学
21	DBJ43/T 305-2014	地下工程混凝土耐久性技术规程 Technical specifications of concrete durability for underground construction		J12711-2014	湖南大学
22	DBJ43/T 306-2014	湖南省住宅工程质量通病防治技术规程 Hunan Provincial regulations of common residential project quality defects		J12817-2014	湖南省建设工程质量安全监督管理总站
23	DBJ43/T 307-2014	烧结装饰砖夹芯保温外墙应用技术规程 Technical specification for application of cavity wall filled with thermal insulation and fired facing bricks		J12816-2014	长沙理工大学
24	DBJ43/T 309-2015	陶粒增强泡沫混凝土砌块建筑技术规程 Technical specification for ceramic reinforced foam concrete block masonry building		J12958-2015	长沙理工大学
25	DBJ43/T 310-2015	湖南省EPS模块外保温工程技术规程 Hunan provincial technical specification for exterior thermal insulation project with EPS module		J13067-2015	长沙理工大学
26	DBJ43/T 311-2015	装配式斜支撑节点钢框架结构技术规程 Technical specifications for prefabricated steel frame structure with diagonal bracing joints		J13063-2015	远大可建科技有限公司 北京工业大学
27	DBJ43/T 312-2015	高性能水泥复合砂浆钢筋网加固砌体结构技术规程 Technical specification for strengthening masonry structure with high performance cement composite mortar reinforcement mesh		J13216-2015	湖南大学

湖南省

序号	标准编号	标准名称	被代替标准编号	备案号	主编单位
28	DBJ43/T 313-2015	建筑施工承插型键槽式钢管支架安全技术规程 Technical specification for safety of keyway-quicklocked steel tubular scaffold in construction		J13248-2015	湖南省建筑工程集团总公司 湖南金峰金属构件有限公司
29	DBJ43/T 314-2015	湖南省绿色建筑评价标准 Evaluation standard for green building in Hunan Province	DBJ43/T 004-2010	J11737-2015	湖南省建筑设计院
30	DBJ43/T 316-2016	湖南省公共建筑能耗监测技术规程 Technical specification for monitoring system of public building energy consumption in Hunan Province		J13578-2016	湖南大学 湖南省建筑科学研究院
31	DBJ43/T 318-2016	硅藻泥工程应用技术规程 Technical specification for engineering application of diatom mud		J13683-2017	长沙理工大学 湖南蓝天豚绿色建筑新材料有限公司
32	DBJ43/T 319-2016	装配式斜支撑节点钢框架结构建筑防火技术规程 Technical specifications for building fire protection of prefabricated steel frame structure with diagonal bracing joints		J13684-2017	远大可建科技有限公司 公安部天津消防研究所
33	DBJ43/T 321-2017	陶粒混凝土屋面与楼地面保温工程技术规程 Technical specification for ceramic concrete roofing, flooring and ground thermal insulation		J13777-2017	长沙理工大学

湖北省

序号	标准编号	标准名称	被代替标准编号	备案号	主编单位
1	DB42/T 159-2012	基坑工程技术规程 Technical specification for excavation engineering		J10534-2013	中南勘察设计院(湖北)有限公司
2	DB42/242-2013	建筑地基基础技术规范 Technical code for building foundation		J12580-2014	湖北省建筑科学研究设计院
3	DB42/T 268-2012	蒸压加气混凝土砌块工程技术规程 Technical specification for blocks engineering of autoclaved aerated concrete		J10398-2012	武汉市墙体材料改革办公室
4	DB42/T 344-2013	城镇道路沥青路面施工技术及验收规程 Construction technique and acceptance rules for asphalt pavement of town roads		J12526-2013	武汉市市政工程质量监督站
5	DB42/408-2006	建筑燃气安全技术规程 Gas safety procedures for building construction		J10891-2006	武汉市城市管理局
6	DB42/T 483-2008	低层住宅冷弯薄壁型钢结构技术规程 Technical specification for cold-formed thin-wall steel structures of low-rise residences		J11272-2008	湖北省建设科技发展中心
7	DB42/489-2008	预应力混凝土管桩基础技术规程 Technical specification for pre-stressed hollow concrete pipe pile foundation		J11217-2008	湖北省勘察设计协会技术咨询部
8	DB42/T 503-2008	建筑工程施工文件管理规范			
9	DB42/T 505-2008	湖北省燃气行业服务规范			
10	DB42/T 523-2008	居住小区信息管网设计规程 Standard for information system pipe network design of residential quarter		J11229-2008	中南建筑设计院等
11	DB42/535-2009	建筑工程施工现场安全防护设施技术规程			
12	DB42/T 536-2009	城市规划信息系统空间数据标准			
13	DB42/T 546-2009	埋地塑料排水管道工程技术规程			
14	DB42/T 553-2008	建筑施工现场安全生产管理规程			
15	DB42/T 559-2013	湖北省低能耗居住建筑节能设计标准	DB42/T 559-2009	J12399-2013	
16	DB42/T 636-2010	湖北省住宅工程质量通病防治技术规程			
17	DB42/T 642-2010	建筑节能工程施工文件管理规范			
18	DB42/T 737-2011	城镇燃气场(站)安全检查导则			
19	DB42/T 743-2011	蒸压砂加气混凝土精确砌块墙体自保温系统应用技术规程 Technical specification for application system of concrete blockwall		J12005-2012	湖北省建设科技发展中心
20	DB42/T 744-2011	CL建筑体系技术规程 Technical specification for CL building system		J12006-2012	湖北省建筑科学研究设计院 宝业湖北建工集团有限公司
21	DB42/T 745-2011	蒸压砂加气混凝土精确砌块薄层砂浆干法施工技术规程		J12007-2012	
22	DB42/T 749-2011	湖北建设工程造价应用软件数据交换规范 Data exchange requirement for construction valuation software in Hubei Province		J11971-2012	湖北省建设工程造价管理总站
23	DB42/T 750-2011	家用燃气燃烧器具安装维修服务质量评价规范 Code of service quality assessment for installation and maintain of domestic gas burning appliances		J11961-2012	武汉市燃气热力管理办公室

湖北省

序号	标准编号	标准名称	被代替标准编号	备案号	主编单位
24	DB42/T 823-2012	建设工程造价咨询质量控制规程 Cost construction engineering consulting, quality control standard		J12067-2012	湖北省建设工程造价咨询协会
25	DB42/T 830-2012	基坑管井降水工程技术规范 Technical regulations for dewatering engineering of tube well of foundation pit		J12066-2012	中冶集团武汉勘察研究院有限公司
26	DB42/831-2012	钻孔灌注桩施工技术规程 Construction technical regulation of bored cast-in-place pile		J12065-2012	中冶集团武汉勘察研究院有限公司
27	DB42/T 832-2012	燃气用镀锌钢管滚压圆锥外螺纹接头			
28	DB42/T 875-2013	湖北省城镇地下管线探测技术规程 Technical specification for detecting and surveying underground pipelines and cables in Hubei Province		J12311-2013	湖北省建设信息中心 武汉科岛地理信息工程有限公司
29	DB42/T 898-2013	无线电信号室内覆盖系统建设规范 Construction specification of indoor coverage system for wireless signal		J12352-2013	湖北省无线电管理委员会办公室
30	DB42/T 914-2013	地下连续墙施工技术规程 Construction technical specification for diaphragm wall		J12445-2013	中建三局建设工程股份有限公司 中建三局第二建设工程有限责任公司
31	DB42/T 927-2013	预制混凝土构件拼装塔机基础技术规程 Technical specification for prefabricated concrete block assembled base of tower crane		J12527-2013	湖北省建筑工程管理局
32	DB42/T 970-2014	聚合物水泥防水材料应用技术规程 Technical specification for application of polymer modified cementitious waterproofing materials		J12685-2014	湖北省建设科技发展中心
33	DB42/T 971-2014	城镇桥梁沥青混凝土铺装层 施工技术与验收规程 Technical and acceptance standards for construction of asphalt pavement on town bridge		J12727-2014	武汉市市政建设集团有限公司
34	DB42/T 990-2014	建设工程钢结构施工安全防护设施技术规程 Technical regulations for safety facilities in steel structure constructions		J12769-2014	武汉市城建安全生产管理站
35	DB42/T 1013-2014	ST60塔式支撑脚手架施工技术规范 Construction specifications for ST60 tower scaffolding		J12890-2014	武汉钢铁建工集团有限责任公司
36	DB42/T 1014-2014	城镇道路土壤固化剂稳定混合料基层技术规程 Urban road soil curing agent stabilized mixture road base technical specification		J12801-2014	湖北省武汉市政工程设计研究院有限责任公司
37	DB42/T 1035-2015	热轧U型钢板桩技术规程 Hot rolled U-sheet pile application protocol		J12970-2015	武汉钢铁建工集团有限责任公司
38	DB42/T 1044-2015	装配整体式混凝土剪力墙结构技术规程 Technical specification for assembled precast concrete shear wall structures		J13046-2015	武汉理工大学

湖北省

序号	标准编号	标准名称	被代替标准编号	备案号	主编单位
39	DB42/T 1045-2015	高固含量乳化型中温沥青路面技术规程 Technical rules of high solid content and emulsified based middle warm asphalt pavement		J13047-2015	武汉市政工程设计研究院有限责任公司
40	DB42/T 1046-2015	住宅厨房卫生间集中排气系统 技术规程 Technical specification for centralized exhaust system of residential kitchen and bathroom		J13049-2015	武汉理工大学
41	DB42/T 1049-2015	房产测绘技术规程 Technical specifications of house property surveying and mapping		J13048-2015	武汉市房产测绘中心
42	DB42/T 1051-2015	燃气铝合金衬塑（PE）复合管道工程技术规程 Technical specification for plastic (PE) liner aluminum alloy composite pipeline for gas supply		J13074-2015	湖北省燃气协会
43	DB42/T 1078-2015	湖北省市（县）城乡总体规划编制规程 Specification for municipality (county) urban-rural overall planning of Hubei Province		J13326-2016	湖北省城市规划设计研究院
44	DB42/T 1093-2015	装配式叠合楼盖钢结构建筑技术规程 Technical specification for buildings of steel structure with assembled floor		J13253-2015	中信建筑设计研究总院有限公司等
45	DB42/T 1107-2015	保温装饰板外墙外保温系统工程技术规程 Technical specification for external thermal insulation systems based on insulated decorative panel		J13273-2015	湖北省建设科技发展中心
46	DB42/T 1124-2015	城市园林绿化养护管理质量标准		J13433-2016	湖北省风景园林学会
47	DB42/T 1144-2016	燃气用不锈钢波纹软管安装及验收规范 Cord for installation and acceptance of corrugated hose assembly		J13432-2016	武汉市燃气热力管理办公室
48	DB42/T 1159-2016	湖北省城镇地下管线信息系统技术规范 Technical specification for the urban underground pipeline		J13440-2016	湖北省城乡建设发展中心

广东省

序号	标准编号	标准名称	被代替标准编号	备案号	主编单位
1	DBJ/T 15-7-2007	冷轧变形钢筋混凝土构件技术规程 Technical specification for concrete structure member with cold-rolled deformed bars	DBJ/T 15-7-92	J11164-2008	深圳市土木建筑学会
2	DBJ 15-18-2014	混凝土小型砌块自承重墙体工程技术规程 Technical specification for non-load-bearing small concrete block wall engineering		J12713-2014	广州大学
3	DBJ/T 15-19-2006	建筑防水工程技术规程 Technical specification for waterproof engineering of construction		J10899-2006	广东省工程建设标准化协会 广州市鲁班建筑防水补强有限公司
4	DBJ/T 15-20-2016	建筑基坑工程技术规程 Technical specification for building foundation excavation	DBJ/T 15-20-97	J13753-2017	广东省基础工程集团有限公司 广东省建筑工程集团有限公司
5	DBJ/T 15-22-2008	锤击式预应力混凝土管桩基础技术规程 Specification for driven prestressed concrete tube-pile foundation		J11386-2009	广东省建筑设计研究院
6	DBJ/T 15-25-2000	轻板墙体工程技术规程 Technical specification for lightweight building panel wall		J10045-2000	华南建筑学院（西院）
7	DBJ 15-30-2002	铝合金门窗工程设计、施工及验收规范 Code for design, installation and aceeptance of aluminium alloy door and window engineering		J10207-2002	广东省建筑科学研究院
8	DBJ 15-31-2016	建筑地基基础设计规范 Design code for building foundation	DBJ 15-31-2003	J13311-2016	广州市建筑科学研究院有限公司 华南理工大学建筑设计研究院
9	DBJ/T 15-32-2003	非承重蒸压灰砂砖墙体工程技术规程 Technical specification for non-loda-bearing autoclaved lime-sand brick wall construeciton		J10290-2003	广州大学工程材料研究所 广州市墙体材料革新与建筑节能办公室
10	DBJ 15-34-2004	大空间智能型主动喷水灭火系统设计规范 Code of design for large-space intelligent active conteol sprinkler systems		J10323-2004	广州市设计院
11	DBJ/T 15-35-2004	混凝土后锚固件抗拔和抗剪性能检测技术规程 Technical specification for testing tension and shear behavior of post-installed fastenings used in concrete		J10361-2004	广州市建筑科学研究院
12	DBJ 15-38-2005	建筑地基处理技术规范 Technical code for ground treatment of buildings		J10523-2005	广州市建筑科学研究院
13	DBJ/T 15-39-2005	燃气直燃型机组机房防火设计规范 Design code of fire protection for gas direct-fired chiller and heater room		J10546-2005	广东省公安厅消防局
14	DBJ/T 15-42-2005	消防安全疏散标志设计、施工及验收规范 Code for design, installation and accetance of fire safety escape signs		J10577-2005	广东省公安厅消防局
15	DBJ/T 15-43-2005	非承重蒸压泡沫混凝土砖墙体工程技术规程 Technical specification for non-load-bearing autoclaved foamed concrete brick wall construction		JI10621-2005	广东省建筑材料研究院 江门天风墙体材料有限公司

广东省

序号	标准编号	标准名称	被代替标准编号	备案号	主编单位
16	DBJ/T 15-45-2005	建设工程质量检测管理信息系统技术标准 Technical standard for management information system of costrucion quality test		JI10629-2005	广东省建设工程质量安全监督检测总站 广州粤建三和软件有限公司
17	DBJ 15-47-2005	IG-100气体灭火系统设计、施工及验收规范 Code for design, installation and acceptance of IG-100 fire-extinguishing systems		J10630-2005	广东省公安厅消防局 西门子楼宇科技（天津）有限公司
18	DBJ 15-48-2005	吸气式感烟火灾探测报警系统设计、施工及验收规范 Code for design, installation and acceptance of aspirating smoke detection fire alarm system		J10635-2005	广东省公安厅消防局 广东金冠安保系统工程有限公司
19	DBJ 15-50-2006	《夏热冬暖地区居住建筑节能设计标准》广东省实施细则		J10754-2006	广东省建筑科学研究院
20	DBJ 15-51-2007	《公共建筑节能设计标准》广东省实施细则 Design standard for energy efficiency of public buildings in GuangDong Province		J10999-2007	广东省建筑科学研究院
21	DBJ 15-52-2007	公共和居住建筑太阳能热水系统一体化设计施工及验收规程 Technical specification for integrated design, installation and acceptance of solar water heating system of popular and residential building		J10998-2007	广东省建筑科学研究院
22	DBJ 15-53-2007	混凝土和钢筋混凝土内衬改性聚氯乙烯排水管道工程技术规程 Technical specification of transformed character pvc lining concrete and reinforced concrete pipeline for sewer engineering		J10978-2007	广东省建筑设计研究院
23	DBJ 15-54-2007	广东省工程建设专业人才资源信息数据标准 Data standards for human resource information of professionals in engineering construction		J11007-2007	广东省建设厅办公室 广东省建设信息中心
24	DBJ 15-55-2007	安全控制与报警逃生门锁系统设计、施工及验收规程 Code for design, installation and acceptance of security control and exit device with alarm system		J11053-2007	广东省公安厅消防局
25	DBJ 15-60-2008	建筑地基基础检测规范 code for testing of building foundation		J11189-2008	广东省建筑科学研究院
26	DBJ 15-61-2008	混凝土结构用成型钢筋制品技术规程 Technical specification for fabricated rebar of concrete structures		J11283-2008	广州市建筑科学研究院
27	DBJ 15-62-2008	轻珠混凝土技术规程 Practice code for EPS lightweight aggregate concrete		J11339-2009	广州大学
28	DBJ 15-63-2008	预应力混凝土管桩机械啮合接头技术规程 Specification for mechanical joggle-joint connection of prestressed concrete piles		J11340-2009	广东省建筑设计研究院
29	DBJ/T 15-64-2009	城市地下空间开发利用规划与设计技术规程 Planning and design technical specification for development and utilization of underground space in cities		J11418-2009	广州市建筑科学研究院

广东省

序号	标准编号	标准名称	被代替标准编号	备案号	主编单位
30	DBJ 15-65-2009	广东省建筑节能工程施工质量验收规范 Code for acceptance of energy efficient building construction of Guangdong Province		J11366-2009	广东省建筑科学研究院
31	DBJ/T 15-66-2009	建筑门窗幕墙玻璃贴膜节能技术规程 Technical specification for energy-efficient windows film installation on construction glass		J11377-2009	华南理工大学建筑节能研究中心
32	DBJ/T 15-68-2009	软瓷建筑装饰工程技术规程 Technical specification for flexible ceramics decoration work		J11423-2009	广东省建筑设计研究院 广州福美软瓷有限公司
33	DBJ/T 15-69-2009	建筑节能材料性能评价及检测技术规程 Specification for evaluation & test techniques of energy-saving building material		J11494-2009	广东省建筑材料研究院
34	DBJ/T 15-70-2009	土钉支护技术规程 Specification for soil nailing techniques in building foundation excavation engineering		J11504-2009	广州市建设科学技术委员会办公室
35	DBJ 15-71-2010	城市地下空间检测监测技术标准 Testing and monitoring technical code for underground space in cities		J11596-2010	广州市建筑科学研究院有限公司
36	DBJ/T 15-72-2010	应用冲击压实技术处理旧水泥混凝土路面施工规程 Specifications of impact compaction construction for treatment of existing cement concrete pavement		J11598-2010	华南理工大学 广东省公路管理局
37	DBJ/T 15-73-2010	建筑塔式起重机安装检验评定规程 Regulations for checking and evaluating of building tower crane after erection		J11653-2010	广东省建筑科学研究院
38	DBJ/T 15-74-2010	预拌混凝土生产质量管理技术规程 Technology regulations of quality management for ready-mixed concrete production		J11666-2010	广东省预拌混凝土行业协会
39	DBJ 15-75-2010	广东省建筑反射隔热涂料应用技术规程 Technical specification for external thermal insulation composite system based on reflective thermal insulating coating on building		J11676-2010	华南理工大学建筑节能研究中心
40	DBJ/T 15-76-2010	玻璃钢内衬混凝土组合管应用技术规程 Technical code for composite glass-fiber plasticreinforced concrete pipe		J11664-2010	广东省建筑材料研究院 深圳大华水泥制品有限公司
41	DBJ/T 15-77-2010	电气火灾监控系统设计、施工及验收规范 Code for design、installation and acceptance of alarm and control system for electric fire prevention		J11703-2010	广东省公安厅消防局 福瑞特国际电气（中山）有限公司
42	DBJ/T 15-78-2011	民用建筑能效测评与标识技术规程 Technical specification for civil building energy efficiency evaluation and labeling		J11849-2011	广东省建筑科学研究院
43	DBJ/T 15-79-2011	刚性-亚刚性桩三维高强复合地基技术规程 Technical specification for rigid-semirigid pile three dimension high-strength composite foundation		J11848-2011	中国建筑科学研究院地基基础研究所 建研地基基础工程有限责任公司广东分公司

广东省

序号	标准编号	标准名称	被代替标准编号	备案号	主编单位
44	DBJ/T 15-80-2011	保障性住房建筑规程 Indemnificatory residential building specification		J11854-2011	广东省建筑设计研究院
45	DBJ/T 15-81-2011	建筑混凝土结构耐火设计技术规程 Code for fire resistance design of building concrete structures		J11855-2011	华南理工大学
46	DBJ 15-82-2011	蒸压加气混凝土砌块自承重墙体技术规程 Technical specification for autoclaved aerated concrete block non load bearing wall		J11857-2011	广州大学
47	DBJ/T 15-83-2017	广东省绿色建筑评价标准 Assessment standard for green building in Guangdong Province	DBJ/T 15-83-2011	J11878-2017	广东省建筑科学研究院
48	DBJ/T 15-84-2011	蒸压陶粒混凝土墙板应用技术规程 Technical specification for application of autoclaved ceramisite concrete wall panel		J11905-2011	广州大学 中山建华墙体材料有限公司
49	DBJ/T 15-85-2011	工程质量安全监督数据标准 The data standard for quality and safety supervision of construction project		J11909-2011	广东省建设工程质量安全监督检测总站
50	DBJ/T 15-86-2011	既有建筑物结构安全性检测鉴定技术标准 Technical standard for inspection and appraisal of existing building structures		J11908-2011	广东省建筑科学研究院
51	DBJ/T 15-87-2011	城市桥梁检测技术标准 Technical standard for inspection of urban bridge		J11965-2012	广东省建筑科学研究院
52	DBJ/T 15-88-2011	建筑幕墙可靠性鉴定技术规程 Technical specification for appraisal of reliability of curtain wall		J11964-2012	广东省建筑科学研究院
53	DBJ/T 15-89-2012	《国家机关办公建筑和大型公共建筑能源审计导则》广东省实施细则 《Energy audit guidelines for state organ office buildings and large public buildings》Regulations of Guangdong Province		J12119-2012	广东省建筑科学研究院
54	DBJ/T 15-90-2012	《民用建筑能耗和节能信息统计报表制度》广东省实施细则 《Statistical reporting system for civil energy consumption and efficient information》Regulations of Guangdong Province		J12118-2012	广东省建筑科学研究院
55	DBJ 15-91-2012	既有民用建筑节能改造技术规程 Technical specification for the retrofitting of existing buildings on energy efficiency		J12234-2012	广东省建筑科学研究院
56	DBJ 15-92-2013	高层建筑混凝土结构技术规程 Technical specification for concrete structures of tall building		J12299-2013	华南理工大学建筑设计研究院
57	DBJ 15-93-2013	民用建筑工程室内环境污染控制技术规程 Technical code for indoor environmental pollution control of civil building engineering		J12300-2013	广州市建筑科学研究院有限公司 广州建设工程质量安全检测中心有限公司
58	DBJ/T 15-94-2013	静压预制混凝土桩基础技术规程 Specification for static pressing precasted concrete pile foundation		J12359-2013	广东省土木建筑学会

广东省

序号	标准编号	标准名称	被代替标准编号	备案号	主编单位
59	DBJ 15-95-2013	现浇混凝土空心楼盖结构技术规程 Technical specification for cast-in-situ concrete hollow floor structure		J12301-2013	广东省建筑科学研究院 广东省建筑设计研究院
60	DBJ 15-96-2013	铝合金模板技术规范 Technical code of aluminum alloy formwork		J12376-2013	广东省建筑科学研究院 广东建星建筑工程有限公司
61	DBJ/T 15-97-2013	建筑工程绿色施工评价标准 Evaluation standard for green construction of building		J12435-2013	广州建筑股份有限公司
62	DBJ 15-98-2014	建筑施工承插型套扣式钢管脚手架安全技术规程 Technical specification for safety of nested steel tubular scaffold in construction		J12622-2014	华南理工大学
63	DBJ/T 15-99-2014	建设工程招标投标造价数据标准 Cost data standard for construction engineering bidding		J12646-2014	广东省建设工程造价管理总站
64	DBJ 15-100-2014	建筑种植工程技术规范 Technical code for building planting engineering		J12782-2014	广州大学
65	DBJ 15-101-2014	建筑结构荷载规范 Load code for the design of building structures		J12795-2014	广东省建筑科学研究院
66	DBJ 15-102-2014	钢结构设计规程 Specification for design of steel structures		J12877-2014	广东省钢结构协会
67	DBJ/T 15-103-2014	基桩自平衡法静载试验技术规程 Technical specification for static loading test of self-balanced method of foundation pile		J12789-2014	广州市建筑科学研究院有限公司
68	DBJ/T 15-104-2015	预拌砂浆、混凝土及制品企业试验室管理规范 Laboratory management code for enterprises of ready-mixed mortar, concrete and its products		J13059-2015	广东省散装水泥管理办公室
69	DBJ/T 15-105-2015	广东省绿色住区评价标准 Evaluation standard for green residential area of Guangdong Province		J13033-2015	广东省房地产行业协会
70	DBJ/T 15-106-2015	顶管技术规程 Technical specification of pipe jacking		J13404-2016	广东省基础工程集团有限公司
71	DBJ/T 15-107-2016	装配式混凝土建筑结构技术规程 Technical specification for precast concrete structures of building		J13145-2016	广东省建筑科学研究院集团股份有限公司
72	DBJ/T 15-108-2015	公共建筑能耗限额编制方法 Method for developing energy consumption quota of public and commercial building		J13209-2015	广东省建筑科学研究院集团股份有限公司
73	DBJ/T 15-109-2015	混凝土技术规范 Technical code for concrete		J13154-2016	广州市建筑科学研究院有限公司
74	DBJ/T 15-110-2015	建筑防火及消防设施检测技术规程 Building fire protection and fire control facilities detection technology procedures		J13259-2015	广东省公安消防总队
75	DBJ/T 15-111-2016	预拌砂浆生产与应用技术管理规程 Technical management specification for production and application of ready-mixed mortar		J13384-2016	广东省散装水泥管理办公室

广东省

序号	标准编号	标准名称	被代替标准编号	备案号	主编单位
76	DBJ/T 15-112-2016	集装箱式房屋技术规程 Technical specification for container type houses		J13385-2016	哈尔滨工业大学深圳研究生院
77	DBJ/T 15-113-2016	再生块体混凝土组合结构技术规程 Technical specification for composite structures containing demolished concrete blocks		J13386-2016	华南理工大学
78	DBJ 15-114-2016	地铁节能工程施工质量验收规范 Code for acceptance of energy efficient metro construction		J13387-2016	广东省建筑科学研究院集团股份有限公司 广州地铁集团有限公司
79	DBJ/T 15-115-2016	广东省建设工程交易规范 Rules for Guangdong Provincial construction engineering trading		J13537-2016	广州公共资源交易中心 广东省建设工程造价管理总站
80	DBJ/T 15-117-2016	《预拌混凝土绿色生产及管理技术规程》广东省实施细则 《Technical specification for green production and management of ready-mixed concrete》Regulations of Guangdong Province		J13598-2016	广东省散装水泥管理办公室 广东省预拌混凝土行业协会
81	DBJ/T 15-118-2016	建筑余泥渣土纳场建设技术规范 Technical code for construction waste and clay residue receiving field		J13540-2016	广东省建筑科学研究院集团股份有限公司
82	DBJ/T 15-119-2016	预拌混凝土用机制砂应用技术规程 Technical specifications for application of manufactured sand in premixed concrete		J13755-2017	广东省建筑科学研究院集团股份有限公司
83	DBJ/T 15-120-2017	城市轨道交通既有结构保护技术规范 Technical code for protection of existing structures of urban rail transit		J13775-2017	广州地铁集团有限公司
84	DBJ/T 15-122-2016	建筑风灾水灾破坏等级评定标准 Criteria for damage grade of buildings caused by wind and flood disaster		J13752-2017	广东省建筑科学研究院集团股份有限公司 广东省建设工程质量安全监督检测总站
85	DBJ/T 15-123-2016	陶瓷薄板幕墙工程技术规程 Technical specification for building ceramic sheet board curtain wall engineering		J13754-2017	广东省建筑设计研究院 广东省建筑科学研究院集团股份有限公司 蒙娜丽莎集团股份有限公司
86	DBJ 15-201-91	建筑地基基础施工及验收规程			广东省建筑科研设计所

广西壮族自治区

序号	标准编号	标准名称	被代替标准编号	备案号	主编单位
1	DBJ/T 45-001-2011	广西壮族自治区绿色建筑设计规范 Code for design of green building in Guangxi Zhuang Autonomous Region		J11901-2011	广西华蓝设计（集团）有限公司
2	DBJ/T 45-001-2012	陶粒混凝土自保温砌块技术规范 The technical specification for ceramsite concrete insulation block		J12064-2012	广西华蓝设计（集团）有限公司
3	DBJ/T 45-001-2012	广西壮族自治区保障性住房建设标准 Standard for indemnificatory housing construction of Guangxi Zhuang Autonomous Region		J12400-2013	广西华蓝设计（集团）有限公司
4	DBJ/T 45-001-2015	复合固化土路面基层和底基层设计施工技术标准 Code for structural design and construction of stabilized soil base and subbase		J12949-2015	广西柳州东风化工股份有限公司
5	DBJ/T 45-002-2011	广西壮族自治区岩土工程勘察规范 Code for investigation of geotechnical engineering in Guangxi		J11629-2011	广西华蓝岩土工程有限公司
6	DBJ/T 45-002-2012	机制砂混凝土应用技术规程 Technical specifications for application of machine-made sand in ready-mixed concrete		J12211-2012	广西建筑科学研究设计院 广西大学
7	DBJ 45/002-2014	桂北地区城镇供水防寒抗冻技术规程 Technical specification for against the cold about urban watter supply in Guibei		J12855-2014	广西华蓝设计（集团）有限公司
8	DBJ/T 45-002-2015	公共建筑节能检测标准 Standard for energy efficiency test of public buildings		J13044-2015	广西绿色建筑节能中心有限责任公司
9	DBJ/45-003-2012	广西公共建筑节能设计规范 Design code for energy efficiency of public buildings of Guangxi Zhuang Autonomous Region		J12221-2012	广西华蓝设计（集团）有限公司
10	DBJ/T 45-003-2013	广西壮族自治区国家机关办公建筑综合能耗、电耗定额 The quota of comprehensive energy and electricity consumption for administrative office buildings		J12516-2013	广西壮族自治区建筑科学研究设计院
11	DBJ 45/003-2014	广西壮族自治区城镇生活垃圾卫生填埋场运行、维护以及考核评价标准 Standards of operation, maintenance and assessment on municipal solid waste landfill in Guangxi Zhuang Autonomous Region		J12853-2014	广西华蓝设计（集团）有限公司
12	DBJ/T 45-003-2015	广西壮族自治区城镇污水处理厂污泥产物土地利用技术 Specification for land application of sewage sludge product in Guangxi Autonomous Region		J13130-2015	中国科学院地理科学与资源研究所
13	DBJ 45/003-2015	广西建筑地基基础设计规范 Code for design of building foundation in Guangxi Autonomous Region		J13143-2015	华蓝设计（集团）有限公司

广西壮族自治区

序号	标准编号	标准名称	被代替标准编号	备案号	主编单位
14	DBJ/45-004-2012	居住区供配电设施建设规范 Constructing code for power supply and distribution facilities' in residential districts		J12235-2012	广西电网公司
15	DBJ/T 45-004-2013	污水源热泵系统工程技术规范 Technical code for sewage source heat pump system		J12454-2013	广西壮族自治区建筑科学研究设计院 广西瑞宝利热能科技有限公司
16	DBJ/T 45-004-2015	城乡道路半导体照明工程技术规范 Technical code for urban and rural road semiconductor lighting engineering		J13129-2015	广西城市建设协会
17	DBJ/45-005-2012	广西建筑节能工程施工质量验收规范 Code for acceptance of energy efficient building construction		J12305-2013	广西建筑科学研究设计院
18	DBJ 45/005-2013	农村生活污水处理设施建设标准 Construction standard of wastewater treatment facilities for village		J12509-2013	广西华蓝设计（集团）有限公司
19	DBJ/T 45-005-2015	城市生活垃圾焚烧厂运行、维护及评价标准 standard for running and assessment on the municipal solid waste incineration plant		J13131-2015	华蓝设计（集团）有限公司
20	DBJ/45-006-2012	二次供水工程技术规程 Technical specification for secondary water supply engineering		J12295-2013	广西华蓝设计（集团）有限公司
21	DBJ/T 45-006-2013	广西壮族自治区商务办公建筑综合能耗、电耗定额 The quota of comprehensive energy and electricity consumption for commercial office buildings		J12515-2013	广西壮族自治区建筑科学研究设计院
22	DBJ/T 45-006-2015	建筑幕墙安全性能检测评估技术规程 Testing and evaluation technical code for building curtain walls safety performance		J13198-2015	广西建筑工程质量检测中心
23	DBJ/T 45-007-2012	先张法预应力混凝土管桩基础技术规程 Technical specification for pretensioned pre-stressed spun concrete pile foundation in Guangxi		J12306-2013	广西城乡规划设计院和邕江大学
24	DBJ/T 45-007-2013	广西壮族自治区星级饭店建筑综合能耗、电耗定额 The quota of comprehensive energy and electricity consumption for star-rated hotels in Guangxi Autonomous Region		J12514-2013	广西壮族自治区建筑科学研究设计院
25	DBJ/T 45-007-2015	旋挖钻孔灌注桩施工技术规程 Construction technical specification for churning driven cast-in-place pile		J13236-2015	广西建工集团基础建设有限公司
26	DBJ/T 45-008-2013	广西壮族自治区商场建筑综合能耗、电耗定额 The quota of comprehensive energy and electricity consumption for markets in Guangxi Autonomous Region		J12513-2013	广西壮族自治区建筑科学研究设计院
27	DBJ45/T 008-2015	广西建筑外门窗遮阳设计应用指南 Guide for design and application of Guangxi building external doors and windows shading		J13237-2015	广西建筑科学研究设计院

广西壮族自治区

序号	标准编号	标准名称	被代替标准编号	备案号	主编单位
28	DBJ/T 45-009-2013	广西壮族自治区医疗卫生建筑综合能耗、电耗定额 The quota of comprehensive energy and electricity consumption for hospitals in Guangxi Autonomous Region		J12512-2013	广西壮族自治区建筑科学研究设计院
29	DBJ/T 45-010-2013	广西文化建筑综合能耗、电耗定额 The quota of comprehensive energy and electricity consumption for cultural buildings in Guangxi		J12511-2013	广西壮族自治区建筑科学研究设计院
30	DBJ45/T 010-2015	组合铝合金模板应用技术规程 Technical specification for application of composite aluminum alloy formwork		J13285-2016	南南铝业股份有限公司
31	DBJ 45/011-2012	烧结砖单位产品能耗限额标准 The norm of energy consumption per unit product of fired brick		J12328-2013	广西壮族自治区建材产品质量监督检验站
32	DBJ/T 45-011-2013	广西壮族自治区普通高等院校建筑综合能耗、电耗定额		J12510-2013	广西壮族自治区建筑科学研究设计院
33	DBJ45/T 011-2015	建筑基坑工程监测技术规程 Technical code for monitoring of building foundation excavation engineering		J13281-2015	南宁市勘察测绘地理信息院
34	DBJ/T 45-012-2013	广西建设项目交通影响评价技术标准 Technical standards of traffic impact analysis of construction projects of Guangxi Zhuang Autonomous Region		J12559-2014	广西华蓝设计（集团）有限公司
35	DBJ/T 45-012-2016	建筑边坡工程技术规程 Technical code for building slope engineering		J13319-2016	华蓝设计（集团）有限公司
36	DBJ/T 45-013-2013	广西农村生态建筑设计规范 Design code for rural ecological building of Guangxi Zhuang Autonomous Region		J12560-2014	广西华蓝设计（集团）有限公司
37	DBJ/T 45-013-2016	低影响开发雨水控制及利用工程设计规范 Code for design of LID stormwater management and harvest engineering		J13320-2016	华蓝设计（集团）有限公司
38	DBJ/T 45-014-2016	地下工程防水技术规程 Technical specification for waterproofing of undergroundworks		J13321-2016	华蓝设计（集团）有限公司
39	DBJ/T 45-014-2013	广西烧结页岩空心砖和空心砌块技术规程 Technical specification for fired shale hollow bricks and blocks		J12561-2014	广西壮族自治区建筑科学研究设计院
40	DBJ/T 45-015-2016	广西城市道路人行道设计规程 Design rules for urban road sidewalk		J13322-2016	南宁市城乡规划设计研究院
41	DBJ/T 45-016-2016	城市桥梁检测评定技术规程 Specification for inspection and evaluation of city bridge		J13337-2016	广西壮族自治区交通规划勘察设计研究院
42	DBJ/T 45-017-2016	城市道路路面设计及施工技术规范 Technical specification for pavement design and construction of urban road		J13338-2016	广西壮族自治区交通规划勘察设计研究院
43	DBJ/T 45-018-2016	城市道路排水管渠设计与施工技术规程 Technical specification for design and construction of urban road sewerage pipeline		J13427-2016	南宁市城乡规划设计研究院

广西壮族自治区

序号	标准编号	标准名称	被代替标准编号	备案号	主编单位
44	DBJ/T 45-019-2016	餐厨垃圾处理技术规范 Technical code for food waste treatment		J13323-2016	华蓝设计（集团）有限公司
45	BDJ/T 45-020-2016	绿色建筑评价标准 Guangxi green building evaluation		J11388-2016	广西壮族自治区建筑科学研究设计院
46	BDJ/T 45-021-2016	城市道路平面交叉口规划与设计规范 Planning and design criteria of at-grade intersections on urban streets		J13448-2016	华蓝设计（集团）有限公司
47	BDJ/T 45-022-2016	城市道路照明工程施工及验收技术规程 Technical regulation for construction and inspection of urban road lighting engineering		J13449-2016	华蓝设计（集团）有限公司
48	DB45/221-2007	广西壮族自治区居住建筑节能设计标准 Design standard for energy efficiency of residential buildings	DB45/221-2005	J11441-2009	广西壮族自治区建筑科学研究设计院
49	DB45/T 392-2007	公共建筑节能设计规范 Design standard for energy efficiency of public buildings in Guangxi Zhuang Autonomous Region		J11191-2008	广西华蓝设计（集团）有限公司（原广西建筑综合设计研究院）
50	DB45/T 393-2007	民用建筑节能检验规范 Energy efficiency inspection standard for civil buildings		J11192-2008	广西壮族自治区建筑工程质量检测中心
51	DB45/T 394-2007	通风与空调系统性能检测规范 Standard for performance testing of ventilation and air conditioning system		J11193-2008	广西壮族自治区建筑工程质量检测中心
52	DB45/T 395-2007	民用建筑与太阳能热水系统一体化应用技术规范 Technical code for solar water heating system integrated with civil building		J11194-2008	南宁市桑普技术发展有限公司 广西大学物理科学与工程技术学院
53	DB45/T 562-2008	居住区供配电设施建设规范 Code for power supply distribution facilities' constructing of residential districts		J11391-2009	广西电网公司
54	DB45/T 618-2009	建筑施工模板及作业平台钢管支架构造安全技术规范 Safety and technical code for the structure of crutches with steel tubes of construction template and working platform		J11564-2010	广西壮族自治区建设工程质量安全监督总站

重庆市

序号	标准编号	标准名称	被代替标准编号	备案号	主编单位
1	DBJ 50/T-036-2014	建筑智能化系统工程设计文件编制深度规范 Engineering design document depth specification of building intelligent system		J10435-2014	重庆市建设技术发展中心 中国医药集团重庆医药设计院
2	DBJ 50/T-039-2015	绿色生态住宅（绿色建筑）小区建设技术规程 Technical specification for eco-residential district (green building)	DBJ/T 50-039-2007	J10535-2015	重庆市建设技术发展中心（重庆市建筑节能中心） 中机中联工程有限公司
3	DBJ/T 50-040-2007	住宅性能评定技术标准 Technical standard for performance assessment of residential building	DBJ/T 50-040-2005	J10536-2008	重庆市建设技术发展中心
4	DBJ 50/T-046-2013	外墙涂料涂饰工程施工及验收规程 Specification for construction and acceptance of exterior wall surface decoration		J10706-2013	重庆市建筑科学研究院 重庆市建设技术发展中心
5	DBJ 50-047-2015	建筑地基基础设计规范 Code for design of building foundation		J13150-2016	中冶赛迪工程技术股份有限公司 重庆市设计院 重庆市土木建筑学会
6	DBJ 50-052-2016	公共建筑节能（绿色建筑）设计标准 Design standards on public building energy saving (green buildings)		J10850-2013	重庆市建设技术发展中心
7	DBJ 50-054-2006	大型商业建筑设计防火规范 Code for fire protection design of large-scale commercial buildings		J10882-2013	机械工业第三设计研究院 重庆市公安局消防局
8	DBJ 50-055-2006	蒸压加气混凝土砌块应用技术规程 Specification for application of aerated concrete blocks		J10912-2006	重庆市建设技术发展中心
9	DBJ 50-056-2006	重庆市住宅建筑群电信用户驻地网建设规范 Specification for application of aerated concrete blocks		J10911-2006	重庆信科设计有限公司（原重庆邮电学院勘察设计院） 重庆市通信管理局通信工程质量监督站
10	DBJ 50-057-2006	重庆市回弹法检测砼抗压强度技术规程		J10955-2007	重庆市建筑科学研究院
11	DBJ 50-058-2006	钢筋混凝土短肢剪力墙、异形柱结构技术规程			重庆市设计院 中国建筑技术集团有限公司（重庆分公司）
12	DBJ 50-059-2006	公共建筑节能工程施工质量验收规程		J10958-2007	重庆市建设技术发展中心 广厦重庆第一建筑（集团）公司
13	DBJ 50-060-2006	建筑防雷工程施工质量控制与验收规程 Code of acceptance and control of construction quality for structure against lightning			重庆市防雷中心
14	DBJ/T 50-061-2007	预拌砂浆生产与应用技术规程 Technical specification for production and application of ready-mixed mortar		J11023-2007	重庆市砼协会
15	DBJ/T 50-062-2007	干混砂浆生产与应用技术规程 Technical specification for production and application of dry-mixed mortar		J11024-2007	重庆市砼协会
16	DBJ/T 50-063-2007	建筑外墙饰面涂饰翻新技术规程 Specification for retrofitting of construction exterior wall surface decoration		J11025-2007	重庆市建设技术发展中心

重庆市

序号	标准编号	标准名称	被代替标准编号	备案号	主编单位
17	DBJ 50-064-2007	城市道路交通规划及路线设计规范 Standard specifications for traffic plan and route design of urban roads		J11026-2007	林同棪国际（重庆）工程咨询有限公司
18	DBJ 50-65-2007	民用建筑门窗安装及验收规程 Rules for install and acceptance of doors and windows for civil buildings		J11046-2007	重庆市建设技术发展中心
19	DBJ50/T-066-2014	绿色建筑评价标准 Evaluation standard for green building	DBJ/T 50-66-2009	J11047-2014	重庆市建筑节能协会绿色建筑专业委员会 万科（重庆）房地产有限公司
20	DBJ/T 50-066-2007	绿色建筑标准 Green building standard		J11048-2007	重庆市建设技术发展中心 重庆龙湖地产发展有限公司
21	DBJ/T 50-067-2007	种植屋面技术规程 Technical specification for planted roofs		J11047-2007	重庆市建设技术新
22	DBJ/T 50-068-2007	清水住宅工程质量验收标准 Standard for quality acceptance of plain residential engineering		J11094-2007	重庆市建设技术发展中心
23	DBJ/T 50-069-2007	居住建筑节能工程施工质量验收规程 Specification for constructional quality acceptance of energy	DBJ/T 50-045-2005	J10645-2007	重庆市建设技术发展中心
24	DBJ 50-071-2010	居住建筑节能65%设计标准 Design standard for energy efficiency 65% of residential building	DBJ/T 50-071-2007	J11571-2010	重庆市建设技术发展中心
25	DBJ 50-072-2007	建筑施工升降机安装与拆卸技术规程 Technical specification of assembling and striping for the builder's hoist		J11122-2008	重庆市土木建筑学会建筑施工机械专业委员会
26	DBJ/T 50-073-2008	市政工程清水混凝土施工技术规程 Technical specification for fair-faced concrete construction of municipal engineering		J11142-2008	重庆交通大学
27	DBJ/T 50-074-2008	住宅工程质量通病控制技术规程 Technical specification for quality control of common failing of housing engineering		J11183-2008	重庆市建设技术发展中心
28	BDJ/T 50-075-2008	挤塑聚苯乙烯石膏复合板外墙内保温系统应用技术规程 Technical specification for application of composite rigid extruded polystyrene form board and plasterboard of interior thermal insulation system on wall		J11203-2008	重庆市建设技术发展中心
29	DBJ/T 50-076-2008	建筑反射隔热涂料外墙保温系统技术规程 Technical specification for external thermal insulation composite system based on reflective thermal insulating coating on building		J11204-2008	重庆市建设技术发展中心
30	DBJ 50-077-2009	建筑施工现场管理标准建筑施工现场管理标准		J11311-2008	重庆市建筑业管理办公室 重庆市建设技术发展中心
31	DBJ 50-078-2008	重庆市城市道路工程施工质量验收规范 Code for acceptance of construction quality of city road engineering of Chongqing		J11224-2008	重庆市建设工程质量监督总站
32	DBJ 50-079-2008	小套型住宅设计规范 Design code for small dwelling size residence		J11227-2008	重庆博建筑设计有限公司
33	DBJ/T 50-80-2008	成品住宅装修工程技术规程		J11262-2008	重庆市建设技术发展中心

重庆市

序号	标准编号	标准名称	被代替标准编号	备案号	主编单位
34	DBJ/T 50-82-2008	重庆市住宅小区智能化系统工程技术规范 Technical code for intelligent system engineering of residential community in Chongqing		J11270-2008	重庆市建筑节能协会
35	DBJ/T 50-83-2008	民用建筑太阳能热水系统一体化应用技术规程 Techinical specification for application of integrated solar water heating system of civil building		J11297-2008	重庆市建设技术发展中心
36	DBJ/T 50-84-2008	河床渗滤取水与水源热泵系统联合应用技术规程 Technical specifications for combining technology application of fetching water by filtration in river bed and water-source heat pump system		J11349-2009	重庆市中设市政工程设计有限公司
37	DBJ 50/T-086-2016	重庆市城市桥梁工程施工质量验收规范 Code for acceptance of construction quality of city bridge engineering of Chongqing		J11341-2009	重庆市建设工程质量监督总站
38	DBJ 50/T-087-2016	建筑施工外挂防护架安全技术规范 Technical code for safety of outside hanging protective frame in construction	DBJ/T 50-87-2008	J11348-2016	重庆市建设技术发展中心 重庆安谐建筑脚手架有限公司
39	DBJ/T 50-88-2008	建筑玻璃隔热膜工程技术规程 Technical specification for architectural glass thermal insulating film engineering		J11356-2009	重庆市建设技术发展中心
40	DBJ/T 50-89-2009	节能彩钢门窗应用技术规程 Technical specification for application of energy efficiency color coated steel window and door		J11378-2009	重庆市建设技术发展中心
41	DBJ/T 50-90-2009	社区公共服务设施配置标准 Standard for community public facilities		J11379-2009	重庆市规划研究中心
42	DBJ/T 50-91-2009	造价软件数据交换标准 Standard for data exchange of construction cost software		J11411-2009	重庆市建设工程造价总站
43	DBJ/T 50-092-2009	跨座式单轨交通防雷技术规范 Technical specifications for straddle monorail transit against lightning		J11410-2009	重庆市防雷中心
44	DBJ/T 50-093-2009	特细砂砌筑砂浆配合比设计规程 Specification for mix-proportion design of super-fine sand masonry mortar		J11412-2009	广厦重庆一建
45	DBJ/T 50-094-2009	住宅小区智能化系统工程验收规范 Code for acceptance of community intelligent system		J11413-2009	重庆市建筑节能中心
46	DBJ/T 50-095-2009	多孔混凝土河道护坡及坡面绿化施工技术规程 Technical code for construction of slope protection and greening with porous concrete		J11416-2009	重庆市建筑科学研究院
47	DBJ/T 50-096-2009	居住建筑围护结构节能应用技术规程 Technical specification for appliflication of energy efficiency on envelope of residential building		J11417-2009	重庆市建筑科学研究院

重庆市

序号	标准编号	标准名称	被代替标准编号	备案号	主编单位
48	DBJ 50-97-2009	餐饮娱乐住宿泵船防火规范 Code for fire protection of dining entertainment accommodation pontoon		J11406-2009	重庆市公安局消防局
49	DBJ/T 50-098-2009	城市绿化养护质量标准 Quality standard for management of landscape greening		J11448-2009	重庆市园林绿化科学研究所
50	DBJ/T 50-99-2010	预拌机制砂混凝土技术规程 Technical specification for ready mixed artificial sand concrete		J11557-2010	重庆市混凝土协会
51	DBJ/T 50-100-2009	建筑边坡工程施工质量验收规范 Code for acceptance of construction quality of building slope engineering		J11558-2010	重庆市建筑科学研究院
52	DBJ/T 50-101-2010	装配式超载自动报警型钢卸料平台技术		J11602-2010	重庆第三建设有限责任公司
53	DBJ/T 50-105-2010	城市地下管线综合管廊建设技术规程 Technical specification for construction of utility tunnel		J11656-2010	重庆市市政设计研究院
54	DBJ/T 50-106-2010	重庆市三峡库区跨江桥梁船撞设计指南 Guide specification for vessel collision design of river-crossing bridges in three gorges reservoir of Chongqing City		J11678-2010	招商局重庆交通科研设计院有限公司
55	DBJ 50-108-2010	城镇给水排水构筑物及管道工程施工质量验收规范 Code of construction quality inspectionfor urban water supply and drainage structure and pipeline works		J11683-2010	重庆市工程建设质量监督总站
56	DBJ/T 50-109-2010	燃气用衬塑（PE）铝合金管道工程技术规程 Applying technical specification of plastic (PE) liner aluminum alloy composited pipeline for gas supply		J11720-2010	重庆市燃气行业协会
57	DBJ 50/T-110-2010	停水自闭阀应用技术规程 Technical regulations for water supply suspending self-closing valve		J11736-2010	重庆市设计院
58	DBJ 50-111-2010	重庆市保障性住房装修设计标准 Decoration design code for indemnificatory residential buildings		J11756-2010	重庆大学
59	DBJ 50-113-2010	成品住宅装修工程技术规程 Technical specification for finished residential building decoration		J11752-2010	重庆市建设技术发展中心
60	DBJ/T 50-119-2011	改性无机粉建筑装饰片材工程技术规程 Technical specification for modified inorganic powderBuilding decorative material (MCM) construction		J11784-2011	重庆市建设技术发展中心
61	DBJ 50-123-2010	建筑护栏技术规程 Technical specification for building guardrail		J11787-2011	重庆市工程建设质量监督总站
62	DBJ 50-125-2011	建筑地基基础工程施工质量验收规范 Code for acceptance of construction quality of building foundation engineering		J11845-2011	重庆市土建筑学会

重庆市

序号	标准编号	标准名称	被代替标准编号	备案号	主编单位
63	DBJ 50-127-2011	非承重节能型烧结页岩空心砌块墙体工程技术规程 Technical specification for blame bearing wall engineering of energy-saving sintering shale hollow block		J11865-2011	重庆市建设技术发展中心
64	DBJ 50-128-2011	城镇道路附属设施工程施工质量验收规范 Code for acceptance of construction quality of city road ancillary facilities engineering		J11888-2011	重庆市工程建设质量监督总站
65	DBJ 50/T-130-2011	建筑立面装饰设计技术导则 The technical guide for building facade decoration design		J11935-2011	重庆市建设技术发展中心
66	DBJ 50/T-131-2011	城镇人行道设计指南 Sidewalk design guidelines of city and town		J11934-2011	重庆中设工程设计公司
67	DBJ 50/T-132-2011	仿幕墙涂料涂饰系统应用技术规程 Specification for application technical of imitation curtain wall finishing system		J11940-2011	重庆市建筑节能协会
68	DBJ 50/T-133-2011	公共租赁房设计标准 Design standards of public rental housing		J11946-2011	重庆市设计院
69	DBJ 50-134-2012	重庆市市政基础设施工程预应力施工质量验收规范 Chongqing city municipal foundation services engineering prestress construction		J11536-2011	重庆市建设工程质量监督总站
70	DBJ 50/T-135-2012	绿色建筑设计规范 Design code for green building		J11967-2012	重庆市设计院
71	DBJ 50/T-136-2012	建筑地基基础检测技术规范 Technical code for testing of building foundation		J11969-2012	重庆市建筑科学研究院
72	DBJ 50/T-137-2012	建筑边坡工程检测技术规范 Technical code for inspection of building slope engineering		J11968-2012	重庆市建筑科学研究院
73	DBJ 50/T-138-2012	建筑隔声门窗工程技术规程 The building sound insulation doors and windows application technique regulations		J12030-2012	重庆市建筑节能协会
74	DBJ 50/T-139-2012	低碳建筑评价标准 Evaluation standard of low-carbon buildings		J12042-2012	重庆大学
75	DBJ 50/T-141-2012	岩棉板薄抹灰外墙外保温系统应用技术规程 Technical specification for application of rock wool external thermal insulation		J12058-2012	中煤科工集团重庆设计研究院
76	DBJ 50/T-142-2012	轻型斜拉式脚手架应用技术规程 Technical specification for light-weight cable-stayed scaffold		J12136-2012	重庆市建设技术发展中心 重庆安谐建筑脚手架有限公司
77	DBJ 50/T-143-2012	无机复合烧结页岩空心砖应用技术规程 Technical regulations for inorganic compound sintered shale hollow bricks		J12096-2012	重庆市建设技术发展中心
78	DBJ 50/T-144-2012	复合酚醛泡沫板薄抹灰外墙外保温系统应用技术规程		J12121-2012	中煤科工集团重庆设计研究院
79	DBJ 50/T-146-2012	注塑型塑料检查井应用技术规程 Technical specification application of plastic inspection chamber for sewerage		J12152-2012	重庆市设计院

重庆市

序号	标准编号	标准名称	被代替标准编号	备案号	主编单位
80	DBJ 50/T-147-2012	重庆市住宅电气设计标准 Code for electrical design for Chongqing		J12149-2012	重庆市设计院
81	DBJ 50/T-148-2012	夹砂玻璃钢（GRP）塑料（PE）复合顶管管道技术规程		J12148-2012	重庆市康伯特塑料有限公司
82	DBJ 50/T-149-2012	挤压成型预应力混凝土空心板生产技术操作规程		J12151-2012	重庆市建筑科学研究院 重庆一建建设集团有限公司
83	DBJ 50/T-150-2012	混凝土用机制砂质量及检验方法标准 Standard for technical requirements and test method of manufactured-sand for concrete		J12150-2012	重庆市建筑科学研究院 北城致远集团有限公司
84	DBJ 50/T-151-2012	全轻混凝土建筑地面保温工程技术规程 Technical specification of full lightweight aggregate concrete thermal insulation		J12168-2012	中煤科工集团重庆设计研究院
85	DBJ 50/T-152-2012	硅酸铝棉板建筑外保温系统应用技术规程 Application technical specification of aluminium silicate cotton board exterior		J12219-2012	重庆市勘察设计协会
86	DBJ/T 50-153-2012	公共建筑能耗监测系统技术规程 Technical specification for metering system of energy consumption of public buildings		J12230-2012	重庆市建设技术发展中心 重庆大学
87	DBJ 50/T-154-2012	行人道透水混凝土应用技术规程 Technical specification for application of pervious concrete in pedestrian walk		J12250-2013	重庆建工住宅建设有限公司
88	DBJ 50/T-155-2012	建筑工人安全操作规程 Specification for safety operation of construction workers		J12252-2013	重庆建工住宅建设有限公司 重庆科技学院
89	DBJ 50-156-2012	旋挖成孔灌注桩工程技术规程 Technical specification for rotary drilling cast-in-place pile		J12222-2012	重庆市建设工程质量监督总站 重庆市建筑科学研究院
90	DBJ 50-157-2013	重庆市房屋建筑与市政基础设施工程现场施工从业人员配备标准 Workers standard for housing construction and municipal infrastructure engineering		J12233-2012	重庆市建设岗位培训中心 重庆城建控股（集团）有限责任公司
91	DBJ 50/T-158-2013	复合硬泡聚氨酯板建筑外保温系统应用技术规程		J12308-2013	中煤科工集团重庆设计研究院
92	DBJ 50/T-159-2013	难燃型挤塑聚苯板建筑外保温系统应用技术规程		J12309-2013	中煤科工集团重庆设计研究院
93	DBJ 50/T-160-2013	难燃型膨胀聚苯板建筑外保温系统应用技术规程		J12298-2013	中煤科工集团重庆设计研究院
94	DBJ 50/T-161-2013	二氧化硅微粉真空隔热保温板建筑保温系统应用技术规程		J12310-2013	重庆市勘察设计协会
95	DBJ 50/T-162-2013	岩棉保温装饰复合板外墙外保温系统应用技术规程 Technical specification for application of decorative composite panels composite board exterior wall		J12290-2013	重庆市勘察设计协会
96	DBJ 50/T-163-2013	公共建筑节能改造应用技术规程 Applied technical code for the retrofitting of public building on energy efficiency		J12307-2013	重庆市设计院 重庆星能建筑节能技术发展有限公司

重庆市

序号	标准编号	标准名称	被代替标准编号	备案号	主编单位
97	DBJ 50-164-2013	民用建筑电线电缆防火设计规范 Code for fireproofing design of wires and cables used in civil buildings		J12286-2013	重庆市公安局消防局 重庆市设计院
98	DBJ 50/T-165-2013	建筑外立面遮阳设施应用技术规程 Technical regulations for solar shading facility on external elevation of buildings		J12337-2013	重庆市建设技术发展中心
99	DBJ 50/T-166-2013	绿色施工管理规程 Management specification for green construction		J12336-2013	重庆建工住宅建设有限公司
100	DBJ 50/T-167-2013	建筑外立面空调室外机位技术规程 Technical specification for building exterior wall air conditioning outdoor seats		J12418-2013	重庆市建设技术发展中心
101	DBJ 50-168-2013	房屋建筑与市政基础设施工程施工模板支撑体系安全技术规范		J12441-2013	重庆城建控股（集团）有限责任公司 重庆建工第九建设有限公司
102	DBJ 50/T-169-2013	混合砂混凝土应用技术规程 Specification of application for concrete made from mixed sand		J10345-2013	重庆市建筑科学研究院 重庆建工新型建材有限公司
103	DBJ 50/T-171-2013	重庆市房屋建筑与市政基础设施工程现场施工专业人员职业标准 Technician occupational standards for site construction of building and municipal infrastructure engineering		J12480-2013	重庆建工第九建设有限公司 重庆市建达职业培训学校
104	DBJ 50/T-172-2013	铝及铝合金管熔化焊对接接头X射线检测工艺技术规程 Technical specification for X-rays test butt welded joints in aluminium and aluminium alloy pipes		J12501-2013	重庆工业设备安装集团有限公司
105	DBJ 50/T-173-2013	塑料片材空铺法屋面防水应用技术规程 Application technology regulation on roof waterproofing by empty paving method with plastic sheet		J12507-2013	重庆市建设技术发展中心
106	DBJ 50/T-175-2014	建筑智能化系统设计规范 Design code for building intelligent system		J12553-2014	重庆市建设技术发展中心 重庆市设计院
107	DBJ 50/T-176-2014	综合医院通风设计规范 Code of design on ventilation in general hospital buildings		J12558-2014	重庆海润节能研究院
108	DBJ 50/T-177-2014	重庆市房屋建筑与市政基础设施工程现场施工技术工人职业技能标准（I） Occupational skill standard of site construction skilled workers for building and municipal infrastructure engineering of Chongqing（I）		J12568-2014	重庆建工第九建设有限公司 重庆市建达职业培训学校
109	DBJ 50/T-178-2014	重庆市城镇道路平面交叉口设计规范 Technical specifications for intersection design on urban roads for Chongqing		J12569-2014	重庆市市政设计研究院
110	DBJ 50/T-179-2014	石灰石粉在水泥混凝土中应用技术规程 Technical specification for application of limestone powder in concrete		J12570-2014	重庆大学 重庆睿亮建材有限公司

重庆市

序号	标准编号	标准名称	被代替标准编号	备案号	主编单位
111	DBJ 50/T-180-2014	预拌混凝土绿色生产管理规程 Management specification for green production of ready-mixed concrete		J12571-2014	重庆建工新型建材有限公司 重庆大学
112	DBJ 50/T-181-2014	绿色照明技术规程 Technical specification for green lights		J12630-2014	重庆市建筑节能协会
113	DBJ 50/T-182-2014	重庆市砖砌体结构房屋装配式构造柱技术规程 Specification for fabricated structural columns of brick masonry buildings in Chongqing		J12632-2014	重庆大学
114	DBJ 50/T-183-2014	可再生能源建筑应用项目系统能效检测标准 Standard for test on system energy efficiency of application projects in renewable energy buildings		J12631-2014	重庆大学
115	DBJ 50-184-2014	建筑施工插槽式钢管模板支撑架安全技术规范 Technical code for safety of slot steel tubular formwork support in construction		J12592-2014	重庆建工第九建设有限公司 群力发（北京）科技开发有限公司
116	DBJ 50/T-185-2014	改性发泡水泥保温板建筑保温系统应用技术规程 Technical specification for application of modified foam cement insulation board thermal insulation system on building		J12642-2014	中煤科工集团重庆设计研究院有限公司 重庆思贝肯节能技术开发有限公司
117	DBJ 50/T-186-2014	装配式住宅建筑设备技术规程 Technical specifications for equipment of assembled buildings		J12672-2014	重庆市设计院
118	DBJ 50/T-187-2014	重庆市住宅用水一户一表设计、施工及验收技术规范 Chongqing city code for design, construction and acceptance of building plumbing installation		J12673-2014	重庆市给水工程设计有限公司
119	DBJ 50/T-188-2014	建筑外墙外保温系统用柔性饰面块材应用技术规程 Application technical specification of decorating flexible block for external wall thermal insulation system on building		J12674-2014	重庆市绿色建筑专业委员会 重庆市绿色建筑技术促进中心
120	DBJ 50/T-189-2014	地下工程地质环境保护技术规范 Technical code for geological environment protection of underground engineering		J12697-2014	重庆地质矿产研究院
121	DBJ 50/T-190-2014	装配式混凝土住宅构件生产与验收技术规程 Code for production and acceptance of concrete structures of precast concrete buildings		J12696-2014	重庆建工新型建材有限公司 重庆建工住宅建设有限公司
122	DBJ 50/T-191-2014	装配式住宅构件生产和安装信息化 The information technology guide for production and installation of prefabricated house		J12694-2014	重庆建工工业有限公司 重庆大学
123	DBJ 50/T-192-2014	装配式混凝土住宅结构施工及质量验收规程 Code for construction and quality acceptance of concrete structures of assembled buildings		J12695-2014	重庆市建设工程质量监督总站 重庆建工住宅建设有限公司
124	DBJ 50-193-2014	装配式混凝土住宅建筑结构设计规程 Design specification for architectural and structures of precast concrete buildings		J12683-2014	重庆市设计院 重庆大学

重庆市

序号	标准编号	标准名称	被代替标准编号	备案号	主编单位
125	DBJ 50/T-194-2014	设施栽培园林植物病虫害防治技术规范 Technical specification for pests control techniques in protected garden plants cultivation		J12748-2014	重庆市风景园林科学研究院
126	DBJ 50/T-195-2014	高强混凝土抗压强度检测技术规程 Technical specification for strength testing of high strength concrete		J12747-2014	重庆市建筑科学研究院
127	DBJ 50/T-196-2014	重庆市城乡建设领域基础数据标准 Basic data standards for Chongqing urban and rural construction		J12745-2014	重庆市建设信息中心 重庆大学
128	DBJ 50/T-197-2014	重庆市城乡建设领域信息安全规范 Information security standards for Chongqing urban and rural construction		J12746-2014	重庆市建设信息中心 重庆大学
129	DBJ 50/T-198-2014	建筑外墙外保温系统用饰面砂浆应用技术规程 Application technical specification of decorative render and plaster for external wall thermal insulation on building		J12744-2014	重庆市勘察设计协会
130	DBJ 50-200-2014	建筑桩基础设计与施工验收规范 Code for design and construction acceptance of building pile foundations		J12720-2014	重庆市设计院
131	DBJ 50-201-2014	焊接箍筋应用技术规程 Technical specification for application of welding stirrup		J12780-2014	中冶建工集团有限公司 重庆建工第九建设有限公司
132	DBJ 50/T-202-2014	无机干粉建筑涂料应用技术规程 Technical specification for architectural inorganic powder coatings		J12799-2014	重庆建工住宅建设有限公司 上海墙特节能材料有限公司
133	DBJ 50/T-203-2014	绿色低碳生态城区评价标准 Evaluation standard for green low-carbon eco-district development in Chongqing		J12824-2014	重庆大学 重庆市绿色建筑专业委员会
134	DBJ 50/T-204-2014	城市道路路面维护评价标准 Technical code of maintenance evaluation for urban road		J12836-2014	重庆市建筑科学研究院
135	DBJ 50/T-205-2014	碱矿渣混凝土应用技术规程 Technique specification for application of alkali-activated slag concrete		J12839-2014	重庆市建筑科学研究院 重庆大学
136	DBJ 50/T-206-2014	模块化同层排水及节水系统应用技术规程 Technical specification for application of modular same-floor drainage & water-saving system		J12852-2014	中煤科工集团重庆设计研究院有限公司
137	DBJ 50/T-209-2014	玻化微珠无机保温板建筑保温系统应用技术规程 Technical specification for application of vitrified beads inorganic insulation board thermal insulation system on building		J12902-2015	中煤科工集团重庆设计研究院有限公司
138	DBJ 50/T-210-2014	电气火灾监控系统设计、施工及验收规范 Code for design, installation and acceptance of alarm and monitoring system for electric fire prevention		J12903-2015	重庆市公安局消防局

重庆市

序号	标准编号	标准名称	被代替标准编号	备案号	主编单位
139	DBJ 50/T-211-2014	绿色建筑检测标准 Standard for inspection of green building		J12904-2015	重庆市建筑科学研究院 重庆市绿色建筑专业委员会
140	DBJ 50/T-212-2015	机制排烟气道系统应用技术规程 Mechanism of exhaust flue gas system application procedures		J12944-2015	重庆市建设技术发展中心 重庆市建设工程质量监督总站
141	DBJ 50/T-213-2015	重庆市建设工程造价技术经济指标采集与发布标准 Collection and publication criteria for technical and economic indicators of construction project cost in Chongqing Municipality		J12972-2015	重庆市建设工程造价管理总站
142	DBJ 50/T-214-2015	绿色建筑设计标准 Design standard of green buildings		J12973-2015	中煤科工集团重庆设计研究院有限公司
143	DBJ 50/T-215-2015	浆固散体材料桩复合地基技术规程 Technical specification for grouted granular material pile composite foundation		J13060-2015	重庆大学 中建三局集团有限公司
144	DBJ 50/T-216-2015	建筑施工轮盘插销式钢管模板支撑架安全技术规范 Technical code for safety of disk-pin joined steel tubular formwork support in construction		J13061-2015	重庆建工第九建设有限公司 中冶建工集团有限公司
145	DBJ 50/T-217-2015	装配式住宅部品标准 Assembled housing parts standard		J13062-2015	重庆市建设技术发展中心
146	DBJ 50/T-219-2015	自承重水泥钢丝网架膨胀珍珠岩墙板应用技术规程 Technical specification for application of non-load-bearing wallboard with mesh cement expanded pearlite sandwich panel		J13083-2015	重庆市绿色建筑技术促进中心
147	DBJ 50/T-220-2015	房屋建筑工程质量保修规程 Specification for quality warranty of building		J13082-2015	重庆建工住宅建设有限公司 重庆建工第二建设有限公司
148	DBJ 50/T-221-2015	行人道透水路面混凝土应用技术规程		J13166-2015	重庆市建筑业协会 重庆建工集团股份有限公司
149	DBJ 50/T-222-2015	屋面保温隔热工程施工技术规程 Construction technical specification of thermal preservation andheat insulation for roof		J13165-2015	重庆建工住宅建设有限公司 重庆建工第七建筑工程有限责任公司
150	DBJ 50/T-223-2015	塔式起重机装配式预应力混凝土基础技术规程 Technical specification for prefabricated concrete foundation of tower cranes		J13167-2015	重庆市建筑业协会机械管理与租赁分会 重庆建工第十一建筑工程有限责任公司
151	DBJ 50/T-224-2015	玻化微珠真空绝热芯材复合无机板薄抹灰外墙外保温系统应用技术规程 Technical specification for application of vacuum adiabatic core material composite inorganic plate		J13184-2015	重庆市绿色建筑技术促进中心 重庆市勘察设计协会
152	DBJ 50/T-225-2015	玻璃纤维增强水泥（GRC）制品应用技术规程 Technical specification for application of glassfibre reinforced cement products		J13235-2015	重庆市渝北区建设工程质量监督站 重庆市建设工程质量监督总站
153	DBJ 50/T-226-2015	混凝土小型空心砌块砌体工程施工及验收规程 Concrete small hollow block masonry construction and acceptance procedures		J10222-2015	重庆市建筑科学研究院 重庆建工第四建设有限责任公司

重庆市

序号	标准编号	标准名称	被代替标准编号	备案号	主编单位
154	DBJ 50/T-227-2015	建设领域创新型企业评价标准 Evaluation criteria for innovative enterprises in the construction field		J13269-2015	重庆市建设技术发展中心
155	DBJ 50/T-228-2015	建设工程绿色施工规范 Specification for green construction of construction		J13305-2016	重庆建工集团股份有限公司 重庆对外建设（集团）有限公司
156	DBJ 50/T-229-2015	建筑地基处理技术规范 Technical code for ground treatment of buildings		J13306-2016	重庆市建筑科学研究院 重庆市建设工程质量检验测试中心
157	DBJ 50/T-230-2015	绿色建材评价标准 Evaluation standard for green building materials		J13307-2016	重庆市绿色建筑技术促进中心
158	DBJ 50/T-231-2015	绿色医院建筑评价标准 Evaluation standard for green hospital building		J13308-2016	重庆海润节能研究院 重庆市建筑节能协会
159	DBJ 50/T-232-2016	建设工程监理工作规程 Supervision procedure for construction project		J13324-2016	重庆联盛建设项目管理有限公司 重庆市建筑科学研究院
160	DBJ 50/T-233-2016	保温装饰复合板外墙外保温系统应用技术规程 Thermal insulation decorative composite board in exterior wall external insulation system technical specification for application		J13325-2016	重庆市建设技术发展中心 重庆建工第七建筑工程有限责任公司
161	DBJ 50/T-235-2016	后张法预应力孔道灌浆应用技术规程 Technical specification for application of post-prestressed grouting		J13435-2016	重庆大学 重庆迪翔建材有限公司
162	DBJ 50/T-236-2016	Z型混凝土复合保温砌块自承重墙体工程技术规程 Type Z self insulation concrete composite block filling of bearing wall engineering technical specification		J13436-2016	重庆市建设技术发展中心 重庆建工第八建设有限责任公司
163	DBJ 50-255-2017	公共建筑节能（绿色建筑）工程施工质量验收规范 Code for acceptance of energy efficient publicbuilding (greenbuilding) construction		J13144-2015	重庆市建设技术发展中心（重庆市建筑节能中心） 重庆市绿色建筑技术促进中心
164	DB 50-5021-2002	塔式起重机安装与拆卸技术规范 Technical specification for installation and dismantlement of tower crane		J12027-2012	重庆市土木学会建筑机械专业委员会
165		地表水水源热泵系统设计标准 Design standard for surface water-source heat pump system		J11772-2011	重庆市建设技术发展中心
166		城市隧道工程施工质量验收规范 Code for construction and quality acceptance of tunnel works in city		J11670-2010	重庆市工程建设质量监督总站
167		跨越式施工支架技术规程 Technical specification for striding construction supporting bracket		J11757-2010	重庆城建集团
168		建筑施工升降机报废规程 Specification for builder's hoist to be scrapped		J11646-2010	重庆市土木建筑学会建筑机械专业委员会
169		重庆市建筑智能化系统工程施工规范 Code for installation of building intellgent system in Chongqing		J11834-2011	重庆市建筑业协会建筑智能化工程专业委员会

重庆市

序号	标准编号	标准名称	被代替标准编号	备案号	主编单位
170		重庆市细水雾灭火系统技术规范 Technical standard of water mist fire protection system of Chongqing		J12860-2014	重庆市公安局消防局 上海同泰火安科技有限公司
171		地埋管地源热泵系统技术规程 Technical regulation for ground-coupled heat pump system		J12719-2014	重庆大学 重庆市设计院
172		建筑边坡工程安全性鉴定规范 Code for appraisal of safety of building slope engineering		J12460-2013	重庆市建筑科学研究院
173		旋转挤压灌注桩技术规程 Technical specification for rotating extrusion cast-in-situ pile		J12841-2014	重庆建工集团股份有限公司 重庆卓典建设工程有限公司
174		地表水水源热泵系统施工质量验收标准 Standard of acceptance for quality of surface water-source heat pump system		J11759-2010	重庆市建设技术发展中心
175		成品住宅装修工程质量验收规范 Constructional quality acceptance for finished residential building decoration		J11751-2010	重庆市建设技术发展中心
176		无机保温砂浆建筑保温系统应用技术规程 Technical specification for application of inorganic insulation mortar thermal insulation system on building		J11609-2010	重庆市建设技术发展中心
177		居住建筑节能50%设计标准 Design standard for energy efficiency 50% of residential building		J11572-2010	重庆市建设技术发展中心
178		地表水水源热泵系统适应性评估标准 Assessment standard for surface water-source heat pump system		J11773-2011	重庆市建设技术发展中心
179		民用建筑电动汽车充电设备配套设施设计规范 Design specifications for civil building electric vehicles' charging facilities construction		J13042-2015	重庆市建设技术发展中心 重庆同乘工程咨询设计有限责任公司
180		市政工程地质勘察规范 Code for geological investigation of municipal engineering		J12493-2013	重庆市涪陵区建筑勘察设计质量审查中心
181		重庆市建设工程档案编制验收标准 Chongqing city construction acceptance criteria for the preparation of files		J11899-2011	重庆市城市建设档案馆
182		地表水水源热泵系统运行管理技术规程 Technical specification for management of surface water-source heat pump system operation		J11758-2010	重庆市建设技术发展中心
183		重庆市住宅建筑群电信用户驻地网建设规范（修订） Standard for construction of customer premise telecommunication network for residential campus in Chongqing		J11863-2011	重庆市通信管理局

重庆市

序号	标准编号	标准名称	被代替标准编号	备案号	主编单位
184		市政工程边坡及挡护结构施工质量验收规范 Code for acceptance of construction quality of municipal slope and retaining structure engineering		J11864-2011	送重庆市建设工程质量监督总站
185		重庆市住宅信报箱建设规范 Constructional code of residential mail-box of Chongqing		J11310-2009	重庆市邮政管理局
186		公共建筑采暖、通风与空调系统节能运行管理标准 Standard for energy efficiency operation and management of heating, ventilation and air-conditioning system of public building		J11309-2008	重庆市建设技术发展中心

四川省

序号	标准编号	标准名称	被代替标准编号	备案号	主编单位
1	DBJ51/T 001-2011	烧结复合自保温砖和砌块墙体保温系统技术规程 Technical specification for wall insulation system of fired composite self-insulation hollow brick and block		J11970-2012	四川省建材工业科学研究院
2	DBJ51/T 002-2011	烧结自保温砖和砌块墙体保温系统技术规程 Technical specification for wall insulation system of fired self-insulation brick and block		J11975-2012	四川省建材工业科学研究院 成都市墙材革新建筑节能办公室
3	DBJ51/T 003-2012	灾区过渡安置点防火规范 Code for fire protection and prevertion of interim settlements of disaster areas		J11976-2012	四川省公安消防总队
4	DBJ 51/004-2012	四川省住宅建筑通信配套光纤入户工程技术规范 Technical code for fiber to the home of communication engineering for residential in Sichuan Province		J12008-2012	中国建筑西南设计研究院有限公司 四川通信科研规划设计有限责任公司
5	DBJ 51/005-2012	城市建筑二次供水工程技术规程		J12035-2012	中国市政工程西南设计研究总院 四川省建筑设计院
6	DBJ 51/006-2012	成都市地源热泵系统施工质量验收规程 Code of acceptance for construction of ground source heat pump system in Chengdu		J12078-2012	四川省建筑科学研究院
7	DB51/T 007-2012	成都市地源热泵系统性能工程评价标准 Code of performance assessment of ground source heat pump system in Chengdu		J12063-2012	四川省建筑科学研究院
8	DBJ 20-7-2013	钢筋电渣压力焊技术规程 Technical specification for electroslag pressure welding of reinforcing steel bar		J12505-2013	四川省建筑科学研究院
9	DBJ51/T 008-2015	四川省建筑工业化混凝土预制构件制作、安装及质量验收规程 Specification for manufacture, erection and acceptance inspection of precasted concrete members in Sichuan Province		J12079-2016	四川省建筑科学研究院 四川华西绿舍建材有限公司
10	DBJ51/T 009-2012	四川省绿色建筑评价标准 Evaluation standard for green building in Sichuan Province		J12097-2012	四川省建筑科学研究院
11	DBJ51/T 010-2012	四川省民用建筑节能工程施工工艺规程 Technical specification of construction for engineering of civil building energy efficiency of Sichuan		J12106-2012	四川建筑职业技术学院 成都建筑工程集团总公司
12	DBJ51/T 011-2012	成都市地源热泵系统运行管理规程 Technical specification for operation and management of ground-source heat pump system in Chengdu		J12197-2012	中国建筑西南设计研究院有限公司
13	DBJ 51/012-2012	成都市地源热泵系统设计技术规程 Technical specification for design of ground-source heat pump system in Chengdu		J12204-2012	中国建筑西南设计研究院有限公司

四川省

序号	标准编号	标准名称	被代替标准编号	备案号	主编单位
14	DBJ51/T 013-2012	酚醛泡沫保温板外墙外保温系统技术规程 Technical specification of external thermal insulating rendering systems on walls made of rigid phenolic foam board for thermal insulation		J12248-2013	四川省建材工业科学研究院
15	DBJ51/T 014-2013	四川省建筑地基基础检测技术规程 Technical code for testing of foundation soil and building foundation in Sichuan Province		J12372-2013	四川省建设工程质量安全监督总站
16	DBJ 51/015-2013	四川省成品住宅装修工程技术标准 Technical standard for finished residential building decoration in Sichuan Province		J12450-2013	成都市建设工程质量监督站 四川省建筑科学研究院
17	DBJ 51/016-2013	四川省农村居住建筑抗震技术规程 Seismic technical specification for rural residential building in Sichuan Province		J12461-2013	四川省建筑科学研究院
18	DBJ51/T 017-2013	四川省民用建筑节能检测评估标准 Standard of energy efficiency test and evaluation for civil buildings in Sichuan Province		J12666-2014	四川省建设科技协会 四川省建筑科学研究院
19	DBJ51/T 018-2013	回弹法检测高强混凝土抗压强度技术规程 Technical specification for inspecting of high strength concrete compressive strength by rebound method		J12543-2014	四川省建筑科学研究院 四川华西绿舍建材有限公司
20	DBJ51/T 019-2013	四川省被动式太阳能建筑设计规范 Technical code for passive solar buildings design in Sichuan Province		J12583-2014	中国建筑西南设计研究院有限公司
21	DBJ51/T 020-2013	四川省绿色学校设计标准 Design code for green school in Sichuan Province		J12584-2014	中国建筑西南设计研究院有限公司
22	DBJ51/T 021-2013	建筑反射隔热涂料应用技术规程 Technical specification for application of architectural reflective thermal insulation coatings		J12542-2014	四川省建筑科学研究院 四川省建设科技协会
23	DBJ51/T 022-2013	旋挖成孔灌注桩施工安全技术规程 Technical specification for construction saftyofrotary drilled and bored pile		J12544-2014	成都市建设工程施工安全监督站 中节能建设工程设计院有限公司
24	DBJ51/T 023-2014	燃气用卡压粘结式薄壁不锈钢管道工程技术规程 Compression and bonding joint thin wall stainless steel gas pipes		J12585-2014	四川省燃气器具产品质量监督检验站
25	DBJ51/T 024-2014	装配整体式混凝土结构设计规程 Design specification for precast reinforced concrete structure		J12586-2014	四川省建筑科学研究院
26	DBJ51/T 025-2014	保温装饰复合板应用技术规程 Technical specification for application of insulated decorative composite panel		J12753-2014	四川省建筑科学研究院 成都市墙材革新建筑节能办公室
27	DBJ51/T 026-2014	建筑施工塔式起重机及施工升降机报废标准 The standard for construction tower cranes and builder's hoists to be scrapped		J12635-2014	四川省建筑科学研究院 四川省建筑工程质量检测中心
28	DBJ51/T 027-2014	建筑工程绿色施工评价与验收规程 Evaluation and acceptance specification for green construction of building		J12687-2014	成都建筑工程集团总公司

四川省

序号	标准编号	标准名称	被代替标准编号	备案号	主编单位
29	DBJ51/T 028-2014	四川省工程建设从业人员资源信息数据标准 Standard for basic data of staff and workers management of engineering construction field of Sichuan Province		J12807-2014	四川省建设科技发展中心
30	DBJ51/T 029-2014	四川省房屋建筑与市政基础设施建设项目管理基础数据标准 Standard for basic data of project management of engineering construction field of Sichuan Province		J12805-2014	四川省建设科技发展中心
31	DBJ51/T 030-2014	四川省工程建设从业企业资源信息数据标准 Standard for basic data of enterprise management of engineering construction field of Sichuan Province		J12806-2014	四川省建设科技发展中心
32	DBJ51/T 031-2014	预应力结构设计与施工技术规程 Technical specification for design and construction of prestressed structures		J12803-2014	四川省建筑科学研究院
33	DBJ51/T 032-2014	城镇道路排水工程施工安全技术规程 Safety technology rules for road drainage works in city and town construction		J12802-2014	中国市政工程西南设计研究总院有限公司
34	DBJ51/T 033-2014	四川省既有建筑电梯增设及改造技术规程 Technical specification for elevator adding or modifying in existing buildings in Sichuan Province		J12879-2014	四川省建筑设计研究院
35	DBJ51/T 034-2014	建筑用能合同能源管理技术规程 Technical code for energy performance contracting of building energy consumption system		J12878-2014	西南交通大学
36	DBJ51/T 035-2014	挤塑聚苯板建筑保温工程技术规程 Technical specification for thermal insulation engineering of building based on extruded polystyrene panel		J12921-2015	四川省建设科技协会
37	DBJ51/T 036-2015	四川省建筑工程现场安全文明施工标准化技术规程 Sichuan Province construction site safety civilization construction standard technical specification		J12918-2015	成都建筑工程集团总公司 四川省建设工程质量安全监督总站
38	DBJ51/T 037-2015	四川省绿色建筑设计标准 Design standard for green building in Sichuan Province		J12919-2015	中国建筑西南设计研究院有限公司
39	DBJ51/T 038-2015	四川省装配整体式住宅建筑设计规程 Design specification for assembled precast residential buildings in Sichuan Province		J12920-2015	四川省建筑设计研究院
40	DBJ51/T 039-2015	四川省民用建筑太阳能热水系统评价标准 Evaluation standard for solar water heating system of civil buildings in Sichuan Province		J13075-2015	西南交通大学
41	DBJ51/T 040-2015	四川省工程建设项目招标代理操作规程 Operational rules on construction projects procurement tendering agency in Sichuan Province		J13114-2015	四川省建设工程招标投标管理总站

四川省

序号	标准编号	标准名称	被代替标准编号	备案号	主编单位
42	DBJ51/T 041-2015	四川省建筑节能门窗应用技术规程 Technical specification for application of building window and door on energy efficiency in Sichuan Province		J13162-2015	四川省建筑科学研究院
43	DBJ51/T 042-2015	四川省建筑工程岩棉制品保温系统技术规程 Technical specification for thermal insulation composite systems of architectural engineering based on rock wool products in Sichuan Province		J13163-2015	四川省建材工业科学研究院
44	DBJ51/T 043-2015	民用建筑机械通风效果测试与评价标准 Standard of the measurement and evaluation for efficiency of civil building mechanical ventilation		J13161-2015	西南交通大学
45	DBJ51/T 044-2015	建筑边坡工程施工质量验收规范 Code for acceptance of constructionquality of building slope engineering in Sichuan Province		J13139-2015	四川省建筑科学研究院 成都市建工科学研究设计院
46	DBJ51/T 045-2015	四川省基桩承载力自平衡法测试技术规程 Technical code for self-balanced method of testing foundation pile bearing capacity in Sichuan Province		J13164-2015	四川省建筑科学研究院
47	DBJ51/T 046-2015	四川省建筑施工承插型钢管支模架安全技术规程 Technical code for safety of disk lock steel tubular scaffold in construction in Sichuan Province		J13171-2015	中国华西企业股份有限公司 成都市建设工程施工安全监督站
48	DBJ51/T 047-2015	四川省建筑工程设计信息模型交付标准 Delivery standard of building information model for design stage in Sichuan Province		J13199-2015	四川省建筑设计研究院
49	DBJ51/T 048-2015	四川省建设工程造价电子数据标准 Electronic data standard for cost of construction projects in Sichuan Province		J13189-2015	四川省建设工程造价管理总站
50	DBJ51/T 049-2015	四川省回弹法检测砖砌体中烧结普通砖抗压强度技术规程 Technical specification for inspecting of common sintered brick in baked brick compressive by rebound method in Sichuan Province		J13219-2015	四川省建筑科学研究院
51	DBJ51/T 050-2015	四川省回弹法检测砖砌体中砌筑砂浆抗压强度技术规程 Technical specification for inspecting of masonry mortar in baked brick compressive by rebound method in Sichuan Province		J13220-2015	四川省建筑科学研究院
52	DBJ51/T 051-2015	四川省水泥基泡沫保温板建筑保温工程技术规程 Technical specification for foamed cement insulation panel in Sichuan Province		J13293-2016	四川省建筑科学研究院 四川省建设科技协会
53	DBJ 51/052-2015	四川省养老院建筑设计规范 Design code for buildings of home for the aged in Sichuan Province		J13142-2015	四川省建筑设计研究院

四川省

序号	标准编号	标准名称	被代替标准编号	备案号	主编单位
54	DBJ51/T 053-2015	四川省智能建筑设计规范 Code for design of intelligent building in Sichuan Province		J13303-2016	中国建筑西南设计研究院有限公司 成都市建筑设计研究院
55	DBJ51/T 054-2015	四川省装配式混凝土结构工程施工与质量验收规程 Specification for construction and quality acceptance of precast concrete structures in Sichuan Province		J13329-2016	成都市土木建筑学会 成都建筑工程集团总公司
56	DBJ 51/055-2016	四川省高寒地区民用建筑供暖通风设计标准 Design standard for heating and ventilation of civil buildings in Sichuan alpine-cold zone		J13304-2016	中国建筑西南设计研究院有限公司
57	DBJ51/T 056-2016	四川省建筑工程绿色施工规程 Regulations for green construction of building in Sichuan Province		J13431-2016	成都市土木建筑学会 成都市第六建筑工程公司
58	DBJ51/T 057-2016	四川省住宅物业管理规程 Specification for residential property management in Sichuan Province		J13471-2016	成都市物业管理协会
59	DBJ51/T 058-2016	四川省公共建筑节能改造技术规程 Technical code for the retrofitting of public building on energy efficiency in Sichuan Province		J13499-2016	四川省建筑科学研究院
60	DBJ51/T 059-2016	四川省再生骨料混凝土及制品应用技术规程 Technical specification for application of recycled aggregate concrete and its products in Sichuan Province		J13519-2016	四川省建材工业科学研究院 成都市墙材革新建筑节能办公室
61	DB51/93-2013	振动（冲击）沉管灌注桩施工及验收规程 Code for construction and acceptance of vibration (impact) driven cast-in-place piles		J12489-2013	四川省建筑科学研究院
62	DB51/T 5012-2013	白蚁防治技术规程 The technical regulation for termite control		J12541-2014	成都市白蚁防治研究所
63	DB51/5016-98	四川省城市园林绿化技术操作规程			成都市园林局
64	DB51/T 5026-2001	成都地区建筑地基基础设计规范 Design code for building foundation of Chengdu Region		J10103-2001	成都市建筑设计研究院
65	DB51/5027-2012	四川省居住建筑节能设计标准 Design standard for energy efficiency of residential buildings in Sichuan Province		J10147-2012	中国建筑西南设计研究院有限公司
66	DB51/T 5032-2005	住宅供水"一户一表"设计、施工及验收技术规程 Technical specification of designing, construction and acceptance in residence water supply of ONE HOUSE ONE WATER-METER		J10569-2005	四川省城镇供排水协会
67	DB51/5033-2014	建筑节能工程施工质量验收规程 Specification for acceptance of energy efficient building construction		J12620-2014	四川省建筑科学研究院 成都市墙材革新建筑节能办公室
68	DB51/T 5034-2012	燃气用衬塑（PE）、衬不锈钢铝合金管道工程技术规程 Applying technical specification of plastic (PE) liner/stainless steel liner aluminum alloy composite pipeline for gas supply		J10772-2012	四川省燃气协会

四川省

序号	标准编号	标准名称	被代替标准编号	备案号	主编单位
69	DB51/T 5035-2012	燃气管道环压连接技术规程 Technical specification for ring compression connection for gas pipes		J10977-2012	四川省燃气协会
70	DB51/T 5036-2007	屋面工程施工工艺规程 Technical specification for roof engineering of construction		J11008-2007	四川建筑职业技术学院 四川省华西集团有限公司 四川省建设工程质量安全监督总站
71	DB51/T 5037-2007	防水工程施工工艺规程 Technical specification of construction for waterproofing engineering		J11009-2007	四川建筑职业技术学院 四川省华西集团有限公司 四川省建设工程质量安全监督总站
72	DB51/T 5038-2007	地面工程施工工艺规程 Technical specification of construction for floor engineering		J11010-2007	四川建筑职业技术学院 四川省华西集团有限公司 四川省建设工程质量安全监督总站
73	DB51/T 5039-2016	四川省砌体结构工程施工工艺规程 Technical specification of construction for masonry structures engineering in Sichuan Province		J11011-2016	四川建筑职业技术学院 四川华西集团有限公司
74	DB51/T 5040-2007	智能建筑工程施工工艺规程 Technical specification of construction for intelligent building engineering		J11012-2007	四川建筑职业技术学院 四川省华西集团有限公司 四川省建设工程质量安全监督总站
75	DB51/T 5041-2007	室外排水用高密度聚乙烯检查井工程技术规程 Technical specification of HDPE inspection chamber engineering for outdoor drainage		J11041-2007	中国市政工程西南设计研究院
76	DB51/T 5042-2007	复合保温石膏板内保温系统工程技术规程 Technical and engineering specification for composite plasterboard of interior thermal insulation system		J11037-2007	中国建筑西南设计研究院 四川省建设科技协会
77	DB51/T 5043-2007	建筑给水内筋嵌入式衬塑钢管管道工程技术规程 Technical specification of steel pipes with protruded line on inwall-liend plastics pipeline engineering for water supply in building		J11072-2007	四川省建筑设计院
78	DB51/T 5046-2014	混凝土结构工程施工工艺规程 Technical specification of construction for concrete engineering		J12872-2014	四川建筑职业技术学院
79	DB51/T 5047-2007	建筑电气工程施工工艺规程 Technical specification of construction in building-electricity engineering		J11084-2007	四川建筑职业技术学院 四川华西集团有限公司 四川省建设工程质量安全监督总站
80	DB51/T 5048-2007	地基与基础工程施工工艺规程 Technical specification of construction for subgrade and foundation		J11085-2007	四川建筑职业技术学院 四川华西集团有限公司 四川省建设工程质量安全监督总站
81	DB51/T 5049-2007	通风与空调工程施工工艺规程 Technical specification of construction for ventilation and air conditioning engineering		J11086-2007	四川建筑职业技术学院 四川华西集团有限公司 四川省建设工程质量安全监督总站

四川省

序号	标准编号	标准名称	被代替标准编号	备案号	主编单位
82	DB51/T 5051-2007	钢结构工程施工工艺规程 Technical specification of construction for steel engineering		J11127-2008	四川建筑职业技术学院 四川华西集团有限公司 四川省建设工程质量安全监督总站
83	DB51/T 5052-2007	建筑给水排水与采暖工程施工工艺规程 Technical specification of building water supply drainage and heating engineering		J11128-2008	四川建筑职业技术学院 四川华西集团有限公司 四川省建设工程质量安全监督总站
84	DB51/T 5053-2007	建筑装饰装修工程施工工艺规程 Technical specification of construction for building decoration		J11129-2008	四川建筑职业技术学院 四川华西集团有限公司 四川省建设工程质量安全监督总站
85	DB51/T 5054-2007	建筑给水薄壁不锈钢管管道工程技术规程 Technical specification of light gauge stainless steel pipeline engineering for building water supply		J11130-2008	中国建筑西南设计研究院
86	DB51/T 5055-2008	室外给水球墨铸铁管管道工程技术规程 Technical specification for ductile iron pipeline of outdoor water supply engineering		J11156-2008	中国市政工程西南设计研究院
87	DB51/T 5056-2008	室外给水钢丝网骨架塑料复合管管道工程技术规程 Technical specification of steel wire mesh and plastic composite pipeline for outdoor water supply engineering		J11157-2008	中国市政工程西南设计研究院
88	DB51/T 5057-2008	城市道路高分子复合材料检查井盖、水箅技术规程 Technical specification of polymer composite manhole lid and water grate for city road		J11174-2008	成都市市政工程协会
89	DB51/T 5058-2014	四川省抗震设防超限高层建筑工程界定标准 Appraisal standard for out-of-code tall building of seismic fortification in Sichuan Province		J12804-2014	四川省建筑设计研究院
90	DB51/T 5059-2015	四川省建筑抗震鉴定与加固技术规程 Technical specification for seismic appraisement and strengthening of building in Sichuan Province	DB51/T 5059-2008	J11251-2015	西南交通大学 四川省建筑科学研究院
91	DB51/T 5060-2013	预拌砂浆生产与应用技术规程 Technical specification for manufacture and application of ready-mixed mortar		J11246-2013	四川省建材工业科学研究院 成都市散装水泥办公室
92	DB51/T 5061-2015	水泥基复合膨胀玻化微珠建筑保温系统技术规程 Technical specification of thermal insulating rendering systems of buildings made of cement based mixed with expanded and vitrified tiny bead		J11301-2015	四川省建材工业科学研究院
93	DB51/T 5062-2013	EPS钢丝网架板现浇混凝土外墙外保温系统技术规程		J11316-2013	成都建筑工程集团总公司
94	DB51/T 5063-2009	在用建筑塔式起重机安全性鉴定标准 The standard for the appraisal of safe-state of tower-cranes in service in construction site		J11345-2009	四川省建筑科学研究院

四川省

序号	标准编号	标准名称	被代替标准编号	备案号	主编单位
95	DB51/T 5065-2009	建筑外窗、遮阳及天窗节能设计规程 Design specification for energy efficiency of building external windows, shading and skylights		J11530-2010	四川省建筑科学研究院
96	DB51/T 5066-2010	居住建筑油烟气集中排放系统应用技术规程 Technical specification for gas fumes centralized drainage system of residential building		J11538-2010	四川省建设工程质量安全监督总站
97	DB51/T 5067-2010	四川省地源热泵系统工程技术实施细则 Technical implementation details for Sichuan ground-source heat pump systems engineering in Sichuan Province		J11549-2010	四川省地质工程勘察院
98	DB51/T 5068-2010	既有玻璃幕墙安全使用性能检测鉴定技术规程 Technical specification for safety performance testing and appraisal of existing glass curtain walls		J11621-2010	四川省建筑科学研究院
99	DB51/5070-2010	先张法预应力高强混凝土管桩基础技术规程 Technical specification code for prestressed high concrete pipe pile foundation		J11712-2010	成都市建设工程质量监督站
100	DB51/T 5071-2011	蒸压加气混凝土砌块墙体自保温工程技术规程 Technical specification for self-insulation engineering of autoclaved aerated concrete blocks wall		J11794-2011	四川华西集团有限公司 成都市墙材革新建筑节能办公室
101	DB51/T 5072-2011	成都地区基坑工程安全技术规范 Technical code for retaining of foundation excavations in Chengdu region		J11927-2011	中国建筑西南勘察设计研究院有限公司 成都市建设工程施工安全监督站

云南省

序号	标准编号	标准名称	被代替标准编号	备案号	主编单位
1	DBJ 53/T-2-2016	普通混凝土配制技术规程 Technical specification for mix proportion design and production of ordinary concrete		J13565-2016	云南省建筑科学研究院
2	DBJ 53/T-3-2016	砌筑砂浆配制技术规程 Technical specification for mix proportion design and production of monsonry mortar		J13566-2016	云南省建筑科学研究院
3	DBJ 53-13-2012	烟草建筑消防设计规范 Code for fire protection design of tobacco buildings		J12285-2013	云南省消防协会
4	DBJ 53-14-2004	公路隧道消防技术规程 Technical specification for fire protection of highway tunnel			云南省消防协会
5	DBJ 53-15-2004	综合布线系统工程施工及验收技术规程 Technical specification for construction and acceptance of integrated wiring system engineering			云南省建筑科学研究院
6	DBJ 53/T-17-2007	云南省住宅小区及商住楼通信设施建设规范 Standard for construction of communication facilities for residential district and business-living building in Yunnan		J11146-2008	云南邮电规划设计院有限公司
7	DBJ 53-18-2008	太阳能热水系统与建筑一体化设计施工技术规程 Technical specification for integrated solar water heating system of building-design and installation		J11137-2008	昆明官房建筑设计有限公司
8	DBJ 53/T-19-2007	加芯搅拌桩技术规程 Technical specification for concrete core mixing pile		J11132-2008	云南大地工程开发公司
9	DBJ 53/T-20-2007	矿渣微粉混凝土（砂浆）技术规程 Applied technical specification of concrete and mortar with ground granulated blast furnace slag		J11147-2008	云南省混凝土协会
10	DBJ 53/T-21-2007	钢筋混凝土保护层塑料限位卡应用技术规程 Techical specication for plasclip of reinforced concrete cover		J11153-2008	云南官房建筑集团股份有限公司
11	DBJ 53/T-22-2008	先张法预应力混凝土管桩应用技术规程 Technical specification for application of pretensioned prestressed concrete pipe pile			云南建筑技术发展中心
12	DBJ 53/T-23-2014	云南省建筑工程施工质量验收统一规程 Yunnan Province specification for acceptance of building construction		J13111-2015	云南省工程质量监督管理站
13	DBJ 53/T-24-2008	云南省建筑工程质量优良等级评定标准 Yunnan Province construction engineering quality grade evaluation standard			云南省工程质量监督管理站
14	DBJ 53/T-25-2010	塑料排水检查井应用技术规程 Technical specification for application of plastic inspection well		J11999-2012	云南省市政工程质量检测站

云南省

序号	标准编号	标准名称	被代替标准编号	备案号	主编单位
15	DBJ 53/T-26-2010	建筑工程应用500MPa热轧带肋钢筋技术规程 Specification for hot rolled ribbed steel bar construction technology application 500MPa		J11998-2012	武钢集团昆明钢铁股份有限公司
16	DBJ 53/T-27-2010	室内无线电信号覆盖系统建设规范 Code for design of indoor wireless signal coverage system		J11997-2012	云南省城乡规划设计研究院
17	DBJ 53/T-28-2010	云南省液化燃气瓶装供应安全技术规程 Technical code for safety of bottled gas supply in Yunnan		J11996-2012	云南省市政工程质量检测站
18	DBJ 53/T-29-2010	云南省城镇照明工程安全生产规程 Yunnan Province safety rules in the production procedures of urban lighting project		J11995-2012	云南省市政工程质量检测站
19	DBJ 53/T-30-2010	变电站消防技术规程 Technical specification of fire protection for substations		J11994-2012	云南省公安消防总队（昆明市滇池路219号
20	DBJ 53/T-31-2011	云南省城镇污水处理厂运行维护及安全技术规程 Technical specification for operation, maintenance and safety of municipal waste		J11993-2012	云南省市政工程质量检测站
21	DBJ 53/T-32-2011	云南省城镇污水处理厂运行维护及安全评定标准 Evaluation standards for operation, maintenance and safety of municipal wastewater		J11992-2012	云南省市政工程质量检测站
22	DBJ 53/T-33-2011	云南省城镇排水设施运行维护及安全技术规程 Technical specification for operation, maintenance and safety of sewer facilitie		J11991-2012	云南省市政工程质量检测站
23	DBJ 53/T-34-2011	云南省城镇排水设施运行维护及安全评定标准 Standard for operation, maintenance and safety evaluation of urban drainage facilities in Yunnan		J11990-2012	云南省市政工程质量检测站
24	DBJ 53/T-35-2011	磷渣微粉混凝土技术规程 Technical specification for phosphorous slag power concrete		J11989-2012	云南省建筑科学研究院
25	DBJ 53/T-36-2011	云南省市政基础设施工程施工质量验收统一规程 Yunnan Province specification for acceptance of municipal infrastructure project		J11988-2012	云南省市政工程质量检测站
26	DBJ 53/T-37-2011	云南省市政基础设施工程施工质量优良等级评定标准 Yunnan municipal infrastructure engineering construction quality evaluation criteria		J11987-2012	云南省市政工程质量检测站
27	DBJ 53/T-38-2011	云南省建设工程造价成果文件数据标准 Data standard for construction cost documents and files of Yunnan Province		J11986-2012	云南省建设工程造价管理协会
28	DBJ 53/T-39-2011	云南省民用建筑节能设计标准 Design standard for energy efficiency of civil buildings		J11995-2012	云南省安泰建设工程施工图设计文件审查中心

云南省

序号	标准编号	标准名称	被代替标准编号	备案号	主编单位
29	DBJ 53/T-40-2011	云南省城镇园林工程施工质量验收规程 Specification for acceptance of construction quality of urban landscape engineering in Yunnan		J13110-2015	云南省市政工程质量检测站
30	DBJ 53/T-41-2011	云南省城镇园林工程安全生产规程 Yunnan Province safety specification of landscape engineering in cities and towns		J13109-2015	云南省市政工程质量检测站
31	DBJ 53/T-42-2011	东川河砂配制混凝土及应用技术规程 Technical specification for application of dongchuan sand in concrete		J13105-2015	云南建工混凝土有限公司
32	DBJ 53/T-43-2011	砂衬齿形桩应用技术规程 Technical specification for application of sand-lined pile			云南省建筑工程设计院
33	DBJ 53/T-44-2011	云南省建筑工程资料管理规程 Specification for building engineering document management of Yunnan Province		J13104-2015	云南省工程质量监督管理站
34	DBJ 53/T-45-2011	云南省建设工程档案编制技术规程 Technical specification for construction projet archive filing in Yunnan		J13103-2015	云南省工程质量监督管理站
35	DBJ 53/T-46-2012	云南省城镇道路及夜景照明工程施工及验收规程 Specification for construction and inspection of Yunnan Province urban road and nightscape lighting engineering		J13102-2015	云南省市政工程质量检测站
36	DBJ 53/T-47-2012	建筑工程叠层橡胶隔震支座性能要求和检验规范 Code for performance requirement and test of laminated rubber seismic isolation bearing for buildings		J13099-2015	云南震安减震科技股份有限公司
37	DBJ 53/T-48-2012	建筑工程叠层橡胶隔震支座施工及验收规范 Code for construction and acceptance of laminated rubber seismic isolation bearing for buildings		J13098-2015	云南震安减震科技股份有限公司
38	DBJ 53/T-49-2015	云南省绿色建筑评价标准 Yunnan Provicial evaluation standard for green building		J13112-2015	云南省建筑科学研究院
39	DBJ 53/T-50-2013	云南省建筑工程结构实体检测技术规程 Technical specification for construction engineering structural entity detection of Yunnan Province		J13097-2015	云南省建设工程质量检测中心有限公司
40	DBJ 53/T-51-2013	云南省建设工程质量检测信息监管系统技术规程 Yunnan construction engineering quality inspection infrkrmation management system technical specification		J13096-2015	云南省工程质量监督管理站
41	DBJ 53/T-52-2013	回弹法检测混凝土抗压强度技术规程 Technical specification for inspecting of concrete compressive strength by rebound method		J13095-2015	云南特斯泰工程检测鉴定有限公司
42	DBJ 53/T-53-2013	超声回弹综合法检测混凝土强度技术规程 Technical specification for detecting strength of concrete by ultrasonic-rebound combined method		J13094-2015	云南特斯泰工程检测鉴定有限公司

云南省

序号	标准编号	标准名称	被代替标准编号	备案号	主编单位
43	DBJ 53/T-54-2013	云南省房地产档案管理技术规程 Regulation of Yunnan Province real estate archives management technology		J13093-2015	云南省住房和城乡建设厅城镇档案工作办公室
44	DBJ 53/T-55-2013	云南省城市管线探测技术规程 Yunnan technical specification for detecting and surveying urban pipelines and cables		J13092-2015	云南省市政工程质量检测站
45	DBJ 53/T-56-2013	云南省河道治理及疏浚工程施工质量验收规程 Yunnan Province specification for acceptance of constructional quality of river rehabilitation and dredging		J13101-2015	云南省市政工程质量检测站
46	DBJ 53/T-57-2013	云南省市政基础设施工程资料管理规程 Yunnan provincial specification for data and information management of municipal infrastructure projects		J13100-2015	云南省市政工程质量检测站
47	DBJ 53/T-58-2013	云南省建设工程造价计价规则及机械仪器仪表台班费用定额		J12613-2014	云南省工程建设技术经济室 云南省建设工程造价管理协会
48	DBJ 53/T-59-2013	云南省市政工程消耗量定额		J12612-2014	云南省工程建设技术经济室 云南省建设工程造价管理协会
49	DBJ 53/T-60-2013	云南省园林绿化工程消耗量定额		J12611-2014	云南省工程建设技术经济室 云南省建设工程造价管理协会
50	DBJ 53/T-61-2013	云南省房屋建筑与装饰工程消耗量定额		J12610-2014	云南省工程建设技术经济室 云南省建设工程造价管理协会
51	DBJ 53/T-62-2013	云南省城市轨道交通工程消耗量定额		J12609-2014	云南省工程建设技术经济室 云南省建设工程造价管理协会
52	DBJ 53/T-63-2013	云南省通用安装工程消耗量定额		J12608-2014	云南省工程建设技术经济室 云南省建设工程造价管理协会
53	DBJ 53/T-64-2013	云南省房屋修缮及仿古建筑工程消耗量定额		J12607-2014	云南省工程建设技术经济室 云南省建设工程造价管理协会
54	DBJ 53/T-65-2014	云南省预制装配式钢筋混凝土检查井制作安装与质量验收规程 Yunnan Provincial specification for fabrication assembly and quality acceptance of prefabricated reinforced concrete manhole		J13091-2015	云南省市政工程质量检测站
55	DBJ 53/T-66-2014	既有建筑结构安全性检测与鉴定技术标准 Technical standard for inspection and assessment for structural safety of existing buildings		J13090-2015	云南省建筑科学研究院
56	DBJ 53/T-67-2014	建筑基坑工程监测技术规程 Technical specification for monitoring of building		J13089-2015	云南省建筑科学研究院
57	DBJ 53/T-68-2014	铝合金模板应用技术规程 Application technical code of aluminum alloy formwork		J13088-2015	云南科保模架有限责任公司
58	DBJ 53/T-69-2014	云南省建筑与市政基础设施工程施工现场专业（管理）人员配备标准 Construction site professional (managerial) staffing standards for building and municipal infrastructure engineering		J13087-2015	云南省工程建设技术经济室

云南省

序号	标准编号	标准名称	被代替标准编号	备案号	主编单位
59	DBJ 53/T-70-2015	建筑隔震工程专用标识技术规程 Technical specification of sign and noting system of basis-isolated project		J13086-2015	云南震安减震科技股份有限公司
60	DBJ 53/T-71-2015	云南省建筑基坑支护技术规程 Technical specification for retaining and protection of building foundation excavations		J13085-2015	云南省建筑科学研究院
61	DBJ 53/T-72-2015	额定电压35kV及以下铝合金电力电缆应用技术规程 Technical application specification of aluminum alloy power cables with rated voltage up to 35kV		J13084-2015	昆明市建筑设计研究院有限责任公司
62	DBJ 53/T-73-2015	云南省施工现场标准员管理规程 Management code for construction site standardization engineer in Yunnan Province		J13562-2016	云南省工程建设技术经济室
63	DBJ 53/T-74-2015	建筑基坑钢板桩支护技术规程 Technical specification for retaining and protection of steel sheet piling in building foundation excavations		J13563-2016	云南省建筑科学研究院
64	DBJ 53/T-75-2015	石屑在混凝土中的应用技术规程 Technical specification for application of limestone chippings in concrete		J13564-2016	云南省建筑科学研究院
65	DBJ 53/T-76-2016	建筑施工轮扣式钢管支架安全技术规程 Technical specification for steel tubular scaffold flexible lock with coupler connected in construction		J13567-2016	云南建工第五建设有限公司
66	DBJ 53/T-77-2016	云南省城镇园林绿化养护技术规程 Urban landscape architecture project in Yunnan Province maintenance and technical regulation		J13568-2016	云南省市政工程质量检测站
67	DBJ 53/T-78-2016	云南省城镇古树名木及后续资源管养技术规程 Technical regulation for urban ancient and famous trees and reserve resources maintenance in Yunnan Province		J13569-2016	云南省市政工程质量检测站
68	DBJ 53/T-79-2016	超缓凝混凝土配制及应用技术规程 Technical specifications for application of ultra-retardation concrete			云南省建设投资控股集团有限公司

贵州省

序号	标准编号	标准名称	被代替标准编号	备案号	主编单位
1	DBJ 22-01-87	贵州省建筑气象参数标准			修编中
2	DBJ 22-04-92	贵阳地区庭院及户内煤气管道技术暂行规定			修编中
3	DBJ 22-11-2000	人工挖孔灌注桩施工技术规程			修编中
4	DBJ 52-016-2010	山砂混凝土技术规程 Technical specification of rock sand concrete	DB 22-016-95	J11802-2011	贵州中建建筑科研设计院有限公司
5	DBJ52/T 017-2014	回弹法检测山砂混凝土抗压强度技术规程 Local standards of Guizhou Province determine of compression strength of rock-sand concrete by rebound hammer method technical specifications		J12555-2014	贵州中建建筑科研设计院有限公司
6	DBJ 22-33-2000	贵州省Ⅷ、Ⅸ内河航道通航标准			修编中
7	DBJ 22-35-2001	建筑给水用聚丙烯（PPR）管道应用技术规程 Engineering technical code of water supply polypropylee pipeline for buildings			贵州省建筑设计研究院
8	DBJ 22-37-2001	型钢井架物料提升机安全技术规程 Safety technicial specification of shape steel headframe stuff hoist			贵州建工集团总公司
9	DBJ 22-39-2002	室外埋地聚乙烯（PE）给水管道工程技术规程 Technical code for buried polyethylene pipeline for water supply engineering			贵州省建筑设计研究院
10	DBJ 22-40-2002	贵州省燃气管道安装工程质量检验评定标准			修编中
11	DBJ 22-41-2004	贯入法检测山砂砌筑砂浆抗压强度技术规程			修编中
12	DBJ 22-42-2004	钻取小芯样法检测山砂混凝土强度技术规程 The technicial specification of pit sand concrete strength testing by small core-drilling method			贵州中建建筑科研设计院有限公司
13	DBJ 22-43-2004	超声回弹综合法检测山砂混凝土强度技术规程			修编中
14	DBJ 22-44-2005	回弹法检测高强山砂混凝土强度技术规程			修编中
15	DBJ 22-45-2004	贵州建筑地基基础设计规范			贵州省建筑设计研究院
16	DBJ 22-46-2004	贵州建筑岩土工程技术规范 Technical code for geotechnical engineering of building of Guizhou			贵州省建筑设计研究院 原贵州工业大学
17	DBJ 22-47-2005	建筑施工落地双排外竹脚手架技术规程 Technical code for double arrange external bamboo scaffold for ground			贵州建工集团第九建筑工程公司
18	DBJ 22-48-2005	钢筋混凝土空腹夹层板楼盖结构技术规程 Technicial specification for reinforced concrete open-web sandwich plate structures			贵州大学空间结构研究所
19	DBJ 52-49-2008	贵州省居住建筑节能设计标准 Design standard for energy efficiency of residential buildings of Guizhou	DB 22/49-2005	J10665-2009	贵州省建筑设计研究院
20	DBJ52/T 50-2006	贵州省城市容貌标准 The townscape criterion of Guizhou Province		J10953-2007	贵州省环境卫生协会
21	DBJ 52-51-2007	闪光对焊箍筋施工技术规程 Technical specification for flash butt welding stirrup		J10939-2007	贵州省建设工程质量监督总站

贵州省

序号	标准编号	标准名称	被代替标准编号	备案号	主编单位
22	DBJ 52-53-2008	城镇山体公园化绿地设计规范		J11352-2009	贵州省园林协会
23	DBJ52/T 055-2015	贵州省高速公路机制砂高性能混凝土技术规程 Specification for highway manufactured-sand high-performance concrete in Guizhou Province		J11361-2015	贵州高速公路集团有限公司 同济大学 贵州省交通规划勘察设计研究院股份有限公司
24	DBJ52/T 057-2011	蒸压磷渣硅酸盐砖建筑技术规程 Technical code for autoclaved phosphrous slag silicate brick masonry buildings		J11894-2011	贵州开磷设计研究院
25	DBJ52/T 058-2012	喀斯特地区灌木护坡技术规范 Technical code for protecting slope with bush in karst area		J12092-2012	贵州科农生态环保科技有限责任公司
26	DBJ52/T 060-2012	贵州省绿色小城镇建设评价标准 Evaluation standard of green construction of small towns in Guizhou Province		J12147-2012	贵州省城乡规划设计研究院
27	DBJ 52-062-2013	贵州省坡地民用建筑设计防火规范 Code for fire protection design of hillside buildings of Guizhou		J12263-2013	贵州省公安消防总队 贵州省建筑设计研究院
28	DBJ52/T 063-2013	建筑节能工程施工质量验收规范贵州省实施细则 Actualize methods of code for acceptance of energy efficient building construction in Guizhou		J12393-2013	贵州省建筑科学研究检测中心 贵州中建建筑科研设计院有限公司
29	DBJ52/T 064-2013	地源热泵系统施工质量验收技术规程 Technical specification for acceptance of ground source heat pump system		J12392-2013	贵州省建筑科学研究检测中心
30	DBJ52/T 065-2013	绿色建筑评价标准（试行） Evaluation standard for green building (for trial implementation)		J12453-2013	贵阳市建筑设计院有限公司
31	DBJ/T 52-066-2014	钢丝网架夹芯整体墙技术规程 Technical specification for wire rack sandwich wall		J12827-2014	贵州省建筑设计研究院 贵州显顺建材有限公司
32	DBJ52/T 067-2015	贵州省城镇燃气用二甲醚应用技术标准 Technical standard for application of dimethyl ether for city gas in Guizhou Province		J12954-2015	贵州省城乡规划设计研究院
33	DBJ52/T 068-2015	灌注桩成孔质量检测技术规程 Technical specification for quality inspection of bored piles		J13169-2015	贵州道兴建设工程检测有限责任公司 贵州中建建筑科研设计院有限公司
34	DBJ52/T 069-2015	建筑工程绿色施工管理规程 Management specification for green construction of building		J13170-2015	贵州建工集团有限公司 贵州中建建筑科研设计院有限公司 贵阳市住房和城乡建设局
35	DBJ52/T 070-2015	建筑反射隔热涂料应用技术规程 Technical specification for application of thermal insulation systems of architectural reflective thermal insulation coatings		J13229-2015	贵州省建材产品质量监督检验院
36	DBJ52/T 071-2016	住宅厨房、卫生间排气道系统技术规程 Kitchen and toilet exhaust ventilation system technical specification		J13157-2016	贵州大学 贵州省计量测试院 贵州中建建筑科研设计院有限公司

贵州省

序号	标准编号	标准名称	被代替标准编号	备案号	主编单位
37	DBJ52/T 072-2015	贵州省种植屋面冷却应用规程 Technical specification for application of planting roof in Guizhou		J13278-2015	中国建筑科学研究院 贵州中建建筑科研设计院有限公司
38	DBJ52/T 073-2015	空气源热泵系统现场测试规程 Code for field test of air source heat pump system		J13279-2015	北京工业大学 中国建筑科学研究院 贵州中建建筑科研设计院有限公司
39	DBJ52/T 074-2016	贵州省建设工程质量检测技术管理规范（房屋建筑和市政基础设施） Local standards of Guizhou Province code of testing tcechnology management（Building and municipal infrastructure engineering quality）		J13280-2016	贵州省建筑科学研究检测中心 贵州省建设工程质量安全监督总站 贵州省建设工程质量检测协会
40	DBJ52/T 075-2016	贵州省公共建筑能耗监测技术规范 Technical specifications of energy consumption monitoring for public building of Guizhou Province		J13327-2016	贵州省建筑科学研究检测中心 贵州中建建筑科研设计院有限公司
41	DBJ52/T 076-2016	贵州省建设工程质量检测报告编制规程（房屋建筑和市政基础设施部分） Local standards of Guizhou Province determine of test reports（Building and municipal infrastructure engineering quality）		J13335-2016	贵州省建筑科学研究检测中心 贵阳永兴建设工程质量检测有限公司 贵州省建设工程质量检测协会 贵州省建设工程质量安全监督总站
42	DBJ52/T 077-2016	贵州省民用建筑绿色设计规范（试行） Code for green design of civil building in Guizhou Province（for trial implementation）		J13356-2016	贵州省建筑设计研究院 贵州中建建筑科研设计院有限公司 贵州省城乡规划设计院 贵阳市建筑设计院有限公司
43	DBJ52/T 078-2016	贵州省绿色生态城区评价标准 Assessment standard for green eco-district in Guizhou Province		J13438-2016	贵州省城乡规划设计研究院
44	DBJ52/T 079-2016	基桩承载力自平衡检测技术规程 Self-balanced load pile foundation technical specification for testing		J13526-2016	贵州中建建筑科研设计院有限公司 贵州道兴建设工程检测有限责任公司

陕西省

序号		标准编号	标准名称	被代替标准编号	备案号	主编单位
1	9	DBJ 61-9-2008	强夯法处理湿陷性黄土地基技术规程 Technical specifications for collapsible loess ground treatment by dynamic compaction		J11247-2008	陕西省建筑科学研究院
2	46	DBJ/T 61-46-2007	回弹法检测泵送混凝土抗压强度技术规程 Technical specificationg for inspecting of pumped compressive strength by rebound method		J11052-2007	陕西省建筑科学研究院
3	47	DBJ 61-47-2008	地铁盾构隧道预制混凝土管片 Prefabricated reinforced concrete segment for metro tunnel		J11170-2008	西安地下铁道有限责任公司
4	47	DBJ 61/T-47-2014	轨道交通预应力混凝土预制梁施工及验收标准 Construction and acceptance criteria for prestressed concrete prefabricated beam of rail transport		J12538-2014	西安市地下铁道有限责任公司 西安科技大学
5	48	DBJ 61-48-2008	陕西省房屋建筑震后重建抗震设防暂行规定 Shaanxi temporary regulation for aseismic design of building reconstructiong in seismic area		J11253-2008	中建西北院
6	49	DBJ/T 61-49-2008	建筑施工拱构型门式钢管脚手架安全技术规程 Technical specification for safety of steel tubular scaffold with arch gabled frames in construction		J11248-2008	西安建筑科技大学
7	51	DBJ 61-51-2009	村庄规划技术规范 Technical code for village planning		J11421-2009	陕西中晟规划设计研究院
8	52	DBJ 61-52-2009	城镇道路和建筑物无障碍设施施工质量验收规程 Specification for construction quality acceptance of barrier-free facilities for urban roads and buildings		J11422-2009	西安市市政设计研究院 西安市建筑设计研究院
9	53	DBJ/T 61-53-2009	芳纶纤维布加固钢筋混凝土结构技术规程 Technical specification for strengthening reinforced concrete structures with aramid fiber sheets		J11446-2009	长安大学
10	54	DBJ/T 61-54-2009	箍筋闪光对焊技术规程 Technical specification for flash butt welding of stirrup		J11600-2010	陕西建工集团有限公司
11	55	DBJ/T 61-55-2009	外墙外保温技术规程——胶粉聚苯颗粒复合型保温系统（第一部分） Technical specification for external thermal insulation on walls Part 1：Composited thermal insulation systems made of rendering with mineral binder and using expanded polystyrene granule as aggregate		J11619-2010	陕西省建筑标准设计办公室
12	56	DBJ/T 61-56-2009	火探管式气体灭火系统技术规范 Code for technology of gas fire-extinguishing system uesd fire trace tube		J11601-2010	陕西省公安厅消防局

陕西省

序号		标准编号	标准名称	被代替标准编号	备案号	主编单位
13	57	DBJ 61-57-2010	建筑场地墓坑探查与处理技术规程 Technical specification for exploration and treatment of grave-pits on construction sites		J11803-2011	陕西建工集团总公司 西北综合勘察设计研究院 中国建筑西北设计研究院
14	58	DBJ/T 61-58-2010	施工现场临时性建筑物应用技术规程 Technical specification for temporary work operation		J11823-2011	陕西省建设工程质量安全监督总站
15	59	DBJ/T 61-59-2010	民用建筑有线电视系统工程技术规程 Code for CATV system engineering technology of civil buildings		J11824-2011	陕西广电网络传媒股份有限公司 中国建筑西北设计研究院有限公司 陕西省建筑设计研究院有限责任公司
16	60	DBJ 61-50-2008	城市绿化养护技术规程 The city afforestation & maintenance techniques rules and regulations		J11249-2008	陕西唐荣园林建设集团有限公司
17	60	DBJ/T 61-60-2011	西安市公共建筑节能设计标准 Design standard for energy efficiency of public buildings in xi'an		J11838-2011	西安市城乡建设委员会
18	63	DBJ/T 61-63-2011	住宅建筑门窗应用技术规范 Technical specification for doors and windows of housing construction		J12044-2012	陕西省产品质量监督检验所
19	64	DBJ/T 61-64 2011	液化天然气汽车加气站设计与施工规范 Code for design and construction of automotive liquified natural gas filling station		J12045-2012	陕西省燃气设计院 陕西省消防总队
20	65	DBJ-61-65-2011	西安市居住建筑节能设计标准 Design standard for energy efficiency of residential buildings in Xian		J10943-2007	中国建筑西北设计研究院 西安市建筑设计研究院 西安市建设工程质量检测中心
21	65	DBJ 61-65-2011	居住建筑节能设计标准 Design standard for energy efficiency of residential buildings		J12026-2012	陕西省建筑标准设计办公室 西安市城乡建设委员会 西安建筑科技大学
22	66	DBJ/T 61-66-2011	陕南地区移民搬迁公共设施建设标准 Relocation of public facilities construction standard for south of Shaanxi Province		J12022-2012	陕西省城乡规划设计研究院 陕西省建筑标准设计办公室
23	67	DBJ/T 61-67-2011	陕北地区移民自建住房技术规范 Relocation selfbuilding housing technical specifications of Shanbei Immigration		J12021-2012	中联西北工程设计研究院 陕西省建筑标准设计办公室
24	68	DBJ/T 61-68-2012	既有村镇住宅抗震加固技术规程 Standard of seismic strengthening technology for existing residence in villages		J12073-2012	长安大学 陕西省住房和城乡建设厅
25	69	DBJ 61/T69-2012	西安市既有公共建筑节能改造技术规范 Technical specification for existing public building energy conservation transform		J12178-2012	西安市城乡建设委员会
26	70	DBJ 61 70-2012	西安市民用建筑太阳能热水系统应用技术规范 Technical code for solar water heating system of xi'an civil buildings		J12202-2012	西安市城乡建设委员会 中国建筑西北设计研究院有限公司
27	71	DBJ 61/T71-2012	西安市既有居住建筑节能改造技术规范		J12220-2012	西安市城乡建设委员会

陕西省

序号		标准编号	标准名称	被代替标准编号	备案号	主编单位
28	72	DBJ 61/T72-2012	城市道路交通管理设施设置技术规范 第1部分：总则 Technical code for urban road traffic management facilities Part 1: General provisions		J12253-2013	西安市政设计研究院有限公司
29	73	DBJ 61/T73-2013	高强混凝土非标准试件抗压强度换算方法 Conversion method of the nonstandard specimens compressive strength of high strength concrete		J12365-2013	陕西省建筑科学研究院
30	74	DBJ 61/T74-2013	卫生间模块化同层排水节水系统应用技术规程 Technical specification for modular same-floor drainage & water-saving system of bathrooms		J12434-2013	陕西省建筑节能与墙体材料改革办公室 中国建筑西北设计研究院有限公司
31	75	DBJ 61/T75-2013	岩棉板外墙外保温系统应用技术规范 Application technology code of external thermal insulation systems based on rock wool board		J12457-2013	西安市城乡建设委员会
32	76	DBJ 61/T76-2013	农村基础设施技术规范 Technical code for rural infrastructure		J12467-2013	西安建筑科技大学
33	77	DBJ 61/T-77-2013	西安市汽车库、停车场设计防火规范 Code for fire protection design of garage and parking area of Xi'an		J12539-2014	西安市城乡建设委员会 西安市公安消防支队 中国建筑西北设计研究院有限公司
34	78	DBJ 61-78-2013	西安市民用建筑太阳能光伏系统应用技术规范 Technical code for application of solar photovoltaic system of Xi'an civil buildings		J12626-2014	西安市城乡建设委员会 陕西省建筑科学研究院
35	79	DBJ 61/T-79-2014	地铁盾构隧道预制管片施工及验收标准 Prefabricated and acceptance criteria for lining segment of metro shield tunnel		J12540-2014	西安市地下铁道有限责任公司 西安科技大学
36	80	DBJ 61/T-80-2014	公共建筑绿色设计标准 Standard for green design of public buildings		J12595-2014	中国建筑西北设计研究院有限公司
37	81	DBJ 61/T81-2014	居住建筑绿色设计标准 Design standard for green residential building		J12598-2014	西安建筑科技大学建筑设计研究院 西安建筑科技大学
38	82	DBJ 61/T-82-2014	可再生能源建筑应用项目验收规程 Code for acceptance of application of renewable energy in buildings		J12597-2014	陕西省建筑科学研究院
39	83	DBJ 61/T83-2014	绿色生态居住小区建设评价标准（试行） Evaluation standard for ecological residential district designs		J12658-2014	中联西北工程设计研究院
40	84	DBJ 61/T-84-2014	建筑与小区雨水利用技术规程 Engineering technical code for rain utilization in building and sub-district		J12596-2014	中国建筑西北设计研究院有限公司 西安市建筑设计研究院
41	85	DBJ 61/T-85-2014	西安市普通预拌砂浆生产与应用技术规程 Technical specification for manufacture and application of ordinary ready-mixed mortar in Xi'an		J12657-2014	西安市城乡建设委员会 长安大学

陕西省

序号		标准编号	标准名称	被代替标准编号	备案号	主编单位
42	86	DBJ 61/T86-2014	泡沫混凝土板外墙外保温系统应用技术规程 Technical specification for external thermal insulation systems of composite foam concrete panel		J12669-2014	陕西省建筑科学研究院 陕西省建筑节能与墙体材料改革办公室
43	87	DBJ 61/T87-2014	装配整体式混凝土结构技术规程（暂行） Technical specification for assembled monolithic concrete structure (Trial)		J12667-2014	西安建工（集团）有限责任公司 西安建筑科技大学
44	88	DBJ 61/T88-2014	再生混凝土结构技术规程 Technical specification for recycled concrete structures		J12668-2014	西安建筑科技大学 西安市建筑工程总公司
45	89	DBJ 61/T-89-2014	轻质蒸压砂加气混凝土砌块及板材技术规程 Technical specification for lightweight sand autoclaved aerated concrete blocks and plates		J12707-2014	陕西省建筑科学研究院 长安大学 陕西凝远绿色建材实业有限责任公司
46	90	DBJ 61/T90-2014	彩钢夹芯板临时建筑防火技术规程 Technical specification for fire protection of color steel sandwich panel temporary building		J12793-2014	陕西省公安消防总队 陕西省建筑标准设计办公室
47	91	DBJ 61/T91-2014	西安地区农村居住建筑节能技术规范 Code for energy efficiency of rural residential buildings in Xi'an		J12883-2014	西安市城乡建设委员会 西安建筑科技大学
48	92	DBJ 61/T92-2014	建筑滑移隔震技术规范 Technical code for sliding isolation of buildings		J12851-2014	陕西省建筑科学研究院
49	93	DBJ 61/T93-2014	坡地民用建筑设计防火规范 Code for fire protection design of hillside buildings		J12908-2015	陕西省公安消防总队 中国建筑西北设计研究院有限公司
50	94	DBJ 61/T94-2015	装配式复合墙结构技术规程 Technical specification for precast composite wall structure		J12968-2015	西安建筑科技大学 中天西北建设投资集团有限公司 中国中铁一局集团有限公司
51	96	DBJ 61/T96-2015	西安市简易自动喷水灭火系统设计规范 Code of design regulations for simple sprinkler systems		J12952-2015	西安市公安消防支队 中国建筑西北设计研究院有限公司
52	97	DBJ 61/T97-2015	西安市公共建筑能耗监测系统技术规范 Technical code for the energy consumption monitoring systems of public buildings in Xi'an		J12969-2015	西安市城乡建设委员会 中国建筑西北设计研究院有限公司
53	98	DBJ 61-98-2015	西安城市轨道交通工程监测技术规范 Code for monitoring of xi'an urban rail transit engineering		J13069-2015	机械工业勘察设计研究院有限公司 西安市地下铁道有限责任公司
54	99	DBJ 61/T99-2015	冷弯薄壁型钢—石膏基砂浆复合墙体技术规程 Technical specification for cold-formed steel framed composite walls with gypsum-based mortar		J13020-2015	西安建筑科技大学
55	100	DBJ 61/T100-2015	工程结构振动测试技术规程 Technical specification for testing engineering structures vibration		J13081-2015	陕西省建筑科学研究院
56	101	DBJ 61/T101-2015	预应力混凝土管桩基础技术规程 Technical specification for pre-stressed concrete pipe pile foundation		J13240-2015	中国有色金属工业西安勘察设计研究院

陕西省

序号		标准编号	标准名称	被代替标准编号	备案号	主编单位
57	102	DBJ 61/T102-2015	沉管夯扩桩技术规程 Technicai specification for pipe sinking compacted base-enlarged pile		J13239-2015	中国有色金属工业西安勘察设计研究院
58	103	DBJ 61-103-2015	DP型烧结多孔砖砌体结构技术规程 Technical specification for masonry structures		J13263-2015	西安墙体材料研究设计院
59	104	DBJ 61/T104-2015	陕西省村镇建筑抗震设防技术规程 Seismic technical specification for building construction in Shaanxi Province		J13301-2016	长安大学
60	105	DBJ 61/T105-2015	建筑基坑支护技术与安全规程 Specification of technology and safety for retaining and protection of building foundation excavations		J13302-2016	西安市建设工程质量安全监督站 陕西省建设工程质量安全监督总站
61	106	DBJ 61/T106-2015	村镇砌体结构民居叠层橡胶支座隔震技术规程 Technical specification for laminated rubber bearing isolators of rural masonry structure dwelling		J13368-2016	长安大学
62	107	DBJ 61/T107-2015	西安市城镇道路太阳能光伏LED路灯照明技术规范 Technical specification of Xi'an town road solar photovoltaic（PV）		J13313-2016	西安市城乡建设委员会 西安市市政公用局
63	108	DBJ 61/T108-2015	陕西省新型农村社区规划建设标准 New rural community planning & construction standard		J13366-2016	陕西省城乡规划设计研究院
64	109	DBJ 61/T109-2015	陕西省村庄规划编制技术规范 Technical specification for village planning of Shaanxi Province LED street lamp lighting		J13314-2016	陕西省城乡规划设计研究院
65	110	DBJ 61/T110-2015	城市公园分级标准 Standard for classification of urban park		J13429-2016	西安市城市管理局
66	111	DBJ 61/T111-2016	西安市幼儿园及儿童活动场所消防安全技术规范		J13367-2016	西安市公安消防支队 中国建筑西北设计研究院有限公司
67	112	DBJ 61/T112-2016	高延性混凝土应用技术规程		J13464-2016	西安建筑科技大学 西安五和土木工程新材料有限公司
68	113	DBJ 61/T113-2016	城市轨道交通隧道穿越地裂缝段技术规范 Technical specification for urban rail transit tunnel crossing ground fracture belt		J13465-2016	长安大学 西安市地下铁道有限责任公司 中铁第一勘察设计院集团有限公司
69	114	DBJ 61/T114-2016	城市轨道交通地下工程防水施工技术规程		J13466-2016	长安大学 西安市地下铁道有限责任公司
70			建筑节能工程施工质量验收规程 Code for quality acceptance of energy efficient building construction		J11076-2007	陕西省建筑科学研究院

甘肃省

序号	标准编号	标准名称	被代替标准编号	备案号	主编单位
1	DB62/T 25-3016-2016	建筑工程资料管理规程 Specification for building engineering document management	DB62/T 25-3016-2005	J10583-2016	甘肃省建设工程安全质量监督管理局 甘肃省工程质量施工安全管理协会
2	DB62/T 25-3040-2009	轻质混凝土吸音块砌体工程施工工艺规程 Construction process for lightweighe-concrete and sound-absorbing brick of engineering		J11375-2009	甘肃省第三建筑工程公司
3	DB62/T 25-3041-2009	钢管混凝土结构技术规程 Technical specification for concrete-filled steel tubular structures		J11395-2009	兰州理工大学 兰州市城市建设设计院
4	DB/T 25-3042-2010	甘肃省农村房屋建设质量验收标准（试行）			甘肃省建筑工程总公司
5	DB62/T 25-3043-2009	砖砌体结构抗震设计规程 Code for seismic design of brick masonry structure		J11514-2009	甘肃建设工程咨询设计有限责任公司
6	DB62/T 25-3044-2009	民用建筑集中采暖供热计量技术规程 Technical specification for heat metering of central heating system		J11515-2009	甘肃省建设科技专家委员会 甘肃省建筑设计研究院 榆中县供热管理站
7	DB62/T 25-3045-2009	大空间智能型主动喷水灭火系统设计规程 Code of design for large-space intelligent active control sprinkler systems		J11520-2009	甘肃省建筑设计研究院
8	DB62/T 25-3046-2009	型钢悬挑扣件式钢管脚手架施工工艺规程 Construction process for steel cantilever fastening steel pipe scaffold		J11532-2010	甘肃省第三建筑工程公司
9	DB62/T 25-3047-2010	城镇生活饮用水二次供水工程技术规程 Technical standard for secondary town potable water supply engineering		J11694-2010	兰州交通大学
10	DB62/T 25-3048-2010	甘肃省城镇规划管理技术规程（试行）		J11754-2010	甘肃省城市科学研究会
11	DB62/T 25-3049-2010	甘肃省民用建筑与太阳能热水系统一体化设计规程 Design standard for building integrated solar water heating system of Gansu civil buildings		J11812-2011	兰州交通大学 甘肃第三建设集团公司
12	DB62/T 25-3050-2010	甘肃省无障碍设施建设标准 The barrier free facilities construction standard of Gansu Province		J11936-2011	甘肃建设工程咨询设计有限责任公司
13	DB62/T 25-3051-2011	砖砌体嵌筋抗震加固技术规程 Technical code for embedded rebar seismic strengthening of brick masonry structure		J11954-2011	甘肃土木工程科学研究院
14	DB62/T 25-3052-2011	风力发电机组基础施工及机组安装技术规程 Foundation construction and installation regulation for wind turbine generator		J11979-2012	甘肃第三建设集团公司
15	DB62/T 25-3053-2011	既有公共建筑节能改造技术规程 Technical specification for the retrofitting of public building on energy efficiency		J11981-2012	甘肃土木工程科学研究院 兰州理工大学
16	DB62/T 25-3054-2011	火探管式感温自启动气体灭火系统技术规程 Technical specification for fireDeTec auto temperature sensible detecting & extinguishing system		J11980-2012	甘肃省建筑设计研究院

甘肃省

序号	标准编号	标准名称	被代替标准编号	备案号	主编单位
17	DB62/T 25-3055-2011	建筑抗震设计规程 Specification for seismic design of buildings		J11982-2012	甘肃建设科技专家委员会
18	DB62/T 25-3056-2012	建筑塑料模板板材 Plastic building form panel		J12040-2012	甘肃第七建设集团股份有限公司
19	DB62/T 25-3057-2012	定型全钢大模板加工技术规程 Technique specification for process of shaped large-area steel formwork		J12041-2012	甘肃第七建设集团股份有限公司
20	DB62/T 25-3058-2012	预拌混凝土质量管理规程 Management specification for quality of ready-mixed concrete		J12039-2012	甘肃第七建设集团股份有限公司
21	DB62/T 25-3059-2012	路桥工地建设标准（试行） Highway & bridge site construction standard		J12054-2012	甘肃路桥建设集团有限公司
22	DB62/T 25-3060-2012	大厚度湿陷性黄土场地工程处理技术规程 Technical specification for ground treatment of collapsible loess with great thickness		J12153-2012	甘肃省土木建筑学会
23	DB62/T 25-3061-2012	自流平地面施工技术规程 Technical specification for self-leveling floor		J12241-2013	甘肃省建设投资（控股）集团总公司 甘肃第三建设集团公司
24	DB62/T 25-3062-2012	太阳能利用与建筑一体化技术规程 Utilization of solar energy and building integration technology standards		J12355-2013	甘肃建筑职业技术学院 甘肃省太阳能建筑专业委员会
25	DB62/T 25-3063-2012	岩土工程勘察规范 Code for investigation of geotechnical engineering		J12356-2013	甘肃中建市政工程勘察设计研究院 甘肃建设工程咨询设计有限责任公司
26	DB62/T 25-3064-2013	绿色建筑评价标准 Assessment standard for green building		J12416-2013	甘肃省建筑科学研究院 甘肃省建材科研设计院
27	DB62/T 25-3065-2013	基桩承载力自平衡检测技术规程 Technical code for self-balanced testing of foundation pile bearing capacity		J12415-2013	甘肃省建筑科学研究院 中国建筑第七工程局有限公司
28	DB62/T 25-3066-2013	泡沫玻璃建筑保温隔热技术规程 Technical specification for the building thermal insulation on foam glass		J12414-2013	甘肃土木工程科学研究院 甘肃建工工程承包有限公司 兰州鹏飞保温隔热有限公司
29	DB62/T 25-3067-2013	居民住宅用电一户一表建设与改造技术规程 Electricity meters construction and retrofit technology regulation for households		J12413-2013	甘肃省电力公司
30	DB62/T 25-3068-2013	断热节能复合砌块墙体保温体系技术规程 Energy-saving insulation block wall heat preservation system technical specification		J12412-2013	西北民族大学（土木工程学院）
31	DB62/T 25-3069-2013	城市园林绿地养护管理标准 Standard for conservation management of urban green space		J12411-2013	兰州市园林绿化局 兰州市园林科学研究所
32	DB62/T 25-3070-2013	兰州地区回弹法检测泵送混凝土抗压强度技术规程 Technical specification for inspecting of pumped concrete compressive strength by rebound method in Lanzhou		J12410-2013	兰州大学（土木工程学院）

甘肃省

序号	标准编号	标准名称	被代替标准编号	备案号	主编单位
33	DB62/T 25-3071-2013	地源热泵系统工程技术规程 Technical specification for ground-source heat pump system		J12409-2013	甘肃建筑职业技术学院 甘肃省建筑设计研究院
34	DB62/T 25-3072-2013	普通清水混凝土施工工艺规程 Technical specification for standard fair-faced concrete construction		J12523-2013	甘肃第七建设集团股份有限公司
35	DB62/T 25-3073-2013	混凝土结构耐久性设计规范 Code for durability design of concrete structures		J12522-2013	甘肃土木工程科学研究院
36	DB62/T 25-3074-2013	岩棉板外墙外保温系统应用技术规程 Application of technical regulations on rock wool board exterior insulation system		J12521-2013	甘肃土木工程科学研究院 甘肃第三建设集团公司
37	DB62/T 25-3075-2013	公路隧道地质雷达检测技术规程 Code for ground penetrating radar detecting technology of highway tunnel		J12531-2014	甘肃省交通科学研究院有限公司
38	DB62/T 25-3076-2013	既有居住建筑节能改造技术规程 Technical specification for energy efficiency retrofitting of existing residential buildings		J12530-2014	甘肃土木工程科学研究院 兰州理工大学
39	DB62/T 25-3077-2014	湿陷性黄土地区抗疏力稳定土路面基层技术规程 Technical specification of anti-collapsing stabilized soil highway roadbases in collapsible loess regions		J12577-2014	甘肃省交通规划勘察设计院有限责任公司 甘肃土木工程科学研究院 陇东学院
40	DB62/T 25-3078-2014	混凝土结构加固施工技术规程 Technical specification for construction of strengthening concrete structures		J12645-2014	甘肃省建筑科学研究院
41	DB62/T 25-3079-2014	被动式太阳能建筑技术规程 Technical code for passive solar building		J12644-2014	甘肃省科学院自然能源研究所 甘肃绿色建筑设计研究院
42	DB62/T 25-3080-2014	庆阳地区回弹法预拌混凝土抗压强度验收技术规程 Qingyang region technical specification for acceptance of premixed concrete compressive strength by rebound method		J12732-2014	兰州理工大学 庆阳市建设局
43	DB62/T 25-3081-2014	绿色建筑施工与验收规范 Code for construction and acceptance of green building		J12733-2014	甘肃省土木工程科学研究院 甘肃第三建设集团公司
44	DB62/T 25-3082-2014	钢筋直螺纹连接技术规程 Technical specification for reinforced straight thread connection		J12734-2014	甘肃省第八建设集团有限责任公司
45	DB62/T 25-3083-2014	HF永久性复合保温模板现浇混凝土建筑保温体系技术规程 Technical specification for cast-in-place concrete composite thermal insulation system of HF external formwork		J12842-2014	西北民族大学
46	DB62/T 25-3084-2014	湿陷性黄土地区建筑灌注桩基技术规程 Technical specification for building cast-in-situ pile foundations in collapsible loess regions		J12880-2014	甘肃众联建设工程科技有限公司 甘肃省建筑设计研究院
47	DB62/T 25-3085-2014	绿色公共建筑能耗标准 Standard for energy use of green public buildings		J12884-2014	甘肃省建筑科学研究院

甘肃省

序号	标准编号	标准名称	被代替标准编号	备案号	主编单位
48	DB62/T 25-3086-2014	绿色居住建筑能耗标准 Standard for energy use of green residential buildings		J12885-2014	甘肃省建筑科学研究院
49	DB62/T 25-3087-2014	绿色公共建筑检测标准 Standard for test of green public building		J12886-2014	甘肃省建材科研设计院
50	DB62/T 25-3088-2014	绿色居住建筑检测标准 Standard for test of green residential building		J12887-2014	甘肃省建材科研设计院
51	DB62/T 25-3089-2014	绿色公共建筑设计标准 Design standard for green public building		J12888-2014	甘肃省建筑设计研究院
52	DB62/T 25-3090-2014	绿色居住建筑设计标准 Design standard for green residential building		J12889-2014	甘肃省建筑设计研究院
53	DB62/T 25-3091-2014	甘肃省城市基础设施专项规划编制导则（试行） Guidelines for urban infrastructure planning of Gansu Province		J12881-2014	中国市政工程西北设计研究院有限公司
54	DB62/T 25-3092-2014	硬泡聚氨酯外墙外保温工程技术规程 Technical specification for rigid polyurethane foam external thermal insulation on walls		J12882-2014	中国聚氨酯工业协会异氰酸酯专业委员会
55	DB62/T 25-3093-2015	民用建筑广播电视网络设施建设标准 Civil architecture in radio and television network facilities construction standards		J12924-2015	甘肃省广播电视网络股份有限公司
56	DB62/T 25-3094-2015	非贮压式超细干粉灭火装置系统技术规程 Technical specification for Non-pressure storage super fine powder extinguish equipment system		J13040-2015	甘肃省建筑设计研究院
57	DB62/T 25-3095-2015	钢丝网复合岩棉板外墙外保温体系应用技术规程 Technical specification for application of steel mesh composite rock wool board external thermal insulation system on walls		J13038-2015	甘肃天鸿金运置业有限公司 甘肃省建筑科学研究院
58	DB62/T 25-3096-2015	发泡陶瓷板保温系统技术规程 Technical specification for application of thermal insulating system of foamed ceramic board		J13125-2015	甘肃省建材科研设计院
59	DB62/T 25-3097-2015	硅钙面层复合保温隔墙板施工及验收规范 Code for construction and acceptance of insnlation composite wall panel with calcium silicate board		J13177-2015	甘肃土木工程科学研究院
60	DB62/T 25-3098-2015	路面废旧料冷再生应用技术规程 Technical specifications of cold recycling for reclaimed asphalt pavement		J13178-2015	甘肃畅陇公路养护技术研究院有限公司
61	DB62/T 25-3099-2015	预应力混凝土管桩基础技术规程 Technical specification for prestressed concrete pipe pile foundation		J13223-2015	甘肃土木工程科学研究院 甘肃省建筑设计研究院
62	DB62/T 25-3100-2015	公路绿化设计规范 Code for design of highway greening		J13226-2015	甘肃省交通规划勘察设计院有限责任公司
63	DB62/T 25-3101-2015	公路绿化施工及质量验收规范 Code for highway greening of construction quality acceptance		J13227-2015	甘肃华运环境建设工程有限公司

甘肃省

序号	标准编号	标准名称	被代替标准编号	备案号	主编单位
64	DB62/T 25-3102-2015	公路绿化养护管理规范 Code for highway greening maintenance management		J13228-2015	甘肃华运环境建设工程有限公司
65	DB62/T 25-3103-2015	公路隧道防火涂料施工质量验收规程 Code for construction quality acceptance of fireproof coating of road tunnel		J13250-2015	甘肃省交通科学研究院有限公司
66	DB62/T 25-3104-2015	建筑边坡工程技术规程 Technical specification for building slope engineering		J13294-2016	兰州交通大学
67	DB62/T 25-3105-2015	火灾应急智能疏散诱导系统技术规程 Technical specification for intelligent fire alarm information displaying and evacuation indicating system		J13295-2016	兰州市城市建设设计院 湖南汇博电子技术有限公司
68	DB62/T 25-3106-2015	地源热泵系统运行管理规程 Regulations for operation management of ground-source heat pump system		J13296-2016	甘肃土木工程科学研究院
69	DB62/T 25-3107-2016	公路路面水泥稳定碎石基层振动法施工技术规范 Technical specification for construction of highway cement stabilized macadam base by vibration test method		J13352-2016	甘肃省远大路业集团有限公司
70	DB62/T 25-3108-2016	低丘缓坡未利用地开发技术规程 Technical specification for reclaimed land of barren hill		J13331-2016	兰州理工大学
71	DB62/T 25-3109-2016	地下工程混凝土结构防腐阻锈防水抗裂技术规程 Technical specification for anticorrosion, rust resistant, waterproof and crack resistant of concrete structure in underground engineering		J13372-2016	甘肃土木工程科学研究院
72	DB62/T 25-3110-2016	防火门监控系统技术规程 Technical code of indicating and control system for fire resistant door		J13373-2016	广州六瑞消防科技有限公司 甘肃省公安消防总队
73	DB6/T 25-3111-2016	建筑基坑工程技术规程 Technical specification for building excavation engineering	DB6/T 25-3001-2000	J10062-2016	甘肃众联建设工程科技有限公司 甘肃中建市政工程勘察设计研究院
74	DB62/T 25-3112-2016	民用建筑与太阳能光伏发电一体化系统技术规程 Technical code for integration of civil buildings and solar pV system		J13419-2016	兰州理工大学
75	DB62/T 25-3113-2016	居住建筑与太阳能供热采暖一体化技术规程 Technical code for integration of residential buildings and solar heating		J13420-2016	兰州理工大学
76	DB62/T 25-3114-2016	建筑与太阳能吸热和反射一体化技术规程 Technical code for the integration of building with solar energy absorption and reflection		J13421-2016	兰州理工大学
77	DB62/T 25-3115-2016	预拌砂浆生产与应用技术规程 Technical specification for manufacture and application of ready-mixed mortar		J13422-2016	甘肃第七建设集团股份有限公司 甘肃土木工程科学研究院

甘肃省

序号	标准编号	标准名称	被代替标准编号	备案号	主编单位
78	DB62/T 25-3116-2016	建筑同层排水系统施工及验收规程 Code of acceptance for construction of same-floor drainage system in buildings		J13451-2016	甘肃第四建设集团有限责任公司
79	DB62/T 25-3117-2016	古建筑油饰彩画施工工艺规程 Construction workmanship code for coating and painting of ancient chinese architecture		J13452-2016	甘肃兴城建设工程有限公司
80	DB62/T 25-3118-2016	湿陷性黄土场地挤密地基技术规程 Technical specification for compacted foundation of collapsible loess site		J13491-2016	甘肃众联建设工程科技有限公司
81	DB62/T 25-3119-2016	生活垃圾卫生填埋场施工技术规程 Technical specification for construction of municipal domestic refus sanitary landfill		J13492-2016	甘肃第二建设集团有限责任公司
82	DB62/T 25-3120-2016	LSP板内嵌轻钢龙骨装配式墙体系统技术规程 Technical specification for LSP board embedded light gauge steel keel assembled wall system		J13557-2016	西北民族大学
83	DB62/T 25-3121-2016	混凝土建筑结构基础隔震技术规程 Technical specification for base seismic-isolation of concrete structures		J13558-2016	甘肃省建筑设计研究院
84	DB62/T 25-3122-2016	全钢附着式升降脚手架安全技术规程 Technical specification for safety of all-steel adhesive lifting scaffol		J13559-2016	甘肃第六建设集团股份有限公司 兰州理工检验技术有限公司
85	DB62/T 25-3123-2016	建筑工程预应力混凝土结构有粘结后张法施工验收规程 Code practice for construction of prestressed concrete structures with bonded posttensioned		J13606-2016	甘肃省第八建设集团有限责任公司
86	DB62/T 25-3124-2016	绿色办公建筑评价标准 Evalution standard for green office building		J13681-2017	甘肃省建筑科学研究院
87	DB62/T 25-3125-2016	预制保温耐热聚乙烯（PE-RTⅡ）低温直埋供热管道技术规范 Technical code for buried polyethylence of raised temperature （PE-RTⅡ）pipeline of prefabricated insulation fir low temperature-heating distribution		J13682-2017	甘肃省建筑设计研究院 兰州市热力总公司
88	DB62/T 25-3126-2016	钢结构检测与鉴定技术规程 Specification for inspection and appraisal of steel structures		J13680-2017	兰州理工大学 甘肃省建设科技专家委员会 钢结构专业委员会

宁夏回族自治区

序号	标准编号	标准名称	被代替标准编号	备案号	主编单位
1	DBJ 28/053-2008	钢结构轻型建筑体系技术导则 The technical rule of steel structures for lightweight building		J11463-2009	宁夏工程建设标准管理中心
2	DBJ 64/054-2015	既有居住建筑节能改造技术规程 Technical specification of renovation of energy efficiency for existing residential buildings	DB64/054-2009	J11466-2015	宁夏工程建设标准管理中心
3	DBJ64/T 056-2015	建筑物围护结构传热系数现场检测标准（热箱法） Field test standard for heat transfer coefficient of building envelope (Hot box method)		J13264-2015	宁夏建筑设计研究院有限公司
4	BDJ 64/T 057-2016	模板早拆体系技术规程 Technical specification for early dismantling formwork system		J13378-2016	宁夏利伟建筑材料新技术研究所
5	DBJ64/T 058-2016	通信基础设施专项规划编制导则 Special planning guide communication infrastructure		J13532-2016	宁夏回族自治区通信管理局
6	DBJ64/T 059-2016	建筑节能门窗工程技术规程 Engineering specification for architechture energy-saving doors and windows		J13696-2016	宁夏建设新技术新产品推广协会
7	DBJ64/T 060-2016	SMC改性沥青路面施工技术规程 Technical specification for construction of SMC modified asphalt pavement		J13695-2016	银川市建设工程综合检测站
8	DB64/265-2006	外墙外保温系统及专用材料质量检验标准 Quality examination standard on special material mortar exterior thermal insulation systems for outer-wall		J11059-2007	宁夏工程建设标准管理中心
9	DB64/T 428-2011	民用建筑设置燃气锅炉房设计防火标准 Fire protection and prevention design standard on gas boiler room in buildings	DB64/T 428-2006	J11841-2011	宁夏回族自治区公安厅消防局
10	DB64/T 498-2007	外墙复合轻质保温板应用技术规程 Applied technical specification of composite light insulation board for outer-wall		J11058-2007	宁夏工程建设标准管理中心
11	DB64/521-2013	居住建筑节能设计标准 Design standard for energy efficiency of residential buildings		J12342-2013	宁夏建设新技术新产品推广协会
12	DB64/558-2009	建筑工人安全操作规程 Safety operation of construction workers		J11521-2009	宁夏建设工程质量安全监督总站
13	DB64/663-2010	EPS空心模块轻钢结构建筑节能体系应用技术规程 Applied technical specification for building energy-saving system of EPS hollow module with light steel structure		J11706-2010	宁夏建筑科学研究院有限公司
14	DB64/664-2015	EPS模块现浇钢筋混凝土外墙外保温应用技术规程 EPS module cast-in-place reinforced concrete external wall thermal insulation application procedures	DB64/664-2010	J11705-2015	宁夏建筑科学研究院有限公司

宁夏回族自治区

序号	标准编号	标准名称	被代替标准编号	备案号	主编单位
15	DB64/665-2015	EPS模块外保温工程技术规程 EPS module external thermal insulation engineering discipline	DB64/665-2012 DB64/665-2010	J12112-2015	宁夏建筑科学研究院有限公司
16	DB64/680-2010	建筑工程安全管理规程 Managment specification of construction engineering safety		J11782-2011	宁夏建设工程质量安全监督总站
17	DB64/T 688-2011	成品住宅套内装修标准 Standard of finished houses decoration		J11840-2011	宁夏建设新技术新产品推广协会
18	DB64/T 689-2011	复合保温钢筋焊接网架混凝土剪力墙（CL建筑体系）技术规程 Thermal insulation composite grid reinforced concrete shear wall welding		J11929-2011	宁夏建筑设计研究院有限公司
19	DB64/696-2011	建筑外墙外保温工程防火技术规程 Building exterior insulation fire technical specification		J11931-2011	宁夏工程建设标准管理中心
20	DB64/T 697-2011	回弹法检测泵送混凝土抗压强度技术规程		J11930-2011	宁夏建筑科学研究院有限公司
21	DB64/707-2011	住宅厨房卫生间防火型变压式排气道应用技术规程 Residential-style kitchen and bathroom exhaust fire-type transformer application		J11928-2011	宁夏建设新技术新产品推广协会
22	DB64/786-2012	EPS模块混凝土剪力墙结构体系应用技术规程 Technical specifications for concrete shear wall structure with EPS modular		J12111-2012	宁夏建筑科学研究院有限公司
23	DB64/787-2012	EPS模块框（钢）架结构工业建筑节能体系技术规程 Technical specifications for industrial building energy-saving with EPS modular		J12110-2012	宁夏建筑科学研究院有限公司
24	DB64/T 795-2012	民用建筑并网光伏发电应用技术规程 Civil construction of grid-connected photovoltaic power generation application		J12194-2012	宁夏建设新技术新产品推广协会
25	DB64/T 954-2014	绿色建筑评价标准 Evaluation standard green building		J12627-2014	宁夏建设新技术新产品推广协会
26	DB64/T 966-2014	抹灰石膏应用技术规程 Plastering gypsum technical specification for application		J12751-2014	宁夏石膏工业协会
27	DB64/T 1056-2014	预制直埋复合塑料保温管道应用技术规程 Prefabricated directly buried composite insulation plastic pipe application procedures		J12945-2015	宁夏青龙塑料管材有限公司
28	DB64/1068-2015	农村住宅节能设计标准 Rural residential energy-saving design standard		J13126-2015	宁夏建筑标准设计办公室

青海省

序号	标准编号	标准名称	被代替标准编号	备案号	主编单位
1	DB63/T 638-2007	粉煤灰砖砌体结构设计与施工技术规程 Technical specification for design and construction of fly ash brick structure		J10970-2007	西宁市土木建筑工程学会
2	DB63/T 663-2007	煤矸石砖及煤矸石多孔砖砌体结构设计与施工技术规程（试行） Technical specification for design and construction of coal gangue brickand coal gangue perforated brick masonry structures		J11068-2007	西宁市土木建筑工程学会
3	DB63/T 664-2007	钢渣普通砖及钢渣多孔砖砌体结构设计与施工技术规程（试行） Technical specification for design and construction of steel slagbrick and steel slag perforated brick masonry structures		J11068-2007	西宁市土木建筑工程学会
4	DB63/T 743-2008	青海省民用建筑太阳能热水系统应用技术规程 Technical specification for solar water heating system of civil buildings		J11230-2008	青海省建筑勘察设计研究院
5	DB63/T 866-2010	民用建筑太阳能利用规划设计规范 Code for planning & design for solar utilization of civil buildings		J11610-2010	中国建筑标准设计研究院 青海省住房和城乡建设厅
6	DB63/T 868-2010	页岩砖及页岩多孔砖砌体结构设计与施工技术规程 Technical specification for design and construction of shalebrick and shale perforated brick masonry structures		J11546-2010	西宁市土木建筑工程学会
7	DB63/T 869-2010	住宅信报箱建设标准 Constructional standard of residential mail-box		J11547-2010	青海省邮政管理局
8	DB63/878-2010	钢筋混凝土复合墙板低层建筑技术规程 Technical specification for low-rise building of reinforced concrete composite wall-panel		J11612-2010	中国建筑科学研究院
9	DB63/T 885-2010	裹体碎石桩法处理地基技术规程 Technical specification for ground treatment by the wrapped gravel columns method		J11628-2010	陕西长嘉实业发展有限公司
10	DB63/T 928-2010	青海省低层民用建筑平板主动式太阳能供热采暖系统应用技术标准 Qinghai Province application technical standard for low-rise residence flat plate active solar heating system		J11667-2010	青海省建筑勘察设计研究院有限公司
11	DB63/T 994-2011	预拌混凝土生产应用技术规程 Premixed concrete production application technology procedures		J11883-2011	青海省建筑建材科学研究院
12	DB63/T 995-2011	预拌砂浆生产应用技术规程 Ready-mixed mortar production application technology procedures		J11885-2011	青海省建筑建材科学研究院
13	DB63/T 996-2011	带消能节点的装配式木结构房屋技术规程 Take away the fabricated timberwork building can node technical procedures		J11884-2011	南京林业大学 东南大学

青海省

序号	标准编号	标准名称	被代替标准编号	备案号	主编单位
14	DB63/T 997-2011	复合保温钢筋焊接网架混凝土剪力墙（CL建筑体系）技术规程 Technical specification for concrete shearwall with composite heat insulation welded steel frame (CL building system)		J11886-2011	青海省建筑勘察设计研究院有限公司
15	DB63/T 1004-2011	青海省既有居住建筑节能改造技术规程 Existing residential building energy-saving technical regulations of Qinghai		J11898-2011	省建筑勘察设计研究院有限公司
16	DB63/T 1111-2012	改性酚醛泡沫防火保温板外墙外保温系统应用技术规程 Phenolic foam fire prevention insulation board exterior insulation system application technology procedures		J12129-2012	青海省规划设计研究院
17	DB63/1141-2012	青海省驻地网通信设施建设规范 Standard for construction of customer premise network for communication facilities in Qinghai		J12134-2012	四川通信科研规划设计有限责任公司
18	DB63/T 1142-2012	固体废弃物免烧普通砖、多孔砖砌体结构设计与施工技术规程 Solid waste unburned ordinary brick, brick masonry structures design and construction		J12135-2012	省建筑勘察设计研究院有限公司
19	DB63/T 1143-2012	青海省受损砌体结构安全性鉴定实施导则 Qinghai Province is impaired in masonry structure safety appraisal implementation		J12243-2013	青海大学土木工程学院
20	DB63/T 1174-2012	散装水泥农村牧区配送站建设技术规范		J12245	青海省建筑建材科学研究院
21	DB63/T 1175-2012	城镇园林绿地养护管理质量标准 Urban landscape green space conservation and management of quality standards		J12244-2013	青海省西宁市园林规划设计院
22	DB63/T 1204-2013	石膏砌块砌体技术规程 Technical specification for gypsum block masonry		J12546-2014	青海恒兴工程设计有限责任公司
23	DB63/T 1205-2013	现浇钢筋混凝土空心楼盖（箱体内模）结构技术规程 Technical specification for cast-in-situ concrete nonow noor structure		J12547-2014	青海省建筑勘察设计研究院有限公司 青海标迪夫科技有限公司
24	DB63/T 1206-2013	城市房屋供配电设施建设技术导则 Guide of supply and distribution facilities construction		J12548-2014	青海省电力公司
25	DB63/T 1306-2014	碳纤维发热线缆地面辐射供暖技术规程 Regulation of heating cable floor radiant heating technology of carbon fiber		J12718-2014	青海东亚工程建设管理咨询有限公司
26	DB63/T 1307-2014	建设工程绿色施工规程 Green construction code of construction		J12848-2014	青海省建筑建材科学研究院 青海一建建筑工程有限责任公司
27	DB63/T 1339-2014	现浇钢筋混凝土结构自保温体系应用技术规程		J13014-2015	青海省建筑建材科学研究院
28	DB63/T 1340-2015	青海省绿色建筑设计标准 Design standard for green building in Qinghai		J13015-2015	西安建筑科技大学 西北工业大学
29	DB63/T 1341-2015	石墨模塑聚苯乙烯泡沫板保温工程技术规程 Technical specification for graphite molded polystyrene foam board insulation		J13016-2015	青海省建设科技开发推广中心
30	DB63/T 1382-2015	住宅工程质量分户验收规程 Acceptance specification for unit quality of housing engineering		J13124-2015	青海省建设工程质量监督总站

新疆维吾尔自治区

序号	标准编号	标准名称	被代替标准编号	备案号	主编单位
1	XJJ 001-2011	严寒和寒冷地区居住建筑节能设计标准实施细则 Design standard for energy efficiency of residential buildings in severe cold and cold zones execution detailed rules	XJJ 001-1999	J11921-2011	新疆维吾尔自治区建筑设计研究院 新疆大学建筑工程学院
2	XJJ 005-2001	家庭装饰装修工程质量验收标准			乌鲁木齐市建筑工程质量监督站
3	XJJ 012-2016	新疆维吾尔自治区实施国家2010（建筑结构）系列规范细则 Xinjiang uygur autonomous region to implement the national 2010（building structure）series of rules and regulations	XJJ 012-2003	J12156-2012	新疆维吾尔自治区建筑设计研究院
4	XJJ 013-2012	城市规划管理技术规定 Technical city planning management regulations	XJJ 013-2004	J10450-2012	新疆维吾尔自治区城乡规划服务中心
5	XJJ 014-2004	农村民居抗震鉴定实施细则			新疆建筑标准设计办公室
6	XJJ 016-2005	建筑地基基础工程施工工艺标准 Construction technology standard in foundation engineering		J10720-2006	新疆生产建设兵团建设工程（集团）有限责任公司
7	XJJ 017-2005	砌体工程施工工艺标准 Construction technology standard in masonry structures		J10721-2006	新疆天一建工投资集团有限责任公司
8	XJJ 018-2005	混凝土结构工程施工工艺标准 Construction technology standard in concrete structures		J10722-2006	新疆兵团建设工程（集团）有限责任公司一建
9	XJJ 019-2005	地下防水工程施工工艺标准 Construction technology standard in underground waterproof engineering		J10723-2006	新疆兵团第四建筑安装工程公司
10	XJJ 020-2005	钢结构工程施工工艺标准 Construction technology standard in stell structures		J10724-2006	新疆建工安装工程有限责任公司
11	XJJ 021-2005	屋面工程施工工艺标准 Construction technology standard in roof engineering		J10725-2006	新疆兵团建设工程（集团）有限责任公司六建
12	XJJ 022-2005	建筑地面工程施工工艺标准 Construction technology standard in ground engineering		J10726-2006	新疆天一建工投资集团有限责任公司
13	XJJ 023-2005	建筑装饰装修工程施工工艺标准 Construction technology standard in decoration of housings		J10727-2006	乌鲁木齐市建筑工程质量监督站 新疆城市建筑装饰工程有限公司 新疆大昌装饰工程有限责任公司 新疆凌云设计工程有限公司 新疆现代装饰工程有限公司 新疆嘉龙装饰装璜工程有限公司 乌鲁木齐大都装饰工程有限公司 新疆羚羊建筑装饰企业有限责任公司
14	XJJ 024-2005	建筑给水排水及采暖工程施工工艺标准 Construction technology standard in water supply and drainage engineering, heating works		J10728-2006	新疆建工集团四建

新疆维吾尔自治区

序号	标准编号	标准名称	被代替标准编号	备案号	主编单位
15	XJJ 025-2005	建筑电气工程施工工艺标准 Construction technology standard in electrical engineering		J10729-2006	新疆建工集团第一建筑工程有限责任公司
16	XJJ 026-2005	通风与空调工程施工工艺标准 Construction technology standard in ventilation and air-conditioning works		J10730-2006	新疆建工集团消防工程有限责任公司 南京中江工程建设有限公司
17	XJJ 027-2005	智能建筑工程施工工艺标准 Construction technology standard in intelligent building systems		J10731-2006	新疆建工集团消防工程有限责任公司
18	XJJ 029-2006	城镇户外广告设施设置规程 Specification for setting of town advertising facilities outdoors		J11019-2007	乌鲁木齐市市政市容管理局
19	XJJ 030-2006	市政基础设施工程施工质量验收统一标准 Unified standard for acceptance of construction quality of municipal infrastructure engineering		J10984-2007	乌鲁木齐市建设工程质量监督站
20	XJJ 034-2017	公共建筑节能设计标准 Design standard for energy efficiency of public buildings region)	XJJ 034-2006	J10997-2017	新疆维吾尔自治区建筑设计研究院
21	XJJ 035-2006	新疆实施国家2001~2004（岩土工程）系列规范细则 Xinjiang detailed rules for implementing 2001-2004 national series of code (geotechnical engineering)		J10983-2007	新疆维吾尔自治区建筑设计研究院
22	XJJ 036-2007	建筑工程资料管理规程 Management specification of construction engineering documentation		J11104-2007	新疆维吾尔自治区建设工程质量监督总站
23	XJJ 037-2008	EPS板薄抹灰外墙外保温施工规程 Construction specification for external insulation system of exterior wall of EPS board with thin plastering		J11255-2008	新疆维吾尔自治区建筑科学研究院
24	XJJ 038-2008	新疆建筑基坑土钉支护统一技术规定 Xinjiang unifying stipulation forsoil nailing retaining and protection of foundation excavations		J11257-2008	新疆岩土工程勘察设计研究院有限责任公司
25	XJJ 039-2008	现浇混凝土EPS板外墙外保温系统施工规程 Construction specification for external insulation system of exterior wall of cast-in-place concrete EPS board		J11256-2008	新疆建工集团第一建筑工程有限责任公司
26	XJJ 040-2009	预拌砂浆应用技术规程 Specification for utility technical of ready-mixed mortar		J11495-2009	新疆维吾尔自治区建筑科学研究院
27	XJJ 041-2011	建筑用塑料窗技术规程 Technical specification for PVC-U windows of building construction		J11805-2011	新疆维吾尔自治区建筑门窗协会 新疆维吾尔自治区建筑科学研究院
28	XJJ 042-2010	住宅小区有线电视配套设施建设标准 Construction standard for facilities of communications in housing district		J11650-2010	新疆通信规划设计院 新疆维吾尔自治区建筑设计研究院
29	XJJ 043-2010	住宅小区通信配套设施建设标准 Constructive standard for necessary installation of cable television in residential district		J11649-2010	新疆广电网络有限责任公司

新疆维吾尔自治区

序号	标准编号	标准名称	被代替标准编号	备案号	主编单位
30	XJJ 044-2010	建设工程施工安全监理规程 Specification for safety construction supervision		J11635-2010	新疆维吾尔自治区建设工程安全监督总站
31	XJJ 045-2011	钢结构防火涂料应用技术规程 Specification for utility technical of fire-retardant coating for steel structures		J11825-2011	自治区消防产品质量监督检验站
32	XJJ 046-2012	城市绿地养护管理标准 Urban green space maintenance management standards		J12224-2012	新疆维吾尔自治区风景园林学会 新疆农业大学林学与园艺学院园林系
33	XJJ 047-2012	村庄规划编制技术规程（试行） Technical specification for village planning to prepare		J12060-2012	新疆维吾尔自治区城乡规划服务中心 新疆城乡规划设计研究院 新疆佳联城建规划设计研究院有限公司
34	XJJ 048-2012	建设项目交通影响评价技术标准 Technical standards of traffic impact analysis of construction project		J12137-2012	乌鲁木齐市城市综合交通项目研究中心 新疆大学 大连理工大学
35	XJJ 049-2012	展览布展防火规程 Technical specification for exhibition exhibition fire protection and prevention		J12138-2012	新疆维吾尔自治区公安消防总队
36	XJJ 050-2012	地下水水源热泵工程技术规程 Technical regulation for groundwater heat pump system		J12260-2013	新疆维吾尔自治区建筑设计研究院
37	XJJ 051-2012	镇（乡）总体规划编制技术规程 Technical specification for town (township) overall planning		J12156-2012	上海同济城市规划设计研究院 上海城市规划设计研究院
38	XJJ 052-2012	直拔法检测混凝土抗压强度技术规程 Technical specification for testing of concrete compressive strength by straight pulling method		J12177-2012	新疆巴州建设工程质量检测中心
39	XJJ 053-2012	发热电缆地面辐射供暖技术规程 Heat cable technical specification for floor radiant heating		J12196-2012	新疆电力科学研究院 自治区建设标准服务中心
40	XJJ 054-2012	剥肋滚轧直螺纹钢筋机械连接工艺标准 Stripping rib rolling thread reinforcing steel bar connection technology standard		J12226-2012	新疆维吾尔自治区建设标准服务中心 宝钢集团新疆八一钢铁有限公司
41	XJJ 055-2012	预拌混凝土生产技术规程 Technical specification for ready-mixed concrete		J12225-2012	新疆维吾尔自治区建筑科学研究院 中建新疆建工（集团）有限公司 乌鲁木齐建材行业协会
42	XJJ 056-2013	住宅物业服务标准 Service standard of residential property		J12375-2013	新疆物业管理专业委员会 新疆建设标准服务中心
43	XJJ 057-2013	建筑工程高强钢筋应用技术导则 Technical guidelines for application of high strength steel bars in construction engineering		J12374-2013	中建新疆建工（集团）有限公司技术中心 新疆建筑设计研究院

新疆维吾尔自治区

序号	标准编号	标准名称	被代替标准编号	备案号	主编单位
44	XJJ 058-2013	电气火灾监控系统技术规程 Technical specification for electrical fire monitoring system		J12373-2013	新疆维吾尔自治区公安消防总队等
45	XJJ 059-2013	住宅区和住宅建筑光纤入户通信工程技术规范 Technical code for fiber to the home communication engineering in residential districtsand residential buildings		J12456-2013	新疆维吾尔自治区通信管理局 工业和信息化部电信规划研究院
46	XJJ 060-2014	胶粉聚苯颗粒复合型外墙外保温工程技术规程 Technical specification for external thermal insulation on outer-walls with composited systems based on mineral binder and expanded polystyrene granule plaster		J12581-2014	北京振利节能环保科技股份有限公司
47	XJJ 061-2014	EPS模块混凝土剪力墙结构工程技术规程 Technical specification for concrete shear wall structural engineer with EPS modules		J12590-2014	新疆建筑设计研究院新疆建筑科学研究院（有限责任公司） 哈尔滨鸿盛房屋节能体系研发中心
48	XJJ 062-2014	建筑护角施工规程 Technical specification for concrete shear wall structural engineer with EPS modules		J12638-2014	中建新疆建工集团第一建筑工程有限公司
49	XJJ/T 063-2014	严寒C区居住建筑节能设计标准 Design standard for energy efficiency of residential buildings in severe cold (C) zones		J12647-2014	新疆维吾尔自治区建筑设计研究院 乌鲁木齐市建筑节能墙体材料革新办公室
50	XJJ 064-2014	建设项目规划选址论证报告编制导则 Site selection report guidelines for the preparation of construction project planning		J12774-2014	新疆维吾尔自治区城乡规划服务中心
51	XJJ 065-2014	火探管式自动灭火装置技术规程 Technical specification for automatic extinguishing equipment with fire trace tube		J12750-2014	新疆维吾尔自治区公安消防总队 陕西西能电力技术有限公司
52	XJJ 066-2014	电热膜供暖技术规程 Technical specification for the electric heating film		J12775-2014	自治区建设标准服务中心 新疆建筑设计研究院 中建新疆建工集团一建
53	XJJ 067-2014	复合保温钢丝焊接网架混凝土剪力墙（CL建筑体系）技术规程 Technical specification for concrete shearwall with welded steel frame and composite thermal insulation (CL building system)		J12798-2014	新疆维吾尔自治区建筑设计研究院 乌鲁木齐建筑设计研究院有限责任公司 新疆生产建设兵团建设工程（集团）有限责任公司 石家庄晶达建筑体系有限公司
54	XJJ 068-2014	民用建筑电气防火设计规范 Code for fireproofing design of electric in civil buildings		J12835-2014	新疆维吾尔自治区建筑设计研究院 新疆维吾尔自治区公安消防总队
55	XJJ 069-2015	硬质酚醛泡沫板薄抹灰外墙外保温工程技术规程 The technical specification for rigid phenolic foam panel external thermal insulation with thin rendering		J12977-2015	新疆维吾尔自治区建筑科学研究院有限责任公司 新疆维吾尔自治区建筑设计研究院 新疆维吾尔自治区酚醛材料行业协会

新疆维吾尔自治区

序号	标准编号	标准名称	被代替标准编号	备案号	主编单位
56	XJJ 070-2015	城市设计技术规程 Technical urban design regulations		J13168-2015	上海市城市规划设计研究院 上海同济城市规划设计研究院 乌鲁木齐市城市规划设计研究院 自治区住房和城乡建设厅城乡规划处
57	XJJ 071-2016	建设工程文件归档技术规程 Technical specification for putting construction project documents into records		J13377-2016	新疆维吾尔自治区城市建设档案馆
58	XJJ 072-2015	建设工程合同备案管理规范 Record management standard on construction project contract		J13262-2015	新疆维吾尔自治区工程造价管理总站
59	XJJ /T 073-2016	寒冷地区居住建筑节能设计标准 Design standard for energy efficiency of residential buildings in cold zones		J13489-2016	新疆维吾尔自治区建筑设计研究院
60	XJJ 074-2016	住宅小区供电设施建设和改造技术标准 The technical standard of power supply facilities construct and reconstruct for residential districts		J13599-2016	国网新疆电力公司乌鲁木齐供电公司
61	XJJ 075-2016	建筑消能减震应用技术规程 Building energy dissipation damping application procedures		J13686-2017	乌鲁木齐建筑设计研究院有限责任公司
62	XJJ 076-2016	再生骨料混凝土技术规程 Technical code on the application of recycled concrete		J13718-2017	新疆大学建筑工程学院

海南省

序号	标准编号	标准名称	被代替标准编号	备案号	主编单位
1	DBJ 02-2006	海南省建筑外门窗抗风压、水密、气密性能控制指标 code for resisting windpressur capacity and water penetration and air permeability of doors and windows		J10875-2006	海南省建设标准定额站
2	DBJ 03-2006	海南省公共建筑节能设计标准 Design standard for energy efficiency of public buildings in Hainan		J10907-2006	海南华磊建筑设计咨询有限公司 海南省勘察设计协会
3	DBJ 04-2006	海南省建筑节能工程施工验收标准（试行） Standard for acceptance of energy efficient building construction in Hainan		J10940-2007	海南省建设标准定额站 海南省建设工程质量安全监督总站
4	DBJ 07-2006	建设工程文明施工标准 The standard for civilized construction of constructive engineering		J10941-2007	海南省建设标准定额站
5	DBJ 08-2007	非承重砌体材料应用技术规程 Technical specification for application of the non-load-bearing masonries		J10989-2007	海南省建设标准定额站
6	DBJ 09-2007	建筑施工优质结构工程评定标准 Evaluating standard for high grade structural engineering of building construction		J11081-2007	海南省建设标准定额站
7	DBJ 46-10-2012	海南省建设工程绿岛杯奖评选标准 Evaluating standard for "the Lv-Dao Cup" award (province excellent quality project) of building engineering	DBJ 10-2007	J12128-2012	海南省建筑业协会
8	DBJ 46-11-2015	农村居住建筑抗震防风技术规程 Technical specification for anti-seismictiy and wind prevention of rural residential buildings	DBJ 11-2008	Jll225-2015	海南省工程抗震办公室 海南省建设标准定额站 海南省建筑设计院
9	DBJ 12-2008	太阳能热水系统与建筑一体化设计施工及验收规程 Code for design and construction and accepting of solar water heating system and building integration		J11367-2009	海南省建设标准定额站
10	DBJ 13-2009	海南省城镇生活垃圾填埋场劳动岗位定员标准（试行）		J11554-2010	海南省住房和城乡建设厅
11	DBJ 14-2009	海南省城乡环境卫生质量标准		J11555-2010	海南省住房和城乡建设厅
12	DBJ 15-2010	海南省城乡容貌标准		J11587-2010	海南省住房和城乡建设厅
13	DBJ 16-2010	海南省公共厕所管理及保洁服务标准		J11625-2010	海南省住房和城乡建设厅
14	DBJ 17-2011	液化天然气（LNG）汽车加气站设计与施工规范 Code for design and construction of automotive liquified natural gas (LNG) filling station		J11871-2011	海南省公安消防总队
15	DBJ 18-2011	预拌混凝土生产及施工技术规程 Technical specification for production and construction of ready-mixed concrete		J11912-2011	海南省建设标准定额站
16	DBJ 19-2011	生活垃圾卫生填埋场运行监管标准 Regulatory standard for operation of municipal solid waste sanitary landfill		J11937-2011	华中科技大学环境科学与工程学院

海南省

序号	标准编号	标准名称	被代替标准编号	备案号	主编单位
17	DBJ 20-2012	电气火灾监控系统设计、施工及验收规范		J12033-2012	海南省公安消防总队
18	DBJ 21-2012	生活垃圾收集转运设施运行监管标准		J12029-2012	华中科技大学环境科学与工程学院
19	DBJ 22-2012	生活垃圾焚烧厂运行监管标准 Standard for operation supervision of municipal solid waste incineration plant		J12037-2012	华中科技大学环境科学与工程学院
20	DBJ 46-023-2012	海南省建筑用反射隔热涂料应用技术规程		J12098-2012	海南省华磊建筑设计咨询有限公司
21	DBJ 46-024-2012	海南省绿色建筑评价标准（试行） Evaluation standard for green building in Hainan		J12127-2012	海南华磊建筑设计咨询有限公司
22	DBJ 46-025-2013	海南省住宅建筑通信设施工程建设标准 Construction standard for communication facilities engineering of residential buildings in Hainan Procince		J12304-2013	海南省通信管理局
23	DBJ 46-026-2013	螺杆灌注桩技术规程 Technical code for half-screw pile		J12334-2013	海南有色工程勘察设计院 海南卓典高科技开发有限公司 海南省建设标准定额站
24	DBJ 46-027-2013	海南省建筑塔式起重机安装使用安全评定规程 The assessment procedure of Hainan Province construction tower crane's installation and applicatons		J12459-2013	海南省建筑机械协会
25	DBJ 46-028-2013	海南省党政机关办公用房维修标准（试行）		J12517-2013	海南省建设标准定额站 海南省建筑设计院
26	DBJ 46-029-2014	海南省公共照明设施养护维修标准 Maintenance standard of public illumination facilities in Hainan Province		J12698-2014	海南省住房和城乡建设厅
27	DBJ 46-030-2014	海南省建设工程造价电子数据标准 Electronic data standards for cost of construction projects in Hainan Province		J12693-2014	成都鹏业软件股份有限公司
28	DBJ 46-031-2014	海南省房屋建筑工程全过程监管信息平台检测数据接口标准（试行） Standard of data interface for whole process surpervision information platform of building construction engineering in Hainan Province		J12792-2014	成都鹏业软件股份有限公司
29	DBJ 46-032-2015	海南省市政道路设施养护维修标准（试行） Maintenance standard of municipal roads facilities in Hainan Province (for trial implementation)		J12925-2015	海口市市政管理局 海口市市政工程设计研究院 海南省建设标准定额站
30	DBJ 46-033-2015			J12926-2015	海口市市政管理局 海口市市政工程设计研究院 海南省建设标准定额站
31	DBJ 46-034-2015	海南省市政排水管渠养护维修标准（试行） Maintenance standard of municipal drainage canal in Hainan Province (for trial implementation)		J12927-2015	海口市市政管理局 海口市市政工程设计研究院 海南省建设标准定额站
32	DBJ 46-035-2015	海南省房屋建筑工程全过程监管信息平台基础数据库（企业、人员、项目、诚信）数据接口标准		J13230-2015	成都鹏业软件股份有限公司

海南省

序号	标准编号	标准名称	被代替标准编号	备案号	主编单位
33	DBJ 46-036-2015	海南省新建住宅小区供配电设施建设技术规范 Technical code for construction of electric power supply facilities in new residential districts of Hainan Province		J13231-2015	海南省建筑设计院 中国南方电网海南电网有限责任公司
34	DBJ 46-037-2016	海南省园林绿化工程施工及验收规范 Code for construction and acceptance of landscaping engineering in Hainan Province		J13333-2016	海南省风景园林协会
35	DBJ 46-038-2016	海南省城镇园林绿地养护管理规范 Code for management of landscape greening in city and town in Hainan Province		J13334-2016	海南省风景园林协会
36	DBJ 46-039-2016	海南省住宅建筑节能和绿色设计标准 Design standard for energy efficiency and green residential building in Hainan		J13388-2016	中国建筑科学研究院海南分院

西藏自治区

序号	标准编号	标准名称	被代替标准编号	备案号	主编单位
1	DBJ 540001-2016	西藏自治区民用建筑节能设计标准 Design standard for energy efficiency of civil buildings		J11036-2016	西藏自治区建筑勘察设计院
2	DBJ 540002-2016	西藏自治区民用建筑供暖通风设计标准 Design standard for heating and ventilation of civil buildings for tibet		J11035-2016	西藏自治区建筑勘察设计院

附：标准英文版目录

序号	标准编号	标准名称	备注
1	GB 50003-2001	Code for design of masonry structures 砌体结构设计规范	中文版为 2011 年版
2	GB 50005-2003	Code for design of timber structures 木结构设计规范（2005 年版）	中文版为 2005 年版
3	GB 50007-2011	Code for design of building foundation 建筑地基基础设计规范	
4	GB 50009-2001	Load code for the design of building structures 建筑结构荷载规范（2006 年版）	中文版为 2012 年版
5	GB 50010-2011	Code for design of concrete structures 混凝土结构设计规范	中文版为 2015 年版
6	GB 50011-2010	Code for seismic design of buildings 建筑抗震设计规范	中文版为 2016 年版
7	GB 50013-2006	Code for design of outdoor water supply engineering 室外给水设计规范	
8	GB 50014-2006	Code for design of outdoor wastewater engineering 室外排水设计规范	中文版为 2016 年版
9	GB 50015-2003	Code for design of building water supply and drainage 建筑给水排水设计规范（2009 年版）	中文版为 2015 年版
10	GB 50016-2006	Code of design on building fire protection and prevention 建筑设计防火规范	中文版为 2014 年版
11	GB 50017-2003	Code for design of steel structures 钢结构设计规范	
12	GB 50018-2002	Technical code of cold-formed thin-wall steel structures 冷弯薄壁型钢结构技术规范	
13	GB 50019-2003	Code for design of heating ventilation and air conditioning 采暖通风与空气调节设计规范	中文版为 2015 年版
14	GB 50021-2001	Code for investigation ofeotechnical engineering 岩土工程勘察规范	中文版为 2009 年版
15	GB 50026-2007	Code for engineering surveying 工程测量规范	
16	GB 50028-2006	Code for design of city gas engineering 城镇燃气设计规范	
17	GB 50034-2004	Standard for lighting design of buildings 建筑照明设计标准	中文版为 2013 年版
18	GB 50041-2008	Code for design of boiler plant 锅炉房设计规范	
19	GB 50045-95	Code for fire protection design of tall buildings 高层民用建筑设计防火规范（2005 年版）	中文版已作废
20	GB 50057-94	Design code for protection of structures against lightning 建筑物防雷设计规范（2000 年版）	中文版为 2010 年版
21	GB 50084-2001	Code of design for sprinkler systems 自动喷水灭火系统设计规范（2005 年版）	
22	GB 50096-2011	Design code for residential buildings 住宅设计规范	

标准英文版目录

序号	标准编号	标准名称	备注
23	GB 50108-2008	Technical code for waterproofing of underground works 地下工程防水技术规范	
24	GB 50116-98	Code for design of automatic fire alarm system 火灾自动报警系统设计规范	中文版为 2013 年版
25	GB 50141-2008	Code for construction and acceptance of water and sewerage structures 给水排水构筑物工程施工及验收规范	
26	GB 50153-2008	Unified standard for reliability design of engineering structures 工程结构可靠性设计统一标准	
27	GB 50157-2003	Code for design of metro 地铁设计规范	中文版为 2013 年版
28	GB 50166-2007	Code for installation and acceptance of fire alarm system 火灾自动报警系统施工及验收规范	
29	GB 50180-93	Code of urban residential areas planning & design 城市居住区规划设计规范	中文版为 2016 年版
30	GB 50189-2005	Design standard for energy efficiency of public buildings 公共建筑节能设计标准	中文版为 2015 年版
31	GB 50202-2002	Code for acceptance of construction quality of building foundation 建筑地基基础工程施工质量验收规范	
32	GB 50203-2002	Code for acceptance of construction quality of masonry engineering 砌体工程施工质量验收规范	中文版为 2011 年版
33	GB 50204-2002	Code for acceptance of constructional quality of concrete structures 混凝土结构工程施工质量验收规范	中文版为 2015 年版
34	GB 50205-2001	Code for acceptance of construction quality of steel structures 钢结构工程施工质量验收规范	
35	GB 50206-2002	Code for construction quality acceptance of timber structures 木结构工程施工质量验收规范	中文版为 2012 年版
36	GB 50207-2002	Code for acceptance of construction quality of roof 屋面工程质量验收规范	中文版为 2012 年版
37	GB 50208-2002	Code for acceptance of construction quality of underground waterproof 地下防水工程质量验收规范	中文版为 2011 年版
38	GB 50209-2002	Code for acceptance of construction quality of building ground 建筑地面工程施工质量验收规范	已有 2010 年版
39	GB 50209-2010	Code for acceptance of construction quality of building ground 建筑地面工程施工质量验收规范	
40	GB 50210-2001	Code for construction quality acceptance of building decoration 建筑装饰装修工程质量验收规范	
41	GB 50229-2006	Code for design of fire protection for fossil fuel power plants and substations 火力发电厂与变电站设计防火规范	
42	GB 50242-2002	Code for acceptance of construction quality of water supply drainage and heating works 建筑给水排水及采暖工程施工质量验收规范	
43	GB 50243-2002	Code of acceptance for construction quality of ventilation and air conditioning works 通风与空调工程施工质量验收规范	中文版为 2016 年版
44	GB 50261-2005	Code for installation and commissioning of sprinkler systems 自动喷水灭火系统施工及验收规范	
45	GB 50268-2008	Code for construction and acceptance of water and sewerage pipeline works 给水排水管道工程施工及验收规范	
46	GB 50300-2001	Unified standard for constructional quality acceptance of building engineering 建筑工程施工质量验收统一标准	中文版为 2013 年版

标准英文版目录

序号	标准编号	标准名称	备注
47	GB 50303-2002	Code of acceptance of construction quality of electrical installation in building 建筑电气工程施工质量验收规范	中文版为2015年版
48	GB 50310-2002	Code for acceptance of installation quality of lifts, escalators and passenger conveyors 电梯工程施工质量验收规范	
49	GB 50311-2007	Code for engineering design of generic cabling system 综合布线系统工程设计规范	中文版为2016年版
50	GB 50312-2007	Code for engineering acceptance of generic cabling system 综合布线系统工程验收规范	中文版为2016年版
51	GB 50314-2006	Standard for design of intelligent building 智能建筑设计标准	中文版为2015年版
52	GB 50343-2004	Technical code for protection against lightning of building electronic information system 建筑物电子信息系统防雷技术规范	中文版为2012年版
53	GB 50348-2004	Technical code for engineering of security and protection system 安全防范工程技术规范	
54	GB 50352-2005	Code for design of civil buildings 民用建筑设计通则	
55	GB 50364-2005	Technical code for solar water heating system of civil buildings 民用建筑太阳能热水系统应用技术规范	
56	GB 50368-2005	Residential building code 住宅建筑规范	
57	GB/T 50375-2006	Evaluating standard for excellent quality of building engineering 建筑工程施工质量评价标准	中文版为2016年版
58	GB/T 50378-2006	Evaluation standard for green building 绿色建筑评价标准	中文版为2014年版
59	GB 50411-2007	Code for acceptance of energy efficient building construction 建筑节能工程施工质量验收规范	
60	GB 50490-2009	Technical code of urban rail transit 城市轨道交通技术规范	
61	GB 50550-2010	Code for acceptance of construction quality of strengthening building structures 建筑结构加固工程施工质量验收规范	
62	GB 50555-2010	Standard for water saving design in civil building 民用建筑节水设计标准	
63	GB 50574-2010	Uniform technical code for wall materials used in buildings 墙体材料应用统一技术规范	
64	GB 50578-2010	Code for constructional quality acceptance of urban rail transit signaling engineering 城市轨道交通信号工程施工质量验收规范	
65	GB 50606-2010	Code for installation of intelligent building systems 智能建筑工程施工规范	
66	GB 50617-2010	Code for construction and acceptance of electrical lighting installation in building 建筑电气照明装置施工与验收规范	
67	GB 50640-2010	Evaluation standard for green construction of building 建筑工程绿色施工评价标准	
68	GB 50666-2011	Code for construction of concrete structures 混凝土结构工程施工规范	
69	GB/T 50668-2011	Standard for energy efficient building assessment 节能建筑评价标准	
70	GB 50688-2011	Code for design of urban road traffic facility 城市道路交通设施设计规范	
71	GB 50702-2011	Code for design of strengthening masonry structures 砌体结构加固设计规范	

标准英文版目录

序号	标准编号	标准名称	备注
72	GB 50755-2012	Code for construction of steel structures 钢结构工程施工规范	
73	JGJ 3-2002	Technical specification for concrete structures of tall building 高层建筑混凝土结构技术规程	中文版为2010年版
74	JGJ 16-2008	Code for electrical design for civil building 民用建筑电气设计规范	
75	JGJ 81-2002	Technical specification for welding of steel structure of building 建筑钢结构焊接技术规程	中文版已作废
76	JGJ 94-2008	Technical code for building pile foundations 建筑桩基技术规范	
77	JGJ 102-2003	Technical code for glass curtain wall engineering 玻璃幕墙工程技术规范	
78	JGJ 107-2003	General technical specification forechanical splicing of bars 钢筋机械连接通用技术规程	中文版为2016年版
79	JGJ 113-2003	Technical specification for application of architectural glass 建筑玻璃应用技术规程	中文版为2015年版
80	JGJ 120-99	Technical specification for retaining and protection of building foundation excavations 建筑基坑支护技术规程	中文版为2012年版
81	JGJ 133-2001	Technical code for metal and stoneurtain walls engineering 金属与石材幕墙工程技术规范	
82	GB 50393-2008	Code for anti-corrosion engineering of steel oil tanks 钢质石油储罐防腐蚀工程技术规范	
83	GB 50517-2010	Code for construction quality acceptance of metallic piping in petrochemical engineering 石油化工金属管道工程施工质量验收规范	
84	GB 50690-2011	Code for construction quality acceptance of non-metallic piping engineering in petrochemical engineering 石油化工非金属管道工程施工质量验收规范	
85	SH/T 3011-2011	Specification for design of process plant layout in petrochemical engineering 石油化工工艺装置布置设计规范	
86	SH 3012-2011	Specification for design of metallic piping layout in petrochemical engineering 石油化工金属管道布置设计规范	
87	SH/T 3034-2012	Specification for design of water supply and wastewater piping in petro chemical industry 石油化工给水排水管道设计规范	
88	SH/T 3059-2012	Specification for piping material design and selection in petrochemical industry 石油化工管道设计器材选用规范	
89	SH/T 3088-2012	Technical specification for tower trays in petrochemical industry 石油化工塔盘技术规范	
90	SH/T 3096-2012	Material selection guideline for design of equipment and piping in units processing sulfur crude oils 高硫原油加工装置设备和管道设计选材导则	
91	SH/T 3098-2011	Speclfieation for design of column in petrochemical industry 石油化工塔器设计规范	
92	SH/T 3129-2012	Material selection cuideline for design of equipment ang piping in unirs processing acid crude oils 高酸原油加工装置设备和管道设计选材导则	
93	SH/T 3139-2011	Technical specification of centrifugal pumps for heavy duty services in petrochemical engineering 石油化工重载荷离心泵工程技术规范	
94	SH/T 3140-2011	Technical specification of centrifugal pumps for medium and light duty services in petrochemical engineering 石油化工中、轻载荷离心泵工程技术规范	

标准英文版目录

序号	标准编号	标准名称	备注
95	SH/T 3143-2012	Technical specification for reciprocating compressor in petrochemical industry 石油化工往复压缩机工程技术规范	
96	SH/T 3171-2011	Technical specification for flexible couplings in petrochemical industry 石油化工挠性联轴器工程技术规范	
97	SH/T 3405-2012	Series of steel pipe size in petrochemical industry 石油化工钢管尺寸系列	
98	SH/T 3408-2012	Steel butt-welding pipe fittings in petrochemical industry 石油化工钢制对焊管件	
99	SH/T 3410-2012	Forged steel socket-welded and threaded fittings in petrochemical industry 石油化工锻钢制承插焊和螺纹管件	
100	SH/T 3004-2011	Design specification for heating, ventilation and air conditioning in petrochemical engineering 石油化工采暖通风与空气调节设计规范	
101	SH/T 3006-2012	Design specification for control room in petrochemical industry 石油化工控制室设计规范	
102	SH/T 3162-2011	Technical specification of liquid ring vacuum pumps and compressors in petrochemical engineering 石油化工液环真空泵和压缩机工程技术规范	
103	SH/T 3163-2011	Specification for classification of static equipment in petrochemical engineering 石油化工静设备分类标准	
104	SH/T 3165-2011	Design specification of material handling in petrochemical engineering 石油化工粉体工程设计规范	
105	SH/T 3166-2011	Design specification for flue gas & air ducts of tubular heater in petrochemical engineering 石油化工管式炉烟风道结构设计规范	
106	SH/T 3167-2012	Steel welded, low-pressure storage tanks 钢制焊接低压储罐	
107	SH/T 3169-2012	Specification for general layout of oil and gas pipeline station 长输油气管道站场布置规范	
108	SH/T 3516-2012	Specification for construction and acceptance of axial compressor-flue gas expander energy recovery unit in flow catalytic cracking unit 催化裂化装置轴流压缩机-烟气轮机能量回收机组施工及验收规范	
109	SH/T 3536-2011	Specification for lifting in petrochemical industry 石油化工工程起重施工规范	
110	SH/T 3548-2011	Acceptance specificationfor construction quality of anticorrosive coating in petrochemical industry 石油化工涂料防腐蚀工程施工质量验收规范	
111	SH/T 3550-2012	Specification for preparation of technical documentation in petrochemical construction projects 石油化工建设工程项目施工技术文件编制规范	
112	SH/T 3004-2011	Design specification for heating, ventilation and air conditioning in petrochemical engineering 石油化工采暖通风与空气调节设计规范	
113	SH/T 3006-2012	Design specification for control room in petrochemical industry 石油化工控制室设计规范	
114	GB 50078-2008	Code for construction ang acceptance of chimney engineering 烟囱工程施工及验收规范	
115	GB 50372-2006	Code for installation acceptance of metallurgical mach ineryironmaking equipment engineering 炼铁机械设备工程安装验收规范	
116	GB 50377-2006	Code for acceptance of mineral processing equipment installation engineering 选矿机械设备工程安装验收规范	

标准英文版目录

序号	标准编号	标准名称	备注
117	GB 50387-2006	Code for engineering installation acceptance of metallurgical mechanical hydraulic, lubrication and pneumatic equipment 冶金机械液压、润滑和气动设备工程安装验收规范	
118	GB 50397-2007	Code for acceptance of electrical equipment installation engineering in metallurgy 冶金电气设备工程安装验收规范	
119	GB 50402-2007	Code for installattion acceptance of sintering mechanical equipment engineering 烧结机械设备工程安装验收规范	
120	GB 50403-2007	Code for engineering installation acceptance of steel-making mechanical equipment 炼钢机械设备工程安装验收规范	
121	GB 50432-2007	Code for design of coking technology 炼焦工艺设计规范	
122	GB 50468-2008	Code for design of welded pipe process 焊管工艺设计规范	
123	GB 50486-2009	Code for design of industrial furnaces in iron & steel works 钢铁厂工业炉设计规范	
124	GB 50491-2009	Code for design of iron pellet engineering 铁矿球团工程设计规范	
125	GB 50541-2009	Code for design of raw material yard of iron and steel plants 钢铁企业原料场工艺设计规范	
126	GB 50551-2010	Code for quality acceptance of palletizing mechanical equipment installation engineering 球团机械设备安装工程质量验收规范	
127	GB 50580-2010	Code for design of continuous casting engineering 连铸工程设计规范	
128	GB 50603-2010	Code for design of general layout and transportation for iron & steel enterprise 钢铁企业总图运输设计规范	
129	GB 50607-2010	Design code for pulverized coal injection of blast furnace 高炉喷吹煤粉工程设计规范	
130	GB 50612-2010	Code for technological design of metallurgical concentrator 冶金矿山选矿厂工艺设计规范	
131	GB 50629-2010	Design code for rolling process of plate and strip mill 板带轧钢工艺设计规范	
132	GB 50679-2011	Code for installation acceptance of metallurgical machinery ironmaking equipment engineering 炼铁机械设备安装规范	
133	GB 50696-2011	Code for design of metallurgical equipment foundations of iron and steel enterprises 钢铁企业冶金设备基础设计规范	
134	GB 50713-2011	Code for design of finishing process of flat steel 板带精整工艺设计规范	
135	GB 50714-2011	Code for design of steel pipe coating workshop process 钢管涂层车间工艺设计规范	
136	GB 50723-2011	Code for installation acceptance of sintering mechanical equipment engineering 烧结机械设备安装规范	
137	GB 50730-2011	Code for engineering construction of metallurgical mechanical hydraulic, lubrication and pneumatic equipment 冶金机械液压、润滑和气动设备工程施工规范	
138	GB 50735-2011	Code for design of ferroalloy process and equipment 铁合金工艺及设备设计规范	
139	GB 50742-2012	Code for installation of steel-making mechanical equipment 炼钢机械设备安装规范	

标准英文版目录

序号	标准编号	标准名称	备注
140	GB/T 50744-2011	Code for acceptance of engineering installation for mechanical equipment of rolling mill 轧机机械设备安装规范	
141	GB 50754-2012	Code for design of steel pipe extrusion engineering 挤压钢管工程设计规范	
142	GB 50825-2013	Code for quality acceptance of reheating furnaces in iron and steel works 钢铁厂加热炉工程质量验收规范	
143	GB 50830-2013	Code for design of metal mine 冶金矿山采矿设计规范	
144	GB 50840-2012	Code for construction and acceptance of slurry pipeline engineering 矿浆管线施工及验收规范	
145	GB 50930-2013	Code for design of cold rolling strip plant 冷轧带钢工厂设计规范	
146	GB 50967-2014	Code for installation of coking and chemical mechanical equipment 焦化机械设备安装规范	
147	GB/T 51075-2015	Code for mineral processing equipment installation engineering 选矿机械设备工程安装规范	
148	GB 50060-2008	Code for design of high voltage electrical installation (3～110kV) 3～110kV 高压配电装置设计规范	
149	GB/T 50062-2008	Code for design of relaying protection and automatic device of electric power installations 电力装置的继电保护和自动装置设计规范	
150	GB/T 50063-2008	Code for design of electrical measuring device of power system 电力装置的电测量仪表装置设计规范	
151	GB 50147-2010	Electric equipment installation engineering code for construction and acceptance of high-voltage 电气装置安装工程　高压电器施工及验收规范	
152	GB 50148-2010	Electric equipment installation engineering code for construction and acceptance of power transformers oil reactor and mutual inductor 电气装置安装工程　电力变压器、油浸电抗器、互感器施工及验收规范	
153	GB 50149-2010	Electric equipment installation engineering code for construction and acceptance of busbar equipments 电气装置安装工程　电气装置安装工程　母线装置施工及验收规范	
154	GB 50150-2006	Electric equipment installation engineering standard for hand-over test of elelctric equipment 电气装置安装工程　电气设备交接试验标准	
155	GB 50168-2006	Electric equipment installation engineering code for construction and acceptance of cable system 电气装置安装工程　电缆线路施工及验收规范	
156	GB 50169-2006	Electric equipment installation engineering code for construction and acceptance of grounding devices 电气装置安装工程　接地装置施工及验收规范	
157	GB 50170-2006	Electric equipment installation engineering code for construction and acceptance of rotating electrical machines 电气装置安装工程　旋转电机施工及验收规范	
158	GB 50171-2012	Electric equipment installation engineering code for construction and acceptance of panel, cabinet and secondary circuit 电气装置安装工程　盘、柜及二次回路接线施工及验收规范	
159	GB 50172-2012	Electric equipment installation engineering code for construction and acceptance of batteries 电气装置安装工程　蓄电池施工及验收规范	
160	GB 50217-2007	Code for design of cables of electric engineering 电力工程电缆设计规范	
161	GB 50227-2008	Code for design of installation of shunt capacitors 并联电容器装置设计规范	

标准英文版目录

序号	标准编号	标准名称	备注
162	GB 50233-2005	Code for construction and acceptance of 110~500kV overhead transmission line 110~500kV 架空送电线路施工及验收规范	中文版为 2014 年版本
163	GB 50545-2010	Code for design of 110kV~750kV overhead transmission line 110kV~750kV 架空输电线路设计规范	
164	GB 50548-2010	Code for investigation and surveying of 330kV~750kV overhead transmission line 330kV~750kV 架空输电线路勘测规范	
165	GB/T 50572-2010	Code for seismic investigation and evaluation of nuclear power plants 核电厂工程地震调查与评价规范	
166	GB 50613-2010	Code for planning and design of urban distribution network 城市配电网规划设计规范	
167	GB/T 50619-2010	Code for design of electrical measuring device of power system 火力发电厂海水淡化工程设计规范	
168	GB 50633-2010	Technical code for engineering surveying of nuclear power station 核电厂工程测量技术规范	
169	GB 50660-2011	Code for design of fossil fired power plant 大中型火力发电厂设计规范	
170	GB/T 50663-2011	Technical code for engineering hydrology for nuclear power plant 核电厂工程水文技术规范	
171	DL/T 5083-2010	Specification of prestressing tendon construction for hydropower and water resources project (English version for approval) 水电水利工程预应力锚索施工技术规范	
172	DL/T 5161.1-2002	Specification for construction quality checkout and evaluation of electric equipment installation Part 1: General rules 电气装置安装工程 质量检验及评定规程 第1部分：通则	
173	DL/T 5161.2-2002	Specification for construction quality checkout and evaluation of electric equipment installation Part 2: High-voltage electric power equipments 电气装置安装工程 质量检验及评定规程 第2部分：高压电器施工质量检验	
174	DL/T 5161.3-2002	Specification for construction quality checkout and evaluation of electric equipment installation Part 3: Power transformers, oil-immersed type reactors and instrument transformers 电气装置安装工程 质量检验及评定规程 第3部分：电力变压器、油浸电抗器、互感器施工质量检验	
175	DL/T 5161.5-2002	Specification for construction quality checkout and evaluation of electric equipment installation Part 5: Power cable lines 电气装置安装工程 质量检验及评定规程 第5部分：电缆线路施工质量检验	
176	DL/T 5161.6-2002	Specification for construction quality checkout and evaluation of electric equipment installation Part 6: Grounding devices 电气装置安装工程 质量检验及评定规程 第6部分：接地装置施工质量检验	
177	DL/T 5161.7-2002	Specification for construction quality checkout and evaluation of electric equipment installation Part 7: Rotating electrical machines 电气装置安装工程 质量检验及评定规程 第7部分：旋转电机施工质量检验	
178	DL/T 5161.8-2002	Specification for construction quality checkout and evaluation of electric equipment installation Part 8: Panels, cabinets and secondary circuit wirings 电气装置安装工程 质量检验及评定规程 第8部分：盘、柜及二次回路结线施工质量检验	
179	DL/T 5161.9-2002	Specification for construction quality checkout and evaluation of electric equipment installation Part 9: Batteries 电气装置安装工程 质量检验及评定规程 第9部分：蓄电池施工质量检验	

标准英文版目录

序号	标准编号	标准名称	备注
180	DL/T 5161.10-2002	Specification for construction quality checkout and evaluation of electric equipment installation Part 10: 35kV and below over-head power line installation 电气装置安装工程 质量检验及评定规程 第10部分：35kV及以下架空电力线路施工质量检验	
181	DL/T 5161.11-2002	Specification for construction quality checkout and evaluation of electric equipment installation Part 11: Elevator 电气装置安装工程 质量检验及评定规程 第11部分：电梯电气装置施工质量检验	
182	DL/T 5161.12-2002	Specification for construction quality checkout and evaluation of electric equipment installation Part 12: Low-voltage apparatus 电气装置安装工程 质量检验及评定规程 第12部分：低压电器施工质量检验	
183	DL/T 5161.13-2002	Specification for construction quality checkout and evaluation of electric equipment installation Part 13: Power converter equipments 电气装置安装工程 质量检验及评定规程 第13部分：电力交流设备施工质量检验	
184	DL/T 5161.14-2002	Specification for construction quality checkout and evaluation of electric equipment installation Part 14: Crane 电气装置安装工程 质量检验及评定规程 第14部分：起重机电气装置施工质量检验	
185	DL/T 5161.15-2002	Specification for construction quality checkout and evaluation of electric equipment installation Part 15: Electric equipments in explosive or inflammable hazardous environment 电气装置安装工程 质量检验及评定规程 第15部分：爆炸及火灾危险环境电气装置施工质量检验	
186	DL/T 5161.16-2002	Specification for construction quality checkout and evaluation of electric equipment installation Part 16: 1kV and under feeder cable engineering 电气装置安装工程 质量检验及评定规程 第16部分：1kV及以下配线工程施工质量检验	
187	DL/T 5161.17-2002	Specification for construction quality checkout and evaluation of electric equipment installation Part 17: Electric lighting devices 电气装置安装工程 质量检验及评定规程 第17部分：电气照明装置施工质量检验	
188	DL/T 5168-2002	Specification for construction quality checkout and evaluation of 110kV~500kV overhead transmission lines 110kV~500kV架空电力线路工程施工质量及评定规程	中文版为2016年版本
189	DL 5190.2-2012	Technical specification for thermal power erection and construction Part 2: Boiler unit 电力建设施工技术规范 第2部分：锅炉机组	
190	DL 5190.4-2012	The technical specification specification for of erection of thermal electric power power erection and construction Part 4: Instrumentation and control 电力建设施工技术规范 第4部分：热工仪表及控制装置	
191	DL 5190.5-2012	Technical specification for thermal power erection and construction Part 5: Piping & system 电力建设施工技术规范 第5部分：管道及系统	
192	DL 5190.8-2012	Technical specification for thermal power erection and construction Part 8: Processing preparation 电力建设施工技术规范 第8部分：加工配制	
193	DL/T 5207-2005	Technical specification for abrasion and cavitation resistance of concrete in hydraulic structures 水工建筑物抗冲磨防空蚀技术规范	
194	DL/T 5210.2-2009	Code for construction quality acceptance and evaluation of electric power construct Part 2: Boiler unit 电力建设施工质量验收及评价规程 第2部分：锅炉机组	

标准英文版目录

序号	标准编号	标准名称	备注
195	DL/T 5210.3-2009	Code for construction quality acceptance and evaluation of electric power construct Part 3: Turbine generator unit 电力建设施工质量验收及评价规程 第3部分：汽轮发电机组	
196	DL/T 5210.4-2009	Code for construction quality acceptance and evaluation of electric power construct-Part 4: Instrumentation and controlcontrol 电力建设施工质量验收及评价规程 第4部分：热工仪表及控制装置	
197	DL/T 5210.5-2009	Code for construction quality acceptance and evaluation of electric power construction Part 5: Piping &. system 电力建设施工质量验收及评价规程 第5部分：管道及系统	
198	DL/T 5210.6-2009	Code for construction quality acceptance and evaluation of electric power construct Part 6: The equipment and the system with the water treatment and hydrogen generation 电力建设施工质量验收及评价规程 第6部分：水处理剂制 氢设备和系统	
199	DL/T 5210.8-2009	Code for construction quality acceptance and evaluation of electric power construct Part 8: Processing preparation 电力建设施工质量验收及评价规程 第8部分：加工配制	
200	DL/T 5231-2010	Specification for construction and acceptance of earth electrode in ±800kV and below DC transmission ±800kV及以下直流输电接地极施工及验收规程	
201	DL/T 5232-2010	Specification for construction and acceptance of electric equipments in ±800kV and below DC converter stations ±800kV及以下直流换流站电气装置安装工程施工及验收规程	
202	DL/T 5233-2010	Specification for construction quality checkout and evaluation of electric equipments in ±800kV and below DC converter stations ±800kV及以下直流换流站电气装置施工质量检验及评定规程	
203	DL/T 5235-2010	Specification for construction and acceptance of ±800kV and under DC overhead transmission lines ±800kV及以下直流架空输电线路工程施工及验收规范	
204	DL/T 5236-2010	Specification for construction quality checkout and evaluation of ±800kV and under DC overhead transmission lines ±800kV及以下直流架空输电线路工程施工质量检验及评定规程	
205	DL/T 5241-2010	Technical specifications for durability of hydraulic concrete 水工混凝土耐久性技术规范	
206	DL/T 5251-2010	Technical code for detection and evaluation of hydraulic concrete structure 水工混凝土建筑物缺陷监测和评估技术规程	
207	DL/T 5257-2010	Specification for construction acceptance of thermal power plant flue gas denitration project 火电厂烟气脱硝工程施工验收技术规程	
208	DL/T 5342-2006	Guide for construction technology for the assembly and erection of the steel towers of 750kV overhead transmission lines 750kV架空送电线路铁塔组立施工工艺导则	
209	DL/T 5343-2006	Guide for tension stringing construction technology of 750kV overhead transmission lines 750kV架空送电线路张力架线施工工艺导则	
210	DL/T 5406-2010	Specifications for chemical grouting of hydraulic structures (English Version For Approval) 水工建筑物化学灌浆施工规范	
211	DL/T 5417-2009	Specification for construction quality inspection and assessment of thermal power plant flue gas desulphurization 火电厂烟气脱硫工程施工质量验收及评定规程	
212	DL/T 5418-2009	Code for construction and acceptance of flue gas desulphurization absorber of thermal power plant 火电厂烟气脱硫吸收塔施工及验收规程	

标准英文版目录

序号	标准编号	标准名称	备注
213	DL/T 5426-2009	Specification for system design of ±800kV UHVDC ±800kV 高压直流输电系统成套设计规程	
214	DL/T 5434-2009	The code of power construction project management 电力建设工程监理规范	
215	DL/T 5437-2009	Code for fossil power construction project from the unit commissioning to completed acceptance 火力发电建设工程启动试运及验收规程	